Linear Algebra

Chefchaouen, Morocco

Toshitsune Miyake

Linear Algebra

From the Beginnings to the Jordan Normal
Forms

Toshitsune Miyake
Hokkaido University
Sapporo, Japan

ISBN 978-981-16-6996-5   ISBN 978-981-16-6994-1   (eBook)
https://doi.org/10.1007/978-981-16-6994-1

This Springer imprint is published by the registered company Springer Nature Singapore Pte Ltd.
The registered company address is: 152 Beach Road, #21-01/04 Gateway East, Singapore 189721,
Singapore

Preface

This book is based on my book *Linear Algebra, From the Beginnings to the Jordan Normal Forms*, written in Japanese.

One of the characteristics of modern mathematics is that we extract common ways of thinking in various fields of mathematics and apply the abstract ideas back to each field. Linear algebra is considered as an abstraction of basic tools in many fields of mathematics. Though linear algebra is a branch of algebra, basic ideas such as vector spaces or linear transformations are quite important concepts of not only algebra but also analysis and geometry. We may say that we cannot study mathematics without knowledge of linear algebra. We consider that explaining linear algebra too abstractly would be unkind and unsuitable for beginners. We describe general definitions and theorems and explain the contents with many examples and exercises. We have attempted to write the book under the following principles:

(1) The purpose of the present book is to give the fundamental concepts of linear algebra. We give plenty of examples and avoid only stating facts or results. Since calculations are very important for the study of linear algebra, we adopt the proofs of theorems suitable for calculations.

(2) We start from the definition of matrices, and we aim to state and explain somewhat advanced topics such as Hermitian inner products and Jordan normal forms at the end.

(3) We give lots of concrete numerical examples and exercises.

(4) We choose the exercises by which the readers can check their understanding of the subjects. We avoid overly difficult exercises.

The topics we discussed in the present book are as follows.

In Chap. 1, we define matrices and explain that matrices are nothing but the abstractions of tables. We define the matrix operations. We explain that the matrix operations are abstractions of operations of tables by giving concrete tables. We offer many kinds of matrices, which will be discussed later. We also connect matrix equations and systems of linear equations.

In Chap. 2, we deal with systems of linear equations in detail. The merit of the Gauss elimination method is that we do not need to verify that the obtained solutions satisfy the original systems of linear equations. Reductions of matrices are defined. We explain that the solution of systems of linear equations by the Gauss elimination method is nothing but the solution of corresponding matrix equations by reductions of matrices. In this book, the row reductions of matrices play very important roles in calculations. We also define regular matrices and inverse matrices. Conditions of matrices to be regular matrices are stated.

In Chap. 3, we define the determinants of matrices and explain properties of determinants. We also calculate the determinants of quite many matrices. Determinants of matrices play important roles in the theory of linear algebra and are very useful for calculations. We also discuss resultants of polynomials as an application of determinants of matrices.

In Chap. 4, we define vector spaces and explain their properties. The linear independence is defined. It is a very important concept in the theory of linear algebra. We define bases and dimensions of vector spaces. We mainly consider the spaces K^n consisting of column vectors of degree n. Since the readers may not be familiar with abstract fields, we take the real number field and the complex number field for the basic field K of vector spaces, especially in examples. Therefore, the readers may consider that the basic field K is the real number field or the complex number field. Though we mainly discuss the spaces K^n, we also consider other vector spaces. As examples of abstract vector spaces, we offer the spaces of polynomials whose degrees are at most n.

In Chap. 5, we deal with linear mappings and their representation matrices. We also discuss eigenvalues, eigenvectors, eigenspaces and diagonalizations of linear transformations and square matrices. We also explain the characteristic polynomials and the minimal polynomials of square matrices and linear transformations.

In Chap. 6, we define and explain inner products of vector spaces over the real number field and orthonormal bases of inner product spaces. We explain orthogonal transformations, orthogonal matrices and diagonalization of symmetric matrices. We also discuss quadratic forms and their canonical forms.

In Chap. 7, we explain Hermitian inner products of vector spaces over the complex number field. Beginning from the definition of Hermitian inner product spaces, we discuss Hermitian transformations (matrices), unitary transformations (matrices) and normal transformations (matrices). We end with spectral resolutions of normal transformations, diagonalizations of normal matrices by unitary matrices (Toeplitz theorem) and also positive definite Hermitian matrices.

In Chap. 8, we explain Jordan normal forms. We show that all linear transformations of vector spaces over the complex number field have the matrices of Jordan normal forms as representation matrices. Though not all square matrices are diagonalizable, all complex square matrices are similar to the matrices of Jordan normal forms, which are quite close to diagonal matrices. The matrices of Jordan normal forms are not only theoretically interesting but also very useful for applications.

I would like to express my deep and sincere gratitude to my former colleague Prof. Yoshitaka Maeda who kindly read the manuscript and gave me many helpful comments and suggestions. Without him, the present book would not have been accomplished.

Sapporo, Japan   Toshitsune Miyake
May 2022

Contents

Notations

$A = \left(a_{ij}\right)_{m \times n} = \left(a_{ij}\right)$ A matrix with the (i, j) component a_{ij} (§1.1, p. 1)

$A, B, C, \cdots$ Matrices (§ 1.1, p. 1)

${}^t A$ The transposed matrix of A (§ 1.1, p. 4)

$-A$ Minus A (§ 1.1, p. 5)

A^r The r-th power of a square matrix A (§ 1.2, p. 12)

$$A = \begin{pmatrix} A_{11} & A_{12} & \cdots & A_{1t} \\ A_{21} & A_{22} & \cdots & A_{2t} \\ \cdots & \cdots & \cdots & \cdots \\ A_{s1} & A_{s2} & \cdots & A_{st} \end{pmatrix}$$
A partition of a matrix A (§ 1.3, p. 17)

$$= \begin{pmatrix} A_{11} & \cdots & A_{1t} \\ \vdots & & \vdots \\ A_{s1} & \cdots & A_{st} \end{pmatrix}$$

$A = \left(a_1 \cdots a_n \right)$ The column partition of a matrix A (§ 1.3, p. 21)

$$A = \begin{pmatrix} {}^t b_1 \\ \vdots \\ {}^t b_m \end{pmatrix}$$
The row partition of a matrix A (§ 1.3, p. 21)

$\left(A \mid b \right)$ The augmented matrix of a system of linear equations $Ax = b$ (§ 1.4, p. 25)

A^{-1} The inverse matrix of a regular matrix A (§ 2.4, p. 55)

A_n The alternating group of degree n (§ 3.1, p. 69)

$\tilde{A}$ The cofactor matrix of a square matrix A (§ 3.4, p. 91)

| $\lvert A \rvert$, $\lvert (a_{ij}) \rvert$, $\det(A)$, $\begin{vmatrix} a_{11} & \cdots & a_{1n} \\ \vdots & & \vdots \\ a_{n1} & \cdots & a_{nn} \end{vmatrix}$ | The determinant of a square matrix $A = (a_{ij}) = \begin{pmatrix} a_{11} & \cdots & a_{1n} \\ \vdots & & \vdots \\ a_{n1} & \cdots & a_{nn} \end{pmatrix}$ (§ 3.2, p. 72) |

A_{ij} The square matrix obtained by removing the i-th row and the j-th column from a square matrix A (§ 3.4, p. 88)

$\lvert A(f(x), g(x)) \rvert$ The resultant of polynomials $f(x)$ and $g(x)$; Sylvester's determinant of polynomials $f(x)$ and $g(x)$ (§ 3.5, p. 96)

$A_1 + \cdots + A_r$ The sum of r matrices $A_1, \cdots, A_r$ of same type (§ 1.2, p. 8)

$A_1 \oplus \cdots \oplus A_r$ The direct sum of r square matrices $A_1, \cdots, A_r$ (§ 5.4, p. 178)

A_T The representation matrix of a linear transformation T (§ 5.6, p. 201)

$\overline{A}$ The complex conjugate of a complex matrix A (§ 6.3, p. 226)

$A[x]$ The quadratic form associated with a square matrix A (§ 6.4, p. 235)

A_k The primary submatrix of a square matrix A (§ 6.4, p. 235)

A^* The adjoint matrix of a square matrix A (§ 7.1, p. 251)

$\sqrt[k]{A}$ The k-th root of a Hermitian matrix A (§ 7.2, p. 269)

$\sqrt{A}$ The square root of a Hermitian matrix A (§ 7.2, p. 269)

$\mathbf{C}$ The complex number field (§ 4.1, p. 108)

$D(f(x))$ The discriminant of a polynomial $f(x)$ (§ 3.5, p. 99)

$\dim(V)$ The dimension of a vector space V (§ 4.4, p. 134)

E, E_n The unit matrix of degree n (§ 1.1, p. 4)

E_{ij} The (i, j)-matrix unit (§ 5.6, p. 200)

$\{E_{11}, \cdots, E_{mn}\}$ The standard basis of $M_{m \times n}(K)$ (§ 5.6, p. 200)

G A group (§ 3.1, p. 68)

$\mathrm{Hom}\,(U, V)$ The space of linear mappings of vector spaces U into V (§ 5.6, p. 201)

$\mathrm{Im}(A)$ The image of a matrix A (§ 5.1, p. 141)

$\mathrm{Im}(T)$ The image of a linear mapping T (§ 5.1, p. 140)

I_U The identity transformation of U (§ 5.1, p. 145)

$J(\lambda; n)$	The Jordan cell of degree n with the eigenvalue λ (§ 8.2, p. 281)
K	A field (§ 4.1, p. 107)
K^n	The vector space consisting of all the column vectors of degree n with components in K (§ 4.1, p. 109)
$\mathrm{Ker}(A)$	The kernel of a matrix A (§ 5.1, p. 141)
$\mathrm{Ker}(T)$	The kernel of a linear mapping T (§ 5.1, p. 140)
$M_{m \times n}(K)$	The space consisting of $m \times n$ matrices over K (§ 5.6, p. 200)
$P_{V/W}$	The orthogonal projection of a vector space V into a subspace W (§ 7.1, p. 248)
Q_σ	The orthogonal matrix obtained by a permutation σ (§ 6.4, p. 235)
$R(f(x), g(x))$	The resultant of polynomials $f(x)$ and $g(x)$; Sylvester's determinant of polynomials $f(x)$ and $g(x)$ (§ 3.5, p. 100)
$\mathbf{R}$	The set of real numbers (§ 2.3, p. 48); the real number field (§ 4.1, p. 108)
$\mathbf{R}[x]$	The set of all polynomials with real coefficients (§ 4.1, p. 109)
$\mathbf{R}[x]_n$	The set of all polynomials with real coefficients of degree n or less (§ 4.1, p. 109)
S_n	The set of all the permutations of n elements (§ 3.1, p. 71); the symmetric group of degree n (§ 3.1, p. 68)
S_n^+	The set of all even permutations of n elements (§ 3.1, p. 68); the alternating group of degree n (§ 3.1, p. 69)
S_n^-	The set of all odd permutations of n elements (§ 3.1, p. 68)
T_A	The linear mapping associated with a matrix A (§ 5.1, p. 140); the linear transformation associated with a square matrix A (§ 5.2, p. 159)
$T_1 \oplus \cdots \oplus T_r$	The direct sum of linear transformations $T_1, \cdots, T_r$ (§ 5.4, p. 178)
T^{-1}	The inverse linear mapping of T; the inverse isomorphism of T; the inverse automorphism of T; the inverse transformation of T (§ 5.1, p. 146)
T^i	The i-th product of a linear transformation T (§ 5.4, p. 184)
T^*	The dual mapping of T; the dual transformation of T (§ 5.6, p. 204); the adjoint transformation of a linear transformation T of an inner product

	space (§ 6.3, p. 224); the adjoint transformation of a linear transformation T of a Hermitian inner product space (§ 7.1, p. 246)
$\sqrt[k]{T}$	The k-th root of a Hermitian transformation T (§ 7.2, p. 269)
$\sqrt{T}$	The square root of a Hermitian transformation T (§ 7.2, p. 269)
$T(U)$	The image of a subspace U by a linear mapping T (§ 5.1, p. 145)
$U \cong V$	A vector space U is isomorphic to a vector space V (§ 5.1, p. 146)
$U^* = \mathrm{Hom}(U, K)$	the dual space of a vector space U (§ 5.6, p. 202)
V	A vector space (§ 4.1, p. 108)
V/W	The quotient space (factor space) of a vector space V by a subspace W (§ 5.6, p. 208)
$W(\lambda; A)$	The eigenspace of a square matrix A with the eigenvalue λ (§ 5.3, p. 162)
$\tilde{W}(\lambda; A)$	The generalized eigenspace of a square matrix A with the eigenvalue λ (§ 8.1, p. 280)
$W(\lambda; T)$	The eigenspace of a linear transformation T with the eigenvalue λ (§ 5.3, p. 162)
$\tilde{W}(\lambda; T)$	The generalized eigenspace of a linear transformation T with the eigenvalue λ (§ 8.1, p. 279)
$W_1 \oplus \cdots \oplus W_r$	The direct sum of subspaces $W_1, \cdots, W_r$ (§ 5.4, p. 175)
$W_1 + \cdots + W_r$	The sum of subspaces $W_1, \cdots, W_r$ (§ 5.4, p. 175)
$W_1 \perp W_2$	Subspaces W_1 and W_2 are orthogonal (§ 6.1, p. 215, § 7.1, p. 247)
$W^\perp$	The orthogonal complement of a subspace W (§ 6.1, p. 216, § 7.1, p. 247)
$\#(X)$	The number of elements in a finite set X (§ 3.1, p. 68)
$X/\sim$	The quotient set of a set X by an equivalent relation $\sim$ (§ 5.6, p. 207)
$\begin{pmatrix} a_1 \\ \vdots \\ a_n \end{pmatrix}$	A vertical (column) vector of degree n (§ 1.1, p. 2, p. 3, § 4.1, p. 124)
$\begin{pmatrix} a_1 & \cdots & a_n \end{pmatrix}$	A horizontal (row) vector of degree n (§ 1.1, p. 2)
a^{-1}	The inverse element of an element $a(\neq 0)$ of a field (§ 4.1, p. 106)

$(\mathbf{a}, \mathbf{b})$	The inner product of column vectors $\mathbf{a}$ and $\mathbf{b}$ (§ 6.1, p. 212, §7.1, p. 246)
$\|\mathbf{a}\|$	The norm of a column vector $\mathbf{a}$ (§ 6.1, p. 213, § 7.1, p. 247)
$\mathbf{a}, \mathbf{b}, \mathbf{c}, \cdots$	Numerical vectors (§ 1.1, p. 2); column vectors (§ 1.1, p. 5)
$^t\mathbf{a}, ^t\mathbf{b}, ^t\mathbf{c}, \cdots$	Transposed vectors of numerical vectors $\mathbf{a}, \mathbf{b}, \mathbf{c}, \cdots$; row vectors (§ 1.1, p. 5)
$\tilde{a}_{ij}$	The (i, j) cofactor of a square matrix $A = \left(a_{ij}\right)$ (§ 3.4, p. 90)
$\text{cl}(a)$	The equivalence class containing a (§ 5.6, p. 207)
e	The unit element of a group (§ 3.1, p. 68)
$e_1, \cdots, e_n$	The unit vectors of degree n (§ 2.4, p. 56; § 4.2, p. 129)
$\{e_1, \cdots, e_n\}$	The standard basis of K^n (§ 4.4, p. 131)
$f'(x), f''(x)$	The derivative and the second derivative of a polynomial $f(x)$, respectively (§ 4.1, p. 108)
$f_{00}(x)$	The constant zero polynomial, which is the zero vector of $R[x]$ and $R[x]_n$ (§ 4.1, p. 108)
$f(A)$	The polynomial of a square matrix A (§ 5.3, p. 167)
$f(T)$	The polynomial of a linear transformation T (§ 5.4, p. 184)
$\{f_1, \cdots, f_n\}$	The dual basis of a basis $\{u_1, \cdots, u_n\}$ (§ 5.6, p. 202)
g^{-1}	The inverse element of an element g of a group G (§ 3.1, p. 68)
$g_A(t)$	The characteristic polynomial of a square matrix A (§ 5.3, p. 168)
$g_T(t)$	The characteristic polynomial of a linear transformation T (§ 5.3, p. 169)
(i, j)	The transposition of elements i and j (§ 3.1, p. 63)
$\left(i_1 i_2 \cdots i_r\right)$	The cyclic permutation (§ 3.1, p. 63)
$i = \sqrt{-1}$	The imaginary unit (§ 3.5, p. 106)
$l(\sigma)$	The length of a permutation σ (§ 3.1, p. 65)
$\text{null}(A)$	The nullity of a matrix A (§ 5.1, p. 141)
$\text{null}(T)$	The nullity of a linear mapping T (§ 5.1, p. 141)
$p_A(t)$	The minimal polynomial of a square matrix A (§ 5.4, p. 185)
$p_T(t)$	The minimal polynomial of a linear transformation T (§ 5.4, p. 184)

p_u	The linear functional associated with u (§ 6.3, p. 225)
$q(x_1, \cdots, x_n)$	A quadratic form (§ 6.4, p. 234)
rank(A)	The rank of a matrix A (§ 2.2, p. 42)
rank(T)	The rank of a linear mapping T (§ 5.1, p. 141)
rank (X)	The rank of a set X of vectors (§ 4.3, p. 124)
sgn(σ)	The signature of a permutation σ (§ 3.1, p. 67)
sgn($q(x_1, \cdots, x_n)$)	The signature of a quadratic form $q(x_1, \cdots, x_n)$ (§ 6.4, p. 239)
tr(A)	The trace of a square matrix A (§ 5.5, p. 199)
$-\mathbf{u}$	The inverse vector of $\mathbf{u}$ (§ 4.1, p. 108)
$< \mathbf{u}_1, \cdots, \mathbf{u}_r >$	The subspace generated by vectors $\mathbf{u}_1, \cdots, \mathbf{u}_r$ (§ 4.4, p. 134)
$(\mathbf{u}, \mathbf{v})$	The inner product of vectors $\mathbf{u}$ and $\mathbf{v}$ (§ 6.1, p. 212); the Hermitian inner product of vectors $\mathbf{u}$ and $\mathbf{v}$ (§ 7.1, p. 245)
$\|\mathbf{u}\|$	The norm of a vector $\mathbf{u}$; the length of a vector $\mathbf{u}$ (§ 6.1, p. 213, § 7.1, p. 246)
$\mathbf{u} \perp \mathbf{v}$	Vectors $\mathbf{u}$ and $\mathbf{v}$ are orthogonal (§ 6.1, p. 216, § 7.1, p. 247)
$O, O_{m,n}$	The zero matrix of size (m, n) (§ 1.1, p. 2)
O_n	The zero matrix of degree n (§ 1.1, p. 4)
$O, O_{U,V}$	The zero mapping of a vector space U into a vector space V (§ 5.1, p. 145)
O_U	The zero transformation of a vector space U (§ 5.1, p. 146)
$\mathbf{0}, \mathbf{0}_n$	The numerical zero vector of degree n (§ 1.1, p. 2); the zero column vector of degree n (§ 1.1, p. 5)
${}^t\mathbf{0}, {}^t\mathbf{0}_n$	The transpose of the zero vector $\mathbf{0}$ or $\mathbf{0}_n$ of degree n (§ 1.1, p. 5); the zero row vector of degree n (§ 1.1, p. 5)
$\mathbf{0}, \mathbf{0}_V$	The zero vector of a vector space V (§ 4.1, p. 109)
$\{1, x, \cdots, x^n\}$	the standard basis of $R[x]_n$ (§ 4.4 p. 131)
$\Delta(x_1, x_2, \cdots, x_n)$	The difference product (§ 3.1, p. 66)
Φ	The isomorphism of Hom(U, V) into $M_{m \times n}(K)$ (§ 5.6, p. 201)
abs(α)	The absolute value of a complex number α (§ 3.5, p. 106)
$\bar{\alpha}$	The complex conjugate of a complex number α (§ 6.3, p. 226)
δ_{ij}	The Kronecker delta (§ 1.1, p. 7)
ε	The unit permutation (§ 3.1, p. 62)

$\sigma = \begin{pmatrix} 1\,2\,\cdots\,n \\ k_1\,k_2\,\cdots\,k_n \end{pmatrix}$ 　　　A permutation (§ 3.1, p. 61)

σ^{-1} 　　　The inverse permutation of a permutation σ (§ 3.1, p. 63)

$\sigma f(x_1, \cdots, x_n)$ 　　　The polynomial acted on by a permutation σ (§ 3.1, p. 65)

$\sim$ 　　　An equivalence relation (§ 5.6, p. 206)

$\equiv$ 　　　A congruence relation (§ 5.6, p. 206)

$m \equiv n \pmod{N}$ 　　　m and n are congruent modulo N (§ 5.6, p. 206)

Chapter 1
Matrices

1.1 Matrices

It would be very convenient if we could treat a bundle of several numbers as one number. For such a purpose, we consider matrices and numerical vectors. We shall begin with defining matrices.

Matrices We arrange mn numbers a_{ij} ($i = 1, 2, \ldots, m$; $j = 1, 2, \ldots, n$) into a rectangle of length m and width n and put them in parentheses. We call such objects *matrices*. Matrices are usually denoted by capital letters A, B, C, ... For a matrix

$$A = \begin{pmatrix} a_{11} & a_{12} & \cdots & a_{1n} \\ a_{21} & a_{22} & \cdots & a_{2n} \\ & & \cdots\cdots & \\ a_{m1} & a_{m2} & \cdots & a_{mn} \end{pmatrix},$$

the pair of numbers (m, n) is called the *size* of A. The size of A is also denoted by $m \times n$. The matrix A is called a matrix of size (m, n) or simply an $m \times n$ matrix. We call the number a_{ij} the (i, j) *component* of A for $i = 1, 2, \ldots, m$ and $j = 1, 2, \ldots, n$. The matrix A is often denoted by $A = (a_{ij})_{m \times n}$ or $A = (a_{ij})$ for brevity.

Example 1 $A = \begin{pmatrix} 1 & 2 & 5 \\ -4 & 6 & -3 \end{pmatrix}$ is a 2 × 3 matrix. The (1, 2) component of A is 2 and the (2, 3) component of A is −3.

Equality of matrices We say that two matrices $A = (a_{ij})_{m \times n}$ and $B = (b_{ij})_{s \times t}$ are *equal* if the following two conditions are satisfied:

(i) Their sizes are equal, or $m = s$ and $n = t$.

(ii) The components of A and B are equal, or $a_{ij} = b_{ij}$ for $i = 1, \ldots, m$ and $j = 1, \ldots, n$.

We often identify a 1 × 1 matrix (a) with the number a.

© The Author(s), under exclusive license to Springer Nature Singapore Pte Ltd. 2022
T. Miyake, *Linear Algebra*, https://doi.org/10.1007/978-981-16-6994-1_1

Zero matrices The $m \times n$ matrix whose components are all zeros is called the *zero matrix* of size (m, n) and is denoted by O or $O_{m,n}$.

Example 2 The zero matrix of size $(2, 3)$ is

$$O = O_{2,3} = \begin{pmatrix} 0 & 0 & 0 \\ 0 & 0 & 0 \end{pmatrix}.$$

Numerical vectors We call $m \times 1$ matrices *vertical vectors of degree m*, and $1 \times n$ matrices *horizontal vectors of degree n*. Both vertical vectors and horizontal vectors are collectively called *numerical vectors*. We denote numerical vectors by **a**, **b**, **c**, …. We say that several numerical vectors are of the *same type* when they are either all vertical vectors or all horizontal vectors.

Example 3 The 3×1 matrix $\begin{pmatrix} 2 \\ -3 \\ 0 \end{pmatrix}$ is a vertical vector of degree 3.

Example 4 Numerical vectors $\begin{pmatrix} 1 & 0 & -3 & 2 \end{pmatrix}$ and $\begin{pmatrix} 3 & -1 & 2 \end{pmatrix}$ are of the same type since both of them are horizontal vectors.

Zero vectors The *vertical zero vector* (resp. *horizontal zero vector*) of degree n is the zero matrix of size $(n, 1)$ (resp. the zero matrix of size $(1, n)$). They are called collectively *zero vectors* and usually denoted by $\mathbf{0}_n$ or $\mathbf{0}$.

Row vectors of matrices Let A be an $m \times n$ matrix:

$$A = \begin{pmatrix} a_{11} & a_{12} & \cdots & a_{1n} \\ a_{21} & a_{22} & \cdots & a_{2n} \\ & & \cdots\cdots & \\ a_{m1} & a_{m2} & \cdots & a_{mn} \end{pmatrix}.$$

The horizontal vectors $\begin{pmatrix} a_{i1} & a_{i2} & \cdots & a_{in} \end{pmatrix}$ of degree n for $i = 1, 2, \ldots, m$ are called *rows* of A, or *row vectors* of A. Rows (resp. row vectors) of A are called the first row (resp. the first row vector), the second row (resp. the second row vector), …, and the m-th row (resp. the m-th row vector) in order from the top to the bottom. We identify horizontal vectors and row vectors and call horizontal vectors also *row vectors*. Horizontal zero vectors are called *zero row vectors*.

Column vectors of matrices The vertical vectors $\begin{pmatrix} a_{1j} \\ a_{2j} \\ \vdots \\ a_{mj} \end{pmatrix}$ of degree m for $j = 1, 2, \ldots, n$ are called the *columns* of A, or *column vectors* of A. Columns (resp. column vectors) of A are called the first column (resp. the first column vector), the second column (resp. the second column vector), …, and the n-th column (resp. the

n-th column vector) from the left to the right in order. We identify vertical vectors and column vectors and call vertical vectors also *column vectors*. Vertical zero vectors are called *zero column vectors*.

Example 5 Let $A = \begin{pmatrix} 1 & -2 & 3 \\ 2 & 5 & -2 \end{pmatrix}$. Then the first row of A is $\begin{pmatrix} 1 & -2 & 3 \end{pmatrix}$ and the second column of A is $\begin{pmatrix} -2 \\ 5 \end{pmatrix}$.

Example 6 Matrices are considered as abstractions of tables. For example, take a table of nutrients:

	Protein	Lipid	Starch
Beef	.28	.5	.05
Flour	.09	.02	.76 (lb.)

In this case, the matrix corresponding to this table is

$$\begin{pmatrix} .28 & .5 & .05 \\ .09 & .02 & .76 \end{pmatrix}.$$

The table below has the same meaning as the above table:

	Beef	Flour
Protein	.28	.09
Lipid	.5	.02
Starch	.05	.76 (lb.)

But the corresponding matrices

$$\begin{pmatrix} .28 & .5 & .05 \\ .09 & .02 & .76 \end{pmatrix} \quad \text{and} \quad \begin{pmatrix} .28 & .09 \\ .5 & .02 \\ .05 & .76 \end{pmatrix}$$

are different.

Square matrices We call $n \times n$ matrices *square matrices of degree n*. For a square matrix $A = (a_{ij})_{n \times n}$, we call the components $a_{11}, a_{22}, \ldots, a_{nn}$ *diagonal components* of the matrix. We denote the zero matrix $O_{n,n}$ by O_n, which is called the *zero matrix of degree n*

Example 7 A matrix $A = \begin{pmatrix} -3 & 2 & 1 \\ 0 & -1 & 2 \\ 1 & 5 & 0 \end{pmatrix}$ is a square matrix of degree 3, and the diagonal components of A are $-3, -1, 0$.

Example 8 The zero matrix of degree 3 is

$$O_3 = O_{3,3} = \begin{pmatrix} 0\,0\,0 \\ 0\,0\,0 \\ 0\,0\,0 \end{pmatrix}.$$

Diagonal matrices A square matrix is called a *diagonal matrix* if all components other than diagonal components are zeros.

Example 9 A matrix $A = \begin{pmatrix} 2 & 0\,0 \\ 0 & -3\,0 \\ 0 & 0\,6 \end{pmatrix}$ is a diagonal matrix of degree 3. The diagonal components of A are 2, -3, 6.

Scalar matrices Let A be a diagonal matrix of degree n. When all diagonal components of A are the same, the diagonal matrix A is called a *scalar matrix*.

Example 10 A matrix $A = \begin{pmatrix} 4\,0\,0 \\ 0\,4\,0 \\ 0\,0\,4 \end{pmatrix}$ is a scalar matrix of degree 3.

Unit matrices A scalar matrix of degree n is called the *unit matrix* of degree n and denoted by E or E_n if all diagonal components are 1.

Example 11 The unit matrix of degree 3 is $E = E_3 = \begin{pmatrix} 1\,0\,0 \\ 0\,1\,0 \\ 0\,0\,1 \end{pmatrix}$.

Transposed matrices Let A be an $m \times n$ matrix. When we exchange the rows and the columns of A, we say that we take the *transposition* of A, and the matrix obtained by exchanging the rows and the columns of A is called the *transposed matrix* of A. The transposed matrix of A is denoted by ${}^tA$. The transposed matrix ${}^tA$ is an $n \times m$ matrix. The (i, j) component of ${}^tA$ is nothing but the (j, i) component of A. We easily see that if we take the transposition of the transposed matrix ${}^tA$, then it turns out to be the original matrix A, namely ${}^t({}^tA) = A$.

Example 12 If $A = \begin{pmatrix} 3\,2 \\ 1\,3 \\ -4\,2 \end{pmatrix}$, then ${}^tA = \begin{pmatrix} 3 & 1 & -4 \\ 2 & 3 & 2 \end{pmatrix}$, which is a 2×3 matrix.

Transposition of numerical vectors If $\mathbf{a}$ is a column vector of degree n, then ${}^t\mathbf{a}$ is a row vector of degree n. To distinguish column vectors and row vectors, we use the letters $\mathbf{a}, \mathbf{b}, \mathbf{c}, \ldots$ for column vectors and the letters ${}^t\mathbf{a}, {}^t\mathbf{b}, {}^t\mathbf{c}, \ldots$ for row vectors. We often denote the zero column vector of degree n by $\mathbf{0}$ or $\mathbf{0}_n$, and the zero row vector of degree n by ${}^t\mathbf{0}$ or ${}^t\mathbf{0}_n$ to distinguish zero column vectors and zero row vectors.

Symmetric matrices A square matrix $A = (a_{ij})$ of degree n is called a *symmetric matrix* if A satisfies the conditions

$$a_{ij} = a_{ji} \quad \text{for} \quad i, j = 1, \ldots, n.$$

Therefore a square matrix A is a symmetric matrix if and only if $A = {}^t A$.

Example 13 A square matrix $A = \begin{pmatrix} 2 & 3 & -1 \\ 3 & 2 & 0 \\ -1 & 0 & 7 \end{pmatrix}$ is a symmetric matrix of degree 3.

Alternating matrices For a matrix $A = (a_{ij})$, we let $-A = (-a_{ij})$. The matrix $-A$ is called the *minus A*. We call a square matrix $A = (a_{ij})$ of degree n an *alternating matrix* if ${}^t A = -A$, namely

$$a_{ij} = -a_{ji} \quad \text{for} \quad i, j = 1, \ldots, n.$$

Example 14 A square matrix $A = \begin{pmatrix} 0 & 3 & -1 \\ -3 & 0 & 5 \\ 1 & -5 & 0 \end{pmatrix}$ is an alternating matrix of degree 3.

Exercise 1.1 Answer the following questions for the matrix

$$A = \begin{pmatrix} -1 & 2 & 6 & -4 \\ 3 & 0 & 12 & 0 \\ 1 & 4 & 0 & 7 \end{pmatrix} :$$

(1) What is the size of A?
(2) What are the $(2, 1)$ component and the $(3, 4)$ component?
(3) What are the second row and the third column?
(4) What is the transposed matrix of A?

Answers. (1) The size of A is $(3, 4)$ or 3×4.

(2) The $(2, 1)$ component is 3. The $(3, 4)$ component is 7.

(3) The second row is $\begin{pmatrix} 3 & 0 & 12 & 0 \end{pmatrix}$. The third column is $\begin{pmatrix} 6 \\ 12 \\ 0 \end{pmatrix}$.

(4) ${}^t A = \begin{pmatrix} -1 & 3 & 1 \\ 2 & 0 & 4 \\ 6 & 12 & 0 \\ -4 & 0 & 7 \end{pmatrix}$. $\|$

Exercise 1.2 Let A be a square matrix of degree n. Show that if a matrix A is both symmetric and alternating, then A is a zero matrix.

Answer. Let $A = (a_{ij})$. Since A is a symmetric matrix, we have

$$a_{ij} = a_{ji} \quad \text{for} \quad i, j = 1, \ldots, n.$$

On the other hand, as A is also an alternating matrix, we have

$$a_{ij} = -a_{ji} \quad \text{for} \quad i, j = 1, \ldots, n.$$

Then we have $a_{ji} = a_{ij} = -a_{ji}$ for $i, j = 1, \ldots, n$. Therefore

$$a_{ji} = 0 \quad \text{for} \quad i, j = 1, \ldots, n.$$

Thus we see that $A = O_n$. ‖

Kronecker delta For positive integers i and j, we define

$$\delta_{i,j} = \begin{cases} 1 \ (i = j), \\ 0 \ (i \neq j). \end{cases}$$

The function δ is called the *Kronecker delta*.

Example 15 $\delta_{2,3} = \delta_{1,3} = \delta_{2,1} = 0$ and $\delta_{1,1} = \delta_{2,2} = \delta_{3,3} = 1$.

Example 16 The square matrix $A = (i\,\delta_{i,j+1} - j\,\delta_{i+1,j})_{3\times3}$ is

$$A = \begin{pmatrix} 0 & -2 & 0 \\ 2 & 0 & -3 \\ 0 & 3 & 0 \end{pmatrix}.$$

Example 17 The unit matrix E_n is expressed as $E_n = (\delta_{i,j})_{n \times n}$.

Triangular matrices When a square matrix $A = (a_{ij})$ of degree n satisfies $a_{ij} = 0$ for $1 \leq j < i \leq n$, A is called an *upper triangular matrix*. Similarly, a square matrix $A = (a_{ij})$ of degree n satisfying $a_{ij} = 0$ for $1 \leq i < j \leq n$ is called a *lower triangular matrix*. Upper triangular matrices and lower triangular matrices are collectively called *triangular matrices*.

Example 18 A matrix $A = \begin{pmatrix} 2 & -1 & 9 \\ 0 & -3 & 1 \\ 0 & 0 & 5 \end{pmatrix}$ is an upper triangular matrix of degree 3.

Example 19 A matrix $A = \begin{pmatrix} -1 & 0 & 0 & 0 \\ 3 & -2 & 0 & 0 \\ 5 & 0 & 1 & 0 \\ 0 & 7 & 2 & 1 \end{pmatrix}$ is a lower triangular matrix of degree 4.

Example 20 If a matrix A is both an upper triangular matrix and a lower triangular matrix, then A is a diagonal matrix. In fact, since $A = (a_{ij})$ is an upper triangular matrix, $a_{ij} = 0$ for $1 \leq j < i \leq n$. As A is a lower triangular matrix, $a_{ij} = 0$ for $1 \leq i < j \leq n$. Then only diagonal components may be non-zeros. Therefore A is a diagonal matrix.

Exercises (Sect. 1.1)

1 Answer the following questions for the matrix $A = \begin{pmatrix} -1 & 3 & -2 \\ 1 & 2 & 5 \end{pmatrix}$:

(1) What is the size of A ?
(2) What is the $(1, 3)$ component of A ?
(3) What is the first row of A ?
(4) What is the third column of A ?
(5) What is the transposed matrix of A ?

2 Answer the following questions for the matrix $A = \begin{pmatrix} 2 & 4 & -3 & 8 \\ 3 & -1 & 2 & -5 \\ 18 & 0 & 2 & 12 \end{pmatrix}$:

(1) What is the size of A ?
(2) What is the $(3, 2)$ component of A ?
(3) What is the second row of A ?
(4) What is the third column of A ?
(5) What is the transposed matrix of A ?

3 Answer the following questions for the matrix $A = \begin{pmatrix} 4 & -1 & 2 \\ -1 & 2 & 0 \\ 9 & 7 & -1 \\ 5 & 0 & 3 \end{pmatrix}$:

(1) What is the size of A ?
(2) What is the $(4, 1)$ component of A ?
(3) What is the third row of A ?
(4) What is the second column of A ?
(5) What is the transposed matrix of A ?

4 Find the numbers a, b, c, d so that they satisfy the following equalities:

(1) $\begin{pmatrix} a-1 & 2 \\ b & c+2 \end{pmatrix} = \begin{pmatrix} 3 & 2d \\ 4 & 7 \end{pmatrix}$.

(2) $\begin{pmatrix} a & b-1 \\ c+2 & 1 \end{pmatrix} = {}^t\begin{pmatrix} 2 & b \\ 2c & d \end{pmatrix}$.

5 When the following matrices A are symmetric matrices, find the numbers a, b, c and the matrices A:

(1) $A = \begin{pmatrix} 1 & 2c+1 & 3 \\ a & -2 & c \\ b & a-2 & 0 \end{pmatrix}$.
(2) $A = \begin{pmatrix} 2 & b-2 & 1 \\ a & 3 & c \\ b-2 & a+1 & 5 \end{pmatrix}$.

(3) $A = \begin{pmatrix} a & 4 & 2 \\ a+b & 2a-1 & b-2 \\ 2b & c & 2c+1 \end{pmatrix}$.

6   When the following matrices A are alternating matrices, find the numbers a, b, c, d and the matrices A:

(1) $A = \begin{pmatrix} 0 & 2c+1 & 3 \\ a & b-2 & c \\ c & d-2 & 0 \end{pmatrix}$.

(2) $A = \begin{pmatrix} 0 & a+1 & -1 \\ b & 3-b & d \\ 1 & c-1 & c \end{pmatrix}$.

(3) $A = \begin{pmatrix} a & a+2b & c+1 \\ -4 & 0 & c \\ 3 & d-2 & b-2 \end{pmatrix}$.

7   Find the following square matrices of degree 3:

(1) $A = \left(\delta_{i,j} + 2\delta_{i+1,j} \right)$.   (2) $A = \left(\delta_{i,j} + \delta_{i,4-j} \right)$.

(3) $A = \left(2\delta_{i,j} + \delta_{i+2,2j+1} \right)$.

8   Show that every diagonal component of an alternating matrix is zero.

1.2   Matrix Operations

In this section, we shall define matrix operations.

Addition and subtraction of matrices   We can define addition and subtraction of matrices *only when the sizes of two matrices are equal*. For two $m \times n$ matrices $A = \left(a_{ij} \right)_{m \times n}$ and $B = \left(b_{ij} \right)_{m \times n}$, we define the *addition* $A + B$ of A and B by adding the (i, j) components of A and B for each (i, j), and define the *subtraction* $A - B$ of B from A by subtracting the (i, j) component of B from the (i, j) component of A for each (i, j):

$$A + B = \left(a_{ij} + b_{ij} \right)_{m \times n}, \; A - B = \left(a_{ij} - b_{ij} \right)_{m \times n}.$$

The matrix $A + B$ is called the *sum* of matrices A and B.

Example 1 $\begin{pmatrix} 1 & -2 & 8 \\ 2 & 5 & -1 \end{pmatrix} + \begin{pmatrix} 3 & 5 & 1 \\ -4 & -1 & 2 \end{pmatrix} = \begin{pmatrix} 4 & 3 & 9 \\ -2 & 4 & 1 \end{pmatrix}$.

Example 2 $\begin{pmatrix} 1 & -2 & 8 \\ 2 & 5 & -1 \end{pmatrix} - \begin{pmatrix} 3 & 5 & 1 \\ -4 & -1 & 2 \end{pmatrix} = \begin{pmatrix} -2 & -7 & 7 \\ 6 & 6 & -3 \end{pmatrix}$.

Example 3   Two families went to two concerts. The tables below show payments for each concert and the total payments for two concerts:

First Concert

	Tickets	Lunches
Smiths	100	85
Johns	110	100

($)

Second Concert

	Tickets	Lunches
Smiths	150	100
Johns	180	130

($)

Total Payments

	Tickets	Lunches
Smiths	250	185
Johns	290	230

($)

Now, let A, B and C be the matrices corresponding to those tables. Then C is equal to $A + B$, the sum of A and B. In fact,

$$A + B = \begin{pmatrix} 100 & 85 \\ 110 & 100 \end{pmatrix} + \begin{pmatrix} 150 & 100 \\ 180 & 130 \end{pmatrix} = \begin{pmatrix} 250 & 185 \\ 290 & 230 \end{pmatrix} = C.$$

Scalar multiplication Let A be an $m \times n$ matrix. For a number c, we define the matrix cA by multiplying each component of A by c:

$$c\left(a_{ij}\right)_{m \times n} = \left(ca_{ij}\right)_{m \times n}.$$

Example 4 For a matrix A, we have $-A = (-1)A$. The left-hand side is the minus A and the right-hand side is the matrix A multiplied by -1.

Example 5 $3\begin{pmatrix} 1 & -2 & 8 \\ 2 & 5 & -1 \end{pmatrix} = \begin{pmatrix} 3 & -6 & 24 \\ 6 & 15 & -3 \end{pmatrix}.$

Example 6 The table of nutrients below is cited in Example 6 of Sect. 1.1:

	Protein	Lipid	Starch
Beef	.28	.5	.05
Flour	.09	.02	.76

(lb.)

To obtain the table of nutrients for 2 lbs., we have only to multiply the corresponding matrix by 2. As for the tables of nutrients, we have

$$2\begin{pmatrix} .28 & .5 & .05 \\ .09 & .02 & .76 \end{pmatrix} = \begin{pmatrix} .56 & 1.0 & .10 \\ .18 & .04 & 1.52 \end{pmatrix}.$$

Matrix multiplication Take two matrices A and B. The matrices A and B can be multiplied *only when the number of columns of A is equal to the number of rows of B*. In other words, AB can be defined only when $A = \left(a_{ij}\right)$ is an $m \times n$ matrix and $B = \left(b_{jk}\right)$ is an $n \times r$ matrix. When this condition is satisfied, we define AB by the $m \times r$ matrix $\left(c_{ik}\right)_{m \times r}$ of which the (i, k) component is given by

$$c_{ik} = \sum_{j=1}^{n} a_{ij} b_{jk} = a_{i1} b_{1k} + \cdots + a_{in} b_{nk}$$

for $i = 1, \ldots, m$ and $k = 1, \ldots, r$. The $m \times r$ matrix $AB = \left(c_{ik} \right)_{m \times r}$ is called the *product* of A and B.

Example 7 $\begin{pmatrix} 2 & 1 & -3 \\ 1 & -5 & 2 \end{pmatrix} \begin{pmatrix} 3 & 1 & 0 \\ 2 & 0 & -1 \\ -1 & 4 & 1 \end{pmatrix} = \begin{pmatrix} 11 & -10 & -4 \\ -9 & 9 & 7 \end{pmatrix}.$

Example 8 $\begin{pmatrix} -1 \\ 2 \end{pmatrix} (3 \ -2) = \begin{pmatrix} -3 & 2 \\ 6 & -4 \end{pmatrix}.$

Example 9 We shall show the meaning of multiplications of matrices by an example. Two families went to a concert. The tables below show the numbers of people in each family and the prices:

Numbers of people

	Adults	Students	Children
Smiths	2	1	2
Johns	1	2	3

Prices

	Tickets	Lunches	
Adults	25	20	
Students	20	25	
Children	15	10	($)

Now, the total payment by each family is:

Total payments

	Tickets	Lunches	
Smiths	100	85	
Johns	110	100	($)

The table of the payments by the families can be obtained by matrix multiplication. Let A, B and C be the matrices corresponding to the tables of numbers of people, the prices and the payments, respectively:

$$A = \begin{pmatrix} 2 & 1 & 2 \\ 1 & 2 & 3 \end{pmatrix}, \ B = \begin{pmatrix} 25 & 20 \\ 20 & 25 \\ 15 & 10 \end{pmatrix}, \ C = \begin{pmatrix} 100 & 85 \\ 110 & 100 \end{pmatrix}.$$

For example, the payment by the Smiths for the lunch (the (1, 2) component of C) can be calculated as

$$85 = 2 \times 20 + 1 \times 25 + 2 \times 10 = \begin{pmatrix} 2 & 1 & 2 \end{pmatrix} \begin{pmatrix} 20 \\ 25 \\ 10 \end{pmatrix}.$$

We can also find the other components of C similarly, and we have

$$AB = \begin{pmatrix} 2 & 1 & 2 \\ 1 & 2 & 3 \end{pmatrix} \begin{pmatrix} 25 & 20 \\ 20 & 25 \\ 15 & 10 \end{pmatrix} = \begin{pmatrix} 100 & 85 \\ 110 & 100 \end{pmatrix} = C.$$

Properties of the matrix operations The matrix operations such as addition, subtraction and multiplication have similar properties to the operations of numbers except for the following two points.

(1) Addition, subtraction and multiplication cannot always be defined for any two matrices.

(2) For two matrices A and B, even if the products AB and BA are defined, AB is not necessarily equal to BA.

Commutativity of matrices If both AB and BA are defined and $AB = BA$ holds, then we say that the matrices A and B are *commutative*.

Example 10 Let $A = \begin{pmatrix} 5 & 4 \\ -2 & -1 \end{pmatrix}$ and $B = \begin{pmatrix} 1 & -2 \\ 1 & 4 \end{pmatrix}$. Then

$$AB = \begin{pmatrix} 9 & 6 \\ -3 & 0 \end{pmatrix} \text{ and } BA = \begin{pmatrix} 9 & 6 \\ -3 & 0 \end{pmatrix}.$$

Since $AB = BA$, the matrices A and B are commutative.

Example 11 Let $A = \begin{pmatrix} 2 & 1 \\ 3 & -1 \end{pmatrix}$ and $B = \begin{pmatrix} 3 & 1 \\ -1 & 2 \end{pmatrix}$. Then

$$AB = \begin{pmatrix} 5 & 4 \\ 10 & 1 \end{pmatrix} \text{ and } BA = \begin{pmatrix} 9 & 2 \\ 4 & -3 \end{pmatrix}.$$

Since $AB \neq BA$, the matrices A and B are not commutative.

Other properties of the operations of numbers are satisfied for matrix operations. We state here those properties. But those equalities below hold only when the operations can be defined. Here A, B, C are matrices, O is a zero matrix, E is a unit matrix and a, b are scalars (i.e. numbers).

Properties of addition:

$$A + B = B + A, \quad A + O = A, \quad (A + B) + C = A + (B + C).$$

Properties of scalar multiplication:

$$0A = O, \quad 1A = A, \quad a(bA) = (ab)A, \quad (aA)B = a(AB).$$

Properties of matrix multiplication:

$$AE = A, \quad EA = A, \quad AO = O, \quad OA = O, \quad (AB)C = A(BC).$$

Distribution law:

$$a(A + B) = aA + aB, \quad (a+b)A = aA + bA,$$
$$A(B + C) = AB + AC, \quad (A + B)C = AC + BC.$$

(3) Let $A_1, \ldots, A_r$ be r matrices of the same size. Let $A_k = \left(a_{ij}^{(k)} \right)$ for $k = 1, \ldots, r$. Then we define the *sum* of $A_1, \ldots, A_r$ by

$$A = A_1 + \cdots + A_r = \left(a_{ij}^{(1)} + \cdots + a_{ij}^{(r)} \right).$$

(4) Let $A_1, A_2, \ldots, A_r$ be matrices. Assume that the matrix multiplication of adjacent matrices can be defined. We define the *product* of $A_1, A_2, \ldots, A_r$ by

$$A_1 A_2 \cdots A_r = \underbrace{(\cdots (A_1 A_2) A_3) \cdots)}_{r-2} A_r.$$

We note that $A_1 A_2 \cdots A_r$ is independent of the order of multiplications as long as the order of matrices is not changed. For example,

$$((A_1 A_2) A_3) A_4 = (A_1 A_2)(A_3 A_4) = A_1(A_2(A_3 A_4)).$$

This can be seen by the associative law of multiplication of matrices which is given after (5) below.

(5) Let A be a square matrix of degree n. For a positive integer r, the *r-th power* $A^r = \underbrace{AA \cdots A}_{r}$ of A can be defined and is independent of the order of multiplications. We understand that $A^0 = E_n$.

Associative law of matrix multiplication: Let A be an $m \times n$ matrix, B an $n \times r$ matrix and C an $r \times s$ matrix. Then the equality

(∗) $$(AB)C = A(BC)$$

holds. The equality (∗) is called the *associative law of matrix multiplication*. We shall show the associative law (∗). Let

$$A = \left(a_{ij} \right)_{m \times n}, \quad B = \left(b_{jk} \right)_{n \times r}, \quad C = \left(c_{kl} \right)_{r \times s}.$$

We note that both matrices $(AB)C$ and $A(BC)$ have the same size (m, s). We are going to find each component of $(AB)C$. We let

$$AB = \left(u_{ik} \right)_{m \times r} \text{ and } (AB)C = \left(v_{il} \right)_{m \times s}$$

and also

$$BC = \left(p_{jl} \right)_{n \times s} \text{ and } A(BC) = \left(q_{il} \right)_{m \times s}.$$

To see the equality $(AB)C = A(BC)$, we have only to see that $v_{il} = q_{il}$ for any $i = 1, \ldots, m$ and $l = 1, \ldots, s$. First we have

$$u_{ik} = a_{i1}b_{1k} + a_{i2}b_{2k} + \cdots + a_{in}b_{nk}$$

and

$$
\begin{aligned}
v_{il} &= u_{i1}c_{1l} + u_{i2}c_{2l} + \cdots + u_{ir}c_{rl} \\
&= (a_{i1}b_{11} + a_{i2}b_{21} + \cdots + a_{in}b_{n1})c_{1l} + \cdots + (a_{i1}b_{1r} + a_{i2}b_{2r} + \cdots + a_{in}b_{nr})c_{rl} \\
&= \left(\sum_{j=1}^{n} a_{ij}b_{j1} \right) c_{1l} + \left(\sum_{j=1}^{n} a_{ij}b_{j2} \right) c_{2l} + \cdots + \left(\sum_{j=1}^{n} a_{ij}b_{jr} \right) c_{rl} \\
&= \sum_{k=1}^{r} \left(\sum_{j=1}^{n} a_{ij}b_{jk} \right) c_{kl} = \sum_{k=1}^{r} \sum_{j=1}^{n} a_{ij}b_{jk}c_{kl}
\end{aligned}
$$

for any $i = 1, \ldots, m$ and $l = 1, \ldots, s$. Next, we have

$$p_{jl} = b_{j1}c_{1l} + b_{j2}c_{2l} + \cdots + b_{jr}c_{rl}$$

and

$$
\begin{aligned}
q_{il} &= a_{i1}p_{1l} + a_{i2}p_{2l} + \cdots + a_{in}p_{nl} \\
&= a_{i1}(b_{11}c_{1l} + b_{12}c_{2l} + \cdots + b_{1r}c_{rl}) + \cdots + a_{in}(b_{n1}c_{1l} + b_{n2}c_{2l} + \cdots + b_{nr}c_{rl}) \\
&= a_{i1} \left(\sum_{k=1}^{r} b_{1k}c_{kl} \right) + a_{i2} \left(\sum_{k=1}^{r} b_{2k}c_{kl} \right) + \cdots + a_{in} \left(\sum_{k=1}^{r} b_{nk}c_{kl} \right) \\
&= \sum_{j=1}^{n} a_{ij} \left(\sum_{k=1}^{r} b_{jk}c_{kl} \right) \\
&= \sum_{j=1}^{n} \sum_{k=1}^{r} a_{ij}b_{jk}c_{kl} = \sum_{k=1}^{r} \sum_{j=1}^{n} a_{ij}b_{jk}c_{kl}.
\end{aligned}
$$

Thus we obtain the equality

$$v_{il} = q_{il} \text{ for any } i = 1, \dots, m \text{ and } l = 1, \dots, s.$$

Therefore the associative law $(*)$ of matrix multiplication holds.

Nilpotent matrices A square matrix A of degree n is called a *nilpotent matrix* if there exists a positive integer m satisfying $A^m = O_n$.

Example 12 Let $A = \begin{pmatrix} 0 & 1 & 0 \\ 0 & 0 & 1 \\ 0 & 0 & 0 \end{pmatrix}$. Then we have

$$A^2 = \begin{pmatrix} 0 & 1 & 0 \\ 0 & 0 & 1 \\ 0 & 0 & 0 \end{pmatrix} \begin{pmatrix} 0 & 1 & 0 \\ 0 & 0 & 1 \\ 0 & 0 & 0 \end{pmatrix} = \begin{pmatrix} 0 & 0 & 1 \\ 0 & 0 & 0 \\ 0 & 0 & 0 \end{pmatrix},$$

$$A^3 = A^2 A = \begin{pmatrix} 0 & 0 & 1 \\ 0 & 0 & 0 \\ 0 & 0 & 0 \end{pmatrix} \begin{pmatrix} 0 & 1 & 0 \\ 0 & 0 & 1 \\ 0 & 0 & 0 \end{pmatrix} = \begin{pmatrix} 0 & 0 & 0 \\ 0 & 0 & 0 \\ 0 & 0 & 0 \end{pmatrix}.$$

Therefore A is a nilpotent matrix.

Exercise 1.3 Find all combinations of two matrices for which we can define multiplication, and find the products of those matrices.

$$A = \begin{pmatrix} 2 & 0 & 1 \\ 0 & 1 & 3 \end{pmatrix}, \quad B = \begin{pmatrix} 1 \\ 2 \\ 1 \end{pmatrix}, \quad C = (1\ 0\ 4), \quad D = \begin{pmatrix} 3 & -1 & 0 \\ 1 & 2 & 1 \\ 0 & 5 & -1 \end{pmatrix}.$$

Answer. The multiplication of matrices X and Y can be defined if and only if the number of columns of X is equal to the number of rows of Y. So, the possible products are AB, AD, BC, CB, CD, DB and $D^2 = DD$.

$$AB = \begin{pmatrix} 3 \\ 5 \end{pmatrix}, \quad AD = \begin{pmatrix} 6 & 3 & -1 \\ 1 & 17 & -2 \end{pmatrix}, \quad BC = \begin{pmatrix} 1 & 0 & 4 \\ 2 & 0 & 8 \\ 1 & 0 & 4 \end{pmatrix},$$

$$CB = (5) = 5, \quad CD = (3\ 19\ -4), \quad DB = \begin{pmatrix} 1 \\ 6 \\ 9 \end{pmatrix},$$

$$D^2 = DD = \begin{pmatrix} 8 & -5 & -1 \\ 5 & 8 & 1 \\ 5 & 5 & 6 \end{pmatrix}. \ \|$$

Exercise 1.4 Find if the following two statements are true or not by giving proofs or counterexamples:

(1) If A is a square matrix of degree n, then

$$A^2 + 3A + 2E_n = (A + 2E_n)(A + E_n).$$

(2) If A and B are square matrices of the same degree, then

$$(A + B)^2 = A^2 + 2AB + B^2.$$

Answers. (1) True. In fact

$$\begin{aligned}(A + 2E_n)(A + E_n) &= A(A + E_n) + 2E_n(A + E_n)\\ &= A^2 + A + 2(A + E_n) = A^2 + 3A + 2E_n.\end{aligned}$$

(2) False. We easily see that

$$(A + B)^2 = (A + B)(A + B) = A^2 + AB + BA + B^2.$$

Therefore the equality holds if and only if $AB = BA$, i.e. A and B are commutative. For example, if $A = \begin{pmatrix} 1 & 1 \\ 0 & 1 \end{pmatrix}$ and $B = \begin{pmatrix} 0 & -1 \\ 1 & 0 \end{pmatrix}$, then we have

$$(A + B)^2 = \begin{pmatrix} 1 & 0 \\ 2 & 1 \end{pmatrix} \neq A^2 + 2AB + B^2 = \begin{pmatrix} 2 & 0 \\ 2 & 0 \end{pmatrix}. \ \|$$

Properties of addition, scalar multiplication and matrix multiplication with respect to the transposition of matrices

$$\,^t(A + B) = \,^tA + \,^tB, \quad \,^t(cA) = c\,^tA, \quad \,^t(AB) = \,^tB\,^tA.$$

Example 13 Let $A = \begin{pmatrix} 1 & -1 \\ 3 & 1 \end{pmatrix}$ and $B = \begin{pmatrix} 5 & 1 \\ -2 & 3 \end{pmatrix}$. Then we see

$$AB = \begin{pmatrix} 1 & -1 \\ 3 & 1 \end{pmatrix}\begin{pmatrix} 5 & 1 \\ -2 & 3 \end{pmatrix} = \begin{pmatrix} 7 & -2 \\ 13 & 6 \end{pmatrix}.$$

and

$$\,^tB\,^tA = \begin{pmatrix} 5 & -2 \\ 1 & 3 \end{pmatrix}\begin{pmatrix} 1 & 3 \\ -1 & 1 \end{pmatrix} = \begin{pmatrix} 7 & 13 \\ -2 & 6 \end{pmatrix}.$$

Thus we have $\,^t(AB) = \,^tB\,^tA$.

Exercises (Sect. 1.2)
1 Find the products of the following matrices:

(1) $\begin{pmatrix} 3 & 2 \\ -1 & 4 \end{pmatrix}\begin{pmatrix} -2 & 1 \\ 5 & 2 \end{pmatrix}.$

(2) $\begin{pmatrix} 0 & -1 \\ -2 & 3 \end{pmatrix}\begin{pmatrix} 2 & -5 \\ -7 & 4 \end{pmatrix}.$

(3) $\begin{pmatrix} 2 & -1 & 2 \\ 1 & 5 & -2 \end{pmatrix} \begin{pmatrix} 2 & 3 & -2 \\ 0 & -2 & 7 \\ 1 & 1 & 3 \end{pmatrix}$.

(4) $\begin{pmatrix} 2 \\ -1 \\ 4 \end{pmatrix} (3 \ 1 \ -2)$.

(5) $(3 \ 1 \ -2) \begin{pmatrix} 2 \\ -1 \\ 4 \end{pmatrix}$.

(6) $\begin{pmatrix} 0 & 1 & 2 \\ 0 & 0 & 1 \\ 0 & 0 & 0 \end{pmatrix}^3$.

(7) $\begin{pmatrix} 1 & 2 & -1 \\ 1 & 0 & 2 \\ 0 & 1 & -1 \end{pmatrix} \begin{pmatrix} 0 & 1 & 1 \\ 2 & -1 & 3 \\ -1 & 0 & 1 \end{pmatrix}$.

(8) $\begin{pmatrix} -2 & 1 & 2 \\ 0 & 1 & -1 \\ 1 & 0 & 1 \end{pmatrix} \begin{pmatrix} 1 & 0 & 1 \\ 1 & 1 & 0 \\ 2 & 0 & 1 \end{pmatrix}$.

(9) $\begin{pmatrix} 0 & 3 & 0 \\ 1 & 1 & -1 \\ 1 & 0 & 2 \end{pmatrix} \begin{pmatrix} 1 & 1 & 0 \\ 0 & -1 & 0 \\ -1 & 1 & 0 \end{pmatrix}$.

(10) $\begin{pmatrix} 1 & 0 & 0 & -1 \\ 0 & 1 & 0 & 1 \\ 0 & 0 & 1 & 1 \\ -1 & 1 & 0 & 0 \end{pmatrix} \begin{pmatrix} 0 & 1 & 1 & 0 \\ -1 & 0 & 1 & 1 \\ 1 & 1 & 0 & 0 \\ 0 & -1 & 0 & 1 \end{pmatrix}$.

(11) $\begin{pmatrix} 0 & 1 & 1 & -1 \\ -1 & 0 & 0 & 1 \\ 1 & 0 & 1 & -1 \end{pmatrix} \begin{pmatrix} 2 & 0 & 0 \\ 1 & 1 & 0 \\ 0 & 1 & 1 \\ 0 & -2 & 1 \end{pmatrix}$.

(12) $\begin{pmatrix} 0 & 1 & -1 \\ 1 & 2 & 0 \\ -2 & 1 & 0 \\ 0 & 1 & -1 \end{pmatrix} \begin{pmatrix} 1 & 0 & 1 & 0 \\ -2 & 0 & 0 & -1 \\ 0 & 1 & -1 & 0 \end{pmatrix}$.

(13) $\begin{pmatrix} 2 & 3 & -1 \\ 0 & 5 & 4 \\ -1 & 0 & -2 \end{pmatrix} \left\{ \begin{pmatrix} 0 & 5 & 9 \\ 3 & -2 & 8 \\ -1 & 8 & 1 \end{pmatrix} - 2 \begin{pmatrix} -1 & 0 & 1 \\ 3 & 2 & 3 \\ -4 & 2 & -1 \end{pmatrix} \right\}$.

2 Find all combinations of two of the following matrices for which we can define multiplication, and also find their products:

$$A = \begin{pmatrix} 2 \\ 1 \\ -1 \end{pmatrix}, B = \begin{pmatrix} 3 & 2 \\ 4 & 1 \\ 0 & 1 \end{pmatrix}, C = (2 \ 0 \ 1), D = \begin{pmatrix} 2 & 3 \\ -1 & 4 \end{pmatrix}.$$

3 Find all combinations of two of the following matrices for which we can define multiplication, and also find their products:

$$A = \begin{pmatrix} 2 & -1 \\ 0 & 1 \\ -2 & 0 \end{pmatrix}, B = \begin{pmatrix} 0 \\ 1 \\ 0 \end{pmatrix}, C = \begin{pmatrix} 1 & 0 \\ 1 & 2 \end{pmatrix}, D = (-1 \ 0 \ 1).$$

4 For the following square matrices A, calculate the n-th power A^n:

(1) $\begin{pmatrix} 0 & 1 & 0 \\ 0 & 0 & 1 \\ 0 & 0 & 0 \end{pmatrix}$. (2) $\begin{pmatrix} 0 & 0 & 1 \\ 1 & 0 & 0 \\ 0 & 1 & 0 \end{pmatrix}$. (3) $\begin{pmatrix} a & 0 & 0 \\ 0 & b & 0 \\ 0 & 0 & c \end{pmatrix}$. (4) $\begin{pmatrix} a & b \\ 0 & 1 \end{pmatrix}$.

5 Find if the following matrices are commutative or not:

(1) $\begin{pmatrix} 0 & 1 & 0 \\ 0 & 0 & 1 \\ 0 & 0 & 1 \end{pmatrix}$, $\begin{pmatrix} a & 0 & 0 \\ 1 & a & 0 \\ 0 & 1 & a \end{pmatrix}$. (2) $\begin{pmatrix} a & 0 & 0 \\ 0 & b & 0 \\ 0 & 0 & c \end{pmatrix}$, $\begin{pmatrix} 0 & 0 & 1 \\ 0 & 1 & 0 \\ 1 & 0 & 0 \end{pmatrix}$.

6 Find the numbers a, b, c, d so that they satisfy the following matrix equation:

$$\begin{pmatrix} 2 & 1 \\ 3 & 4 \end{pmatrix} \begin{pmatrix} a & -2 \\ c & 3 \end{pmatrix} = \begin{pmatrix} 1 & b \\ -1 & d \end{pmatrix}.$$

7 Let A be a square matrix of degree n. If $A^m = O_n$, then show that

$$(E_n - A)(E_n + A + \cdots + A^{m-1}) = E_n.$$

8 Show that if A and B are commutative nilpotent matrices of degree n, then the matrices AB and $A + B$ are also nilpotent matrices.

9 Show that any square matrix is uniquely expressed as the sum of a symmetric matrix and an alternating matrix.

1.3 Partitions of Matrices

Partitions of matrices Division of matrices into several small matrices is very useful since we may treat small matrices as if they are numbers. For such a purpose, we divide matrices into small matrices by rows and columns. Such divisions are called *partitions of matrices*. When an $m \times n$ matrix A is divided in the following way:

$$A = \begin{array}{c} \\ m_1\{ \\ m_2\{ \\ \vdots \\ m_s\{ \end{array} \overbrace{\begin{array}{|c|c|c|c|}}^{n_1} \overbrace{}^{n_2} \cdots \overbrace{}^{n_t} \left(\begin{array}{c|c|c|c} A_{11} & A_{12} & \cdots & A_{1t} \\ \hline A_{21} & A_{22} & \cdots & A_{2t} \\ \hline \cdots & \cdots & \cdots & \cdots \\ \hline A_{s1} & A_{s2} & \cdots & A_{st} \end{array}\right),$$

we say that A is *partitioned* into $s \times t$ small matrices. Here A_{kl} is an $m_k \times n_l$ matrix for $k = 1, \ldots, s$ and $l = 1, \ldots, t$. We note that the numbers of rows of $A_{k1}, \ldots, A_{kt}$ are the same for $k = 1, \ldots, s$, and the numbers of columns of $A_{1l}, \ldots, A_{sl}$ are the same for $l = 1, 2, \ldots, t$. We call it a *partition of a matrix A*.

Example 1

$$A = \begin{pmatrix} 2 & 3 & 0 \\ 1 & -2 & 0 \\ \hline 5 & 3 & -9 \end{pmatrix} = \begin{pmatrix} A_{11} & A_{12} \\ A_{21} & A_{22} \end{pmatrix}. \quad \text{Here}$$

$$A_{11} = \begin{pmatrix} 2 & 3 \\ 1 & -2 \end{pmatrix}, A_{12} = \begin{pmatrix} 0 \\ 0 \end{pmatrix}, A_{21} = \begin{pmatrix} 5 & 3 \end{pmatrix}, A_{22} = \begin{pmatrix} -9 \end{pmatrix}.$$

Matrix multiplication using partitions Let A be a matrix partitioned into $s \times t$ small matrices and B a matrix partitioned into $t \times u$ small matrices in the following ways:

$$A = \begin{pmatrix} \overbrace{A_{11}}^{n_1} & \overbrace{A_{12}}^{n_2} & \cdots & \overbrace{A_{1t}}^{n_t} \\ \hline A_{21} & A_{22} & \cdots & A_{2t} \\ \hline \cdots & \cdots & \cdots & \cdots \\ \hline A_{s1} & A_{s2} & \cdots & A_{st} \end{pmatrix}, \quad B = \begin{matrix} n_1\{ \\ n_2\{ \\ \vdots \\ n_t\{ \end{matrix} \begin{pmatrix} B_{11} & B_{12} & \cdots & B_{1u} \\ \hline B_{21} & B_{22} & \cdots & B_{2u} \\ \hline \cdots & \cdots & \cdots & \cdots \\ \hline B_{t1} & B_{t2} & \cdots & B_{tu} \end{pmatrix}.$$

We assume that the number of columns of A_{ik} is equal to the number of rows of B_{kj} for each i, j. Then we can calculate the product AB as if small matrices A_{ik}, B_{kj} are numbers:

$$AB = \begin{pmatrix} C_{11} & C_{12} & \cdots & C_{1u} \\ C_{21} & C_{22} & \cdots & C_{2u} \\ \cdots & \cdots & \cdots \\ C_{s1} & C_{s2} & \cdots & C_{su} \end{pmatrix}, C_{ij} = A_{i1}B_{1j} + A_{i2}B_{2j} + \cdots + A_{it}B_{tj}.$$

Since the proof in a general case is too long, we give the proof in a special case in Example 2 below.

Example 2 We consider the case that both A and B are square matrices of degree 4 which are partitioned into small 2×2 matrices:

$$A = \begin{pmatrix} A_{11} & A_{12} \\ A_{21} & A_{22} \end{pmatrix} = \begin{pmatrix} a_{11} & a_{12} & a_{13} & a_{14} \\ a_{21} & a_{22} & a_{23} & a_{24} \\ \hline a_{31} & a_{32} & a_{33} & a_{34} \\ a_{41} & a_{42} & a_{43} & a_{44} \end{pmatrix}$$

and

$$B = \begin{pmatrix} B_{11} & B_{12} \\ B_{21} & B_{22} \end{pmatrix} = \begin{pmatrix} b_{11} & b_{12} & b_{13} & b_{14} \\ b_{21} & b_{22} & b_{23} & b_{24} \\ \hline b_{31} & b_{32} & b_{33} & b_{34} \\ b_{41} & b_{42} & b_{43} & b_{44} \end{pmatrix}.$$

Then we can calculate AB using partitioned small matrices:

$$AB = \begin{pmatrix} A_{11}B_{11} + A_{12}B_{21} & A_{11}B_{12} + A_{12}B_{22} \\ A_{21}B_{11} + A_{22}B_{21} & A_{21}B_{12} + A_{22}B_{22} \end{pmatrix}.$$

In fact, we have

$$A_{11}B_{11} + A_{12}B_{21}$$
$$= \begin{pmatrix} a_{11} & a_{12} \\ a_{21} & a_{22} \end{pmatrix}\begin{pmatrix} b_{11} & b_{12} \\ b_{21} & b_{22} \end{pmatrix} + \begin{pmatrix} a_{13} & a_{14} \\ a_{23} & a_{24} \end{pmatrix}\begin{pmatrix} b_{31} & b_{32} \\ b_{41} & b_{42} \end{pmatrix}$$
$$= \begin{pmatrix} a_{11}b_{11} + a_{12}b_{21} + a_{13}b_{31} + a_{14}b_{41} & a_{11}b_{12} + a_{12}b_{22} + a_{13}b_{32} + a_{14}b_{42} \\ a_{21}b_{11} + a_{22}b_{21} + a_{23}b_{31} + a_{24}b_{41} & a_{21}b_{12} + a_{22}b_{22} + a_{23}b_{32} + a_{24}b_{42} \end{pmatrix},$$
$$A_{11}B_{12} + A_{12}B_{22}$$
$$= \begin{pmatrix} a_{11} & a_{12} \\ a_{21} & a_{22} \end{pmatrix}\begin{pmatrix} b_{13} & b_{14} \\ b_{23} & b_{24} \end{pmatrix} + \begin{pmatrix} a_{13} & a_{14} \\ a_{23} & a_{24} \end{pmatrix}\begin{pmatrix} b_{33} & b_{34} \\ b_{43} & b_{44} \end{pmatrix}$$
$$= \begin{pmatrix} a_{11}b_{13} + a_{12}b_{23} + a_{13}b_{33} + a_{14}b_{43} & a_{11}b_{14} + a_{12}b_{24} + a_{13}b_{34} + a_{14}b_{44} \\ a_{21}b_{13} + a_{22}b_{23} + a_{23}b_{33} + a_{24}b_{43} & a_{21}b_{14} + a_{22}b_{24} + a_{23}b_{34} + a_{24}b_{44} \end{pmatrix},$$

$$A_{21}B_{11} + A_{22}B_{21}$$
$$= \begin{pmatrix} a_{31} & a_{32} \\ a_{41} & a_{42} \end{pmatrix}\begin{pmatrix} b_{11} & b_{12} \\ b_{21} & b_{22} \end{pmatrix} + \begin{pmatrix} a_{33} & a_{34} \\ a_{43} & a_{44} \end{pmatrix}\begin{pmatrix} b_{31} & b_{32} \\ b_{41} & b_{42} \end{pmatrix}$$
$$= \begin{pmatrix} a_{31}b_{11} + a_{32}b_{21} + a_{33}b_{31} + a_{34}b_{41} & a_{31}b_{12} + a_{32}b_{22} + a_{33}b_{32} + a_{34}b_{42} \\ a_{41}b_{11} + a_{42}b_{21} + a_{43}b_{31} + a_{44}b_{41} & a_{41}b_{12} + a_{42}b_{22} + a_{43}b_{32} + a_{44}b_{42} \end{pmatrix},$$
$$A_{21}B_{12} + A_{22}B_{22}$$
$$= \begin{pmatrix} a_{31} & a_{32} \\ a_{41} & a_{42} \end{pmatrix}\begin{pmatrix} b_{13} & b_{14} \\ b_{23} & b_{24} \end{pmatrix} + \begin{pmatrix} a_{33} & a_{34} \\ a_{43} & a_{44} \end{pmatrix}\begin{pmatrix} b_{33} & b_{34} \\ b_{43} & b_{44} \end{pmatrix}$$
$$= \begin{pmatrix} a_{31}b_{13} + a_{32}b_{23} + a_{33}b_{33} + a_{34}b_{43} & a_{31}b_{14} + a_{32}b_{24} + a_{33}b_{34} + a_{34}b_{44} \\ a_{41}b_{13} + a_{42}b_{23} + a_{43}b_{33} + a_{44}b_{43} & a_{41}b_{14} + a_{42}b_{24} + a_{43}b_{34} + a_{44}b_{44} \end{pmatrix}.$$

Therefore we obtain that the product AB calculated using partitioned small matrices is equal to the product AB obtained by the definition of the multiplication.

Exercise 1.5 Calculate the product AB of the matrices A and B using the given partitions of matrices:

$$A = \begin{pmatrix} 1 & 2 & 4 & 5 & 1 \\ 0 & 2 & 1 & 1 & 3 \\ 3 & 0 & 2 & 0 & 1 \end{pmatrix}, \qquad B = \begin{pmatrix} -1 & 2 \\ 0 & 3 \\ 1 & -2 \\ -1 & 1 \\ 2 & -3 \end{pmatrix}.$$

Answer.

$$AB = \begin{pmatrix} (1\ 2)\begin{pmatrix} -1 & 2 \\ 0 & 3 \end{pmatrix} + (4)(1\ -2) + (5\ 1)\begin{pmatrix} -1 & 1 \\ 2 & -3 \end{pmatrix} \\ \begin{pmatrix} 0 & 2 \\ 3 & 0 \end{pmatrix}\begin{pmatrix} -1 & 2 \\ 0 & 3 \end{pmatrix} + \begin{pmatrix} 1 \\ 2 \end{pmatrix}(1\ -2) + \begin{pmatrix} 1 & 3 \\ 0 & 1 \end{pmatrix}\begin{pmatrix} -1 & 1 \\ 2 & -3 \end{pmatrix} \end{pmatrix}$$

$$
= \left(\begin{array}{c} (-1\ 8) + (4\ -8) + (-3\ 2) \\ \left(\begin{array}{cc} 0 & 6 \\ -3 & 6 \end{array} \right) + \left(\begin{array}{cc} 1 & -2 \\ 2 & -4 \end{array} \right) + \left(\begin{array}{cc} 5 & -8 \\ 2 & -3 \end{array} \right) \end{array} \right)
$$

$$
= \left(\begin{array}{cc} 0 & 2 \\ 6 & -4 \\ 1 & -1 \end{array} \right). \quad \|
$$

When there are zero matrices among partitioned small matrices, products of matrices can be easily calculated by using partitions of matrices.

Exercise 1.6 Let A_1 and B_1 be square matrices of degree m, and A_2 and B_2 square matrices of degree n. Then show the following equality:

$$
\left(\begin{array}{cc} A_1 & O_{m,n} \\ O_{n,m} & A_2 \end{array} \right) \left(\begin{array}{cc} B_1 & O_{m,n} \\ O_{n,m} & B_2 \end{array} \right) = \left(\begin{array}{cc} A_1 B_1 & O_{m,n} \\ O_{n,m} & A_2 B_2 \end{array} \right).
$$

Answer.

$$
\left(\begin{array}{cc} A_1 & O_{m,n} \\ O_{n,m} & A_2 \end{array} \right) \left(\begin{array}{cc} B_1 & O_{m,n} \\ O_{n,m} & B_2 \end{array} \right)
$$

$$
= \left(\begin{array}{cc} A_1 B_1 + O_{m,n} O_{n,m} & A_1 O_{m,n} + O_{m,n} B_2 \\ O_{n,m} B_1 + A_2 O_{n,m} & O_{n,m} O_{m,n} + A_2 B_2 \end{array} \right)
$$

$$
= \left(\begin{array}{cc} A_1 B_1 & O_{m,n} \\ O_{n,m} & A_2 B_2 \end{array} \right). \quad \|
$$

The partition of matrices is very efficient in the following cases:
(1) There are zero matrices among partitioned small matrices.
(2) A matrix is partitioned into column vectors.
(3) A matrix is partitioned into row vectors.
The efficiency (1) is seen in Exercise 1.6 above. In Exercise 1.6, we consider only the case that the matrices are partitioned into four small matrices. But we easily see that similar considerations are applicable to other cases. If matrices have zero matrices among small matrices, the determinants of the matrices, which will be defined in Chap. 3, are also easily calculated.

Now, we consider the partitioning of a matrix into column vectors mentioned in (2) and also the partitions of a matrix into row vectors mentioned in (3). Let A be an $m \times n$ matrix:

$$
A = \left(\begin{array}{cccc} a_{11} & a_{12} & \cdots & a_{1n} \\ a_{21} & a_{22} & \cdots & a_{2n} \\ & \cdots\cdots & & \\ a_{m1} & a_{m2} & \cdots & a_{mn} \end{array} \right).
$$

Column partitions of matrices We divide an $m \times n$ matrix $A = (a_{ij})$ into n column vectors in the following way:

$$A = (\mathbf{a}_1 \; \mathbf{a}_2 \; \cdots \; \mathbf{a}_n),$$

$$\mathbf{a}_1 = \begin{pmatrix} a_{11} \\ a_{21} \\ \vdots \\ a_{m1} \end{pmatrix}, \; \mathbf{a}_2 = \begin{pmatrix} a_{12} \\ a_{22} \\ \vdots \\ a_{m2} \end{pmatrix}, \ldots, \mathbf{a}_n = \begin{pmatrix} a_{1n} \\ a_{2n} \\ \vdots \\ a_{mn} \end{pmatrix}.$$

We call this partition of a matrix A into column vectors the *column partition* of A.

Example 3 (Column partition of a matrix.)

$$A = \begin{pmatrix} 1 & 3 & 4 & 4 \\ 2 & 1 & 0 & -1 \\ 1 & 0 & 5 & 0 \end{pmatrix} = (\mathbf{a}_1 \; \mathbf{a}_2 \; \mathbf{a}_3 \; \mathbf{a}_4),$$

$$\mathbf{a}_1 = \begin{pmatrix} 1 \\ 2 \\ 1 \end{pmatrix}, \; \mathbf{a}_2 = \begin{pmatrix} 3 \\ 1 \\ 0 \end{pmatrix}, \; \mathbf{a}_3 = \begin{pmatrix} 4 \\ 0 \\ 5 \end{pmatrix}, \; \mathbf{a}_4 = \begin{pmatrix} 4 \\ -1 \\ 0 \end{pmatrix}.$$

Row partitions of matrices We also divide an $m \times n$ matrix $A = (a_{ij})$ into m row vectors in the following way:

$$A = \begin{pmatrix} {}^t\mathbf{b}_1 \\ {}^t\mathbf{b}_2 \\ \vdots \\ {}^t\mathbf{b}_m \end{pmatrix}, \qquad \begin{aligned} {}^t\mathbf{b}_1 &= (a_{11} \; a_{12} \; \cdots \; a_{1n}), \\ {}^t\mathbf{b}_2 &= (a_{21} \; a_{22} \; \cdots \; a_{2n}), \\ &\cdots \\ {}^t\mathbf{b}_m &= (a_{m1} \; a_{m2} \; \cdots \; a_{mn}). \end{aligned}$$

We call this partition of a matrix A into row vectors the *row partition* of A.

Example 4 (Row partition of a matrix.)

$$A = \begin{pmatrix} 1 & 3 & 4 & 4 \\ 2 & 1 & 0 & -1 \\ 1 & 0 & 5 & 0 \end{pmatrix} = \begin{pmatrix} {}^t\mathbf{b}_1 \\ {}^t\mathbf{b}_2 \\ {}^t\mathbf{b}_3 \end{pmatrix}, \qquad \begin{aligned} {}^t\mathbf{b}_1 &= (1 \; 3 \; 4 \; 4), \\ {}^t\mathbf{b}_2 &= (2 \; 1 \; 0 \; -1), \\ {}^t\mathbf{b}_3 &= (1 \; 0 \; 5 \; 0). \end{aligned}$$

Since numerical vectors are special matrices, we can define the multiplication of numerical vectors as that of matrices. Let ${}^t\mathbf{a}$ be a row vector of degree n and $\mathbf{b}$ a column vector of degree n. If we let

$$^t\mathbf{a} = \begin{pmatrix} a_1 & a_2 & \cdots & a_n \end{pmatrix} \quad \text{and} \quad \mathbf{b} = \begin{pmatrix} b_1 \\ b_2 \\ \vdots \\ b_n \end{pmatrix},$$

then

$$^t\mathbf{ab} = \begin{pmatrix} a_1 & a_2 & \cdots & a_n \end{pmatrix} \begin{pmatrix} b_1 \\ b_2 \\ \vdots \\ b_n \end{pmatrix} = a_1 b_1 + a_2 b_2 + \cdots + a_n b_n.$$

Using these products of vectors, we can rewrite the multiplication of matrices. Let
$A = \begin{pmatrix} ^t\mathbf{a}_1 \\ ^t\mathbf{a}_1 \\ \vdots \\ ^t\mathbf{a}_m \end{pmatrix}$ be the row partition of an $m \times n$ matrix A and $B = \begin{pmatrix} \mathbf{b}_1 & \mathbf{b}_2 & \cdots & \mathbf{b}_r \end{pmatrix}$ the
column partition of an $n \times r$ matrix B. Then

$$AB = \begin{pmatrix} ^t\mathbf{a}_1\mathbf{b}_1 & ^t\mathbf{a}_1\mathbf{b}_2 & \cdots & ^t\mathbf{a}_1\mathbf{b}_r \\ ^t\mathbf{a}_2\mathbf{b}_1 & ^t\mathbf{a}_2\mathbf{b}_2 & \cdots & ^t\mathbf{a}_2\mathbf{b}_r \\ \cdots\cdots \\ ^t\mathbf{a}_m\mathbf{b}_1 & ^t\mathbf{a}_m\mathbf{b}_2 & \cdots & ^t\mathbf{a}_m\mathbf{b}_r \end{pmatrix} = \begin{pmatrix} ^t\mathbf{a}_1 B \\ ^t\mathbf{a}_2 B \\ \vdots \\ ^t\mathbf{a}_m B \end{pmatrix}$$

$$= \begin{pmatrix} A\mathbf{b}_1 & A\mathbf{b}_2 & \cdots & A\mathbf{b}_r \end{pmatrix}.$$

Example 5 For square matrices $A = \begin{pmatrix} ^t\mathbf{a}_1 \\ ^t\mathbf{a}_2 \\ ^t\mathbf{a}_3 \end{pmatrix} = \begin{pmatrix} 1 & 2 & -1 \\ 0 & 3 & -2 \\ 5 & 0 & 6 \end{pmatrix}$ and $B =$
$\begin{pmatrix} \mathbf{b}_1 & \mathbf{b}_2 & \mathbf{b}_2 \end{pmatrix} = \begin{pmatrix} -2 & 0 & 3 \\ 1 & 2 & 4 \\ -3 & 1 & 2 \end{pmatrix}$, we have

$$C = AB = \begin{pmatrix} c_{ij} \end{pmatrix}, c_{ij} = {}^t\mathbf{a}_i\mathbf{b}_j.$$

For example, $c_{23} = {}^t\mathbf{a}_2\mathbf{b}_3 = \begin{pmatrix} 0 & 3 & -2 \end{pmatrix} \begin{pmatrix} 3 \\ 4 \\ 2 \end{pmatrix} = 8.$

Exercises (Sect. 1.3)
1 Find the products of the following pairs of matrices using their given partitions:

(1) $\left(\begin{array}{cc|cc} 2 & 1 & 1 & 0 \\ 4 & 3 & 0 & 1 \\ \hline 0 & 0 & 1 & 2 \\ 0 & 0 & 0 & 1 \end{array}\right)\left(\begin{array}{cc|cc} 1 & 1 & 1 & 0 \\ 3 & 2 & 0 & 1 \\ \hline 0 & 0 & 2 & 1 \\ 0 & 0 & 1 & 0 \end{array}\right).$

(2) $\left(\begin{array}{cc|cc} 3 & -1 & 0 & 0 \\ 1 & 3 & 0 & 0 \\ \hline 1 & 0 & 1 & 2 \\ 0 & 1 & -1 & 1 \end{array}\right)\left(\begin{array}{cc|cc} 1 & 1 & 0 & 0 \\ 3 & 2 & 0 & 0 \\ \hline 1 & 0 & 2 & 1 \\ 0 & 1 & 1 & 0 \end{array}\right).$

(3) $\left(\begin{array}{c|cc|c} -1 & 0 & 0 & 0 \\ \hline 0 & -3 & 1 & 0 \\ 0 & 1 & 1 & 0 \\ \hline 0 & 0 & 0 & 1 \end{array}\right)\left(\begin{array}{c|cc|c} 1 & 0 & 0 & 0 \\ \hline 0 & 1 & 0 & 0 \\ 0 & 2 & -2 & 0 \\ \hline 0 & 0 & 0 & 2 \end{array}\right).$

(4) $\left(\begin{array}{c|cc|c} 2 & 0 & 0 & 1 \\ \hline 0 & 2 & 1 & 0 \\ 0 & -1 & 2 & 0 \\ \hline 0 & 0 & 0 & 1 \end{array}\right)\left(\begin{array}{c|cc|c} 1 & 0 & 0 & 2 \\ \hline 0 & 1 & 4 & 0 \\ 0 & -1 & 0 & 0 \\ \hline 0 & 0 & 0 & -3 \end{array}\right).$

2 Let $A = \begin{pmatrix} \mathbf{a}_1 & \mathbf{a}_2 & \mathbf{a}_3 \end{pmatrix}$ be the column partition of a matrix A. Express the following matrices using column vectors $\mathbf{a}_1$, $\mathbf{a}_2$, $\mathbf{a}_3$ of A:

(1) $A\begin{pmatrix} 2 \\ 1 \\ 3 \end{pmatrix}.$

(2) $A\begin{pmatrix} -1 \\ 4 \\ -2 \end{pmatrix}.$

3 Let $A = \begin{pmatrix} {}^t\mathbf{b}_1 \\ {}^t\mathbf{b}_2 \\ {}^t\mathbf{b}_3 \end{pmatrix}$ be the row partition of a matrix A. Express the following matrices using row vectors ${}^t\mathbf{b}_1$, ${}^t\mathbf{b}_2$, ${}^t\mathbf{b}_3$ of A:

(1) $\begin{pmatrix} 3 & -2 & 4 \end{pmatrix} A.$

(2) $\begin{pmatrix} 1 & 5 & -1 \end{pmatrix} A.$

4 Let $A = \begin{pmatrix} \mathbf{a}_1 & \mathbf{a}_2 & \mathbf{a}_3 \end{pmatrix}$ be the column partition of a matrix A. Express the following matrices using column vectors $\mathbf{a}_1$, $\mathbf{a}_2$, $\mathbf{a}_3$ of A:

(1) $A\begin{pmatrix} 3 & 1 \\ 0 & -4 \\ 1 & 1 \end{pmatrix}.$

(2) $A\begin{pmatrix} -2 & 0 & 3 \\ 1 & -1 & 2 \\ 0 & 1 & -2 \end{pmatrix}.$

5 Let $A = \begin{pmatrix} {}^t\mathbf{b}_1 \\ {}^t\mathbf{b}_2 \\ {}^t\mathbf{b}_3 \end{pmatrix}$ be the row partition of a matrix A. Express the following matrices using row vectors ${}^t\mathbf{b}_1$, ${}^t\mathbf{b}_2$, ${}^t\mathbf{b}_3$ of A:

(1) $\begin{pmatrix} 1 & -1 & 1 \\ 1 & 0 & 1 \end{pmatrix} A.$

(2) $\begin{pmatrix} 2 & 1 & -1 \\ 0 & 3 & 1 \\ 1 & 1 & 0 \end{pmatrix} A.$

6 Let A_1 and B_1 be square matrices of degree m, and A_2 and B_2 square matrices of degree n. If A_1 and B_1 are commutative and A_2 and B_2 are commutative, then show that $A = \begin{pmatrix} A_1 & O_{m,n} \\ O_{n,m} & A_2 \end{pmatrix}$ and $B = \begin{pmatrix} B_1 & O_{m,n} \\ O_{n,m} & B_2 \end{pmatrix}$ are also commutative.

7 When A is an $m \times n$ matrix, find the k-th power $\begin{pmatrix} E_m & A \\ O_{n,m} & E_n \end{pmatrix}^k$ for a positive integer k.

1.4   Matrices and Systems of Linear Equations

We begin with the following matrix equation.

Exercise 1.7   Solve the following matrix equation:

$$\begin{pmatrix} 2 & 3 \\ 1 & -4 \end{pmatrix} \begin{pmatrix} x \\ y \end{pmatrix} = \begin{pmatrix} 7 \\ 9 \end{pmatrix}.$$

Answer. Multiply the two matrices on the left-hand side of the equality, then we have

$$\begin{pmatrix} 2x + 3y \\ x - 4y \end{pmatrix} = \begin{pmatrix} 7 \\ 9 \end{pmatrix}.$$

Therefore solving the matrix equation is equivalent to solving the following system of linear equations:

$$\begin{cases} 2x + 3y = 7, \\ x - 4y = 9. \end{cases}$$

It is easy to solve this system of linear equations, and we have the answer

$$x = 5, y = -1. \parallel$$

In Exercise 1.7, we see that a matrix equation is nothing but a system of linear equations. Conversely, we shall show that systems of linear equations are considered as matrix equations.

Coefficient matrices of systems of linear equations   For a system of linear equations of n variables consisting of m linear equations:

$$(*) \quad \begin{cases} a_{11}x_1 + a_{12}x_2 + \cdots + a_{1n}x_n = b_1, \\ a_{21}x_1 + a_{22}x_2 + \cdots + a_{2n}x_n = b_2, \\ \quad\quad \cdots\cdots \\ a_{m1}x_1 + a_{m2}x_2 + \cdots + a_{mn}x_n = b_m, \end{cases}$$

we let

$$A = \begin{pmatrix} a_{11} & a_{12} & \cdots & a_{1n} \\ a_{21} & a_{22} & \cdots & a_{2n} \\ & & \cdots\cdots & \\ a_{m1} & a_{m2} & \cdots & a_{mn} \end{pmatrix}, \quad \mathbf{x} = \begin{pmatrix} x_1 \\ x_2 \\ \vdots \\ x_n \end{pmatrix}, \quad \mathbf{b} = \begin{pmatrix} b_1 \\ b_2 \\ \vdots \\ b_m \end{pmatrix}.$$

We call the matrix A the *coefficient matrix* and $\mathbf{x}$ the *variable vector* of the system of linear equations.

Augmented matrices We call the $m \times (n+1)$ matrix

$$(A|\mathbf{b}) = \begin{pmatrix} a_{11} & a_{12} & \cdots & a_{1n} & b_1 \\ a_{21} & a_{22} & \cdots & a_{2n} & b_2 \\ & & \cdots\cdots & & \vdots \\ a_{m1} & a_{m2} & \cdots & a_{mn} & b_m \end{pmatrix}$$

the *augmented matrix* of the system of linear equations $(*)$. As we see in Exercise 1.7, the solution of the system of linear equations $(*)$ is nothing but the solution of the matrix equation

$$(**) \qquad\qquad A\mathbf{x} = \mathbf{b}.$$

Then we identify the system of linear equations $(*)$ and the matrix equation $(**)$, and call the matrix equations systems of linear equations. Therefore the coefficient matrix of the system of linear equations $A\mathbf{x} = \mathbf{b}$ is

$$A$$

and the augmented matrix of $A\mathbf{x} = \mathbf{b}$ is

$$(A|\mathbf{b}).$$

Example 1 We consider a system of linear equations:

$$\begin{cases} 2x_1 + x_2 - x_3 = 2, \\ x_1 - 2x_2 + 4x_3 = 3, \\ -2x_1 + x_2 + 3x_3 = 2. \end{cases}$$

This system of linear equations is expressed as the matrix equation

$$\begin{pmatrix} 2 & 1 & -1 \\ 1 & -2 & 4 \\ -2 & 1 & 3 \end{pmatrix} \begin{pmatrix} x_1 \\ x_2 \\ x_3 \end{pmatrix} = \begin{pmatrix} 2 \\ 3 \\ 2 \end{pmatrix}.$$

Example 2 We consider the system of linear equations from Example 1. The coefficient matrix is

$$\begin{pmatrix} 2 & 1 & -1 \\ 1 & -2 & 4 \\ -2 & 1 & 3 \end{pmatrix}$$

and the augmented matrix is

$$\left(\begin{array}{ccc|c} 2 & 1 & -1 & 2 \\ 1 & -2 & 4 & 3 \\ -2 & 1 & 3 & 2 \end{array} \right).$$

Exercise 1.8 For a system of linear equations:

$$\begin{cases} 3x_1 - 2x_2 + x_3 + 4x_4 = 7, \\ x_1 \qquad - 3x_3 + x_4 = 5, \\ 2x_1 - x_2 + 9x_3 \qquad = 0, \end{cases}$$

answer the following questions:

(1) Express the system of linear equations as a matrix equation.
(2) Find the coefficient matrix and the augmented matrix.

Answers. (1) The system of linear equations is expressed as the matrix equation

$$\begin{pmatrix} 3 & -2 & 1 & 4 \\ 1 & 0 & -3 & 1 \\ 2 & -1 & 9 & 0 \end{pmatrix} \begin{pmatrix} x_1 \\ x_2 \\ x_3 \\ x_4 \end{pmatrix} = \begin{pmatrix} 7 \\ 5 \\ 0 \end{pmatrix}.$$

(2) The coefficient matrix is $\begin{pmatrix} 3 & -2 & 1 & 4 \\ 1 & 0 & -3 & 1 \\ 2 & -1 & 9 & 0 \end{pmatrix}.$

The augmented matrix is $\left(\begin{array}{cccc|c} 3 & -2 & 1 & 4 & 7 \\ 1 & 0 & -3 & 1 & 5 \\ 2 & -1 & 9 & 0 & 0 \end{array} \right).$ ‖

Exercise 1.9 Consider the matrix equation

$$\begin{pmatrix} 2 & -1 & 4 \\ -1 & 2 & 5 \end{pmatrix} \begin{pmatrix} x_1 \\ x_2 \\ x_3 \end{pmatrix} = \begin{pmatrix} -1 \\ -5 \end{pmatrix}.$$

(1) What are the coefficient matrix and the augmented matrix?
(2) Express the matrix equation as a system of linear equations.

Answers. (1) The coefficient matrix and the augmented matrix are

$$\begin{pmatrix} 2 & -1 & 4 \\ -1 & 2 & 5 \end{pmatrix} \quad \text{and} \quad \left(\begin{array}{ccc|c} 2 & -1 & 4 & -1 \\ -1 & 2 & 5 & -5 \end{array} \right).$$

(2) $\begin{cases} 2x_1 - x_2 + 4x_3 = -1, \\ -x_1 + 2x_2 + 5x_3 = -5. \end{cases}$ ‖

Linear combinations of numerical vectors Since numerical vectors are special cases of matrices, we can add numerical vectors and multiply numbers to them as matrices. Take n numerical vectors $\mathbf{a}_1, \mathbf{a}_2, \ldots, \mathbf{a}_n$ of the same type and the same degree. The numerical vector

$$c_1 \mathbf{a}_1 + c_2 \mathbf{a}_2 + \cdots + c_n \mathbf{a}_n \quad (c_i : \text{numbers})$$

is called the *linear combination* of $\mathbf{a}_1, \mathbf{a}_2, \ldots, \mathbf{a}_n$.

Example 3 The column vector $\begin{pmatrix} 2 \\ 3 \\ -1 \end{pmatrix}$ is expressed as a linear combination of the column vectors $\begin{pmatrix} 1 \\ 0 \\ 0 \end{pmatrix}, \begin{pmatrix} 0 \\ 1 \\ 0 \end{pmatrix}$ and $\begin{pmatrix} 0 \\ 0 \\ 1 \end{pmatrix}$. In fact, we have

$$\begin{pmatrix} 2 \\ 3 \\ -1 \end{pmatrix} = 2 \begin{pmatrix} 1 \\ 0 \\ 0 \end{pmatrix} + 3 \begin{pmatrix} 0 \\ 1 \\ 0 \end{pmatrix} + (-1) \begin{pmatrix} 0 \\ 0 \\ 1 \end{pmatrix}$$

$$= 2 \begin{pmatrix} 1 \\ 0 \\ 0 \end{pmatrix} + 3 \begin{pmatrix} 0 \\ 1 \\ 0 \end{pmatrix} - \begin{pmatrix} 0 \\ 0 \\ 1 \end{pmatrix}.$$

Example 4 The row vector $\begin{pmatrix} 3 & -2 & 5 \end{pmatrix}$ is expressed as a linear combination of the row vectors $\begin{pmatrix} 1 & 0 & 0 \end{pmatrix}, \begin{pmatrix} 0 & 1 & 0 \end{pmatrix}$ and $\begin{pmatrix} 0 & 1 & 0 \end{pmatrix}$. In fact, we have

$$\begin{pmatrix} 3 & -2 & 5 \end{pmatrix} = 3 \begin{pmatrix} 1 & 0 & 0 \end{pmatrix} + (-2) \begin{pmatrix} 0 & 1 & 0 \end{pmatrix} + 5 \begin{pmatrix} 0 & 0 & 1 \end{pmatrix}$$

$$= 3 \begin{pmatrix} 1 & 0 & 0 \end{pmatrix} - 2 \begin{pmatrix} 0 & 1 & 0 \end{pmatrix} + 5 \begin{pmatrix} 0 & 0 & 1 \end{pmatrix}.$$

Linear combinations of column vectors and matrix equations Let $A =$
$\begin{pmatrix} a_{11} & \cdots & a_{1n} \\ \vdots & & \vdots \\ a_{m1} & \cdots & a_{mn} \end{pmatrix}$ be an $m \times n$ matrix and $A = (\mathbf{a}_1 \cdots \mathbf{a}_n)$ the column partition of

A. For a variable vector $\mathbf{x} = \begin{pmatrix} x_1 \\ \vdots \\ x_n \end{pmatrix}$, $A\mathbf{x}$ is expressed as

$$A\mathbf{x} = (\mathbf{a}_1 \cdots \mathbf{a}_n) \begin{pmatrix} x_1 \\ \vdots \\ x_n \end{pmatrix}$$
$$= x_1 \mathbf{a}_1 + \cdots + x_n \mathbf{a}_n.$$

Let $\mathbf{b} = \begin{pmatrix} b_1 \\ \vdots \\ b_m \end{pmatrix}$ be a column vector of degree m. Then solving the linear equation
$A\mathbf{x} = \mathbf{b}$ is equivalent to finding numbers $x_1, \ldots, x_n$ satisfying

$$x_1 \mathbf{a}_1 + \cdots + x_n \mathbf{a}_n = \mathbf{b}.$$

Therefore using the same notation as above, the following three equations are equivalent:

(1) $\begin{cases} a_{11}x_1 + \cdots + a_{1n}x_n = b_1, \\ a_{21}x_1 + \cdots + a_{2n}x_n = b_2, \\ \quad \cdots\cdots \\ a_{m1}x_1 + \cdots + a_{mn}x_n = b_m. \end{cases}$

(2) $A\mathbf{x} = \mathbf{b}.$

(3) $x_1 \mathbf{a}_1 + \cdots + x_n \mathbf{a}_n = \mathbf{b}.$

Exercise 1.10 Express $\begin{pmatrix} 1 \\ 3 \end{pmatrix}$ as a linear combination of $\begin{pmatrix} 1 \\ 2 \end{pmatrix}$ and $\begin{pmatrix} 2 \\ 3 \end{pmatrix}$.

Answer. Let

$$\begin{pmatrix} 1 \\ 3 \end{pmatrix} = x_1 \begin{pmatrix} 1 \\ 2 \end{pmatrix} + x_2 \begin{pmatrix} 2 \\ 3 \end{pmatrix}.$$

Then we have the matrix equation

$$\begin{pmatrix} 1 & 2 \\ 2 & 3 \end{pmatrix} \begin{pmatrix} x_1 \\ x_2 \end{pmatrix} = \begin{pmatrix} 1 \\ 3 \end{pmatrix},$$

which is equivalent to the system of linear equations:

$$\begin{cases} x_1 + 2x_2 = 1 \cdots\cdots \text{①}, \\ 2x_1 + 3x_2 = 3 \cdots\cdots \text{②} \end{cases}$$

Calculating ② − ① × 2, we have

$$-x_2 = 1 \quad \text{or} \quad x_2 = -1.$$

Substituting the value $x_2 = -1$ for the equation ①, we obtain $x_1 = 3$ and

$$\begin{pmatrix} 1 \\ 3 \end{pmatrix} = 3 \begin{pmatrix} 1 \\ 2 \end{pmatrix} - \begin{pmatrix} 2 \\ 3 \end{pmatrix}. \ \|$$

We study systems of linear equations systematically in the next chapter.

Example 5 We shall express the row vector $(1\ 3)$ as a linear combination of the row vectors $(1\ 2)$ and $(2\ 3)$. Let

$$(1\ 3) = x_1 (1\ 2) + x_2 (2\ 3).$$

Then we have the system of linear equations:

$$\begin{cases} x_1 + 2x_2 = 1, \\ 2x_1 + 3x_2 = 3. \end{cases}$$

This system of linear equations is the same system of linear equations in Exercise 1.10 and the solution is $x_1 = 3$, $x_2 = -1$. Then

$$(1\ 3) = 3(1\ 2) - (2\ 3).$$

Expression of linear combinations of numerical vectors Let $\mathbf{u}_1, \ldots, \mathbf{u}_m$ be numerical vectors of the same type and the same degree. We can express linear combinations by a formal matrix multiplication in the following way:

$$a_1 \mathbf{u}_1 + \cdots + a_m \mathbf{u}_m = (\mathbf{u}_1 \ \cdots \ \mathbf{u}_m) \begin{pmatrix} a_1 \\ \vdots \\ a_m \end{pmatrix}.$$

Here the right-hand side is the formal matrix multiplication of the formal row vector $(\mathbf{u}_1 \ \cdots \ \mathbf{u}_m)$ of degree m consisting of vectors $\mathbf{u}_1, \ldots, \mathbf{u}_m$ and the column vector $\begin{pmatrix} a_1 \\ \vdots \\ a_m \end{pmatrix}$ of degree m.

Example 6 $2\mathbf{u}_1 + 3\mathbf{u}_2 = (\mathbf{u}_1 \ \mathbf{u}_2) \begin{pmatrix} 2 \\ 3 \end{pmatrix}.$

Expression of vectors by formal multiplication of matrices Let $\mathbf{u}_1, \ldots, \mathbf{u}_m$ be numerical vectors of the same type and the same degree. We express n linear combinations

$$\begin{cases} \mathbf{v}_1 = a_{11}\mathbf{u}_1 + \cdots + a_{m1}\mathbf{u}_m, \\ \quad \cdots \\ \mathbf{v}_n = a_{1n}\mathbf{u}_1 + \cdots + a_{mn}\mathbf{u}_m \end{cases}$$

in the following way:

$$\begin{pmatrix} \mathbf{v}_1 \cdots \mathbf{v}_n \end{pmatrix} = \begin{pmatrix} \mathbf{u}_1 \cdots \mathbf{u}_m \end{pmatrix} \begin{pmatrix} a_{11} \cdots a_{1n} \\ \cdots \\ a_{m1} \cdots a_{mn} \end{pmatrix}.$$

Here the right-hand side is the formal matrix multiplication of the formal row vector $\begin{pmatrix} \mathbf{u}_1 \cdots \mathbf{u}_m \end{pmatrix}$ of degree m consisting of vectors $\mathbf{u}_1, \ldots, \mathbf{u}_m$ and the matrix $\begin{pmatrix} a_{11} \cdots a_{1n} \\ \cdots \\ a_{m1} \cdots a_{mn} \end{pmatrix}.$

Example 7 $\begin{pmatrix} 3\mathbf{u}_1 - \mathbf{u}_2 & 4\mathbf{u}_1 + 2\mathbf{u}_2 \end{pmatrix} = \begin{pmatrix} \mathbf{u}_1 & \mathbf{u}_2 \end{pmatrix} \begin{pmatrix} 3 & 4 \\ -1 & 2 \end{pmatrix}.$

Example 8

$$\left(\begin{pmatrix} -1 \\ 2 \end{pmatrix} \begin{pmatrix} 3 \\ -1 \end{pmatrix} \right) \begin{pmatrix} 1 & 2 \\ -1 & 3 \end{pmatrix}$$

$$= \left(\begin{pmatrix} -1 \\ 2 \end{pmatrix} - \begin{pmatrix} 3 \\ -1 \end{pmatrix} \quad 2\begin{pmatrix} -1 \\ 2 \end{pmatrix} + 3\begin{pmatrix} 3 \\ -1 \end{pmatrix} \right) = \left(\begin{pmatrix} -4 \\ 3 \end{pmatrix} \begin{pmatrix} 7 \\ 1 \end{pmatrix} \right).$$

Theorem 1.4.1 *Let* $\mathbf{u}_1, \ldots, \mathbf{u}_m$; $\mathbf{v}_1, \ldots, \mathbf{v}_n$; $\mathbf{w}$ *be numerical vectors of the same type and the same degree. We assume the following:*

(1) The vector $\mathbf{w}$ *is a linear combination of the vectors* $\mathbf{v}_1, \ldots, \mathbf{v}_n$.

(2) Each vector $\mathbf{v}_j$ *is a linear combination of the vectors* $\mathbf{u}_1, \ldots, \mathbf{u}_m$ *for* $j = 1, \ldots, n$.

Then $\mathbf{w}$ *is a linear combination of* $\mathbf{u}_1, \ldots, \mathbf{u}_m$.

Proof Since $\mathbf{w}$ is a linear combination of $\mathbf{v}_1, \ldots, \mathbf{v}_n$, $\mathbf{w}$ is expressed as

$$\mathbf{w} = b_1\mathbf{v}_1 + \cdots + b_n\mathbf{v}_n.$$

Since each vector of $\mathbf{v}_1, \ldots, \mathbf{v}_n$ is a linear combination of $\mathbf{u}_1, \ldots, \mathbf{u}_m$, we see

$$\begin{pmatrix} \mathbf{v}_1 \cdots \mathbf{v}_n \end{pmatrix} = \begin{pmatrix} \mathbf{u}_1 \cdots \mathbf{u}_m \end{pmatrix} \begin{pmatrix} a_{11} \cdots a_{1n} \\ \cdots \\ a_{m1} \cdots a_{mn} \end{pmatrix}.$$

Then we have

$$\mathbf{w} = b_1\mathbf{v}_1 + \cdots + b_n\mathbf{v}_n = \begin{pmatrix} \mathbf{v}_1 & \cdots & \mathbf{v}_n \end{pmatrix} \begin{pmatrix} b_1 \\ \vdots \\ b_n \end{pmatrix}$$

$$= \begin{pmatrix} \mathbf{u}_1 & \cdots & \mathbf{u}_m \end{pmatrix} \begin{pmatrix} a_{11} & \cdots & a_{1n} \\ & \cdots & \\ a_{m1} & \cdots & a_{mn} \end{pmatrix} \begin{pmatrix} b_1 \\ \vdots \\ b_n \end{pmatrix}$$

$$= \begin{pmatrix} \mathbf{u}_1 & \cdots & \mathbf{u}_m \end{pmatrix} \begin{pmatrix} a_{11}b_1 + \cdots + a_{1n}b_n \\ \vdots \\ a_{m1}b_1 + \cdots + a_{mn}b_n \end{pmatrix}$$

$$= (a_{11}b_1 + \cdots + a_{1n}b_n)\mathbf{u}_1 + \cdots + (a_{m1}b_1 + \cdots + a_{mn}b_n)\mathbf{u}_m$$

Therefore $\mathbf{w}$ is a linear combination of $\mathbf{u}_1, \ldots, \mathbf{u}_m$. ∥

Exercises (Sect. 1.4)

1 Express the following systems of linear equations as matrix equations. Also, find the coefficient matrices and the augmented matrices:

(1) $\begin{cases} 2x_1 + 3x_2 = -1, \\ x_1 - x_2 = 2. \end{cases}$

(2) $\begin{cases} x_1 + 2x_2 - x_3 = 2, \\ -x_1 + 3x_3 = 8, \\ x_2 - 2x_3 = -4. \end{cases}$

(3) $\begin{cases} 2x_1 + 3x_3 = 8, \\ x_1 - x_2 - 2x_3 = -1, \\ 2x_1 + x_2 + x_3 = 3. \end{cases}$

2 Express the following matrix equations as systems of linear equations:

(1) $\begin{pmatrix} 2 & 1 & 3 \\ 0 & -1 & 2 \\ 1 & 0 & -1 \end{pmatrix} \begin{pmatrix} x_1 \\ x_2 \\ x_3 \end{pmatrix} = \begin{pmatrix} 1 \\ 2 \\ -2 \end{pmatrix}.$

(2) $\begin{pmatrix} 3 & 0 & 1 \\ 1 & -1 & 2 \end{pmatrix} \begin{pmatrix} x_1 \\ x_2 \\ x_3 \end{pmatrix} = \begin{pmatrix} -1 \\ 0 \end{pmatrix}.$

(3) $\begin{pmatrix} 1 & -1 & 2 & 1 \\ 1 & 1 & 3 & -2 \\ 0 & 2 & -1 & 1 \end{pmatrix} \begin{pmatrix} x_1 \\ x_2 \\ x_3 \\ x_4 \end{pmatrix} = \begin{pmatrix} 0 \\ -2 \\ 2 \end{pmatrix}.$

3 In the following, is column vector $\mathbf{a}$ a linear combination of column vectors $\mathbf{b}_1, \mathbf{b}_2$? If it is true, then express $\mathbf{a}$ as a linear combination of $\mathbf{b}_1$ and $\mathbf{b}_2$:

(1) $\mathbf{a} = \begin{pmatrix} -2 \\ 1 \end{pmatrix}$, $\mathbf{b}_1 = \begin{pmatrix} 3 \\ -1 \end{pmatrix}$, $\mathbf{b}_2 = \begin{pmatrix} 1 \\ 1 \end{pmatrix}$.

(2) $\mathbf{a} = \begin{pmatrix} 1 \\ 2 \\ 1 \end{pmatrix}$, $\mathbf{b}_1 = \begin{pmatrix} 1 \\ 3 \\ 0 \end{pmatrix}$, $\mathbf{b}_2 = \begin{pmatrix} 2 \\ 3 \\ 1 \end{pmatrix}$.

(3) $\mathbf{a} = \begin{pmatrix} 0 \\ 1 \\ 5 \end{pmatrix}$, $\mathbf{b}_1 = \begin{pmatrix} 1 \\ 1 \\ 2 \end{pmatrix}$, $\mathbf{b}_2 = \begin{pmatrix} 2 \\ 1 \\ -1 \end{pmatrix}$.

4 Find the conditions on the numbers a, b so that, the column vector $\mathbf{a}$ is expressed as a linear combination of $\mathbf{b}_1$ and $\mathbf{b}_2$:

(1) $\mathbf{a} = \begin{pmatrix} a \\ 2 \\ 3 \end{pmatrix}$, $\mathbf{b}_1 = \begin{pmatrix} 1 \\ 2 \\ 1 \end{pmatrix}$, $\mathbf{b}_2 = \begin{pmatrix} 2 \\ 3 \\ 1 \end{pmatrix}$.

(2) $\mathbf{a} = \begin{pmatrix} 0 \\ a \\ b \end{pmatrix}$, $\mathbf{b}_1 = \begin{pmatrix} 1 \\ -1 \\ 1 \end{pmatrix}$, $\mathbf{b}_2 = \begin{pmatrix} 2 \\ 1 \\ 3 \end{pmatrix}$.

(3) $\mathbf{a} = \begin{pmatrix} a \\ 1 \\ b \end{pmatrix}$, $\mathbf{b}_1 = \begin{pmatrix} 1 \\ 1 \\ 1 \end{pmatrix}$, $\mathbf{b}_2 = \begin{pmatrix} -1 \\ 2 \\ 1 \end{pmatrix}$.

5 Let $\mathbf{a}_1$, $\mathbf{a}_2$, $\mathbf{b}_1$, $\mathbf{b}_2$, $\mathbf{c}$, $\mathbf{c}_1$, $\mathbf{c}_2$ be numerical vectors of the same type and the same degree.

(1) When

$$\mathbf{c} = \mathbf{b}_1 - 3\mathbf{b}_2 \quad \text{and} \quad \begin{cases} \mathbf{b}_1 = 2\mathbf{a}_1 + 3\mathbf{a}_2, \\ \mathbf{b}_2 = -\mathbf{a}_1 + 4\mathbf{a}_2, \end{cases}$$

express $\mathbf{c}$ as a linear combination of $\mathbf{a}_1$ and $\mathbf{a}_2$.

(2) When

$$\mathbf{c} = 2\mathbf{b}_1 + \mathbf{b}_2 \quad \text{and} \quad \begin{cases} \mathbf{b}_1 = -\mathbf{a}_1 - 4\mathbf{a}_2, \\ \mathbf{b}_2 = 3\mathbf{a}_1 + \mathbf{a}_2, \end{cases}$$

express $\mathbf{c}$ as a linear combination of $\mathbf{a}_1$ and $\mathbf{a}_2$.

(3) When

$$\begin{cases} \mathbf{c}_1 = \mathbf{b}_1 + 2\mathbf{b}_2, \\ \mathbf{c}_2 = -\mathbf{b}_1 + \mathbf{b}_2, \end{cases} \quad \text{and} \quad \begin{cases} \mathbf{b}_1 = 3\mathbf{a}_1 - 2\mathbf{a}_2, \\ \mathbf{b}_2 = -\mathbf{a}_1 + 2\mathbf{a}_2, \end{cases}$$

express $\mathbf{c}_1$ and $\mathbf{c}_2$ as linear combinations of $\mathbf{a}_1$ and $\mathbf{a}_2$.

Chapter 2
Linear Equations

2.1 Reductions of Linear Equations and Matrices

We begin with solving the following system of linear equations:

$$\text{(I)} \quad \begin{cases} 2x + 3y = 8, \\ x + 2y = 5. \end{cases}$$

To solve this system of linear equations, we use only multiplications of constants to linear equations, additions of linear equations and exchanges of two linear equations.

$$\text{(II)} \quad \begin{cases} -y = -2, & \text{①} + \text{②} \times (-2) \\ x + 2y = 5. \end{cases}$$

$$\text{(III)} \quad \begin{cases} -y = -2, \\ x = 1. & \text{②} + \text{①} \times 2 \end{cases}$$

$$\text{(IV)} \quad \begin{cases} x = 1, & \text{①} \leftrightarrow \text{②} \\ -y = -2. \end{cases}$$

$$\text{(V)} \quad \begin{cases} x = 1, \\ y = 2. & \text{②} \times (-1) \end{cases}$$

Here ① and ② stand for the first equation and the second equation of the previous system of linear equations, respectively. The operation $\text{①} + \text{②} \times (-2)$ means to add -2 times the second equation to the first equation, the operation $\text{①} \leftrightarrow \text{②}$ means to exchange the first equation and the second equation, and $\text{②} \times (-1)$ means -1 times the second equation.

Therefore the solution of the system of linear equations (I) is

$$\begin{cases} x = 1, \\ y = 2. \end{cases}$$

T. Miyake, *Linear Algebra*, https://doi.org/10.1007/978-981-16-6994-1_2

From now on, we call a system of linear equations simply a *linear equation* and each linear equation in a system of linear equations an *equation*.

Reductions of linear equations In the above, we used the following three deformations of linear equations:

(i) Multiply one equation by a non-zero constant (IV $\Rightarrow$ V).
(ii) Exchange two equations (III $\Rightarrow$ IV).
(iii) Add a constant times one equation to another equation (I $\Rightarrow$ II and II $\Rightarrow$ III).

We call these three deformations the *reductions* of linear equations.

Readers may wonder why one should always treat two equations together. Why do we not substitute the obtained value of one variable for another equation to get the answer? Such a way to solve linear equations may be good enough for simple linear equations, but it is not so for linear equations consisting of many equations and with many variables. It takes a lot of time and labor to verify if the obtained values satisfy the original linear equation. However the reduction procedures of linear equations are reversible. Therefore the obtained solution is the solution of all linear equations appearing in the procedures. It must especially satisfy the original linear equation. That is the reason why we always consider all equations together. The reversibility of the reduction procedures is shown in the following statements:

(i) The reverse of the multiplication of one equation by a non-zero constant c is the multiplication of the obtained equation by the constant $1/c$.
(ii) The reverse of the exchange of two equations is the exchange of two equations.
(iii) The reverse of the addition of a times the jth equation to the ith equation is the addition of $-a$ times the jth equation to the ith equation.

Gauss elimination method The method of solving linear equations by reductions is called the *Gauss elimination method*, or the *Gauss sweeping-out method*. Since the reduction procedures are reversible, all linear equations appearing in the procedures are equivalent. Then the solutions of linear equations appearing in each procedure are the same and we do not need to verify that the solution we obtained satisfies the original linear equation.

Next, we consider the augmented matrices of linear equations:

$$(\text{I}) \begin{cases} 2x + 3y = 8, \\ x + 2y = 5. \end{cases} \qquad\qquad \begin{pmatrix} 2 & 3 & 8 \\ 1 & 2 & 5 \end{pmatrix}.$$

$$(\text{II}) \begin{cases} -y = -2, \\ x + 2y = 5. \end{cases} \quad ① + ② \times (-2) \quad \begin{pmatrix} 0 & -1 & -2 \\ 1 & 2 & 5 \end{pmatrix}.$$

$$(\text{III}) \begin{cases} -y = -2, \\ x = 1. \end{cases} \quad ② + ① \times 2 \quad \begin{pmatrix} 0 & -1 & -2 \\ 1 & 0 & 1 \end{pmatrix}.$$

$$(\text{IV}) \begin{cases} x = 1, \\ -y = -2. \end{cases} \quad ① \leftrightarrow ② \quad \begin{pmatrix} 1 & 0 & 1 \\ 0 & -1 & -2 \end{pmatrix}.$$

$$(\text{V}) \begin{cases} x = 1, \\ y = 2. \end{cases} \quad ② \times (-1) \quad \begin{pmatrix} 1 & 0 & 1 \\ 0 & 1 & 2 \end{pmatrix}.$$

We consider the deformations of matrices which correspond to the reductions of linear equations. Such deformations of matrices are called the *row-reductions of matrices*. We shall define row-reductions of matrices and solve linear equations by row-reductions of matrices. The matrices adequate for obtaining answers are called the reduced matrices, which will be defined in the next section (Sect. 2.2). We deform the augmented matrices of linear equations into the reduced matrices by row-reductions and solve linear equations.

Row-reductions of matrices We call the following three deformations of matrices the *row-reductions of matrices*, which correspond to the reductions of linear equations. We also call the row-reductions of matrices simply the *reductions of matrices*:

(i) Multiply one row by a non-zero constant.
(ii) Exchange two rows.
(iii) Add a constant times one row to another row.

In principle, we should do only one reduction at one time. However, we may do several reductions at one time when such deformations can be surely done by repeating reductions separately, but we must do them carefully. For example, we may do several reductions at one time in the following two cases:

(1) We add constants times a fixed row to other rows.
(2) We multiple several rows by non-zero constants.

Example 1 We shall solve the following linear equation by reductions of the augmented matrix:

$$\begin{cases} x_1 - 2x_2 - x_3 = 1. \\ x_2 + 2x_3 = -1, \\ 2x_1 \quad\quad + x_3 = 3. \end{cases}$$

The augmented matrix of the linear equation is

$$\left(\begin{array}{ccc|c} 1 & -2 & -1 & 1 \\ 0 & 1 & 2 & -1 \\ 2 & 0 & 1 & 3 \end{array} \right).$$

We shall take reductions of this augmented matrix. In the processes of reductions, ①, ②, ③ indicate the first row, the second row and the third row of the previous matrices, respectively.

$$\begin{array}{ccc|c}
1 & -2 & -1 & 1 \\
0 & 1 & 2 & -1 \\
2 & 0 & 1 & 3 \\
\end{array} \quad \begin{array}{l} ③+①×(-2) \end{array}$$

$$\begin{array}{ccc|c}
1 & -2 & -1 & 1 \\
0 & 1 & 2 & -1 \\
0 & 4 & 3 & 1 \\
\end{array} \quad \begin{array}{l} ①+②×2 \\ ③+②×(-4) \end{array}$$

$$\begin{array}{ccc|c}
1 & 0 & 3 & -1 \\
0 & 1 & 2 & -1 \\
0 & 0 & -5 & 5 \\
\end{array} \quad \begin{array}{l} ③×(-1/5) \end{array}$$

$$\begin{array}{ccc|c}
1 & 0 & 3 & -1 \\
0 & 1 & 2 & -1 \\
0 & 0 & 1 & -1 \\
\end{array} \quad \begin{array}{l} ①+③×(-3) \\ ②+③×(-3) \end{array}$$

$$\begin{array}{ccc|c}
1 & 0 & 0 & 2 \\
0 & 1 & 0 & 1 \\
0 & 0 & 1 & -1 \\
\end{array}$$

The linear equation corresponding to the last augmented matrix is

$$\begin{cases}
x_1 & = 2, \\
& x_2 & = 1, \\
& & x_3 = -1.
\end{cases}$$

And this is the solution of the given linear equation.

Exercise 2.1 Solve the linear equation

$$\begin{cases}
2x_1 + 3x_2 - x_3 = -3, \\
-x_1 + 2x_2 + 2x_3 = 1, \\
x_1 + x_2 - x_3 = -2
\end{cases}$$

by reductions of the augmented matrix.

Answer. In the processes of reductions, ①, ②, ③ indicate the first row, the second row and the third row of the previous matrices, respectively.

2	3	−1	−3
−1	2	2	1
1	1	−1	−2
0	1	1	1	①+③× (−2)
0	3	1	−1	②+③
1	1	−1	−2
1	1	−1	−2
0	3	1	−1	① ↔ ③
0	1	1	1
1	0	−2	−3	①+③× (−1)
0	0	−2	−4	②+③× (−3)
0	1	1	1
1	0	−2	−3
0	0	1	2	②× (−1/2)
0	1	1	1
1	0	−2	−3
0	1	1	1	② ↔ ③
0	0	1	2
1	0	0	1	①+③× 2
0	1	0	−1	②+③× (−1).
0	0	1	2

Hence the answer is

$$\begin{cases} x_1 = 1, \\ x_2 = -1, \quad \| \\ x_3 = 2. \end{cases}$$

Exercises (Sect. 2.1)

1 Solve the following linear equations by the Gauss elimination method:

(1) $\begin{cases} 2x_1 + 3x_2 = -1, \\ x_1 - x_2 = 2. \end{cases}$

(2) $\begin{cases} 3x_1 + 2x_2 = 0, \\ x_1 - 2x_2 = 8. \end{cases}$

(3) $\begin{cases} x_1 - 2x_2 = 1, \\ 2x_1 + x_2 = 7. \end{cases}$

(4) $\begin{cases} x_1 + 3x_2 = 5, \\ x_1 - x_2 = -3. \end{cases}$

(5) $\begin{cases} x_1 + 2x_2 - x_3 = 2, \\ -x_1 + 3x_3 = 8, \\ x_2 - 2x_3 = -4. \end{cases}$

(6) $\begin{cases} x_1 + 2x_2 - 2x_3 = -3, \\ x_1 - x_2 + 3x_3 = 5, \\ -x_1 + 4x_2 - 2x_3 = -7. \end{cases}$

$$(7)\ \begin{cases} x_2 - 2x_3 = -3, \\ x_1 + x_2 + x_3 = 4, \\ x_1 - x_2 + x_3 = 2. \end{cases}$$

$$(8)\ \begin{cases} x_1 + x_2 + x_3 = 0, \\ 2x_2 + x_2 + x_3 = -1, \\ x_1 \qquad + 2x_3 = 1. \end{cases}$$

$$(9)\ \begin{cases} 2x_1 + x_2 + 2x_3 = 2, \\ -x_1 - x_2 + 3x_3 = -6, \\ x_1 - x_2 - x_3 = 0. \end{cases}$$

$$(10)\ \begin{cases} x_1 + x_2 - x_3 + 2x_4 = 6, \\ -x_1 + x_2 + 2x_3 + x_4 = -2, \\ x_1 \qquad + x_3 - x_4 = 0, \\ -x_1 + x_2 - x_3 + 2x_4 = 2. \end{cases}$$

2 Solve the following linear equations by reductions of the augmented matrices:

$$(1)\ \begin{pmatrix} 1 & 1 \\ 1 & -1 \end{pmatrix} \begin{pmatrix} x_1 \\ x_2 \end{pmatrix} = \begin{pmatrix} 3 \\ -1 \end{pmatrix}.$$

$$(2)\ \begin{pmatrix} 3 & 5 \\ 1 & 3 \end{pmatrix} \begin{pmatrix} x_1 \\ x_2 \end{pmatrix} = \begin{pmatrix} 4 \\ 0 \end{pmatrix}.$$

$$(3)\ \begin{pmatrix} 1 & 2 \\ 1 & -1 \end{pmatrix} \begin{pmatrix} x_1 \\ x_2 \end{pmatrix} = \begin{pmatrix} 1 \\ -2 \end{pmatrix}.$$

$$(4)\ \begin{pmatrix} 2 & 1 \\ -1 & 3 \end{pmatrix} \begin{pmatrix} x_1 \\ x_2 \end{pmatrix} = \begin{pmatrix} 5 \\ 1 \end{pmatrix}.$$

$$(5)\ \begin{pmatrix} 2 & 1 & 3 \\ 0 & -1 & 2 \\ 1 & 0 & -1 \end{pmatrix} \begin{pmatrix} x_1 \\ x_2 \\ x_3 \end{pmatrix} = \begin{pmatrix} 1 \\ 2 \\ -2 \end{pmatrix}.$$

$$(6)\ \begin{pmatrix} 2 & 3 & 0 \\ 1 & -1 & 1 \\ 3 & 1 & -3 \end{pmatrix} \begin{pmatrix} x_1 \\ x_2 \\ x_3 \end{pmatrix} = \begin{pmatrix} -1 \\ 4 \\ -4 \end{pmatrix}.$$

$$(7)\ \begin{pmatrix} 1 & 0 & -1 \\ 1 & 1 & 1 \\ 1 & -1 & 0 \end{pmatrix} \begin{pmatrix} x_1 \\ x_2 \\ x_3 \end{pmatrix} = \begin{pmatrix} -2 \\ 2 \\ -3 \end{pmatrix}.$$

$$(8)\ \begin{pmatrix} 2 & 2 & 1 \\ 0 & 1 & 1 \\ -1 & 2 & 0 \end{pmatrix} \begin{pmatrix} x_1 \\ x_2 \\ x_3 \end{pmatrix} = \begin{pmatrix} 6 \\ 1 \\ 0 \end{pmatrix}.$$

$$(9)\ \begin{pmatrix} 2 & 3 & 1 & 1 \\ -2 & 1 & 1 & 2 \\ 1 & 1 & 0 & 1 \\ 1 & -1 & 1 & 2 \end{pmatrix} \begin{pmatrix} x_1 \\ x_2 \\ x_3 \\ x_4 \end{pmatrix} = \begin{pmatrix} -1 \\ -4 \\ -1 \\ 1 \end{pmatrix}.$$

3 Deform the matrix $\begin{pmatrix} 1 & 0 & | & 1 \\ 0 & 1 & | & 2 \end{pmatrix}$ into $\begin{pmatrix} 2 & 3 & | & 8 \\ 1 & 2 & | & 5 \end{pmatrix}$ by reductions (the reverse of (I)$\Rightarrow$(V) on Sect. 2.1).

2.2 Reduced Matrices

We have tried to solve linear equations by reductions of augmented matrices. To solve linear equations, we should deform augmented matrices into matrices for which we can easily obtain answers.

Primary components We call the leftmost non-zero component of a row vector the *primary component* of the row vector. Any zero row vector has no primary component.

Example 1 If the row vector is

$$\begin{pmatrix} 0 & 0 & 2 & -3 \end{pmatrix},$$

then the primary component of the row vector is 2.

Example 2 The zero row vector

$$\begin{pmatrix} 0 & 0 & \cdots & 0 \end{pmatrix}$$

has no primary component.

Reduced matrices The matrix satisfying the following four conditions is called a *row-reduced matrix* or simply a *reduced matrix*:

(I) If there are zero row vectors, then they are under non-zero row vectors.

(II) The primary component of any non-zero row vector is 1.

(III) When we denote the primary component of the ith row by a_{ij_i}, we have

$$j_1 < j_2 < j_3 < \cdots .$$

In other words, if a row is under another row, then the position of the primary component of the row is to the right of the column of the primary component of the upper row.

(IV) If a column contains the primary component of a row, all the other components of the column are zeros. In other words, if a_{ij_i} is the primary component of the ith row, then all the components a_{kj_i} of the j_ith column are zeros for $k \neq i$.

Example 3 The zero matrix $O_{m,n} = \begin{pmatrix} 0 & 0 & \cdots & 0 \\ 0 & 0 & \cdots & 0 \\ \vdots & \vdots & \ddots & \vdots \\ 0 & 0 & \cdots & 0 \end{pmatrix}$ of size (m, n) satisfies the four conditions of the reduced matrix. Therefore all the zero matrices are reduced matrices.

Example 4 The unit matrix $E_n = \begin{pmatrix} 1 & & & 0 \\ & 1 & & \\ & & \ddots & \\ 0 & & & 1 \end{pmatrix}$ of degree n satisfies the four conditions of the reduced matrix. Therefore all the unit matrices are also reduced matrices.

Though reduced matrices seem very complicated, it is not so difficult to understand them. Here we give some examples of reduced matrices.

Example 5 (*Examples of reduced matrices*) In the matrices in this example, ①'s mean the primary components of the rows. We are going to see that the matrix

$$\begin{pmatrix} 0 & ① & 4 & 0 & -1 \\ 0 & 0 & 0 & ① & 6 \\ 0 & 0 & 0 & 0 & 0 \end{pmatrix}$$

is a reduced matrix. In fact:

(I) A zero row vector (the third row) is under non-zero rows (the first row and the second row).

(II) The primary components of non-zero rows are 1.

(III) The primary components are the $(1, 2)$ component and the $(2, 4)$ component and $2 < 4$.

(IV) The columns containing the primary components are the second column and the fourth column. In these columns, components other than the primary components are zeros.

We note that by (III), zeros before the primary components are arranged in a terraced pattern.

The following matrices are also reduced matrices:

$$\begin{pmatrix} 0 & ① & 3 & 0 & 2 \\ 0 & 0 & 0 & 0 & 0 \end{pmatrix}, \qquad \begin{pmatrix} ① & -2 & 0 & 4 & 0 & -1 \\ 0 & 0 & ① & -4 & 1 & 1 \end{pmatrix},$$

$$\begin{pmatrix} ① & 0 & 1 & 4 & 0 & -1 \\ 0 & ① & 7 & -4 & 0 & 1 \\ 0 & 0 & 0 & 0 & ① & 3 \end{pmatrix}, \qquad \begin{pmatrix} 0 & 0 & ① & -4 & 0 & -1 \\ 0 & 0 & 0 & 0 & ① & 3 \\ 0 & 0 & 0 & 0 & 0 & 0 \\ 0 & 0 & 0 & 0 & 0 & 0 \end{pmatrix}.$$

Exercise 2.2 Explain the reason why the following matrices are not reduced matrices, and deform them into reduced matrices by reductions:

$$(1) \begin{pmatrix} 0 & 2 & 1 & 0 & 1 \\ 0 & 0 & 0 & 1 & 2 \\ 0 & 0 & 0 & 0 & 0 \end{pmatrix}. \qquad (2) \begin{pmatrix} 0 & 1 & 0 & 1/3 & 1/2 \\ 0 & 0 & 0 & 0 & 0 \\ 0 & 0 & 1 & -1 & 3 \end{pmatrix}.$$

$$(3) \begin{pmatrix} 1 & 1 & 3 & 1 & 2 \\ 0 & 0 & 1 & 2 & 0 \\ 0 & 0 & 0 & 0 & 1 \end{pmatrix}. \qquad (4) \begin{pmatrix} 0 & 0 & 0 & 1 & 1 \\ 0 & 0 & 1 & 0 & -2 \\ 1 & 3 & 0 & 0 & 2 \end{pmatrix}.$$

Answers. (1) The given matrix does not satisfy condition (II) in the first row. Multiply the first row by $1/2$, then we have the reduced matrix

$$\begin{pmatrix} 0 & 2 & 1 & 0 & 1 \\ 0 & 0 & 0 & 1 & 2 \\ 0 & 0 & 0 & 0 & 0 \end{pmatrix} \longrightarrow \begin{pmatrix} 0 & 1 & 1/2 & 0 & 1/2 \\ 0 & 0 & 0 & 1 & 2 \\ 0 & 0 & 0 & 0 & 0 \end{pmatrix}.$$

(2) The given matrix does not satisfy condition (I). Exchange the second row and the third row, then we have the reduced matrix

$$\begin{pmatrix} 0 & 1 & 0 & 1/3 & 1/2 \\ 0 & 0 & 0 & 0 & 0 \\ 0 & 0 & 1 & -1 & 3 \end{pmatrix} \longrightarrow \begin{pmatrix} 0 & 1 & 0 & 1/3 & 1/2 \\ 0 & 0 & 1 & -1 & 3 \\ 0 & 0 & 0 & 0 & 0 \end{pmatrix}.$$

(3) The given matrix does not satisfy condition (IV) in the third column and the fifth column. We do the following two reductions to the matrix:

(i) Add -3 times the second row to the first row.
(ii) Add -2 times the third row to the first row.

Then we have the reduced matrix

$$\begin{pmatrix} 1 & 1 & 3 & 1 & 2 \\ 0 & 0 & 1 & 2 & 0 \\ 0 & 0 & 0 & 0 & 1 \end{pmatrix} \longrightarrow \begin{pmatrix} 1 & 1 & 0 & -5 & 2 \\ 0 & 0 & 1 & 2 & 0 \\ 0 & 0 & 0 & 0 & 1 \end{pmatrix} \longrightarrow \begin{pmatrix} 1 & 1 & 0 & -5 & 0 \\ 0 & 0 & 1 & 2 & 0 \\ 0 & 0 & 0 & 0 & 1 \end{pmatrix}.$$

(4) The given matrix does not satisfy condition (III). Exchange the first row and the third row, then we have the reduced matrix

$$\begin{pmatrix} 0 & 0 & 0 & 1 & 1 \\ 0 & 0 & 1 & 0 & -2 \\ 1 & 3 & 0 & 0 & 2 \end{pmatrix} \longrightarrow \begin{pmatrix} 1 & 3 & 0 & 0 & 2 \\ 0 & 0 & 1 & 0 & -2 \\ 0 & 0 & 0 & 1 & 1 \end{pmatrix}. \quad \|$$

Theorem 2.2.1 *For matrices, the following statements hold:*

(1) Any matrix A can be deformed into a reduced matrix B by reductions.
(2) The reduced matrix B exists uniquely. We call the reduced matrix B the reduced matrix of A.

Proof (1) We know that zero matrices are reduced matrices (Example 3). Take a non-zero $m \times n$ matrix A. We are going to deform A into a reduced matrix by reductions.

We use induction on n. If $n = 1$, then A is a non-zero $m \times 1$ matrix $\begin{pmatrix} a_{11} \\ a_{21} \\ \vdots \\ a_{m1} \end{pmatrix}$. Since A is not a zero matrix, there exists at least one non-zero component a_{i1}. If $a_{11} = 0$, then exchange the first row and the ith row. Thus we may assume that $a_{11} \neq 0$. Multiplying the first row by $1/a_{11}$, we may also assume that $a_{11} = 1$. We add $-a_{i1}$ times the first row to the ith row for $i = 2, \ldots, m$. Then we have a reduced matrix.

Let $n \geq 2$. We assume that any $m \times (n-1)$ matrix can be deformed into a reduced matrix. Now, we are going to show that an $m \times n$ matrix A can be deformed into a reduced matrix. When the first column is a zero vector, A is expressed as $A = \begin{pmatrix} \mathbf{0} & A_1 \end{pmatrix}$, where $\mathbf{0}$ is the zero column vector and A_1 is an $m \times (n-1)$ matrix. By the induction assumption, A_1 can be deformed into a reduced matrix B_1 by reductions. So A can be deformed into a reduced matrix $B = \begin{pmatrix} \mathbf{0} & B_1 \end{pmatrix}$ by reductions.

Assume that the first column of A is not a zero vector, then we can deform the

first column into $\begin{pmatrix} 1 \\ 0 \\ \vdots \\ 0 \end{pmatrix}$ by reductions as shown above. Therefore A is deformed

into a matrix $\begin{pmatrix} 1 & {}^t\mathbf{b} \\ \mathbf{0}_{m-1} & A_1 \end{pmatrix}$. If $A_1 = O_{m-1,n-1}$, then the matrix $\begin{pmatrix} 1 & {}^t\mathbf{b} \\ \mathbf{0}_{m-1} & A_1 \end{pmatrix} =$
$\begin{pmatrix} 1 & {}^t\mathbf{b} \\ \mathbf{0}_{m-1} & O_{m-1,n-1} \end{pmatrix}$ is a reduced matrix.

We assume that $A_1 \neq O_{m-1,n-1}$. Then by the induction assumption, A_1 can be
deformed into a reduced matrix B_1 by reductions. Therefore A can be deformed into
the matrix $\begin{pmatrix} 1 & {}^t\mathbf{b} \\ \mathbf{0} & B_1 \end{pmatrix}$. Denote ${}^t\mathbf{b} = \begin{pmatrix} b_{12} & \cdots & b_{1n} \end{pmatrix}$. Let ${}^t\mathbf{b}_2, \ldots, {}^t\mathbf{b}_s$ be all the non-zero
rows of B_1. Then there exists the primary component 1 in the row ${}^t\mathbf{b}_j$ for $j = 2, \ldots, s$.
Let the primary component of ${}^t\mathbf{b}_j$ be the (j, m_j) component for $j = 2, \ldots, s$. We
add $-b_{1m_j}$ times the $(j + 1)$th row of $\begin{pmatrix} 1 & {}^t\mathbf{b} \\ \mathbf{0} & B_1 \end{pmatrix}$ to the first row $\begin{pmatrix} 1 & {}^t\mathbf{b} \end{pmatrix}$ of the matrix
$\begin{pmatrix} 1 & {}^t\mathbf{b} \\ \mathbf{0} & B_1 \end{pmatrix}$ in order for $j = 2, \ldots, s$ and obtain $B = \begin{pmatrix} 1 & {}^t\mathbf{b}' \\ \mathbf{0} & B_1 \end{pmatrix}$. Then the matrix B is a
reduced matrix and the matrix A is deformed into the reduced matrix $B = \begin{pmatrix} 1 & {}^t\mathbf{b}' \\ \mathbf{0} & B_1 \end{pmatrix}$.

(2) The uniqueness of the reduced matrix B obtained from A will be proved in
Theorem 4.3.6. ‖

Ranks of matrices By Theorem 2.2.1, we can define the rank of a matrix. Let A be
a matrix and B the reduced matrix of A. Then we define the *rank* of A by

$$\text{rank}(A) = \text{ the number of non-zero rows of } B.$$

Since the number of non-zero rows is equal to the number of the primary components
of B, we can also say

$$\text{rank}(A) = \text{ the number of columns of } B \text{ containing}$$
$$\text{primary components.}$$

Therefore we have the following theorem.

Theorem 2.2.2 *When A is an $m \times n$ matrix, we have*

$$\text{rank}(A) \leq m, \text{rank}(A) \leq n.$$

We shall give an example of reductions of a matrix.

Example 6 We are going to find the reduced matrix of the following matrix A by
reductions and rank (A):

$$A = \begin{pmatrix} 0\,0\,0 & 2 & 3\,2 \\ 0\,3\,6 & -9 & -4\,7 \\ 0\,2\,4 & -6 & -4\,2 \end{pmatrix}.$$

If there are zero rows, we move zero rows under non-zero rows. Next, take the row of which the primary component is leftmost. In this case, take the third row, for example, and exchange the third row and the first row:

$$A \longrightarrow \begin{pmatrix} 0\,2\,4 & -6 & -4\,2 \\ 0\,3\,6 & -9 & -4\,7 \\ 0\,0\,0 & 2 & 3\,2 \end{pmatrix} \quad ① \leftrightarrow ③.$$

Multiply the first row by 1/2:

$$\longrightarrow \begin{pmatrix} 0\,1\,2 & -3 & -2\ 1 \\ 0\,3\,6 & -9 & -4\,7 \\ 0\,0\,0 & 2 & 3\,2 \end{pmatrix} \quad ① \times (1/2).$$

Add −3 times the first row to the second row:

$$\longrightarrow \begin{pmatrix} 0\,1\,2 & -3 & -2\ 1 \\ 0\,0\,0 & 0 & 2\ 4 \\ 0\,0\,0 & 2 & 3\,2 \end{pmatrix} \quad ② + ① \times (-3).$$

Exchange the second row and the third row:

$$\longrightarrow \begin{pmatrix} 0\,1\,2 & -3 & -2\ 1 \\ 0\,0\,0 & 2 & 3\,2 \\ 0\,0\,0 & 0 & 2\ 4 \end{pmatrix} \quad ② \leftrightarrow ③.$$

Multiply both the second row and the third row by 1/2:

$$\longrightarrow \begin{pmatrix} 0\,1\,2 & -3 & -2\ 1 \\ 0\,0\,0 & 1 & 3/2\ 1 \\ 0\,0\,0 & 0 & 1\ 2 \end{pmatrix} \quad \begin{matrix} ② \times (1/2) \\ ③ \times (1/2). \end{matrix}$$

Add 3 times the second row to the first row:

$$\longrightarrow \begin{pmatrix} 0\,1\,2\,0 & 5/2 & 4 \\ 0\,0\,0\,1 & 3/2 & 1 \\ 0\,0\,0\,0 & 1 & 2 \end{pmatrix} \quad ① + ② \times 3.$$

Add −5/2 times the third row to the first row, and add −3/2 times the third row to the second row:

$$\longrightarrow \begin{pmatrix} 0\ 1\ 2\ 0\ 0 -1 \\ 0\ 0\ 0\ 1\ 0 -2 \\ 0\ 0\ 0\ 0\ 1\ \ 2 \end{pmatrix} \quad \begin{matrix} ①+③\times(-5/2) \\ ②+③\times(-3/2). \end{matrix}$$

The last matrix is the reduced matrix of A. Since the rank of A is the number of primary components of the reduced matrix of A, we have $\text{rank}(A) = 3$.

In Example 6, we connect the matrices appearing in the processes of reductions by the symbol $\rightarrow$. But to calculate the reduced matrices, it would be better to express matrices vertically like the matrices in Example 1 and Exercise 2.1 in Sect. 2.1. (See also Exercise 2.3 below.)

Exercise 2.3 Find the reduced matrix of the matrix

$$A = \begin{pmatrix} 0 & 0 & 0 & 0 & 0 & 0 & 0 \\ 0 & 1 & 2 & 5 & -1 & 5 & 1 \\ 0 & -1 & -2 & -2 & -5 & 0 & -4 \\ 0 & 1 & 2 & 2 & 5 & 2 & -2 \end{pmatrix}$$

by reductions.

Answer.

0	0	0	0	0	0	0	
0	1	2	5	−1	5	1	
0	−1	−2	−2	−5	0	−4	
0	1	2	2	5	2	−2	
0	1	2	2	5	2	−2	
0	1	2	5	−1	5	1	$①\leftrightarrow④$
0	−1	−2	−2	−5	0	−4	
0	0	0	0	0	0	0	
0	1	2	2	5	2	−2	
0	0	0	3	−6	3	3	$②+①\times(-1)$
0	0	0	0	0	2	−6	$③+①$
0	0	0	0	0	0	0	
0	1	2	2	5	2	−2	
0	0	0	1	−2	1	1	$②\times(1/3)$
0	0	0	0	0	1	−3	$③\times(1/2)$
0	0	0	0	0	0	0	
0	1	2	0	9	0	−4	$①+②\times(-2)$
0	0	0	1	−2	1	1	
0	0	0	0	0	1	−3	
0	0	0	0	0	0	0	
0	1	2	0	9	0	−4	
0	0	0	1	−2	0	4	$②+③\times(-1).$
0	0	0	0	0	1	−3	
0	0	0	0	0	0	0	

Therefore the reduced matrix of A is

$$
\begin{pmatrix}
0 & 1 & 2 & 0 & 9 & 0 & -4 \\
0 & 0 & 0 & 1 & -2 & 0 & 4 \\
0 & 0 & 0 & 0 & 0 & 1 & -3 \\
0 & 0 & 0 & 0 & 0 & 0 & 0
\end{pmatrix}. \quad \parallel
$$

Exercises (Sect. 2.2)

1 Find if the following matrices are reduced matrices or not, and also find the reduced matrices of the unreduced matrices by reductions:

(1) $\begin{pmatrix} 0 & 0 & 0 \\ 0 & 0 & 1 \end{pmatrix}$.

(2) $\begin{pmatrix} 1 & 2 & -3 \\ 0 & 1 & 1 \\ 0 & 0 & 0 \end{pmatrix}$.

(3) $\begin{pmatrix} 0 & 1 & 0 \\ 0 & 0 & 1 \\ 0 & 0 & 1 \end{pmatrix}$.

(4) $\begin{pmatrix} 1 & 0 & 0 & 1 \\ 0 & 2 & 1 & 0 \\ 0 & 0 & 1 & 1 \end{pmatrix}$.

(5) $\begin{pmatrix} 0 & 1 & 2 & 1 \\ 0 & 0 & 0 & 0 \\ 0 & 0 & 0 & 0 \end{pmatrix}$.

(6) $\begin{pmatrix} 0 & 1 & 0 & 0 \\ 1 & 0 & 0 & 0 \\ 0 & 0 & 1 & 0 \end{pmatrix}$.

(7) $\begin{pmatrix} 1 & 0 & 0 & 1 \\ 0 & 0 & 0 & 1 \\ 0 & 0 & 0 & 0 \end{pmatrix}$.

(8) $\begin{pmatrix} 0 & 1 & 0 & 1 \\ 0 & -1 & 3 & 0 \\ 0 & 0 & 1 & 0 \end{pmatrix}$.

(9) $\begin{pmatrix} 0 & 1 & -1 & 2 \\ 1 & 1 & -1 & 5 \\ 1 & 0 & 0 & 3 \end{pmatrix}$.

(10) $\begin{pmatrix} 0 & 0 & 1 & 0 \\ 1 & 1 & 1 & 0 \\ 0 & 0 & 1 & 2 \end{pmatrix}$.

(11) $\begin{pmatrix} 1 & 0 & 0 & -3 \\ 0 & 2 & 2 & 2 \\ 0 & 0 & 0 & 1 \end{pmatrix}$.

(12) $\begin{pmatrix} 0 & 1 & 0 & 2 & 0 \\ 0 & 2 & 1 & 0 & 0 \\ 0 & 0 & 0 & 2 & -2 \\ 0 & 0 & 0 & 1 & -1 \end{pmatrix}$.

2 Show that all the reduced square matrices of degree 2 are

$$
\begin{pmatrix} 0 & 0 \\ 0 & 0 \end{pmatrix}, \begin{pmatrix} 0 & 1 \\ 0 & 0 \end{pmatrix}, \begin{pmatrix} 1 & 0 \\ 0 & 1 \end{pmatrix}, \begin{pmatrix} 1 & * \\ 0 & 0 \end{pmatrix}.
$$

Here any number can be taken in place of $*$.

3 Find all the reduced 2×4 matrices. Put $*$ into the components where any number can be taken as in Exercise 2 above.

4 Find all the reduced square matrices of degree 3. Put $*$ into the components where any number can be taken as in Exercise 2 above.

5 Find the reduced matrices and the ranks of the following matrices:

(1) $\begin{pmatrix} 2 & 1 \\ 1 & 0 \end{pmatrix}.$

(2) $\begin{pmatrix} 1 & 2 & -3 \\ 1 & 1 & 1 \end{pmatrix}.$

(3) $\begin{pmatrix} 0 & 1 & 0 \\ 1 & 2 & -1 \end{pmatrix}.$

(4) $\begin{pmatrix} 1 & 0 & 2 & 1 \\ 2 & 1 & 1 & 0 \\ 0 & 1 & 1 & 0 \end{pmatrix}.$

(5) $\begin{pmatrix} 0 & 1 & 2 & 1 \\ 0 & 0 & 2 & 0 \\ 1 & 0 & 0 & 3 \end{pmatrix}.$

(6) $\begin{pmatrix} 0 & 1 & 3 & 1 \\ 1 & 0 & 1 & 1 \\ 1 & -2 & -5 & -1 \end{pmatrix}.$

(7) $\begin{pmatrix} 1 & 2 & 3 & 2 \\ 1 & 2 & 1 & 1 \\ 1 & 2 & -1 & 0 \end{pmatrix}.$

(8) $\begin{pmatrix} 0 & 1 & 2 & -1 & 1 \\ 1 & 0 & 2 & 0 & -1 \\ 1 & 0 & 2 & 0 & -1 \\ 1 & 2 & 4 & -2 & 3 \end{pmatrix}.$

(9) $\begin{pmatrix} 1 & 0 & 1 & 2 & 1 \\ 0 & 3 & -3 & -3 & 1 \\ -1 & 1 & -2 & -3 & 1 \\ 0 & 1 & -1 & -1 & -1 \end{pmatrix}.$

(10) $\begin{pmatrix} 1 & 2 & 2 & 1 & 5 \\ 1 & 2 & 1 & 0 & 3 \\ 0 & 0 & -1 & -1 & -2 \\ -1 & -2 & -1 & 0 & -3 \end{pmatrix}.$

(11) $\begin{pmatrix} 1 & 0 & 1 & 3 & 1 \\ 2 & -1 & 0 & 5 & 1 \\ -2 & 1 & 0 & -5 & 1 \\ 0 & 1 & 2 & -1 & 0 \end{pmatrix}.$

(12) $\begin{pmatrix} 2 & 4 & 6 & -1 & 1 \\ 0 & 0 & 0 & 2 & 2 \\ 1 & 2 & 3 & 1 & 2 \\ 1 & 2 & 3 & 3 & 4 \\ 0 & 0 & 0 & 1 & 1 \end{pmatrix}.$

2.3 Solutions of Linear Equations

We are going to solve linear equations by reductions of matrices. Let

$$Ax = b$$

be a linear equation. Here $A = (a_{ij})$ is an $m \times n$ matrix, $\mathbf{x} = \begin{pmatrix} x_1 \\ \vdots \\ x_n \end{pmatrix}$ a variable

vector of degree n and $\mathbf{b} = \begin{pmatrix} b_1 \\ \vdots \\ b_m \end{pmatrix}$ a column vector of degree m. The coefficient

matrix of the linear equation is A and the augmented matrix is $(A|\mathbf{b})$. The rank of
a matrix is equal to the number of columns containing primary components of the
reduced matrix of the matrix. The coefficient matrix A and the augmented matrix
$(A|\mathbf{b})$ can be deformed into their reduced matrices simultaneously by the same

reductions. Therefore the reduced matrix of A and the first $m \times n$ part of the reduced matrix of the augmented matrix $(A|\mathbf{b})$ are the same. Thus we have either

$$\text{rank}\,(A|\mathbf{b}) = \text{rank}(A) \text{ or rank}\,(A|\mathbf{b}) = \text{rank}(A) + 1.$$

Assume rank $(A|\mathbf{b}) = \text{rank}(A) + 1$. In this case, the reduced matrix of the augmented matrix is of the form

$$\begin{pmatrix}
0 \cdots 0 & 1 * 0 * * 0 * & \cdots\cdots * & 0 \\
0 \cdots 0 & 0 0 1 * * 0 * & \cdots\cdots * & 0 \\
\vdots & \vdots\vdots\vdots\vdots\vdots\vdots\vdots\vdots & \vdots & \vdots \\
0 \cdots 0 & 0 0 0 0 0 1 * & \cdots\cdots * & 0 \\
0 \cdots 0 & 0 0 0 0 0 0 0 & \cdots\cdots 0 & 1 \\
0 \cdots 0 & 0 0 0 0 0 0 0 & \cdots\cdots 0 & 0 \\
\vdots & \vdots\vdots\vdots\vdots\vdots\vdots\vdots\vdots & \vdots & \vdots \\
0 \cdots 0 & 0 0 0 0 0 0 0 & \cdots\cdots 0 & 0
\end{pmatrix}.$$

The existence of the row

$$\begin{pmatrix} 0 \cdots 0 & 0 0 0 0 0 0 0 & \cdots\cdots 0 & 1 \end{pmatrix}$$

in the reduced matrix of the augmented matrix $(A|\mathbf{b})$ is quite important. The equation corresponding to this row is

$$0x_1 + 0x_2 + \cdots + 0x_n = 1$$

and there are no numbers $x_1, x_2, \ldots, x_n$ satisfying this equation. Therefore we have no solution of the linear equation $A\mathbf{x} = \mathbf{b}$.

Next, we assume that rank $(A|\mathbf{b}) = \text{rank}(A)$. Then the reduced matrix of $(A|\mathbf{b})$ is of the form

$$\begin{pmatrix}
0 \cdots & 1 * 0 * * 0 * & \cdots * & b_1' \\
0 \cdots & 0 0 1 * * 0 * & \cdots * & b_2' \\
\vdots & \vdots\vdots\vdots\vdots\vdots\vdots\vdots & \vdots & \\
0 \cdots & 0 0 0 0 0 1 * & \cdots * & b_r' \\
0 \cdots & 0 0 0 0 0 0 0 & \cdots 0 & 0 \\
\vdots & \vdots\vdots\vdots\vdots\vdots\vdots\vdots & \vdots & \\
0 \cdots & 0 0 0 0 0 0 0 & \cdots 0 & 0
\end{pmatrix}.$$

If we substitute any numbers for the variables which do not correspond to primary components, then the values of the variables corresponding to primary components are uniquely determined. Thus we obtain solutions.

Example 1 Let us solve the linear equation

$$\begin{pmatrix} 1 & 0 & -1 \\ 0 & 1 & 2 \end{pmatrix} \begin{pmatrix} x_1 \\ x_2 \\ x_3 \end{pmatrix} = \begin{pmatrix} 2 \\ -3 \end{pmatrix}.$$

The augmented matrix $\left(\begin{array}{ccc|c} 1 & 0 & -1 & 2 \\ 0 & 1 & 2 & -3 \end{array}\right)$ is already reduced. In this example, the first variable x_1 and the second variable x_2 correspond to primary components, but the third variable x_3 does not correspond to a primary component. Substitute any number, say c, for the variable x_3; then the linear equation is written as

$$\begin{cases} x_1 \quad - c = 2, \\ \quad x_2 + 2c = -3. \end{cases}$$

Thus we obtain the solution of the linear equation

$$\begin{pmatrix} x_1 \\ x_2 \\ x_3 \end{pmatrix} = \begin{pmatrix} 2 + c \\ -3 - 2c \\ c \end{pmatrix} (c \in \mathbf{R}).$$

Here $\mathbf{R}$ is the set of real numbers.

By the above consideration, we have the following theorem.

Theorem 2.3.1 *A linear equation $A\mathbf{x} = \mathbf{b}$ has at least one solution if and only if*

$$\mathrm{rank}\left(A \mid \mathbf{b}\right) = \mathrm{rank}(A).$$

Exercise 2.4 Solve the following linear equation:

$$\begin{pmatrix} 1 & 0 & -1 & 0 & -2 \\ 0 & 1 & 1 & 0 & 1 \\ -1 & 0 & 1 & 1 & 1 \\ 2 & 1 & -1 & 0 & -3 \end{pmatrix} \begin{pmatrix} x_1 \\ x_2 \\ x_3 \\ x_4 \\ x_5 \end{pmatrix} = \begin{pmatrix} 1 \\ -2 \\ 3 \\ 1 \end{pmatrix}.$$

Answer. We are going to deform the augmented matrix into the reduced matrix by reductions.

$$\begin{array}{ccccc|c}
1 & 0 & -1 & 0 & -2 & 1 \\
0 & 1 & 1 & 0 & 1 & -2 \\
-1 & 0 & 1 & 1 & 1 & 3 \\
2 & 1 & -1 & 0 & -3 & 1 \\
\hline
1 & 0 & -1 & 0 & -2 & 1 \\
0 & 1 & 1 & 0 & 1 & -2 \\
0 & 0 & 0 & 1 & -1 & 4 \\
0 & 1 & 1 & 0 & 1 & -1 \\
\hline
1 & 0 & -1 & 0 & -2 & 1 \\
0 & 1 & 1 & 0 & 1 & -2 \\
0 & 0 & 0 & 1 & -1 & 4 \\
0 & 0 & 0 & 0 & 0 & 1 \\
\hline
1 & 0 & -1 & 0 & -2 & 0 \\
0 & 1 & 1 & 0 & 1 & 0 \\
0 & 0 & 0 & 1 & -1 & 0 \\
0 & 0 & 0 & 0 & 0 & 1 \\
\end{array}$$

Rows: ③+① , ④+①×(−2) ; ④+②×(−1) ; ①+④×(−1), ②+④×2, ③+④×(−4).

Since the rank of the augmented matrix is 4 and the rank of the coefficient matrix is 3, the linear equation does not have any solutions. ‖

Remark. To see that the linear equation has no solutions, we do not necessarily have to deform the augmented matrix into the reduced matrix. In the above case, since the row $(0\ 0\ 0\ 0\ 0|1)$ exists already in the matrix previous to the reduced matrix, the given linear equation has no solution.

Exercise 2.5 Solve the following linear equation:

$$\begin{pmatrix} 1 & -2 & 0 & 3 & 0 \\ 1 & -2 & 1 & 2 & 1 \\ 2 & -4 & 1 & 5 & 2 \end{pmatrix} \begin{pmatrix} x_1 \\ x_2 \\ x_3 \\ x_4 \\ x_5 \end{pmatrix} = \begin{pmatrix} 2 \\ 2 \\ 5 \end{pmatrix}.$$

Answer. We are going to deform the augmented matrix into the reduced matrix by reductions.

$$\begin{array}{ccccc|c}
1 & -2 & 0 & 3 & 0 & 2 \\
1 & -2 & 1 & 2 & 1 & 2 \\
2 & -4 & 1 & 5 & 2 & 5 \\
\hline
1 & -2 & 0 & 3 & 0 & 2 \\
0 & 0 & 1 & -1 & 1 & 0 \\
0 & 0 & 1 & -1 & 2 & 1 \\
\hline
1 & -2 & 0 & 3 & 0 & 2 \\
0 & 0 & 1 & -1 & 1 & 0 \\
0 & 0 & 0 & 0 & 1 & 1 \\
\hline
1 & -2 & 0 & 3 & 0 & 2 \\
0 & 0 & 1 & -1 & 0 & -1 \\
0 & 0 & 0 & 0 & 1 & 1 \\
\end{array}$$

Rows: ②+①×(−1), ③+①×(−2) ; ③+②×(−1) ; ②+③×(−1).

The linear equation corresponding to the reduced matrix of the augmented matrix is

$$\begin{cases} x_1 - 2x_2 \quad + 3x_4 \quad = \quad 2, \\ \qquad\qquad x_3 - x_4 \quad = -1, \\ \qquad\qquad\qquad\qquad x_5 = \quad 1. \end{cases}$$

The variables x_1, x_3, x_5 correspond to primary components, and the variables x_2, x_4 do not correspond to primary components. If we put arbitrary numbers c_1 and c_2 into the variables x_2 and x_4, respectively, then the variables x_1, x_3, x_5 are uniquely determined. Therefore we have

$$\begin{cases} x_1 = 2 + 2c_1 - 3c_2, \\ x_3 = -1 + c_2, \\ x_5 = 1. \end{cases}$$

Thus we have the following answer:

$$\mathbf{x} = \begin{pmatrix} x_1 \\ x_2 \\ x_3 \\ x_4 \\ x_5 \end{pmatrix} = \begin{pmatrix} 2 + 2c_1 - 3c_2 \\ c_1 \\ -1 + c_2 \\ c_2 \\ 1 \end{pmatrix} \quad (c_1, c_2 \in \mathbf{R}). \quad \|$$

In Exercise 2.5, the answer can be expressed as

$$\mathbf{x} = \begin{pmatrix} 2 + 2c_1 - 3c_2 \\ c_1 \\ -1 + c_2 \\ c_2 \\ 1 \end{pmatrix}$$

$$= \begin{pmatrix} 2 \\ 0 \\ -1 \\ 0 \\ 1 \end{pmatrix} + c_1 \begin{pmatrix} 2 \\ 1 \\ 0 \\ 0 \\ 0 \end{pmatrix} + c_2 \begin{pmatrix} -3 \\ 0 \\ 1 \\ 1 \\ 0 \end{pmatrix} \quad (c_1, c_2 \in \mathbf{R}).$$

In this expression, $\mathbf{x}$ is expressed as a linear combination of three column vectors.

Assume that a linear equation $A\mathbf{x} = \mathbf{b}$ has a solution. If there exists in the reduced matrix of A at least one column which has no primary component, we can take an arbitrary number for the corresponding variable and the solution is not uniquely determined. Therefore the solution is uniquely determined if and only if all columns of the reduced matrix of A have primary components. In other words, the rank of A is equal to the number of variables.

Combining this argument with Theorem 2.3.1, we obtain the following theorem.

Theorem 2.3.2 *A linear equation*

$$A\mathbf{x} = \mathbf{b}$$

of n variables has a unique solution if and only if

$$\text{rank}(A) = \text{rank}\left(A\big|\mathbf{b}\right) = n.$$

Homogeneous linear equations We call a linear equation

$$A\mathbf{x} = \mathbf{0}$$

a *homogeneous linear equation*. In other words, a linear equation $A\mathbf{x} = \mathbf{b}$ is called a homogeneous linear equation when $\mathbf{b} = \mathbf{0}$. Any homogeneous linear equation $A\mathbf{x} = \mathbf{0}$ has always a solution $\mathbf{x} = \mathbf{0}$. We call this solution the *trivial solution*. For homogeneous linear equations, we have the following theorem.

Theorem 2.3.3 *Let A be an m × n matrix.*
(1) A homogeneous linear equation

$$A\mathbf{x} = \mathbf{0}$$

has only the trivial solution if and only if

$$\text{rank}(A) = n.$$

(2) If m < n, then $A\mathbf{x} = \mathbf{0}$ has non-trivial solutions.

Proof (1) is the special case of Theorem 2.3.2 since

$$\text{rank}(A) = \text{rank}\left(A\big|\mathbf{0}\right).$$

(2) Since the rank of a matrix is equal or less than the number of rows (Theorem 2.2.2), we see that

$$\text{rank}(A) \leq m < n.$$

Therefore the homogeneous linear equation $A\mathbf{x} = \mathbf{0}$ has non-trivial solutions by (1).
∥

Solutions of homogeneous linear equations We shall solve homogeneous linear equations. Take a homogeneous linear equation

$$A\mathbf{x} = \mathbf{0}.$$

Since the augmented matrix of the homogeneous linear equation is $\left(A\big|\mathbf{0}\right)$, the last column of the matrix appearing in each step of the reductions is always $\mathbf{0}$. So, we may consider only reductions of the coefficient matrix A instead of the augmented matrix $(A|\mathbf{0})$.

Example 2 We are going to solve the following homogeneous linear equation:

$$\begin{pmatrix} 1 & -2 & 3 & -2 \\ -1 & 2 & -2 & 1 \end{pmatrix} \begin{pmatrix} x_1 \\ x_2 \\ x_3 \\ x_4 \end{pmatrix} = \begin{pmatrix} 0 \\ 0 \end{pmatrix}.$$

Since the linear equation is homogeneous, we have only to deform the coefficient matrix of the linear equation:

$$
\begin{array}{cccc}
1 & -2 & 3 & -2 \\
-1 & 2 & -2 & 1 \\
\hline
1 & -2 & 3 & -2 \\
0 & 0 & 1 & -1 \\
\hline
1 & -2 & 0 & 1 \\
0 & 0 & 1 & -1
\end{array}
\qquad
\begin{array}{l}
\\
\\
\\
②+① \\
①+②\times(-3)
\end{array}
$$

Since the linear equation corresponding to the reduced matrix is

$$\begin{cases} x_1 - 2x_2 & + x_4 = 0, \\ & x_3 - x_4 = 0, \end{cases}$$

we have the solution

$$\mathbf{x} = \begin{pmatrix} 2c_1 - c_2 \\ c_1 \\ c_2 \\ c_2 \end{pmatrix} = c_1 \begin{pmatrix} 2 \\ 1 \\ 0 \\ 0 \end{pmatrix} + c_2 \begin{pmatrix} -1 \\ 0 \\ 1 \\ 1 \end{pmatrix} \quad (c_1, c_2 \in \mathbf{R}).$$

Exercise 2.6 Solve the following homogeneous linear equation:

$$\begin{pmatrix} 1 & -2 & 0 & 0 & 1 & 2 \\ 1 & -1 & 1 & 0 & 1 & 0 \\ 2 & -5 & -1 & 0 & 3 & 3 \end{pmatrix} \begin{pmatrix} x_1 \\ x_2 \\ x_3 \\ x_4 \\ x_5 \\ x_6 \end{pmatrix} = \begin{pmatrix} 0 \\ 0 \\ 0 \end{pmatrix}.$$

Answer. Since the linear equation is homogeneous, we have only to deform the coefficient matrix of the linear equation to solve the linear equation. We are going to deform the coefficient matrix:

$$
\begin{array}{rrrrrr}
1 & -2 & 0 & 0 & 1 & 2 \\
1 & -1 & 1 & 0 & 1 & 0 \\
2 & -5 & -1 & 0 & 3 & 3 \\
\hline
1 & -2 & 0 & 0 & 1 & 2 \\
0 & 1 & 1 & 0 & 0 & -2 \\
0 & -1 & -1 & 0 & 1 & -1 \\
\hline
1 & 0 & 2 & 0 & 1 & -2 \\
0 & 1 & 1 & 0 & 0 & -2 \\
0 & 0 & 0 & 0 & 1 & -3 \\
\hline
1 & 0 & 2 & 0 & 0 & 1 \\
0 & 1 & 1 & 0 & 0 & -2 \\
0 & 0 & 0 & 0 & 1 & -3
\end{array}
\qquad
\begin{array}{l}
\\
\\
\\
\\
②+①\times(-1) \\
③+①\times(-2) \\
①+②\times 2 \\
\\
③+② \\
①+③\times(-1)
\end{array}
$$

Since the linear equation corresponding to the reduced matrix is

$$
\begin{cases}
x_1 \quad\quad + 2x_3 \quad\quad + x_6 = 0, \\
\quad\quad x_2 + x_3 \quad\quad - 2x_6 = 0, \\
\quad\quad\quad\quad\quad\quad x_5 - 3x_6 = 0,
\end{cases}
$$

we have the solution

$$
\mathbf{x} =
\begin{pmatrix}
-2c_1 - c_3 \\
-c_1 + 2c_3 \\
c_1 \\
c_2 \\
3c_3 \\
c_3
\end{pmatrix}
= c_1
\begin{pmatrix}
-2 \\ -1 \\ 1 \\ 0 \\ 0 \\ 0
\end{pmatrix}
+ c_2
\begin{pmatrix}
0 \\ 0 \\ 0 \\ 1 \\ 0 \\ 0
\end{pmatrix}
+ c_3
\begin{pmatrix}
-1 \\ 2 \\ 0 \\ 0 \\ 3 \\ 1
\end{pmatrix}
\quad (c_1, c_2, c_3 \in \mathbf{R}). \quad \|
$$

Exercises (Sect. 2.3)

1 Solve the following linear equations:

(1) $\begin{pmatrix} -3 & 3 & 1 \\ 1 & -1 & 2 \end{pmatrix} \begin{pmatrix} x_1 \\ x_2 \\ x_3 \end{pmatrix} = \begin{pmatrix} 7 \\ 0 \end{pmatrix}.$

(2) $\begin{pmatrix} 2 & -1 & 5 \\ 0 & 2 & 2 \\ 1 & 0 & 3 \end{pmatrix} \begin{pmatrix} x_1 \\ x_2 \\ x_3 \end{pmatrix} = \begin{pmatrix} -1 \\ 6 \\ 1 \end{pmatrix}.$

(3) $\begin{pmatrix} 1 & -1 & 1 \\ -1 & 0 & -3 \\ -1 & 2 & 1 \end{pmatrix} \begin{pmatrix} x_1 \\ x_2 \\ x_3 \end{pmatrix} = \begin{pmatrix} 1 \\ 1 \\ 2 \end{pmatrix}.$

(4) $\begin{pmatrix} 2 & -1 & 1 \\ -1 & 1 & -3 \\ 3 & -2 & 4 \end{pmatrix} \begin{pmatrix} x_1 \\ x_2 \\ x_3 \end{pmatrix} = \begin{pmatrix} 2 \\ 0 \\ 3 \end{pmatrix}.$

(5) $\begin{pmatrix} 1 & -1 & 1 & -3 \\ -1 & 0 & -3 & 3 \\ 1 & 2 & 7 & -2 \end{pmatrix} \begin{pmatrix} x_1 \\ x_2 \\ x_3 \\ x_4 \end{pmatrix} = \begin{pmatrix} 3 \\ -4 \\ 5 \end{pmatrix}.$

(6) $\begin{pmatrix} 2 & -3 & -3 & 3 \\ -1 & 1 & 1 & -2 \\ 1 & -2 & -2 & 1 \\ 1 & 2 & 0 & 1 \end{pmatrix} \begin{pmatrix} x_1 \\ x_2 \\ x_3 \\ x_4 \end{pmatrix} = \begin{pmatrix} 2 \\ 0 \\ 1 \\ 2 \end{pmatrix}.$

(7) $\begin{pmatrix} 1 & 1 & 1 & 1 & 0 \\ 0 & 1 & -2 & 0 & 2 \\ 2 & 3 & 0 & 2 & 2 \\ 1 & 0 & 3 & 1 & -2 \end{pmatrix} \begin{pmatrix} x_1 \\ x_2 \\ x_3 \\ x_4 \\ x_5 \end{pmatrix} = \begin{pmatrix} 3 \\ 1 \\ 7 \\ 2 \end{pmatrix}.$

(8) $\begin{pmatrix} 1 & 0 & -2 & 0 & 1 \\ 2 & 1 & -3 & 5 & 0 \\ 0 & 2 & 2 & 0 & -4 \\ 1 & 0 & -2 & 2 & 1 \\ 0 & 1 & 1 & 1 & -2 \end{pmatrix} \begin{pmatrix} x_1 \\ x_2 \\ x_3 \\ x_4 \\ x_5 \end{pmatrix} = \begin{pmatrix} -1 \\ 5 \\ 4 \\ 1 \\ 3 \end{pmatrix}.$

2   Solve the following homogeneous linear equations:

(1) $\begin{pmatrix} 1 & -1 & -1 \\ -1 & 0 & -1 \\ 1 & 2 & 5 \end{pmatrix} \begin{pmatrix} x_1 \\ x_2 \\ x_3 \end{pmatrix} = \begin{pmatrix} 0 \\ 0 \\ 0 \end{pmatrix}.$

(2) $\begin{pmatrix} 2 & -1 & 3 \\ -1 & 1 & -2 \\ 1 & -3 & 4 \end{pmatrix} \begin{pmatrix} x_1 \\ x_2 \\ x_3 \end{pmatrix} = \begin{pmatrix} 0 \\ 0 \\ 0 \end{pmatrix}.$

(3) $\begin{pmatrix} 1 & 0 & 2 & -1 \\ 0 & 1 & -1 & 1 \\ -1 & 3 & -5 & 4 \end{pmatrix} \begin{pmatrix} x_1 \\ x_2 \\ x_3 \\ x_4 \end{pmatrix} = \begin{pmatrix} 0 \\ 0 \\ 0 \end{pmatrix}.$

(4) $\begin{pmatrix} 1 & -2 & 1 & 2 & 3 \\ 1 & -2 & 0 & 1 & 2 \\ -1 & 2 & 1 & 0 & -1 \end{pmatrix} \begin{pmatrix} x_1 \\ x_2 \\ x_3 \\ x_4 \\ x_5 \end{pmatrix} = \begin{pmatrix} 0 \\ 0 \\ 0 \end{pmatrix}.$

3   Find the conditions on the numbers a, b for the following linear equations to have solutions:

(1) $\begin{pmatrix} 2 & 1 & 3 \\ 0 & -1 & 1 \\ 1 & 1 & 1 \end{pmatrix} \begin{pmatrix} x_1 \\ x_2 \\ x_3 \end{pmatrix} = \begin{pmatrix} 1 \\ a \\ b \end{pmatrix}.$

(2) $\begin{pmatrix} 1 & -1 & 1 \\ 1 & 1 & 2 \\ 2 & -2 & a \end{pmatrix} \begin{pmatrix} x_1 \\ x_2 \\ x_3 \end{pmatrix} = \begin{pmatrix} 2 \\ 5 \\ 5 \end{pmatrix}.$

4 Let $\mathbf{x}_0$ be a solution of a linear equation

$$(*)\qquad A\mathbf{x} = \mathbf{b}.$$

Show that if $\mathbf{x}_1$ is a solution of the homogeneous linear equation

$$(**)\qquad A\mathbf{x} = \mathbf{0},$$

then $\mathbf{x}_0 + \mathbf{x}_1$ is a solution of $(*)$. Furthermore, show that any solution of $(*)$ is expressed as $\mathbf{x}_0 + \mathbf{x}_1$ with a solution $\mathbf{x}_1$ of $(**)$.

2.4   Regular Matrices

We consider only square matrices in this section.

Inverse matrices Let A be a square matrix of degree n. A square matrix B of degree n is called an *inverse matrix* of A if it satisfies

$$AB = BA = E_n.$$

If A has an inverse matrix, it is uniquely determined. In fact, if A has two inverse matrices B and C, then

$$B = BE_n = B(AC) = (BA)C = E_nC = C.$$

Therefore we see $B = C$.

Regular matrices When a square matrix A has the inverse matrix, A is called a *regular matrix*, and the inverse matrix of A is denoted by

$$A^{-1}.$$

The next theorem will be proved in Sect. 3.4.

Theorem 2.4.1 *Let A and B be square matrices of degree n. If A and B satisfy the equality $AB = E_n$, then A and B are regular matrices and B is the inverse matrix of A.*

Before calculating the inverse matrix, we state the following theorem which shows equivalent conditions for the regularity of a square matrix.

Theorem 2.4.2 *For a square matrix A of degree n, the following five conditions are equivalent:*
(1) $\mathrm{rank}(A) = n$.
(2) The reduced matrix of A is the unit matrix E_n.

(3) The linear equation $A\mathbf{x} = \mathbf{b}$ has a unique solution for any column vector $\mathbf{b}$ of degree n.

(4) The homogeneous linear equation $A\mathbf{x} = \mathbf{0}$ has only the trivial solution.

(5) The matrix A is a regular matrix.

Proof $(1) \Rightarrow (2)$ Since A is a square matrix of degree n satisfying $\mathrm{rank}(A) = n$, all columns of the reduced matrix of A contain primary components. Therefore they are not zero vectors. This implies that the reduced matrix must be the unit matrix E_n.

$(2) \Rightarrow (3)$ Since the reduced matrix of A is E_n, the reduced matrix of the augmented matrix $(A|\mathbf{b})$ of the linear equation $A\mathbf{x} = \mathbf{b}$ is written as

$$(E_n|\mathbf{b}')$$

with a column vector $\mathbf{b}'$ of degree n. If we write

$$\mathbf{b}' = \begin{pmatrix} b_1' \\ b_2' \\ \vdots \\ b_n' \end{pmatrix},$$

then the solution of the linear equation $A\mathbf{x} = \mathbf{b}$ is

$$x_1 = b_1', x_2 = b_2', \ldots, x_n = b_n'.$$

Therefore the linear equation has a unique solution.

$(3) \Rightarrow (4)$ Condition (4) is the special case of of (3). In fact, if we take $\mathbf{b} = \mathbf{0}$ in (3), then the linear equation $A\mathbf{x} = \mathbf{0}$ has a unique solution $\mathbf{x} = \mathbf{0}$.

$(4) \Rightarrow (1)$ This is nothing but the assertion of Theorem 2.3.3 (1).

Thus we showed the equivalence of conditions (1) through (4).

$(3) \Rightarrow (5)$ We let $\mathbf{e}_1, \mathbf{e}_2, \ldots, \mathbf{e}_n$ be the column vectors of degree n given by

$$(*) \quad \mathbf{e}_1 = \begin{pmatrix} 1 \\ 0 \\ \vdots \\ 0 \end{pmatrix}, \mathbf{e}_2 = \begin{pmatrix} 0 \\ 1 \\ \vdots \\ 0 \end{pmatrix}, \ldots, \mathbf{e}_n = \begin{pmatrix} 0 \\ \vdots \\ 0 \\ 1 \end{pmatrix}.$$

Then the linear equations

$$A\mathbf{x} = \mathbf{e}_1, A\mathbf{x} = \mathbf{e}_2, \ldots, A\mathbf{x} = \mathbf{e}_n$$

have unique solutions by the assumption. We denote the solution of $A\mathbf{x} = \mathbf{e}_i$ by $\mathbf{x} = \mathbf{c}_i$ for $i = 1, \ldots, n$ and let the matrix C be

$$C = (\mathbf{c}_1 \ \mathbf{c}_2 \ \cdots \ \mathbf{c}_n).$$

Then C is a square matrix of degree n and satisfies

$$AC = A \begin{pmatrix} \mathbf{c}_1 & \mathbf{c}_2 & \cdots & \mathbf{c}_n \end{pmatrix}$$
$$= \begin{pmatrix} A\mathbf{c}_1 & A\mathbf{c}_2 & \cdots & A\mathbf{c}_n \end{pmatrix}$$
$$= \begin{pmatrix} \mathbf{e}_1 & \mathbf{e}_2 & \cdots & \mathbf{e}_n \end{pmatrix} = E_n.$$

By Theorem 2.4.1, A is a regular matrix and C is the inverse matrix of A.

(5) $\Rightarrow$ (4) Assume $A\mathbf{x} = \mathbf{0}$. Multiplying the linear equation $A\mathbf{x} = \mathbf{0}$ by the inverse matrix A^{-1} from the left-hand side, we have

$$A^{-1}A\mathbf{x} = A^{-1}\mathbf{0} = \mathbf{0}.$$

We also know $A^{-1}A\mathbf{x} = E_n\mathbf{x} = \mathbf{x}$. Therefore we have $\mathbf{x} = \mathbf{0}$ and this is (4).

Thus we have proved the equivalence of (1) through (5).   ‖

Calculation of inverse matrices The proof of the above theorem shows the way to calculate the inverse matrices. For a regular matrix A of degree n, we solve the following n linear equations:

$$A\mathbf{x} = \mathbf{e}_1, \ A\mathbf{x} = \mathbf{e}_2, \ldots, \ A\mathbf{x} = \mathbf{e}_n.$$

We let $\mathbf{x} = \mathbf{c}_i$ be the solution of the linear equation $A\mathbf{x} = \mathbf{e}_i$ for $i = 1, \ldots, n$. Then as we showed in the proof of (3)$\Rightarrow$(5), we have

$$A^{-1} = \begin{pmatrix} \mathbf{c}_1 & \mathbf{c}_2 & \cdots & \mathbf{c}_n \end{pmatrix}.$$

Since the reduced matrix of $\begin{pmatrix} A | \mathbf{e}_i \end{pmatrix}$ is $\begin{pmatrix} E_n | \mathbf{c}_i \end{pmatrix}$, we can simultaneously deform $\begin{pmatrix} A | \mathbf{e}_i \end{pmatrix}$ by reductions for $i = 1, 2, \ldots, n$, and we obtain

$$\begin{pmatrix} A | E_n \end{pmatrix} = \begin{pmatrix} A | \mathbf{e}_1 & \mathbf{e}_2 & \cdots & \mathbf{e}_n \end{pmatrix}$$
$$\rightarrow \begin{pmatrix} E_n | \mathbf{c}_1 & \mathbf{c}_2 & \cdots & \mathbf{c}_n \end{pmatrix} = \begin{pmatrix} E_n | A^{-1} \end{pmatrix}.$$

Therefore $\begin{pmatrix} E_n | A^{-1} \end{pmatrix}$ is the reduced matrix of $\begin{pmatrix} A | E_n \end{pmatrix}$. We note that the reduction processes of the matrices $\begin{pmatrix} A | \mathbf{e}_i \end{pmatrix}$ into $\begin{pmatrix} E_n | \mathbf{c}_i \end{pmatrix}$ are the same for $i = 1, 2, \ldots, n$. Therefore the inverse matrix A^{-1} appears in the right-hand side of the reduced matrix of $\begin{pmatrix} A | E_n \end{pmatrix}$.

Unit vectors We call the vectors $\mathbf{e}_1, \ldots, \mathbf{e}_n$ defined in $(*)$ above the *unit vetcors* of degree n.

Exercise 2.7 Find the inverse matrix of

$$A = \begin{pmatrix} 1 & 2 & 1 \\ 2 & 3 & 1 \\ 1 & 2 & 2 \end{pmatrix}.$$

Answer. We are going to find the reduced matrix of

$$\begin{pmatrix} A | E_3 \end{pmatrix}.$$

Now, by the calculation

1	2	1	1	0	0
2	3	1	0	1	0
1	2	2	0	0	1
1	2	1	1	0	0
0	−1	−1	−2	1	0
0	0	1	−1	0	1
1	2	1	1	0	0
0	1	1	2	−1	0
0	0	1	−1	0	1
1	0	−1	−3	2	0
0	1	1	2	−1	0
0	0	1	−1	0	1
1	0	0	−4	2	1
0	1	0	3	−1	−1
0	0	1	−1	0	1

②+①×(−2)
③+①×(−1)

②×(−1)

①+②×(−2)

①+③
②+③×(−1).

we obtain

$$A^{-1} = \begin{pmatrix} -4 & 2 & 1 \\ 3 & -1 & -1 \\ -1 & 0 & 1 \end{pmatrix}. \quad \|$$

Remark. Let A be a square matrix of degree n. If the reduced matrix of $(A|E_n)$ is not of the form $(E_n|*)$, then the matrix A is not a regular matrix and A does not have the inverse matrix.

Exercises (Sect. 2.4)

1 Find the inverse matrices of the following matrices:

(1) $\begin{pmatrix} 2 & -1 \\ 1 & -1 \end{pmatrix}$.

(2) $\begin{pmatrix} 1 & -6 \\ -1 & 5 \end{pmatrix}$.

(3) $\begin{pmatrix} 1 & 5 \\ -1 & -3 \end{pmatrix}$.

(4) $\begin{pmatrix} 1 & 0 & -2 \\ 0 & 1 & 0 \\ -1 & 0 & 1 \end{pmatrix}$.

(5) $\begin{pmatrix} 1 & 1 & 1 \\ 1 & 1 & 0 \\ 1 & 0 & 0 \end{pmatrix}$.

(6) $\begin{pmatrix} 2 & -1 & 0 \\ 2 & -1 & -1 \\ 1 & 0 & -1 \end{pmatrix}$.

(7) $\begin{pmatrix} -3 & -6 & 2 \\ 3 & 5 & -2 \\ 1 & 3 & -1 \end{pmatrix}$.

(8) $\begin{pmatrix} 1 & -1 & -3 \\ 1 & 1 & -1 \\ -1 & 1 & 5 \end{pmatrix}$.

(9) $\begin{pmatrix} 1 & 2 & -1 \\ 1 & 1 & 0 \\ 3 & 1 & -1 \end{pmatrix}$.

$$(10) \quad \begin{pmatrix} 1 & 1 & 1 & 1 \\ 0 & 1 & 1 & 1 \\ 0 & 0 & 1 & 1 \\ 0 & 0 & 0 & 1 \end{pmatrix}. \qquad\qquad (11) \quad \begin{pmatrix} 2 & 0 & 1 & 0 \\ 0 & -1 & 1 & -2 \\ 1 & 0 & 1 & 0 \\ 0 & 1 & -1 & 3 \end{pmatrix}.$$

2 Solve the following linear equations using the inverse matrices of the coefficient matrices:

$$(1) \quad \begin{pmatrix} 7 & -2 \\ 3 & -1 \end{pmatrix} \begin{pmatrix} x_1 \\ x_2 \\ x_3 \end{pmatrix} = \begin{pmatrix} 3 \\ 4 \end{pmatrix}.$$

$$(2) \quad \begin{pmatrix} 5 & -2 & 2 \\ 3 & -1 & 2 \\ -2 & 1 & -1 \end{pmatrix} \begin{pmatrix} x_1 \\ x_2 \\ x_3 \end{pmatrix} = \begin{pmatrix} -1 \\ 0 \\ 3 \end{pmatrix}.$$

$$(3) \quad \begin{pmatrix} 2 & 2 & -3 \\ -1 & -1 & 1 \\ 1 & 0 & -1 \end{pmatrix} \begin{pmatrix} x_1 \\ x_2 \\ x_3 \end{pmatrix} = \begin{pmatrix} 1 \\ 2 \\ -3 \end{pmatrix}.$$

$$(4) \quad \begin{pmatrix} 3 & 2 & 1 \\ 3 & 3 & 1 \\ 1 & -2 & 1 \end{pmatrix} \begin{pmatrix} x_1 \\ x_2 \\ x_3 \end{pmatrix} = \begin{pmatrix} 4 \\ 1 \\ 2 \end{pmatrix}.$$

$$(5) \quad \begin{pmatrix} 4 & 1 & -1 \\ 5 & 3 & -1 \\ 1 & 1 & 0 \end{pmatrix} \begin{pmatrix} x_1 \\ x_2 \\ x_3 \end{pmatrix} = \begin{pmatrix} a \\ b \\ c \end{pmatrix}.$$

3 Let A and B be regular matrices of degree n. Show the following assertions:

(1) A^{-1} is also a regular matrix and $(A^{-1})^{-1} = A$.

(2) $^t A$ is also a regular matrix and $(^t A)^{-1} =^t (A^{-1})$. (Therefore we simply denote this matrix by $^t A^{-1}$.)

(3) AB is also a regular matrix and $(AB)^{-1} = B^{-1}A^{-1}$.

4 Assume that A and B are commutative square matrices. Show that the following matrices are also commutative:

(1) A^{-1}, B. \qquad\qquad (2) A^{-1}, B^{-1}. \qquad\qquad (3) $^t A$, $^t B$.

5 Let A be a square matrix. Show that if there exists a non-zero matrix B satisfying $AB = O$, then A is not a regular matrix.

6 Assume that A is a nilpotent matrix of degree n. Show that $E_n + A$ and $E_n - A$ are regular matrices. Also find their inverse matrices.

7 Let A and D be regular matrices of degree m and of degree n, respectively. Let B and C be an arbitrary $m \times n$ matrix and an arbitrary $n \times m$ matrix, respectively. Show that the following matrices X, Y and Z are regular matrices and find X^{-1}, Y^{-1} and Z^{-1}:

$$X = \begin{pmatrix} A & B \\ O_{n,m} & D \end{pmatrix}, Y = \begin{pmatrix} A & O_{m,n} \\ C & D \end{pmatrix}, Z = \begin{pmatrix} B & A \\ D & O_{n,m} \end{pmatrix}.$$

Chapter 3
Determinants

3.1 Permutations

Permutations Before explaining determinants, we begin with defining permutations of finite elements. Let $X_n = \{1, 2, \ldots, n\}$ be a set of n elements $1, 2, \ldots, n$. Replacements of elements in X_n are called *permutations* of n elements. Permutations are considered as mappings of X_n into itself. When a permutation σ maps

$$1 \mapsto k_1, \quad 2 \mapsto k_2, \quad \cdots \quad , n \mapsto k_n$$

we denote the permutation σ by

$$\sigma = \begin{pmatrix} 1 & 2 & \cdots & n \\ k_1 & k_2 & \cdots & k_n \end{pmatrix}.$$

Here the element k_i is the image of i by the permutation σ and

$$\sigma(i) = \begin{pmatrix} 1 & 2 & \cdots & n \\ k_1 & k_2 & \cdots & k_n \end{pmatrix}(i) = k_i.$$

Example 1 If $\sigma = \begin{pmatrix} 1 & 2 & 3 \\ 2 & 3 & 1 \end{pmatrix}$, then $\sigma(2) = \begin{pmatrix} 1 & 2 & 3 \\ 2 & 3 & 1 \end{pmatrix}(2) = 3$.

Example 2 The permutation

$$\sigma = \begin{pmatrix} 1 & 2 & 3 & 4 \\ 3 & 1 & 4 & 2 \end{pmatrix}$$

maps $\{1, 2, 3, 4\}$ into $\{1, 2, 3, 4\}$ according to

$$\sigma(1) = 3, \quad \sigma(2) = 1, \quad \sigma(3) = 4, \quad \sigma(4) = 2.$$

© The Author(s), under exclusive license to Springer Nature Singapore Pte Ltd. 2022
T. Miyake, *Linear Algebra*, https://doi.org/10.1007/978-981-16-6994-1_3

We may change the order of pairs $\left\{\begin{matrix} i \\ k_i \end{matrix}\right\}$ if each pair $\left\{\begin{matrix} i \\ k_i \end{matrix}\right\}$ is unchanged.

Example 3 Take a permutation $\sigma = \begin{pmatrix} 1\,2\,3\,4\,5 \\ 3\,4\,5\,2\,1 \end{pmatrix}$. Then the permutation σ can be also expressed as

$$\begin{pmatrix} 5\,2\,3\,1\,4 \\ 1\,4\,5\,3\,2 \end{pmatrix}.$$

Therefore $\sigma = \begin{pmatrix} 2\,4\,1\,5\,3 \\ 4\,2\,3\,1\,5 \end{pmatrix} = \begin{pmatrix} 3\,1\,2\,5\,4 \\ 5\,3\,4\,1\,2 \end{pmatrix}$, etc..

We may omit the pairs $\left\{\begin{matrix} i \\ i \end{matrix}\right\}$ in the expressions.

Example 4 $\begin{pmatrix} 1\,2\,3\,4\,5 \\ 3\,2\,5\,4\,1 \end{pmatrix} = \begin{pmatrix} 1\,3\,5 \\ 3\,5\,1 \end{pmatrix} = \begin{pmatrix} 5\,1\,3 \\ 1\,3\,5 \end{pmatrix}$.

Products of permutations Let σ and τ be permutations of n elements. We define the *product* $\sigma\tau$ of σ and τ by

$$\sigma\tau(i) = \sigma(\tau(i)) \qquad (i \in X_n).$$

Example 5 Let $\sigma = \begin{pmatrix} 1\,2\,3\,4 \\ 4\,3\,1\,2 \end{pmatrix}$, $\tau = \begin{pmatrix} 1\,2\,3\,4 \\ 2\,3\,4\,1 \end{pmatrix}$. Then we have

$$\sigma\tau(1) = \sigma(\tau(1)) = \sigma(2) = 3,$$
$$\sigma\tau(2) = \sigma(\tau(2)) = \sigma(3) = 1,$$
$$\sigma\tau(3) = \sigma(\tau(3)) = \sigma(4) = 2,$$
$$\sigma\tau(4) = \sigma(\tau(4)) = \sigma(1) = 4.$$

Therefore we have

$$\sigma\tau = \begin{pmatrix} 1\,2\,3\,4 \\ 3\,1\,2\,4 \end{pmatrix} = \begin{pmatrix} 1\,2\,3 \\ 3\,1\,2 \end{pmatrix}.$$

Similarly, we obtain that

$$\tau\sigma = \begin{pmatrix} 1\,2\,3\,4 \\ 1\,4\,2\,3 \end{pmatrix} = \begin{pmatrix} 2\,3\,4 \\ 4\,2\,3 \end{pmatrix}.$$

We note $\sigma\tau \neq \tau\sigma$ in this case.

Unit permutation The permutation which does not change any element in X_n is called the *unit permutation*. The unit permutation is denoted by ε, or

$$\varepsilon = \begin{pmatrix} 1 \cdots n \\ 1 \cdots n \end{pmatrix} \quad \text{for} \quad X_n = \{1, \ldots, n\}.$$

We have $\varepsilon\sigma = \sigma\varepsilon = \sigma$ for any permutation σ of X_n.

Inverse permutations For a permutation

$$\sigma = \begin{pmatrix} 1 & 2 & \cdots & n \\ k_1 & k_2 & \cdots & k_n \end{pmatrix},$$

we define the permutation σ^{-1} by

$$\sigma^{-1} = \begin{pmatrix} k_1 & k_2 & \cdots & k_n \\ 1 & 2 & \cdots & n \end{pmatrix}.$$

The permutation σ^{-1} is called the *inverse permutation* of σ and satisfies

$$\sigma\sigma^{-1} = \sigma^{-1}\sigma = \varepsilon.$$

Example 6 If $\sigma = \begin{pmatrix} 1\,2\,3\,4\,5 \\ 4\,5\,1\,3\,2 \end{pmatrix}$, then we have

$$\sigma^{-1} = \begin{pmatrix} 4\,5\,1\,3\,2 \\ 1\,2\,3\,4\,5 \end{pmatrix} = \begin{pmatrix} 1\,2\,3\,4\,5 \\ 3\,5\,4\,1\,2 \end{pmatrix}.$$

Transpositions The permutation $\begin{pmatrix} i_1 & i_2 \\ i_2 & i_1 \end{pmatrix}$ of two elements i_1 and i_2 is called the *transposition* of i_1 and i_2. We denote the transposition of i_1 and i_2 by $(i_1\ i_2)$. If $\sigma = (i_1\ i_2)$ is a transposition, then

$$\sigma^{-1} = (i_2\ i_1) = \sigma.$$

Example 7 We see that

$$\begin{pmatrix} 3\,5 \\ 5\,3 \end{pmatrix} = (3\ 5) = (5\ 3).$$

Cyclic permutations Let σ be a permutation which permutes r elements $i_1, i_2, \ldots, i_r$ cyclically according to

$$i_1 \mapsto i_2, \quad i_2 \mapsto i_3, \quad \cdots \quad , \quad i_{r-1} \mapsto i_r, \quad i_r \mapsto i_1$$

and does not change any element not in $\{i_1, i_2, \ldots, i_r\}$. The permutation σ is called the *cyclic permutation*. We denote the cyclic permutation σ by

$$\sigma = \left(i_1 \; i_2 \; \cdots \; i_r \right).$$

Example 8 We consider the cyclic permutation $\sigma = \left(1\;3\;7\;5 \right)$ of $X_8 = \{1, 2, 3, 4, 5, 6, 7, 8\}$. Then we have

$$\sigma = \left(1\;3\;7\;5 \right) = \begin{pmatrix} 1\;2\;3\;4\;5\;6\;7\;8 \\ 3\;2\;7\;4\;1\;6\;5\;8 \end{pmatrix}.$$

Example 9 By the cyclic permutation $\sigma = \left(2\;5\;7\;3 \right)$, we have

$$\sigma(2) = 5, \quad \sigma(5) = 7, \quad \sigma(7) = 3, \quad \sigma(3) = 2$$

and

$$\sigma(i) = i \quad \text{for} \quad i \neq 2, 3, 5, 7.$$

Sets of permutations We denote by S_n the set of the permutations of n elements.

Example 10 We easily see that the sets S_1, S_2, S_3 are as follows:

$$S_1 = \{\varepsilon\},$$
$$S_2 = \left\{\varepsilon, \left(1\;2 \right)\right\},$$
$$S_3 = \left\{\varepsilon, \left(1\;2 \right), \left(1\;3 \right), \left(2\;3 \right), \left(1\;2\;3 \right), \left(1\;3\;2 \right)\right\}.$$

Lengths of permutations The number of elements changed by a permutation σ is called the *length* of σ. The length of σ is denoted by $l(\sigma)$. We consider the length of the unit permutation ε as 0. There is no permutation of length 1 and the length of a transposition $\left(i_1 \; i_2 \right) = \begin{pmatrix} i_1 \; i_2 \\ i_2 \; i_1 \end{pmatrix}$ is 2.

Example 11 Let $\sigma = \begin{pmatrix} 1\;2\;3\;4\;5 \\ 1\;4\;3\;5\;2 \end{pmatrix}$. Since $\sigma = \begin{pmatrix} 2\;4\;5 \\ 4\;5\;2 \end{pmatrix}$, σ changes only three elements, 2, 4, 5. Therefore the length of σ is $l(\sigma) = 3$.

Theorem 3.1.1 *Any permutation $\sigma \in S_n$ ($n \geq 2$) can be expressed as a product of transpositions, which is called a transposition product expression of σ.*

Proof We shall prove the theorem by induction on the length of a permutation. If $l(\sigma) = 0$, then $\sigma = \varepsilon$. Since $\varepsilon = \left(1\;2 \right)\left(1\;2 \right)$, the unit permutation ε is a product of transpositions.

Assume that the assertion is true for permutations whose lengths are less than l. We shall show the theorem is true for any permutation of length l. Let $\sigma \in S_n$ be a permutation with $l(\sigma) = l$ denoted by

$$\sigma = \begin{pmatrix} i_1 & i_2 & \cdots & i_l \\ k_1 & k_2 & \cdots & k_l \end{pmatrix}.$$

Here $i_j \neq k_j$ for $j = 1, \ldots, l$. Then the permutation

$$(i_1 \; k_1) \, \sigma$$

does not change i_1. Then the possible elements which $(i_1 \; k_1) \, \sigma$ may change are only $i_2, i_3, \ldots, i_l$. Therefore the length of the permutation

$$(i_1 \; k_1) \, \sigma$$

is less than or equal to $l - 1$. By the induction assumption, the permutation $(i_1 \; k_1) \, \sigma$ is a product of transpositions

$$(i_1 \; k_1) \, \sigma = \sigma_1 \sigma_2 \cdots \sigma_r.$$

By multiplying the transposition $(i_1 \; k_1)$ from the left on both sides of the equality, the permutation σ is a product of transpositions

$$\sigma = (i_1 \; k_1) \, \sigma_1 \sigma_2 \cdots \sigma_r. \quad \|$$

Example 12 The permutation $\sigma = \begin{pmatrix} 1 & 2 & 3 \\ 3 & 1 & 2 \end{pmatrix}$ is expressed as

$$\sigma = \begin{pmatrix} 1 & 2 & 3 \\ 3 & 1 & 2 \end{pmatrix} = (1 \; 2)(1 \; 3) = (1 \; 3)(2 \; 3).$$

The permutation σ in Example 12 is also expressed as

$$\sigma = (2 \; 3)(1 \; 2) = (2 \; 3)(1 \; 2)(1 \; 2)(1 \; 2) = \cdots.$$

Then the numbers of transpositions in expressions of the permutation σ as products of transpositions are not uniquely determined. But the parity of the numbers of transpositions appearing in expressions of σ as products of transpositions is uniquely determined for any $\sigma \in S_n$ (Theorem 3.1.2 below).

Action of permutations on polynomials Let $\sigma \in S_n$ and $f(x_1, x_2, \ldots, x_n)$ a polynomial of n variables. Define the *action* of σ on the polynomial $f(x_1, x_2, \ldots, x_n)$ by

$$\sigma f(x_1, x_2, \ldots, x_n) = f(x_{\sigma(1)}, x_{\sigma(2)}, \ldots, x_{\sigma(n)}).$$

We see that

$$\varepsilon f(x_1, x_2, \ldots, x_n) = f(x_1, x_2, \ldots, x_n).$$

Furthermore, since $\sigma(\tau x_i) = \sigma x_{\tau(i)} = x_{\sigma(\tau(i))} = x_{\sigma\tau(i)} = (\sigma\tau)x_i$, we have

$$\sigma\{\tau f(x_1, x_2, \ldots, x_n)\} = (\sigma\tau)f(x_1, x_2, \ldots, x_n).$$

Example 13 Let $\sigma = \begin{pmatrix} 2 & 3 \end{pmatrix}$ and $\tau = \begin{pmatrix} 1 & 2 \end{pmatrix}$. Then we have $\sigma\tau = \begin{pmatrix} 1 & 2 & 3 \\ 3 & 1 & 2 \end{pmatrix}$. Letting τ and σ successively act on the polynomial $f(x_1, x_2, x_3) = x_1 x_2 + x_3^2$, we obtain

$$\sigma\{\tau f(x_1, x_2, x_3)\} = \sigma(x_2 x_1 + x_3^2) = x_3 x_1 + x_2^2.$$

On the other hand, we let $\sigma\tau$ act on the polynomial $f(x_1, x_2, x_3)$ and obtain

$$(\sigma\tau)f(x_1, x_2, x_3) = (\sigma\tau)(x_1 x_2 + x_3^2) = x_3 x_1 + x_2^2.$$

Therefore $\sigma\{\tau f(x_1, x_2, x_3)\} = (\sigma\tau)f(x_1, x_2, x_3)$.

Difference products For $n \geq 2$, we let

$$\Delta(x_1, x_2, \ldots, x_n) = \prod_{1 \leq i < j \leq n} (x_i - x_j).$$

The polynomial $\Delta(x_1, x_2, \ldots, x_n)$ is called the *difference product* of n variables.

Example 14 If $n = 3$, then the difference product is

$$\Delta(x_1, x_2, x_3) = (x_1 - x_2)(x_1 - x_3)(x_2 - x_3).$$

Example 15 If $n = 4$, then the difference product is

$$\begin{aligned} \Delta(x_1, x_2, x_3, x_4) &= (x_1 - x_2)(x_1 - x_3)(x_1 - x_4) \\ &\quad \times (x_2 - x_3)(x_2 - x_4) \\ &\quad \times (x_3 - x_4). \end{aligned}$$

Theorem 3.1.2 *Let $n \geq 2$. For $\sigma \in S_n$, the parity of numbers of transpositions in a transposition product expression of σ is uniquely determined and independent of the transposition product expressions.*

Proof We are going to see how the difference product $\Delta(x_1, x_2, \ldots, x_n)$ changes by the action of the transposition $\tau = \begin{pmatrix} i & j \end{pmatrix}$ $(i < j)$. By the action of τ, the component $(x_k - x_l)$ changes only when k or l is either i or j. Such components are:

(1) $(x_1 - x_i), \ldots, (x_{i-1} - x_i),$ (2) $(x_1 - x_j), \ldots, (x_{i-1} - x_j),$
(3) $(x_i - x_{i+1}), \ldots, (x_i - x_{j-1}),$ (4) $(x_{i+1} - x_j), \ldots, (x_{j-1} - x_j),$
(5) $(x_i - x_{j+1}), \ldots, (x_i - x_n),$ (6) $(x_j - x_{j+1}), \ldots, (x_j - x_n),$
(7) $(x_i - x_j).$

The other components are not changed by the action of τ. We easily see the following:
  (i) The components in (1) and (2) are exchanged by the action of τ.
  (ii) The components in (5) and (6) are also exchanged by the action of τ.
  (iii) We have

$$\tau(x_i - x_k) = x_j - x_k = -(x_k - x_j)$$

and

$$\tau(x_k - x_j) = x_k - x_i = -(x_i - x_k)$$

for $k = i + 1, \ldots, j - 1$. Then the product of all components in (3) and (4) is unchanged by the action of τ.
  (iv) The component $(x_i - x_j)$ is changed by the action of τ as
$$\tau(x_i - x_j) = x_j - x_i = -(x_i - x_j).$$

Then the action of the transposition τ on $\Delta(x_1, x_2, \ldots, x_n)$ is

$$\tau\Delta(x_1, x_2, \ldots, x_n) = -\Delta(x_1, x_2, \ldots, x_n).$$

Therefore, if the permutation σ is a product $\sigma = \tau_1\tau_2\ldots\tau_r$ of r transpositions $\tau_1, \tau_2, \ldots, \tau_r$, then

$$\sigma\Delta(x_1, x_2, \ldots, x_n) = (-1)^r\Delta(x_1, x_2, \ldots, x_n).$$

If σ is expressed as another product $\sigma = \mu_1\mu_2\ldots\mu_s$ of σ transpositions, then

$$\sigma\Delta(x_1, x_2, \ldots, x_n) = (-1)^s\Delta(x_1, x_2, \ldots, x_n).$$

Thus the equality $(-1)^r\Delta(x_1, x_2, \ldots, x_n) = (-1)^s\Delta(x_1, x_2, \ldots, x_n)$ is obtained. Since $\Delta(x_1, x_2, \ldots, x_n) \neq 0$, we have $(-1)^r = (-1)^s$. Therefore the parity of the numbers of transpositions appearing in transposition product expressions of the permutation σ is uniquely determined. $\parallel$

Signatures of permutations When a permutation σ is a product of r transpositions, we define the *signature* of σ by

$$\text{sgn}(\sigma) = (-1)^r,$$

which is independent of transposition product expressions of σ.

Even permutations, odd permutations A permutation σ is called an *even permutation* if $\text{sgn}(\sigma) = 1$ and an *odd permutation* if $\text{sgn}(\sigma) = -1$. We denote the subset of all even permutations of S_n by S_n^+ and the subset of all odd permutations of S_n by S_n^-.

Example 16 Consider the permutations in S_3. Since

$$(1\,2\,3) = (1\,3)(1\,2), \qquad \text{sgn}\left((1\,2\,3)\right) = 1$$

and

$$(1\,3\,2) = (1\,2)(1\,3), \qquad \text{sgn}\left((1\,3\,2)\right) = 1,$$

we have

$$S_3^+ = \left\{\varepsilon, (1\,2\,3), (1\,3\,2)\right\}$$

and

$$S_3^- = \left\{(1\,2), (1\,3), (2\,3)\right\}.$$

Groups Though knowledge of groups is not necessary to understand the theory of linear algebra, we will have a better understanding of the linear algebra if we know the concept of a group.

Let G be a set with an operation $(f, g) \mapsto fg \in G$ for $f, g \in G$. When the operation of G has the following properties, we call G a *group*.

(1) The *associative law*: $(fg)h = f(gh)$ $(f, g, h \in G)$.

(2) There exists an element $e \in G$ satisfying $ge = eg = g$ for any element $g \in G$. The element e is called the *unit element* of G.

(3) For any element $g \in G$, there exists an element $g^{-1} \in G$ satisfying $gg^{-1} = g^{-1}g = e$. The element g^{-1} is called the *inverse element* of g.

Symmetric groups The set S_n is a group considering the product of permutations as the group operation. The unit permutation ε is the unit element of the group S_n and the inverse permutations are the inverse elements of the group S_n. The group S_n is called the *symmetric group* of degree n.

Numbers of elements in finite sets For a finite set X, the number of elements in X is denoted by $\#(X)$.

Orders of groups Let G be a group. The number $\#(G)$ of the elements in G is called the *order* of G.

Theorem 3.1.3 *Let $n \geq 2$. Then we have the following:*

(1) The order $\#(S_n)$ of the symmetric group S_n of degree n is $n!$.

(2) The elements of the sets S_n^+ and S_n^- correspond to each other by the correspondence

$$S_n^+ \ni \sigma \longmapsto \left(1\ 2\right) \sigma \in S_n^-.$$

So, we have $\#(S_n^+) = \#(S_n^-) = \dfrac{n!}{2}.$

Proof (1) Each permutation $\sigma = \begin{pmatrix} 1 & 2 & \cdots & n \\ k_1 & k_2 & \cdots & k_n \end{pmatrix}$ is determined by the elements $k_1, k_2, \ldots, k_n$. Since we can take any element between 1 and n for k_1, the number of choices of k_1 is n. Any element between 1 and n except k_1 can be taken as the element k_2. Then the number of choices of k_2 is $n-1$. Repeating these choices of k_i for $i = 3, \ldots, n$, we see that the number of possible choices of elements $k_1, k_2, \ldots, k_n$ is $n!$. Therefore we have $\#(S_n) = n!$.

(2) Let $\sigma \in S_n^+$. Since σ is expressed as a product of $2r$ transpositions by the definition of S_n^+, $\left(1\ 2\right)\sigma$ is a product of $2r+1$ transpositions. Thus $\left(1\ 2\right)\sigma$ belongs to S_n^-. Conversely, let $\tau \in S_n^-$. Since τ is expressed as a product of $2r+1$ transpositions, $\left(1\ 2\right)\tau$ is a product of $2r+2$ transpositions and $\left(1\ 2\right)\tau \in S_n^+$. Therefore permutations in S_n^+ and S_n^- correspond to each other by the mapping

$$S_n^+ \ni \sigma \longmapsto \left(1\ 2\right) \sigma \in S_n^-.$$

Thus we have $\#(S_n^+) = \#(S_n^-) = \dfrac{n!}{2}$. ‖

Subgroups When a subset H of a group G containing the unit element e is a group with the operation of G, we call H a *subgroup* of G.

Alternating groups The set S_n^+ is a subgroup of S_n. In fact, the unit permutation ε belongs to S_n^+. Furthermore, if $\sigma, \tau \in S_n^+$, then the number of transpositions appearing in a transposition product expression of $\sigma\tau$ is also even. Thus we have $\sigma\tau \in S_n^+$. The associative law holds trivially. Let $\sigma \in S_n^+$. Since the numbers of transpositions appearing in transposition product expressions of σ and σ^{-1} are equal, we have $\sigma^{-1} \in S_n^+$. Then S_n^+ is a group with the operation of S_n. Therefore S_n^+ is a subgroup of S_n.

We call the subgroup S_n^+ the *alternating group* of degree n. The alternating group of degree n is also denoted by A_n.

Example 17 When $n \geq 2$, the order $\#(A_n)$ of the alternating group A_n is $\dfrac{n!}{2}$.

Example 18 The alternating group A_2 is $\{\varepsilon\}$.

Example 19 The alternating group of degree 3 is

$$A_3 = S_3^+ = \left\{\varepsilon,\ \left(1\ 2\ 3\right),\ \left(1\ 3\ 2\right)\right\}.$$

Exercises (Sect. 3.1)

1 Find the following products of permutations:

(1) $\begin{pmatrix} 1\,2\,3 \\ 3\,1\,2 \end{pmatrix}\begin{pmatrix} 1\,2\,3 \\ 2\,3\,1 \end{pmatrix}$.

(2) $\begin{pmatrix} 1\,2\,3\,4 \\ 3\,4\,1\,2 \end{pmatrix}\begin{pmatrix} 2\,3\,4 \\ 3\,4\,2 \end{pmatrix}$.

(3) $\begin{pmatrix} 1\,2\,3\,4 \\ 3\,4\,2\,1 \end{pmatrix}\begin{pmatrix} 1\,2\,3\,4 \\ 2\,1\,4\,3 \end{pmatrix}$.

(4) $\begin{pmatrix} 1\,2\,3\,4\,5 \\ 5\,4\,2\,1\,3 \end{pmatrix}\begin{pmatrix} 1\,2\,3\,4\,5 \\ 4\,3\,5\,1\,2 \end{pmatrix}$.

(5) $(1\,3)(2\,3)(2\,4)$.

(6) $(1\,2\,3)(1\,4\,2)(2\,4)$.

(7) $(1\,4\,2)(2\,3)(1\,2\,4\,3)(2\,3)$.

(8) $(1\,5\,2\,4)(1\,3\,4)(1\,2\,4\,5)(2\,4\,3)$.

2 Express the following permutations as products of cyclic permutations:

(1) $\begin{pmatrix} 1\,2\,3\,4\,5\,6\,7 \\ 4\,7\,6\,5\,1\,2\,3 \end{pmatrix}$.

(2) $\begin{pmatrix} 1\,2\,3\,4\,5\,6\,7\,8 \\ 3\,1\,5\,8\,2\,4\,6\,7 \end{pmatrix}$.

(3) $\begin{pmatrix} 1\,2\,3\,4\,5\,6\,7\,8 \\ 7\,8\,1\,2\,6\,5\,3\,4 \end{pmatrix}$.

(4) $\begin{pmatrix} 1\,2\,3\,4\,5\,6\,7\,8\,9 \\ 9\,1\,5\,7\,3\,4\,8\,6\,2 \end{pmatrix}$.

(5) $\begin{pmatrix} 1\,2\,3\,4\,5\,6\,7\,8\,9 \\ 8\,5\,7\,9\,3\,1\,2\,6\,4 \end{pmatrix}$.

(6) $\begin{pmatrix} 1\,2\,3\,4\,5\,6\,7\,8\,9\,10 \\ 4\,6\,9\,5\,10\,2\,8\,3\,7\,1 \end{pmatrix}$.

3 Express the following cyclic permutations as products of transpositions:

(1) $(1\,6\,3\,2\,4)$.

(2) $(2\,5\,3\,6\,4\,7)$.

(3) $(1\,2\,5\,3\,7\,6\,4)$.

(4) $(1\,2\,3\,4\,5\,6\,7\,8)$.

4 Show that the signature of a cyclic permutation of length r is $(-1)^{r-1}$.

5 Express the following permutations as products of transpositions and find their signatures:

(1) $\begin{pmatrix} 1\,2\,3\,4 \\ 3\,4\,1\,2 \end{pmatrix}$.

(2) $\begin{pmatrix} 1\,2\,3\,4 \\ 2\,4\,1\,3 \end{pmatrix}$.

(3) $\begin{pmatrix} 1\,2\,3\,4\,5 \\ 5\,4\,2\,3\,1 \end{pmatrix}$.

(4) $\begin{pmatrix} 2\,3\,4\,6\,7 \\ 4\,6\,7\,3\,2 \end{pmatrix}$.

(5) $\begin{pmatrix} 1\ 2\ 3\ 4\ 5\ 6 \\ 6\ 4\ 5\ 2\ 3\ 1 \end{pmatrix}.$ (6) $\begin{pmatrix} 1\ 2\ 3\ 4\ 5\ 6 \\ 6\ 5\ 4\ 3\ 1\ 2 \end{pmatrix}.$

(7) $\begin{pmatrix} 1\ 2\ 3\ 4\ 5\ 6 \\ 6\ 4\ 5\ 1\ 2\ 3 \end{pmatrix}.$ (8) $\begin{pmatrix} 1\ 2\ 3\ 4\ 5\ 6 \\ 4\ 6\ 5\ 1\ 2\ 3 \end{pmatrix}.$

(9) $\begin{pmatrix} 1\ 2\ 3\ 4\ 5\ 6\ 7 \\ 3\ 7\ 5\ 1\ 4\ 2\ 6 \end{pmatrix}.$ (10) $\begin{pmatrix} 1\ 2\ 3\ 4\ 5\ 6\ 7 \\ 2\ 7\ 6\ 5\ 1\ 4\ 3 \end{pmatrix}.$

(11) $\begin{pmatrix} 1\ 2\ 3\ 4\ 5\ 6\ 7\ 8\ 9 \\ 4\ 8\ 5\ 7\ 9\ 2\ 1\ 6\ 3 \end{pmatrix}.$

6 Find the order $\#(S_4)$ of S_4 and all the permutations in S_4.

7 Find the order $\#(A_4)$ of A_4 and all the permutations in A_4.

8 For the following permutations σ and polynomials $f(x_1, x_2, x_3)$, find the polynomials $(\sigma f)(x_1, x_2, x_3)$:

(1) $\sigma = \begin{pmatrix} 1\ 2 \end{pmatrix}$, $f(x_1, x_2, x_3) = x_1 x_2 + 2x_2 + 3x_1 x_3$.

(2) $\sigma = \begin{pmatrix} 2\ 3 \end{pmatrix}$, $f(x_1, x_2, x_3) = 2x_1 x_3 + x_2^2 + x_2 x_3$.

(3) $\sigma = \begin{pmatrix} 1\ 2\ 3 \end{pmatrix}$, $f(x_1, x_2, x_3) = 2x_1 x_3 + x_2^2 + x_2 x_3$.

(4) $\sigma = \begin{pmatrix} 1\ 3\ 2 \end{pmatrix}$, $f(x_1, x_2, x_3) = (x_1 - x_2)(x_1 - x_3)(x_2 - x_3)$.

9 For the following permutations σ and polynomials $f(x_1, x_2, x_3, x_4)$, find the polynomials $(\sigma f)(x_1, x_2, x_3, x_4)$:

(1) $\sigma = \begin{pmatrix} 2\ 3 \end{pmatrix}\begin{pmatrix} 1\ 4 \end{pmatrix}$,
   $f(x_1, x_2, x_3, x_4) = x_1^3 x_2 - x_2 x_3 + 2x_1 x_2 x_4 + 3x_2 x_3 x_4^2$.

(2) $\sigma = \begin{pmatrix} 1\ 3\ 4 \end{pmatrix}$,
   $f(x_1, x_2, x_3, x_4) = x_1 x_2 x_4 - x_2 x_3 - 2x_1^2 x_3 + 3x_1 x_2 x_3 + 4x_2 x_4^2$.

3.2 Determinants and Their Properties

Determinants For a square matrix $A = (a_{ij})$ of degree n, we define the *determinant* of A by

$$|A| = \sum_{\sigma \in S_n} \operatorname{sgn}(\sigma) a_{1\sigma(1)} a_{2\sigma(2)} \cdots a_{n\sigma(n)}.$$

The determinant of A is also expressed in the following way:

$$\begin{vmatrix} a_{11}\ a_{12}\ \cdots\ a_{1n} \\ a_{21}\ a_{22}\ \cdots\ a_{2n} \\ \cdots\cdots\cdots \\ a_{n1}\ a_{n2}\ \cdots\ a_{nn} \end{vmatrix}, \quad \det \begin{pmatrix} a_{11}\ a_{12}\ \cdots\ a_{1n} \\ a_{21}\ a_{22}\ \cdots\ a_{2n} \\ \cdots\cdots\cdots \\ a_{n1}\ a_{n2}\ \cdots\ a_{nn} \end{pmatrix}, \quad \det(A), \quad |(a_{ij})|.$$

Example 1 We know $S_2 = \{\varepsilon, \sigma = \begin{pmatrix} 1\ 2 \end{pmatrix}\}$ by Example 10, Sect. 3.1 and $\text{sgn}(\varepsilon) = 1$, $\text{sgn}(\sigma) = \text{sgn}\begin{pmatrix} 1\ 2 \end{pmatrix} = -1$, so we obtain

$$\begin{vmatrix} a_{11}\ a_{12} \\ a_{21}\ a_{22} \end{vmatrix} = \text{sgn}(\varepsilon)a_{1\varepsilon(1)}a_{2\varepsilon(2)} + \text{sgn}(\sigma)a_{1\sigma(1)}a_{2\sigma(2)}$$

$$= a_{11}a_{22} - a_{12}a_{21}.$$

Example 2 We know

$$S_3 = \left\{ \sigma_0 = \varepsilon,\ \ \sigma_1 = \begin{pmatrix} 1\ 2 \end{pmatrix},\ \ \sigma_2 = \begin{pmatrix} 1\ 3 \end{pmatrix},\ \ \sigma_3 = \begin{pmatrix} 2\ 3 \end{pmatrix}, \right.$$

$$\left. \sigma_4 = \begin{pmatrix} 1\ 2\ 3 \end{pmatrix},\ \ \sigma_5 = \begin{pmatrix} 1\ 3\ 2 \end{pmatrix} \right\}$$

by Example 10, Sect. 3.1. We also know

$$\text{sgn}(\sigma_0) = \text{sgn}(\sigma_4) = \text{sgn}(\sigma_5) = 1,$$
$$\text{sgn}(\sigma_1) = \text{sgn}(\sigma_2) = \text{sgn}(\sigma_3) = -1$$

by Example 16, Sect. 3.1. Therefore we obtain

$$\begin{vmatrix} a_{11}\ a_{12}\ a_{13} \\ a_{21}\ a_{22}\ a_{23} \\ a_{31}\ a_{32}\ a_{33} \end{vmatrix} = \sum_{i=0}^{5} \text{sgn}(\sigma_i)a_{1\sigma_i(1)}a_{2\sigma_i(2)}a_{3\sigma_i(3)}$$

$$= a_{11}a_{22}a_{33} - a_{12}a_{21}a_{33} - a_{13}a_{22}a_{31}$$
$$-a_{11}a_{23}a_{32} + a_{12}a_{23}a_{31} + a_{13}a_{21}a_{32}.$$

Example 3 $\begin{vmatrix} -2\ 3 \\ 5\ 7 \end{vmatrix} = (-2) \cdot 7 - 3 \cdot 5 = -29.$

Example 4

$$\begin{vmatrix} 3\ \ \ 0\ \ \ 2 \\ 0\ -2\ \ \ 1 \\ -1\ \ \ 2\ -3 \end{vmatrix} = 3 \cdot (-2) \cdot (-3) - 0 \cdot 0 \cdot (-3) - 2 \cdot (-2) \cdot (-1)$$

$$- 3 \cdot 1 \cdot 2 + 0 \cdot 1 \cdot (-1) + 2 \cdot 0 \cdot 2$$
$$= 8.$$

The rule of Sarrus The *rule of Sarrus* is a way to calculate determinants of the square matrices of degree 2 or 3. More precisely, the determinants of the square matrices of degree 2 or 3 can be calculated by

summing the products of components from upper left to lower right

and

subtracting the products of components from upper right to lower left.

We must emphasize that the rule of Sarrus is applicable to matrices of only degree 2 or 3, and is not applicable to matrices of degree more than 3.

Case of a matrix of degree 2:

$$\begin{vmatrix} a_{11} & a_{12} \\ a_{21} & a_{22} \end{vmatrix} = a_{11}a_{22} - a_{12}a_{21}.$$

Case of a matrix of degree 3:

$$\begin{vmatrix} a_{11} & a_{12} & a_{13} \\ a_{21} & a_{22} & a_{23} \\ a_{31} & a_{32} & a_{33} \end{vmatrix} = a_{11}a_{22}a_{33} + a_{12}a_{23}a_{31} + a_{13}a_{21}a_{32}$$
$$-\{a_{11}a_{23}a_{32} + a_{12}a_{21}a_{33} + a_{13}a_{22}a_{31}\}.$$

Example 5

$$\begin{vmatrix} 2 & -3 & 1 \\ 3 & 6 & -4 \\ -1 & -2 & 5 \end{vmatrix} = 2 \cdot 6 \cdot 5 + (-3) \cdot (-4) \cdot (-1) + 1 \cdot 3 \cdot (-2)$$
$$- \{2 \cdot (-4) \cdot (-2) + (-3) \cdot 3 \cdot 5 + 1 \cdot 6 \cdot (-1)\}$$
$$= 77.$$

When we calculate determinants of matrices of degree larger than 3, the next theorem is fundamental.

Theorem 3.2.1

$$\begin{vmatrix} a_{11} & a_{12} & \cdots & a_{1n} \\ 0 & a_{22} & \cdots & a_{2n} \\ \vdots & \vdots & & \vdots \\ 0 & a_{n2} & \cdots & a_{nn} \end{vmatrix} = a_{11} \begin{vmatrix} a_{22} & \cdots & a_{2n} \\ \vdots & & \vdots \\ a_{n2} & \cdots & a_{nn} \end{vmatrix} \quad (n \geq 2).$$

Proof We write $a_{i1} = 0$ for $i = 2, 3, \ldots, n$. Let $\sigma \in S_n$. If $\sigma(1) \neq 1$, then there exists $k\,(2 \leq k \leq n)$ satisfying $\sigma(k) = 1$. Since $a_{k\sigma(k)} = a_{k1} = 0$, we see that $a_{1\sigma(1)}a_{2\sigma(2)} \cdots a_{n\sigma(n)} = 0$. This implies that the sum of the terms $a_{1\sigma(1)}a_{2\sigma(2)} \cdots a_{n\sigma(n)}$ satisfying $\sigma(1) \neq 1$ is 0. Therefore we see that

$$\begin{vmatrix} a_{11} & a_{22} & \cdots & a_{2n} \\ 0 & a_{22} & \cdots & a_{2n} \\ \vdots & \vdots & & \vdots \\ 0 & a_{n2} & \cdots & a_{nn} \end{vmatrix} = \sum_{\sigma \in S_n} \mathrm{sgn}(\sigma) a_{1\sigma(1)}a_{2\sigma(2)} \cdots a_{n\sigma(n)}$$
$$= a_{11} \sum_{\sigma \in S_n, \sigma(1)=1} \mathrm{sgn}(\sigma) a_{2\sigma(2)} \cdots a_{n\sigma(n)}.$$

The restrictions of σ satisfying $\sigma(1) = 1$ to $\{2, 3, \ldots, n\}$ are permutations of $\{2, 3, \ldots, n\}$, and all permutations of $\{2, 3, \ldots, n\}$ can be obtained as the restrictions of $\sigma \in S_n$ satisfying $\sigma(1) = 1$. Furthermore, the signatures of σ and the restrictions of σ are equal. Therefore we have

$$= a_{11} \begin{vmatrix} a_{22} & \cdots & a_{2n} \\ \vdots & & \vdots \\ a_{n2} & \cdots & a_{nn} \end{vmatrix}. \quad \|$$

Example 6 $\begin{vmatrix} 3 & 2 & 7 \\ 0 & 1 & -2 \\ 0 & 1 & 3 \end{vmatrix} = 3 \begin{vmatrix} 1 & -2 \\ 1 & 3 \end{vmatrix} = 3(1 \cdot 3 - (-2) \cdot 1) = 15.$

Example 7

$$\begin{vmatrix} 2 & 3 & -1 & 5 \\ 0 & -1 & 0 & -6 \\ 0 & 0 & 5 & 10 \\ 0 & 0 & 0 & 7 \end{vmatrix} = 2 \begin{vmatrix} -1 & 0 & -6 \\ 0 & 5 & 10 \\ 0 & 0 & 7 \end{vmatrix}$$

$$= 2 \cdot (-1) \begin{vmatrix} 5 & 10 \\ 0 & 7 \end{vmatrix} = 2 \cdot (-1) \cdot 5 \cdot 7 = -70.$$

Example 8 The determinant of an upper triangular matrix is obtained as follows:

$$\begin{vmatrix} a_{11} & a_{12} & \cdots\cdots & \cdots & a_{1n} \\ 0 & a_{22} & \cdots\cdots & \cdots & a_{2n} \\ \vdots & & \ddots & \ddots & \vdots \\ \vdots & & & \ddots & \vdots \\ 0 & 0 & \cdots & 0 & a_{nn} \end{vmatrix} = a_{11} \begin{vmatrix} a_{22} & \cdots\cdots & \cdots & a_{2n} \\ 0 & \ddots & \ddots & \vdots \\ \vdots & \ddots & \ddots & \vdots \\ 0 & \cdots & 0 & a_{nn} \end{vmatrix}$$

$$= \cdots = a_{11}a_{22}\cdots a_{nn}.$$

Theorem 3.2.2

$$(1) \quad \begin{vmatrix} a_{11} & \cdots & a_{1n} \\ \vdots & & \vdots \\ ca_{i1} & \cdots & ca_{in} \\ \vdots & & \vdots \\ a_{n1} & \cdots & a_{nn} \end{vmatrix} = c \begin{vmatrix} a_{11} & \cdots & a_{1n} \\ \vdots & & \vdots \\ a_{i1} & \cdots & a_{in} \\ \vdots & & \vdots \\ a_{n1} & \cdots & a_{nn} \end{vmatrix}.$$

$$(2) \quad \begin{vmatrix} a_{11} & \cdots & a_{1n} \\ \vdots & & \vdots \\ a_{i1}+b_{i1} & \cdots & a_{in}+b_{in} \\ \vdots & & \vdots \\ a_{n1} & \cdots & a_{nn} \end{vmatrix} = \begin{vmatrix} a_{11} & \cdots & a_{1n} \\ \vdots & & \vdots \\ a_{i1} & \cdots & a_{in} \\ \vdots & & \vdots \\ a_{n1} & \cdots & a_{nn} \end{vmatrix} + \begin{vmatrix} a_{11} & \cdots & a_{1n} \\ \vdots & & \vdots \\ b_{i1} & \cdots & b_{in} \\ \vdots & & \vdots \\ a_{n1} & \cdots & a_{nn} \end{vmatrix}.$$

Proof

(1) The left-hand side $= \displaystyle\sum_{\sigma \in S_n} \mathrm{sgn}(\sigma)\, a_{1\sigma(1)} a_{2\sigma(2)} \cdots (c a_{i\sigma(i)}) \cdots a_{n\sigma(n)}$

$= c \displaystyle\sum_{\sigma \in S_n} \mathrm{sgn}(\sigma)\, a_{1\sigma(1)} a_{2\sigma(2)} \cdots a_{i\sigma(i)} \cdots a_{n\sigma(n)}$

$=$ the right-hand side.

(2) The left-hand side $= \displaystyle\sum_{\sigma \in S_n} \mathrm{sgn}(\sigma) a_{1\sigma(1)} a_{2\sigma(2)} \cdots (a_{i\sigma(i)} + b_{i\sigma(i)}) \cdots a_{n\sigma(n)}$

$= \displaystyle\sum_{\sigma \in S_n} \mathrm{sgn}(\sigma) a_{1\sigma(1)} a_{2\sigma(2)} \cdots a_{i\sigma(i)} \cdots a_{n\sigma(n)}$

$+ \displaystyle\sum_{\sigma \in S_n} \mathrm{sgn}(\sigma) a_{1\sigma(1)} a_{2\sigma(2)} \cdots b_{i\sigma(i)} \cdots a_{n\sigma(n)}$

$=$ the right-hand side. $\|$

Example 9

$$\begin{vmatrix} -1 & 2 & 0 \\ 2a+3 & 2b-9 & 2c+6 \\ 7 & 2 & 4 \end{vmatrix} = \begin{vmatrix} -1 & 2 & 0 \\ 2a & 2b & 2c \\ 7 & 2 & 4 \end{vmatrix} + \begin{vmatrix} -1 & 2 & 0 \\ 3 & -9 & 6 \\ 7 & 2 & 4 \end{vmatrix}$$

(Theorem 3.2.2 (2))

$$= 2 \begin{vmatrix} -1 & 2 & 0 \\ a & b & c \\ 7 & 2 & 4 \end{vmatrix} + 3 \begin{vmatrix} -1 & 2 & 0 \\ 1 & -3 & 2 \\ 7 & 2 & 4 \end{vmatrix}$$

(Theorem 3.2.2 (1))

$$= 2(-8a - 4b + 16c) + 3 \cdot 36$$
$$= -16a - 8b + 32c + 108.$$

Theorem 3.2.3   *(1) If two rows of a matrix are exchanged, then the determinant is changed in sign:*

$$
\begin{matrix}
i \to \\
\\
j \to \\
\end{matrix}
\begin{vmatrix}
a_{11} & \cdots & a_{1n} \\
\vdots & & \vdots \\
a_{i1} & \cdots & a_{in} \\
\vdots & & \vdots \\
a_{j1} & \cdots & a_{jn} \\
\vdots & & \vdots \\
a_{n1} & \cdots & a_{nn}
\end{vmatrix}
= -
\begin{vmatrix}
a_{11} & \cdots & a_{1n} \\
\vdots & & \vdots \\
a_{j1} & \cdots & a_{jn} \\
\vdots & & \vdots \\
a_{i1} & \cdots & a_{in} \\
\vdots & & \vdots \\
a_{n1} & \cdots & a_{nn}
\end{vmatrix}
\begin{matrix}
\\
\\
\leftarrow i \\
\\
\leftarrow j \\
\end{matrix}
.
$$

(2)   If two rows of a matrix are equal, then the determinant is zero.

Proof (1) For a permutation $\sigma \in S_n$, we put $\tau = \sigma \left(i \ j \right)$. Then we have

$$\tau(i) = \sigma \left(i \ j \right)(i) = \sigma(j), \qquad \tau(j) = \sigma \left(i \ j \right)(j) = \sigma(i),$$

$$\tau(k) = \sigma(k) \quad \text{for} \quad k \neq i, j,$$

and

$$\text{sgn}(\tau) = \text{sgn}\left(\sigma \left(i \ j \right) \right) = -\text{sgn}(\sigma).$$

Since the correspondence $\sigma \mapsto \tau$ of S_n into S_n is one to one, we have

$$\text{the left-hand side} = \sum_{\sigma \in S_n} \text{sgn}(\sigma)\, a_{1\sigma(1)} \cdots a_{i\sigma(i)} \cdots a_{j\sigma(j)} \cdots a_{n\sigma(n)}$$

$$\parallel \qquad\qquad \parallel \qquad \parallel$$

$$= \sum_{\tau \in S_n} (-\text{sgn}(\tau))\, a_{1\tau(1)} \cdots a_{i\tau(j)} \cdots a_{j\tau(i)} \cdots a_{n\tau(n)}$$

$$= -\sum_{\tau \in S_n} \text{sgn}(\tau)\, a_{1\tau(1)} \cdots \underset{\underset{i}{\uparrow}}{a_{j\tau(i)}} \cdots \underset{\underset{j}{\uparrow}}{a_{i\tau(j)}} \cdots a_{n\tau(n)}$$

$$= \text{the right-hand side.}$$

(2)   Let A be a matrix whose two rows are equal. We exchange these two rows. Then the matrix itself does not change. On the other hand, when we exchange the two rows, the determinant of the matrix is changed in sign by (1). Therefore we have

$$\left| A \right| = - \left| A \right|.$$

Thus we obtain $\left| A \right| = 0$. $\parallel$

Example 10
$$\begin{vmatrix} 2 & 3 & -1 \\ 6 & 9 & -3 \\ 1 & 5 & 7 \end{vmatrix} = 3 \begin{vmatrix} 2 & 3 & -1 \\ 2 & 3 & -1 \\ 1 & 5 & 7 \end{vmatrix} = 0.$$
(Theorem 3.2.2 (1))   (Theorem 3.2.3 (2))

Theorem 3.2.4   *If we add a constant times one row to another row, then the determinant is unchanged:*

$$
\begin{vmatrix}
a_{11} & \cdots & a_{1n} \\
\vdots & & \vdots \\
a_{i1} + ca_{j1} & \cdots & a_{in} + ca_{jn} \\
\vdots & & \vdots \\
a_{j1} & \cdots & a_{jn} \\
\vdots & & \vdots \\
a_{n1} & \cdots & a_{nn}
\end{vmatrix}
=
\begin{vmatrix}
a_{11} & \cdots & a_{1n} \\
\vdots & & \vdots \\
a_{i1} & \cdots & a_{in} \\
\vdots & & \vdots \\
a_{j1} & \cdots & a_{jn} \\
\vdots & & \vdots \\
a_{n1} & \cdots & a_{nn}
\end{vmatrix}.
$$

Proof By Theorem 3.2.2 (2) and (1), we have

$$
\begin{vmatrix}
a_{11} & \cdots & a_{1n} \\
\vdots & & \vdots \\
a_{i1} + ca_{j1} & \cdots & a_{in} + ca_{jn} \\
\vdots & & \vdots \\
a_{j1} & \cdots & a_{jn} \\
\vdots & & \vdots \\
a_{n1} & \cdots & a_{nn}
\end{vmatrix}
=
\begin{vmatrix}
a_{11} & \cdots & a_{1n} \\
\vdots & & \vdots \\
a_{i1} & \cdots & a_{in} \\
\vdots & & \vdots \\
a_{j1} & \cdots & a_{jn} \\
\vdots & & \vdots \\
a_{n1} & \cdots & a_{nn}
\end{vmatrix}
+ c
\begin{vmatrix}
a_{11} & \cdots & a_{1n} \\
\vdots & & \vdots \\
a_{j1} & \cdots & a_{jn} \\
\vdots & & \vdots \\
a_{j1} & \cdots & a_{jn} \\
\vdots & & \vdots \\
a_{n1} & \cdots & a_{nn}
\end{vmatrix}.
$$

Since the last determinant is 0 by Theorem 3.2.3 (2), we see that

$$
\begin{vmatrix}
a_{11} & \cdots & a_{1n} \\
\vdots & & \vdots \\
a_{i1} + ca_{j1} & \cdots & a_{in} + ca_{jn} \\
\vdots & & \vdots \\
a_{j1} & \cdots & a_{jn} \\
\vdots & & \vdots \\
a_{n1} & \cdots & a_{nn}
\end{vmatrix}
=
\begin{vmatrix}
a_{11} & \cdots & a_{1n} \\
\vdots & & \vdots \\
a_{i1} & \cdots & a_{in} \\
\vdots & & \vdots \\
a_{j1} & \cdots & a_{jn} \\
\vdots & & \vdots \\
a_{n1} & \cdots & a_{nn}
\end{vmatrix}. \quad \|
$$

The calculations of determinants are quite similar to reductions. We shall calculate determinants of several matrices. In the examples below, ⓘ stands for the ith row of the previous determinant.

Example 11
$$
\begin{vmatrix}
0 & 0 & -1 \\
0 & 4 & 2 \\
3 & -1 & -2
\end{vmatrix}
= -
\begin{vmatrix}
3 & -1 & -2 \\
0 & 4 & 2 \\
0 & 0 & -1
\end{vmatrix}
= 12.
$$
$$
① \leftrightarrow ③ \qquad \text{(Example 8)}
$$

Example 12
$$
\begin{vmatrix}
2 & 4 & 4 \\
-2 & -5 & 3 \\
0 & 1 & -1
\end{vmatrix}
=
\begin{vmatrix}
2 & 4 & 4 \\
0 & -1 & 7 \\
0 & 1 & -1
\end{vmatrix}
= 2
\begin{vmatrix}
-1 & 7 \\
1 & -1
\end{vmatrix}
= 2
\begin{vmatrix}
-1 & 7 \\
0 & 6
\end{vmatrix}
$$
$$
② + ① \qquad \text{(Theorem 3.2.1)} \quad ② + ①
$$
$$
= 2 \cdot (-1) \cdot 6 = -12.
$$

Example 13

$$\begin{vmatrix} 1 & 3 & 4 \\ 2 & 5 & -7 \\ -3 & 2 & -1 \end{vmatrix} = \begin{vmatrix} 1 & 3 & 4 \\ 0 & -1 & -15 \\ -3 & 2 & -1 \end{vmatrix} \quad ②+①\times(-2)$$

$$= \begin{vmatrix} 1 & 3 & 4 \\ 0 & -1 & -15 \\ 0 & 11 & 11 \end{vmatrix} \quad ③+①\times 3$$

$$= \begin{vmatrix} -1 & -15 \\ 11 & 11 \end{vmatrix} \qquad \text{(Theorem 3.2.1)}$$

$$= (-1)\cdot 11 \begin{vmatrix} 1 & 15 \\ 1 & 1 \end{vmatrix} \qquad \text{(Theorem 3.2.2(1))}$$

$$= -11 \begin{vmatrix} 1 & 15 \\ 0 & -14 \end{vmatrix} \quad ②+①\times(-1)$$

$$= (-11)\cdot 1 \cdot (-14) = 154 \quad \text{(Example 8)}.$$

Example 14

$$\begin{vmatrix} 1 & 0 & -2 & 3 \\ 0 & 3 & -2 & 1 \\ -1 & 0 & -1 & 0 \\ -2 & 0 & 1 & 1 \end{vmatrix} = \begin{vmatrix} 1 & 0 & -2 & 3 \\ 0 & 3 & -2 & 1 \\ 0 & 0 & -3 & 3 \\ 0 & 0 & -3 & 7 \end{vmatrix} \quad \begin{matrix} ③+① \\ ④+①\times 2 \end{matrix}$$

$$= 1\cdot 3 \begin{vmatrix} -3 & 3 \\ -3 & 7 \end{vmatrix} \qquad \text{(Theorem 3.2.1)}$$

$$= 3 \begin{vmatrix} -3 & 3 \\ 0 & 4 \end{vmatrix} \qquad ②+①\times(-1)$$

$$= 3\cdot(-3)\cdot 4 = -36 \quad \text{(Example 8)}.$$

Exercises (Sect. 3.2)

1 Find the following determinants by the rule of Sarrus:

(1) $\begin{vmatrix} 1 & 3 \\ 2 & 4 \end{vmatrix}$.
(2) $\begin{vmatrix} 3 & -1 \\ -5 & 2 \end{vmatrix}$.
(3) $\begin{vmatrix} a & b \\ c & d \end{vmatrix}$.

(4) $\begin{vmatrix} 1 & 2 & 0 \\ 0 & -5 & 2 \\ 7 & 0 & 6 \end{vmatrix}$.
(5) $\begin{vmatrix} 0 & 1 & -2 \\ 1 & 3 & 0 \\ 0 & -2 & 5 \end{vmatrix}$.

2 Find the following determinants:

(1) $\begin{vmatrix} 2 & 0 & -4 \\ 0 & 3 & 2 \\ 0 & -1 & -1 \end{vmatrix}$.

(2) $\begin{vmatrix} 2 & 3 & 5 \\ 8 & 13 & -1 \\ 6 & -9 & 6 \end{vmatrix}$.

(3) $\begin{vmatrix} 12 & 16 & 32 \\ -6 & 13 & 4 \\ 15 & 10 & -20 \end{vmatrix}$.

(4) $\begin{vmatrix} 0 & 1 & 4 \\ 0 & -5 & 7 \\ 3 & 2 & 1 \end{vmatrix}$.

(5) $\begin{vmatrix} 0 & 0 & -2 \\ 2 & 5 & -1 \\ 6 & -4 & 7 \end{vmatrix}$.

(6) $\begin{vmatrix} -9 & 6 & 3 \\ -2 & 2 & 4 \\ 5 & -5 & -10 \end{vmatrix}$.

(7) $\begin{vmatrix} 1/4 & 1/6 & 2/3 \\ 1/12 & 1/6 & 1/4 \\ 1/4 & 0 & 1/6 \end{vmatrix}$.

(8) $\begin{vmatrix} 99 & 100 & 101 \\ 100 & 99 & 100 \\ 101 & 101 & 99 \end{vmatrix}$.

(9) $\begin{vmatrix} 2 & -4 & -5 & 3 \\ -6 & 13 & 14 & 1 \\ 1 & -2 & -2 & -8 \\ 2 & -5 & 0 & 5 \end{vmatrix}$.

(10) $\begin{vmatrix} 0 & -3 & -6 & 15 \\ -2 & 5 & 14 & 4 \\ 1 & -3 & -2 & 5 \\ 15 & 10 & 10 & -5 \end{vmatrix}$.

(11) $\begin{vmatrix} 0 & 0 & 0 & 0 & 3 \\ 0 & 2 & 0 & 0 & 5 \\ 0 & 13 & -2 & 0 & -4 \\ 0 & -6 & 1 & 2 & 2 \\ 8 & 1 & 2 & 3 & 4 \end{vmatrix}$.

(12) $\begin{vmatrix} 1 & -1 & -1 & 1 & -1 \\ 1 & -1 & 1 & 1 & 1 \\ 1 & 1 & -1 & 1 & -1 \\ -1 & 1 & 1 & 1 & 1 \\ 1 & 1 & 1 & -1 & -1 \end{vmatrix}$.

(13) $\begin{vmatrix} 1 & 0 & 0 & 1 & 1 \\ 0 & 1 & 0 & 1 & 2 \\ 0 & 0 & 1 & -1 & 0 \\ 2 & 1 & 3 & 1 & 0 \\ 1 & 1 & -2 & 0 & 0 \end{vmatrix}$.

(14) $\begin{vmatrix} 0 & 0 & \cdots & 0 & 1 \\ 0 & 0 & \cdots & 1 & 0 \\ \vdots & \vdots & & \vdots & \vdots \\ 0 & 1 & \cdots & 0 & 0 \\ 1 & 0 & \cdots & 0 & 0 \end{vmatrix}$ (degree $n \geq 2$).

3.3 Properties of Determinants (Continued)

In Sect. 3.2, we calculated determinants by reductions on rows. By the next theorem, we can also use reductions on columns to calculate determinants.

Theorem 3.3.1 *For a square matrix A of degree n, we have*

$$\left| {}^tA \right| = \left| A \right|.$$

Proof If we let $A = \left(a_{ij} \right)$ and ${}^tA = \left(b_{ij} \right)$, then we see $b_{ij} = a_{ji}$. Thus we see that

$$\left| {}^tA \right| = \sum_{\sigma \in S_n} \operatorname{sgn}(\sigma) \, b_{1\sigma(1)} b_{2\sigma(2)} \cdots b_{n\sigma(n)}$$

$$= \sum_{\sigma \in S_n} \operatorname{sgn}(\sigma) \, a_{\sigma(1)1} a_{\sigma(2)2} \cdots a_{\sigma(n)n}.$$

Since the set $\{\,\sigma(1),\ \sigma(2),\ \ldots,\ \sigma(n)\,\}$ is equal to the set $\{\,1,\ 2,\ \ldots,\ n\,\}$ and

$$a_{\sigma(i)i} = a_{j\sigma^{-1}(j)} \quad \text{for} \quad j = \sigma(i),$$

we have

$$a_{\sigma(1)1}a_{\sigma(2)2} \cdots a_{\sigma(n)n} = a_{1\sigma^{-1}(1)}a_{2\sigma^{-1}(2)} \cdots a_{n\sigma^{-1}(n)}$$

by changing the order of the multiplication. Since we easily see that $\operatorname{sgn}(\sigma^{-1}) = \operatorname{sgn}(\sigma)$, we obtain

$$\left|\,{}^{t}A\,\right| = \sum_{\sigma \in S_n} \operatorname{sgn}(\sigma^{-1})\, a_{1\sigma^{-1}(1)}a_{2\sigma^{-1}(2)} \cdots a_{n\sigma^{-1}(n)}.$$

When σ runs over all elements of S_n, σ^{-1} also does over all elements of S_n. By letting $\tau = \sigma^{-1}$, we have

$$= \sum_{\tau \in S_n} \operatorname{sgn}(\tau)\, a_{1\tau(1)}a_{2\tau(2)} \cdots a_{n\tau(n)} = \left|\,A\,\right|. \quad \|$$

Example 1
$$\begin{vmatrix} 3 & -1 & 4 \\ 0 & 2 & -5 \\ -2 & 1 & -1 \end{vmatrix} = \begin{vmatrix} 3 & 0 & -2 \\ -1 & 2 & 1 \\ 4 & -5 & -1 \end{vmatrix}.$$ They are the determinants of a matrix

and its transposed matrix. Then they are equal by Theorem 3.3.1. In fact, both of them are 15.

Theorem 3.3.2

$$\begin{vmatrix} a_{11} & 0 & \cdots & 0 \\ a_{21} & a_{22} & \cdots & a_{2n} \\ \vdots & \vdots & & \vdots \\ a_{n1} & a_{n2} & \cdots & a_{nn} \end{vmatrix} = a_{11} \begin{vmatrix} a_{22} & \cdots & a_{2n} \\ \vdots & & \vdots \\ a_{n2} & \cdots & a_{nn} \end{vmatrix}.$$

Proof

$$\begin{vmatrix} a_{11} & 0 & \cdots & 0 \\ a_{21} & a_{22} & \cdots & a_{2n} \\ \vdots & \vdots & & \vdots \\ a_{n1} & a_{n2} & \cdots & a_{nn} \end{vmatrix} = \begin{vmatrix} a_{11} & a_{21} & \cdots & a_{n1} \\ 0 & a_{22} & \cdots & a_{n2} \\ \vdots & \vdots & & \vdots \\ 0 & a_{2n} & \cdots & a_{nn} \end{vmatrix}$$

(Theorem 3.3.1)

$$= a_{11} \begin{vmatrix} a_{22} & \cdots & a_{n2} \\ \vdots & & \vdots \\ a_{2n} & \cdots & a_{nn} \end{vmatrix} = a_{11} \begin{vmatrix} a_{22} & \cdots & a_{2n} \\ \vdots & & \vdots \\ a_{n2} & \cdots & a_{nn} \end{vmatrix}. \|$$

(Theorem 3.2.1) (Theorem 3.3.1)

Example 2
$$\begin{vmatrix} a_{11} & 0 & \cdots\cdots & 0 \\ a_{21} & a_{22} & 0 & \cdots & 0 \\ a_{31} & a_{32} & \ddots & \ddots & \vdots \\ \vdots & \vdots & \ddots & \ddots & 0 \\ a_{n1} & a_{n2} & \cdots\cdots & \cdots & a_{nn} \end{vmatrix} = a_{11} \begin{vmatrix} a_{22} & 0 & \cdots & 0 \\ a_{32} & \ddots & \ddots & \vdots \\ \vdots & \ddots & \ddots & 0 \\ a_{n2} & \cdots\cdots & \cdots & a_{nn} \end{vmatrix}$$

by Theorem 3.3.2. Applying Theorem 3.3.2 repeatedly, we have
$$= \cdots = a_{11}a_{22}\cdots a_{nn}.$$

Example 3

$$\begin{vmatrix} 3 & 0 & 0 & 0 \\ 4 & -2 & 7 & 2 \\ 6 & 0 & 5 & 0 \\ -1 & 0 & -1 & -1 \end{vmatrix} = 3 \begin{vmatrix} -2 & 7 & 2 \\ 0 & 5 & 0 \\ 0 & -1 & -1 \end{vmatrix} \quad \text{(Theorem 3.3.2)}$$

$$= 3 \cdot (-2) \begin{vmatrix} 5 & 0 \\ -1 & -1 \end{vmatrix} \quad \text{(Theorem 3.2.1)}$$

$$= 3 \cdot (-2) \cdot 5 \cdot (-1) = 30.$$

By the next theorem, calculations of determinants will be easier. Their properties on columns can be proved from the corresponding properties on rows by taking transpositions of matrices.

Theorem 3.3.3 *(1) If we multiply a constant c to a column, then the determinant of the matrix is c times the determinant of the original matrix.*
(2) Assume that one column of a matrix is a sum of two column vectors. Then the determinant of the matrix is the sum of the determinants of the matrices taking each column vectors.
(3) If two columns are exchanged, then the determinant of the matrix is changed in sign.
(4) If two columns are the same, then the determinant of the matrix is zero.
(5) If we add a constant times one column to another column, then the determinant of the matrix is unchanged.

Proof As we mentioned above, the properties of determinants on columns can be proved by the corresponding properties of determinants on rows and Theorem 3.3.1. ∥

In the examples below, the rows are denoted by ①, ②, . . . , and the columns are denoted by (i), (ii), ...

Example 4

$$
\begin{vmatrix}
2 & 0 & -4 & 6 \\
2 & 1 & -5 & 5 \\
-1 & -3 & 4 & -1 \\
4 & 2 & -6 & 7
\end{vmatrix}
=
\begin{vmatrix}
2 & 0 & 0 & 0 \\
2 & 1 & -1 & -1 \\
-1 & -3 & 2 & 2 \\
4 & 2 & 2 & -5
\end{vmatrix}
\quad
\begin{array}{l}
\text{(iii)+(i)} \times 2 \\
\text{(iv)+(i)} \times (-3)
\end{array}
$$

(Theorem 3.3.3 (5))

$$
= 2
\begin{vmatrix}
1 & -1 & -1 \\
-3 & 2 & 2 \\
2 & 2 & -5
\end{vmatrix}
= 2
\begin{vmatrix}
1 & 0 & 0 \\
-3 & -1 & -1 \\
2 & 4 & -3
\end{vmatrix}
\quad
\begin{array}{l}
\text{(ii)+(i)} \\
\text{(iii)+(i)}
\end{array}
$$

(Theorem 3.3.2)   (Theorem 3.3.3 (5))

$$
= 2
\begin{vmatrix}
-1 & -1 \\
4 & -3
\end{vmatrix}
= 2
\begin{vmatrix}
-1 & 0 \\
4 & -7
\end{vmatrix}
\quad \text{(ii)+(i)} \times (-1)
$$

(Theorem 3.3.2)   (Theorem 3.3.3 (5))

$$
= 2 \cdot (-1) \cdot (-7) = 14.
$$

(Example 2)

Example 5

$$
\begin{vmatrix}
3 & 1 & -1 & 6 \\
0 & 0 & 4 & 3 \\
-1 & 7 & 3 & -2 \\
1 & -1 & 0 & 2
\end{vmatrix}
= -
\begin{vmatrix}
1 & -1 & 0 & 2 \\
0 & 0 & 4 & 3 \\
-1 & 7 & 3 & -2 \\
3 & 1 & -1 & 6
\end{vmatrix}
\quad ① \leftrightarrow ④
$$

(Theorem 3.2.3 (1))

$$
= -
\begin{vmatrix}
1 & -1 & 0 & 2 \\
0 & 0 & 4 & 3 \\
0 & 6 & 3 & 0 \\
0 & 4 & -1 & 0
\end{vmatrix}
\quad
\begin{array}{l}
③ + ① \\
④ + ① \times (-3)
\end{array}
$$

(Theorem 3.2.4)

$$
= -
\begin{vmatrix}
0 & 4 & 3 \\
6 & 3 & 0 \\
4 & -1 & 0
\end{vmatrix}
=
\begin{vmatrix}
3 & 4 & 0 \\
0 & 3 & 6 \\
0 & -1 & 4
\end{vmatrix}
\quad \text{(i)} \leftrightarrow \text{(iii)}
$$

(Theorem 3.2.1)   (Theorem 3.3.3 (3))

$$
= 3
\begin{vmatrix}
3 & 6 \\
-1 & 4
\end{vmatrix}
= 3
\begin{vmatrix}
3 & 0 \\
-1 & 6
\end{vmatrix}
\quad \text{(ii)} + \text{(i)} \times (-2)
$$

(Theorem 3.2.1)   (Theorem 3.3.3 (5))

$$
= 3 \cdot 3 \cdot 6 = 54.
$$

(Example 2)

Theorem 3.3.4 *Let A be a square matrix of degree r and D a square matrix of degree s. Then for an $r \times s$ matrix B and an $s \times r$ matrix C, we have*

$$\begin{vmatrix} A & B \\ O_{s,r} & D \end{vmatrix} = \begin{vmatrix} A & O_{r,s} \\ C & D \end{vmatrix} = |A| \cdot |D|.$$

Proof By Theorem 3.3.1, we have only to prove

$$\begin{vmatrix} A & O_{r,s} \\ C & D \end{vmatrix} = |A| \cdot |D|.$$

Let $n = r + s$. We write

$$X = \begin{pmatrix} A & O_{r,s} \\ C & D \end{pmatrix} = (a_{ij}).$$

By definition,

$$|X| = \sum_{\sigma \in S_n} \operatorname{sgn}(\sigma) a_{1\sigma(1)} \cdots a_{r\sigma(r)} a_{r+1\sigma(r+1)} \cdots a_{n\sigma(n)}.$$

We note that $a_{ij} = 0$ if $1 \le i \le r$ and $r + 1 \le j \le n$. Then we see that

$$a_{1\sigma(1)} \cdots a_{r\sigma(r)} a_{r+1\sigma(r+1)} \cdots a_{n\sigma(n)} = 0$$

if there exists a number larger than r among $\{\sigma(1), \sigma(2), \ldots, \sigma(r)\}$. Therefore we have only to take the summation on σ satisfying $\{\sigma(1), \ldots, \sigma(r)\} = \{1, \ldots, r\}$. If σ satisfies $\{\sigma(1), \ldots, \sigma(r)\} = \{1, \ldots, r\}$, then

$$\{\sigma(r+1), \ldots, \sigma(n)\} = \{r+1, \ldots, n\}.$$

We define the permutation τ of $\{1, \ldots, r\}$ and the permutation ρ of $\{r+1, \ldots, n\}$ by

$$\tau = \begin{pmatrix} 1 & \cdots & r \\ \sigma(1) & \cdots & \sigma(r) \end{pmatrix}, \qquad \rho = \begin{pmatrix} r+1 & \cdots & n \\ \sigma(r+1) & \cdots & \sigma(n) \end{pmatrix}.$$

Then we have $\sigma = \tau\rho$. The permutation τ runs over all permutations of r elements $\{1, \ldots, r\}$, and ρ does over all permutations of s elements $\{r+1, \ldots, n\}$. Let S_r be the set of all permutations of $\{1, \ldots, r\}$ and S'_s the set of all permutations of $\{r+1, \ldots, n\}$. Since $\operatorname{sgn}(\sigma) = \operatorname{sgn}(\tau\rho) = \operatorname{sgn}(\tau)\operatorname{sgn}(\rho)$, we obtain

$$|X| = \sum_{\tau \in S_r, \rho \in S'_s} \text{sgn}(\tau)\text{sgn}(\rho)a_{1\tau(1)} \cdots a_{r\tau(r)}a_{r+1\rho(r+1)} \cdots a_{n\rho(n)}$$

$$= \left(\sum_{\tau \in S_r} \text{sgn}(\tau)a_{1\tau(1)} \cdots a_{r\tau(r)}\right)\left(\sum_{\rho \in S'_s} \text{sgn}(\rho)a_{r+1\rho(r+1)} \cdots a_{n\rho(n)}\right)$$

$$= |A| \cdot |D|. \quad \|$$

Example 6

$$\begin{vmatrix} 2 & 7 & 13 & 5 \\ 5 & 3 & 8 & 2 \\ 0 & 0 & 9 & 4 \\ 0 & 0 & -2 & 1 \end{vmatrix} = \begin{vmatrix} 2 & 7 \\ 5 & 3 \end{vmatrix} \cdot \begin{vmatrix} 9 & 4 \\ -2 & 1 \end{vmatrix}$$

$$= (-29) \cdot 17 = -493.$$

Theorem 3.3.5 *For two square matrices A, B of degree n, we have the equality*

$$|AB| = |A| \cdot |B|.$$

Proof We shall calculate the determinant $\begin{vmatrix} A & O_n \\ -E_n & B \end{vmatrix}$ in two ways.

  (i) By Theorem 3.3.4, we have

$$\begin{vmatrix} A & O_n \\ -E_n & B \end{vmatrix} = |A| \cdot |B|.$$

(ii) Let

$$A = \left(a_{ij}\right) \quad \text{and} \quad B = \left(b_{ij}\right).$$

To calculate $\begin{vmatrix} A & O_n \\ -E_n & B \end{vmatrix}$, we add b_{1k} times the first column, b_{2k} times the second column, ..., and b_{nk} times the nth column to the $(n+k)$th column. We repeat these operations for $k = 1, \ldots, n$. Then by Theorem 3.3.3 (5), we have

$$\begin{vmatrix} A & O_n \\ -E_n & B \end{vmatrix} = \begin{vmatrix} A & AB \\ -E_n & O_n \end{vmatrix}.$$

To calculate the right-hand side, we exchange the ith row and the $n+i$th row for $i = 1, \ldots, n$. Then by Theorems 3.3.3 (3) and 3.3.4, we have

$$\begin{vmatrix} A & AB \\ -E_n & O_n \end{vmatrix} = (-1)^n \begin{vmatrix} -E_n & O_n \\ A & AB \end{vmatrix}$$

$$= (-1)^n \, |-E_n| \cdot |AB|$$

$$= |AB|.$$

By (i) and (ii), the theorem is proved.   ‖

Example 7 Let $A = \begin{pmatrix} 2 & -5 \\ 3 & 2 \end{pmatrix}$ and $B = \begin{pmatrix} 2 & 3 \\ -1 & 1 \end{pmatrix}$. Then $AB = \begin{pmatrix} 9 & 1 \\ 4 & 11 \end{pmatrix}$. Since we see that

$$|A| = \begin{vmatrix} 2 & -5 \\ 3 & 2 \end{vmatrix} = 4 + 15 = 19$$

and

$$|B| = \begin{vmatrix} 2 & 3 \\ -1 & 1 \end{vmatrix} = 2 + 3 = 5,$$

we have

$$|AB| = |A| \cdot |B| = 19 \cdot 5 = 95.$$

In fact,

$$|AB| = \begin{vmatrix} 9 & 1 \\ 4 & 11 \end{vmatrix} = 9 \cdot 11 - 1 \cdot 4 = 95.$$

From now on, we denote by (i), (ii), $\cdots$ the first column, the second column, $\cdots$.

Example 8 (*Proof of Theorem* 3.3.5 *in the case* $n = 2$)

$$\begin{vmatrix} A & O_2 \\ -E_2 & B \end{vmatrix} = \begin{vmatrix} a_{11} & a_{12} & 0 & 0 \\ a_{21} & a_{22} & 0 & 0 \\ -1 & 0 & b_{11} & b_{12} \\ 0 & -1 & b_{21} & b_{22} \end{vmatrix} = \begin{vmatrix} a_{11} & a_{12} & a_{11}b_{11} + a_{12}b_{21} & 0 \\ a_{21} & a_{22} & a_{21}b_{11} + a_{22}b_{21} & 0 \\ -1 & 0 & 0 & b_{12} \\ 0 & -1 & 0 & b_{22} \end{vmatrix}$$

$$((\text{iii}) + (\text{i}) \times b_{11} + (\text{ii}) \times b_{21}))$$

$$= \begin{vmatrix} a_{11} & a_{12} & a_{11}b_{11} + a_{12}b_{21} & a_{11}b_{12} + a_{12}b_{22} \\ a_{21} & a_{22} & a_{21}b_{11} + a_{22}b_{21} & a_{21}b_{12} + a_{22}b_{22} \\ -1 & 0 & 0 & 0 \\ 0 & -1 & 0 & 0 \end{vmatrix}.$$

$$((\text{iv}) + (\text{i}) \times b_{12} + (\text{ii}) \times b_{22})$$

$$= (-1)^2 \begin{vmatrix} -1 & 0 & 0 & 0 \\ 0 & -1 & 0 & 0 \\ a_{11} & a_{12} & a_{11}b_{11} + a_{12}b_{21} & a_{11}b_{12} + a_{12}b_{22} \\ a_{21} & a_{22} & a_{21}b_{11} + a_{22}b_{21} & a_{21}b_{12} + a_{22}b_{22} \end{vmatrix}$$

$$= (-1)^2 \begin{vmatrix} -E_2 & O_2 \\ A & AB \end{vmatrix} = (-1)^2 (-1)^2 \, |AB| = |AB|.$$

(Theorem 3.3.4)

Example 9 Let A be a square matrix. Then for any positive integer n, we have

(∗) $$|A^n| = |A|^n.$$

We shall show this by induction on n. If $n = 1$, (∗) is trivial. Assume that (∗) holds for $n - 1$. Then we have

$$|A^n| = |AA^{n-1}| = |A| \cdot |A^{n-1}|$$
$$= |A| \cdot |A|^{n-1} = |A|^n,$$

and (∗) holds for n.

Example 10 We shall show the factorization
$$(ac - bd)^2 + (ad + bc)^2 = (a^2 + b^2)(c^2 + d^2)$$

using determinants of matrices. Consider the equality of matrices
$$\begin{pmatrix} ac - bd & ad + bc \\ -(ad + bc) & ac - bd \end{pmatrix} = \begin{pmatrix} a & b \\ -b & a \end{pmatrix}\begin{pmatrix} c & d \\ -d & c \end{pmatrix}.$$

We shall take the determinants of both sides. The determinant of the left-hand side is

$$\begin{vmatrix} ac - bd & ad + bc \\ -(ad + bc) & ac - bd \end{vmatrix} = (ac - bd)^2 + (ad + bc)^2$$

and that of the right-hand side is

$$\begin{vmatrix} a & b \\ -b & a \end{vmatrix} \cdot \begin{vmatrix} c & d \\ -d & c \end{vmatrix} = (a^2 + b^2)(c^2 + d^2).$$

Thus we have the following factorization:

$$(ac - bd)^2 + (ad + bc)^2 = (a^2 + b^2)(c^2 + d^2).$$

Exercises (Sect. 3.3)
1 Find the following determinants:

(1) $\begin{vmatrix} 5 & -3 & 14 \\ -5 & 6 & 7 \\ 10 & 3 & -7 \end{vmatrix}$.

(2) $\begin{vmatrix} 2 & 16 & 3 \\ 4 & 8 & -6 \\ 8 & 8 & 12 \end{vmatrix}$.

(3) $\begin{vmatrix} 5 & 4 & 7 & 9 \\ -1 & 3 & 9 & -2 \\ 1 & -3 & -8 & 1 \\ 5 & 4 & 2 & 11 \end{vmatrix}$.

(4) $\begin{vmatrix} 1 & -1 & 2 & 1 \\ 2 & -1 & 1 & 2 \\ -1 & 1 & 2 & 1 \\ 2 & 1 & 1 & 1 \end{vmatrix}$.

(5) $\begin{vmatrix} 3 & 1 & 3 & 5 \\ 6 & 2 & 2 & 6 \\ -3 & 1 & 0 & 1 \\ 3 & 1 & 1 & 6 \end{vmatrix}.$

(6) $\begin{vmatrix} -1 & -4 & 3 & 4 \\ 1 & 2 & -3 & -2 \\ 7 & 9 & 4 & 2 \\ -9 & 7 & -3 & 6 \end{vmatrix}.$

(7) $\begin{vmatrix} 7 & -6 & 2 & 1 \\ -7 & 3 & 3 & -1 \\ 21 & 0 & -9 & 3 \\ 7 & -3 & 2 & 3 \end{vmatrix}.$

(8) $\begin{vmatrix} 2 & 0 & 5 & 3 \\ 4 & 3 & -10 & 2 \\ -2 & 6 & 5 & 0 \\ 4 & 3 & 10 & 1 \end{vmatrix}.$

(9) $\begin{vmatrix} 2 & -6 & 2 & 4 \\ -1 & 0 & 4 & 3 \\ 3 & 9 & 6 & -6 \\ 1 & 9 & -10 & 2 \end{vmatrix}.$

(10) $\begin{vmatrix} 0 & 7 & -2 & 3 \\ 2 & 14 & 2 & 6 \\ -3 & 28 & 4 & -6 \\ 2 & 21 & 6 & 3 \end{vmatrix}.$

(11) $\begin{vmatrix} 3 & 5 & 1 & 2 & -1 \\ 2 & 6 & 0 & 9 & 1 \\ 0 & 0 & 7 & 1 & 2 \\ 0 & 0 & 3 & 2 & 5 \\ 0 & 0 & 0 & 0 & -6 \end{vmatrix}.$

(12) $\begin{vmatrix} 5 & 3 & -1 & 2 & 1 \\ 1 & 4 & 0 & 9 & 3 \\ 3 & -6 & -7 & 1 & 2 \\ 2 & 7 & 0 & 0 & 0 \\ 1 & 5 & 0 & 0 & 0 \end{vmatrix}.$

(13) $\begin{vmatrix} 0 & 0 & 2 & 1 & -1 & 1 \\ 0 & 0 & 0 & 0 & 0 & 3 \\ 2 & 3 & 1 & -1 & 1 & 1 \\ 0 & 0 & -3 & 2 & 1 & 4 \\ 4 & -1 & 1 & -1 & 1 & 1 \\ 0 & 0 & 1 & 2 & 5 & 1 \end{vmatrix}.$

(14) $\begin{vmatrix} 7 & 3 & -1 & 4 & 0 & 0 \\ 6 & 2 & 1 & 3 & 1 & 1 \\ 2 & 1 & 0 & 0 & 0 & 0 \\ 2 & 1 & 0 & 0 & -1 & 2 \\ 3 & -1 & 0 & 0 & 0 & 0 \\ 1 & 4 & 2 & -1 & 0 & 0 \end{vmatrix}.$

2 Show that if A is a regular matrix, then $|A| \neq 0$ and $|A^{-1}| = |A|^{-1}$.

3 By calculating the determinants of both sides of

$$\begin{pmatrix} a & b \\ b & a \end{pmatrix}\begin{pmatrix} c & d \\ d & c \end{pmatrix} = \begin{pmatrix} ac+bd & ad+bc \\ bc+ad & bd+ac \end{pmatrix},$$

show the following equality:

$$(a^2 - b^2)(c^2 - d^2) = (ac+bd)^2 - (ad+bc)^2.$$

4 Let A, B and C be square matrices of degree n. Express the determinant $\begin{vmatrix} A & B \\ C & O_n \end{vmatrix}$ using the determinants $|A|$, $|B|$ and $|C|$.

5 Let B be a square matrix of degree r, C a square matrix of degree s and D an $s \times r$ matrix. Express the determinant $\begin{vmatrix} O_{r,s} & B \\ C & D \end{vmatrix}$ using the determinants $|B|$, $|C|$ and $|D|$.

6 Let A and B be square matrices of degree n. Show that

$$\begin{vmatrix} A & B \\ B & A \end{vmatrix} = |A + B| \cdot |A - B|.$$

7 Let m be a positive odd number. Show that if A is a square matrix of degree n and $A^m = E_n$, then $|A| = 1$.

3.4   Cofactor Matrices and Cramer's Rule

Let A be a square matrix of degree $n \geq 2$ and A_{ij} the square matrix of degree $n - 1$ obtained by removing the ith row and the jth column from the matrix A.

Example 1 If $A = \begin{pmatrix} 3 & 1 & -2 \\ 4 & -3 & 0 \\ 2 & 6 & 5 \end{pmatrix}$, then

$$A_{11} = \begin{pmatrix} -3 & 0 \\ 6 & 5 \end{pmatrix}, \quad A_{12} = \begin{pmatrix} 4 & 0 \\ 2 & 5 \end{pmatrix}, \quad A_{13} = \begin{pmatrix} 4 & -3 \\ 2 & 6 \end{pmatrix},$$

$$A_{22} = \begin{pmatrix} 3 & -2 \\ 2 & 5 \end{pmatrix}, \quad A_{31} = \begin{pmatrix} 1 & -2 \\ -3 & 0 \end{pmatrix}, \quad A_{33} = \begin{pmatrix} 3 & 1 \\ 4 & -3 \end{pmatrix}.$$

Cofactor expansions of determinants The jth column $\begin{pmatrix} a_{1j} \\ a_{2j} \\ \vdots \\ a_{nj} \end{pmatrix}$ of a square matrix

$A = (a_{ij})$ of degree $n \geq 2$ is expressed as the sum of n column vectors:

$$\begin{pmatrix} a_{1j} \\ a_{2j} \\ \vdots \\ a_{nj} \end{pmatrix} = \begin{pmatrix} a_{1j} \\ 0 \\ \vdots \\ 0 \end{pmatrix} + \begin{pmatrix} 0 \\ a_{2j} \\ \vdots \\ 0 \end{pmatrix} + \cdots + \begin{pmatrix} 0 \\ 0 \\ \vdots \\ a_{nj} \end{pmatrix}.$$

By Theorem 3.3.3 (2), the determinant $|A|$ is expressed as

$$(*) \qquad |A| = \begin{vmatrix} a_{11} & \cdots & a_{1j} & \cdots & a_{1n} \\ a_{21} & \cdots & 0 & \cdots & a_{2n} \\ \vdots & & \vdots & & \vdots \\ a_{n1} & \cdots & 0 & \cdots & a_{nn} \end{vmatrix} + \begin{vmatrix} a_{11} & \cdots & 0 & \cdots & a_{1n} \\ a_{21} & \cdots & a_{2j} & \cdots & a_{2n} \\ \vdots & & \vdots & & \vdots \\ a_{n1} & \cdots & 0 & \cdots & a_{nn} \end{vmatrix}$$

$$+ \cdots + \begin{vmatrix} a_{11} & \cdots & 0 & \cdots & a_{1n} \\ a_{21} & \cdots & 0 & \cdots & a_{2n} \\ \vdots & & \vdots & & \vdots \\ a_{n1} & \cdots & a_{nj} & \cdots & a_{nn} \end{vmatrix}.$$

We shall calculate the ith determinant of the right-hand side of the equality. First we exchange the ith row and the $(i-1)$th row. Next, we exchange the $(i-1)$th row and the $(i-2)$th row and so on until the original ith row takes the place of the first row. We similarly exchange columns until the jth column takes to the place of the first column:

$$\begin{vmatrix} a_{11} & \cdots & 0 & \cdots & a_{1n} \\ \vdots & & \vdots & & \vdots \\ a_{i1} & \cdots & a_{ij} & \cdots & a_{in} \\ \vdots & & \vdots & & \vdots \\ a_{n1} & \cdots & 0 & \cdots & a_{nn} \end{vmatrix} = (-1)^{i-1} \begin{vmatrix} a_{i1} & \cdots & a_{ij} & \cdots & a_{in} \\ a_{11} & \cdots & 0 & \cdots & a_{1n} \\ \vdots & & \vdots & & \vdots \\ \vdots & & \vdots & & \vdots \\ a_{n1} & \cdots & 0 & \cdots & a_{nn} \end{vmatrix}$$

(move the ith row to the position of the first row, Theorem 3.2.3 (1))

$$= (-1)^{i+j-2} \begin{vmatrix} a_{ij} & a_{i1} & \cdots & a_{in} \\ 0 & a_{11} & \cdots & a_{1n} \\ \vdots & \vdots & & \vdots \\ \vdots & \vdots & & \vdots \\ 0 & a_{n1} & \cdots & a_{nn} \end{vmatrix}$$

(move the jth column to the position of the first column, Theorem 3.3.3 (3))
$$= (-1)^{i+j} a_{ij} |A_{ij}| \qquad \text{(Theorem 3.2.1).}$$
Here A_{ij} is the square matrix of degree $n-1$ obtained by removing the ith row and the jth column from A. Substituting these into the equality $(*)$ for $i = 1, \ldots, n$, we obtain

$$|A| = (-1)^{1+j} a_{1j} |A_{1j}| + \cdots + (-1)^{n+j} a_{nj} |A_{nj}|.$$

We call this equality the *cofactor expansion* of the determinant $|A|$ with respect to the jth column for $j = 1, \ldots, n$.

We can also take the ith row in place of the jth column. By a similar consideration, we obtain

$$|A| = (-1)^{i+1} a_{i1} |A_{i1}| + \cdots + (-1)^{i+n} a_{in} |A_{in}|.$$

We call this equality the *cofactor expansion* of the determinant $|A|$ with respect to the ith row for $i = 1, \ldots, n$.

Example 2 (*The cofactor expansion with respect to the 2nd column*)

$$\begin{vmatrix} 2 & 7 & 4 \\ 3 & 2 & 0 \\ 1 & 5 & 3 \end{vmatrix} = -7 \begin{vmatrix} 3 & 0 \\ 1 & 3 \end{vmatrix} + 2 \begin{vmatrix} 2 & 4 \\ 1 & 3 \end{vmatrix} - 5 \begin{vmatrix} 2 & 4 \\ 3 & 0 \end{vmatrix}.$$

Example 3 (*The cofactor expansion with respect to the 2nd row*)

$$\begin{vmatrix} 4 & 3 & 2 \\ 0 & 0 & 2 \\ 2 & 1 & 3 \end{vmatrix} = -0 \begin{vmatrix} 3 & 2 \\ 1 & 3 \end{vmatrix} + 0 \begin{vmatrix} 4 & 2 \\ 2 & 3 \end{vmatrix} - 2 \begin{vmatrix} 4 & 3 \\ 2 & 1 \end{vmatrix} = -2 \begin{vmatrix} 4 & 3 \\ 2 & 1 \end{vmatrix}.$$

Example 4 (*The cofactor expansion of a square matrix of degree n with respect to the first row*)

$$\begin{vmatrix} a & 0 & \cdots\cdots & 0 & b \\ b & a & 0 & & 0 \\ 0 & b & \ddots & \ddots & \vdots \\ \vdots & \ddots & \ddots & \ddots & 0 \\ \vdots & & \ddots & b & a & 0 \\ 0 & \cdots\cdots & 0 & b & a \end{vmatrix}$$

$$= a \begin{vmatrix} a & 0 & \cdots\cdots & 0 \\ b & \ddots & \ddots & \vdots \\ 0 & \ddots & \ddots & \ddots & \vdots \\ \vdots & \ddots & b & a & 0 \\ 0 & \cdots & 0 & b & a \end{vmatrix} + (-1)^{1+n} b \begin{vmatrix} b & a & 0 & \cdots & 0 \\ 0 & b & \ddots & \ddots & \vdots \\ \vdots & \ddots & \ddots & \ddots & 0 \\ \vdots & & \ddots & b & a \\ 0 & \cdots\cdots & 0 & b \end{vmatrix}$$

$$= a^n + (-1)^{n+1} b^n = a^n - (-b)^n.$$

Cofactors of matrices For a square matrix $A = (a_{ij})$ of degree $n \geq 2$, we define $\tilde{a}_{ij}$ by

$$\tilde{a}_{ij} = (-1)^{i+j} |A_{ji}|,$$

and call it the (i, j) *cofactor* of the matrix A. We must be careful about the indices since the index (i, j) of $\tilde{a}_{ij}$ and the index (j, i) of A_{ji} are in reverse order. Using cofactors of a matrix, the cofactor expansions obtained above can be rewritten into the following forms:

$$(**) \qquad |A| = a_{1j}\tilde{a}_{j1} + \cdots + a_{nj}\tilde{a}_{jn} \qquad (j = 1, \ldots, n),$$
$$(***) \qquad |A| = a_{i1}\tilde{a}_{1i} + \cdots + a_{in}\tilde{a}_{ni} \qquad (i = 1, \ldots, n).$$

Cofactor matrices For a square matrix $A = (a_{ij})$ of degree $n \geq 2$, we define the square matrix $\tilde{A}$ of degree n by

$$\tilde{A} = (\tilde{a}_{ij})$$

whose (i, j) component is the (i, j) cofactor $\tilde{a}_{ij}$ of A. We call $\tilde{A}$ the *cofactor matrix* of A.

Example 5 For a matrix $A = \begin{pmatrix} a & b \\ c & d \end{pmatrix}$, we denote by $\tilde{A} = \begin{pmatrix} \tilde{a}_{11} & \tilde{a}_{12} \\ \tilde{a}_{21} & \tilde{a}_{22} \end{pmatrix}$ the cofactor matrix of A. Then

$$\tilde{a}_{11} = |A_{11}| = d, \qquad\qquad \tilde{a}_{12} = (-1)^3 |A_{21}| = -b,$$
$$\tilde{a}_{21} = (-1)^3 |A_{12}| = -c, \qquad \tilde{a}_{22} = |A_{22}| = a.$$

Therefore we have $\tilde{A} = \begin{pmatrix} d & -b \\ -c & a \end{pmatrix}$.

Theorem 3.4.1 *Let $\tilde{A}$ be the cofactor matrix of a square matrix A of degree $n \geq 2$. Then*

$$A\tilde{A} = \tilde{A}A = d\,E_n \qquad (d = |A|).$$

Proof Let $A = (a_{ij})$, $\tilde{A} = (\tilde{a}_{ij})$ and $A\tilde{A} = (c_{ij})$. Then

$$c_{ij} = a_{i1}\tilde{a}_{1j} + \cdots + a_{in}\tilde{a}_{nj}.$$

If $i = j$, then this is nothing but the cofactor expansion $(***)$. Therefore

$$c_{ii} = a_{i1}\tilde{a}_{1i} + \cdots + a_{in}\tilde{a}_{ni} = |A|.$$

Next, we assume $i \neq j$. Let B be a matrix whose kth row is equal to the kth row of A for $k \neq j$, and the jth row of B is equal to the ith row of A. Since the ith row and the jth row of B are the same, we have $|B| = 0$ by Theorem 3.2.3 (2). Let $B = (b_{kl})$ and $\tilde{B} = (\tilde{b}_{kl})$ the cofactor matrix of B. Then we see that

$$b_{j1} = a_{i1}, \quad \ldots, \quad b_{jn} = a_{in}.$$

Since $B_{j1} = A_{j1}, \ldots, B_{jn} = A_{jn}$, we also see that

$$\tilde{b}_{1j} = \tilde{a}_{1j}, \quad \cdots, \quad \tilde{b}_{nj} = \tilde{a}_{nj}.$$

By the cofactor expansion $(***)$ for the matrix B, we see

$$0 = |B| = b_{j1}\tilde{b}_{1j} + \cdots + b_{jn}\tilde{b}_{nj} = a_{i1}\tilde{a}_{1j} + \cdots + a_{in}\tilde{a}_{nj} = c_{ij}.$$

Therefore we obtain

$$c_{ij} = \begin{cases} |A| = d \ (i = j), \\ \quad 0 \quad (i \neq j). \end{cases}$$

This implies $A\tilde{A} = d\,E_n$. The equality $\tilde{A}A = d\,E_n$ can be proved in the same way by (∗∗). ‖

Theorem 3.4.2 *(1) A square matrix A of degree n is regular if and only if* $|A| \neq 0$.

(2) *For a regular matrix A, we have* $A^{-1} = \dfrac{1}{d}\tilde{A}$ $\left(d = |A|\right)$. *Here* $\tilde{A}$ *is the cofactor matrix of A.*

Proof (1) If a matrix A is regular, then there exists the inverse matrix A^{-1}. Since $AA^{-1} = E_n$, we see $|A| \cdot |A^{-1}| = |AA^{-1}| = |E_n| = 1$. Therefore $|A| \neq 0$. Conversely, we assume $d = |A| \neq 0$. Let $B = (1/d)\tilde{A}$, then

$$AB = (1/d)A\tilde{A} = (1/d)d\,E_n = E_n \quad \text{and} \quad BA = (1/d)\tilde{A}A = (1/d)d\,E_n = E_n$$

by Theorem 3.4.1. Therefore A is a regular matrix and B is the inverse matrix of A. (2) is proved in (1). ‖

We can give the postponed proof of Theorem 2.4.1 by Theorem 3.4.2.

Proof of Theorem 2.4.1. Assume $AB = E_n$. Then we see that

$$|A| \cdot |B| = |AB| = |E_n| = 1.$$

Then we see $|A| \neq 0$. Therefore by Theorem 3.4.2, the matrix A is a regular matrix, and has the inverse matrix A^{-1}. Furthermore, we obtain

$$B = E_n B = (A^{-1}A)B = A^{-1}(AB) = A^{-1}E_n = A^{-1}. \quad ‖$$

Exercise 3.1 Let $A = \begin{pmatrix} 1 & 2 & 3 \\ 1 & 1 & -1 \\ 4 & 1 & 5 \end{pmatrix}$. Find the cofactor matrix $\tilde{A}$ and the inverse matrix A^{-1}.

Answer. We denote the cofactor matrix by $\tilde{A} = \left(\tilde{a}_{ij}\right)$. Then $\tilde{a}_{ij} = (-1)^{i+j}|A_{ji}|$. Here A_{ji} is the square matrix of degree 2 obtained by removing the jth row and the ith column from A.

$$\tilde{a}_{11} = (-1)^{1+1}\begin{vmatrix} 1 & -1 \\ 1 & 5 \end{vmatrix} = 6, \qquad \tilde{a}_{12} = (-1)^{1+2}\begin{vmatrix} 2 & 3 \\ 1 & 5 \end{vmatrix} = -7,$$

$$\tilde{a}_{13} = (-1)^{1+3}\begin{vmatrix} 2 & 3 \\ 1 & -1 \end{vmatrix} = -5, \qquad \tilde{a}_{21} = (-1)^{2+1}\begin{vmatrix} 1 & -1 \\ 4 & 5 \end{vmatrix} = -9,$$

$$\tilde{a}_{22} = (-1)^{2+2}\begin{vmatrix} 1 & 3 \\ 4 & 5 \end{vmatrix} = -7, \qquad \tilde{a}_{23} = (-1)^{2+3}\begin{vmatrix} 1 & 3 \\ 1 & -1 \end{vmatrix} = 4,$$

$$\tilde{a}_{31} = (-1)^{3+1}\begin{vmatrix} 1 & 1 \\ 4 & 1 \end{vmatrix} = -3, \qquad \tilde{a}_{32} = (-1)^{3+2}\begin{vmatrix} 1 & 2 \\ 4 & 1 \end{vmatrix} = 7,$$

$$\tilde{a}_{33} = (-1)^{3+3}\begin{vmatrix} 1 & 2 \\ 1 & 1 \end{vmatrix} = -1, \qquad |A| = -21.$$

Therefore we have

$$\tilde{A} = \begin{pmatrix} 6 & -7 & -5 \\ -9 & -7 & 4 \\ -3 & 7 & -1 \end{pmatrix}$$

and

$$A^{-1} = \frac{1}{-21}\begin{pmatrix} 6 & -7 & -5 \\ -9 & -7 & 4 \\ -3 & 7 & -1 \end{pmatrix}. \quad \|$$

If the coefficient matrix of a linear equation is regular, we can solve the linear equation by Cramer's rule below.

Theorem 3.4.3 (Cramer's rule) *Let A be a regular matrix of degree $n \geq 2$. Denote the column partition of A by $A = \begin{pmatrix} \mathbf{a}_1 & \cdots & \mathbf{a}_n \end{pmatrix}$. Then the solution of the linear equation*

$$A\mathbf{x} = \mathbf{b}$$

is given by

$$\mathbf{x} = \begin{pmatrix} x_1 \\ \vdots \\ x_n \end{pmatrix}, \qquad x_i = \frac{|\mathbf{a}_1 \cdots \overset{i}{\mathbf{b}} \cdots \mathbf{a}_n|}{|A|}.$$

Here $|\mathbf{a}_1 \cdots \overset{i}{\mathbf{b}} \cdots \mathbf{a}_n|$ is the determinant of the matrix where the ith column vector $\mathbf{a}_i$ of A is replaced with the column vector $\mathbf{b}$.

Proof Since A is a regular matrix, we know that there exists a unique solution by Theorem 2.4.2. Let $\mathbf{x}$ be the solution. Then

$$\mathbf{b} = A\mathbf{x} = \begin{pmatrix} \mathbf{a}_1 & \cdots & \mathbf{a}_n \end{pmatrix}\mathbf{x} = x_1\mathbf{a}_1 + \cdots + x_n\mathbf{a}_n.$$

Therefore we have by Theorem 3.3.3 (1), (2), (4)

$$|\, \mathbf{a}_1 \cdots \overset{i}{\mathbf{b}} \cdots \mathbf{a}_n \,| = |\, \mathbf{a}_1 \ \cdots \ \overset{i}{x_1 \mathbf{a}_1} + \cdots + x_n \mathbf{a}_n \ \cdots \ \mathbf{a}_n \,|$$

$$= x_1 |\, \mathbf{a}_1 \ \cdots \ \overset{i}{\mathbf{a}_1} \cdots \ \mathbf{a}_n \,| + \cdots + x_i |\, \mathbf{a}_1 \ \cdots \ \overset{i}{\mathbf{a}_i} \cdots \ \mathbf{a}_n \,| +$$

$$\cdots + x_n |\, \mathbf{a}_1 \ \cdots \ \overset{i}{\mathbf{a}_n} \cdots \ \mathbf{a}_n \,| = x_i |\, \mathbf{a}_1 \ \cdots \ \overset{i}{\mathbf{a}_i} \cdots \ \mathbf{a}_n \,| = x_i |\, A \,|.$$

Since $|\, A \,| \neq 0$, we obtain $x_i = \dfrac{|\, \mathbf{a}_1 \cdots \overset{i}{\mathbf{b}} \cdots \mathbf{a}_n \,|}{|\, A \,|}.$ $\|$

Example 6 We shall solve $\begin{pmatrix} 5 & 1 \\ 3 & 2 \end{pmatrix}\begin{pmatrix} x_1 \\ x_2 \end{pmatrix} = \begin{pmatrix} 2 \\ 0 \end{pmatrix}$. By Cramer's rule (Theorem 3.4.3), we have

$$x_1 = \frac{\begin{vmatrix} 2 & 1 \\ 0 & 2 \end{vmatrix}}{\begin{vmatrix} 5 & 1 \\ 3 & 2 \end{vmatrix}} = \frac{4}{7}, \quad x_2 = \frac{\begin{vmatrix} 5 & 2 \\ 3 & 0 \end{vmatrix}}{\begin{vmatrix} 5 & 1 \\ 3 & 2 \end{vmatrix}} = -\frac{6}{7}. \ (Ans.) \quad \begin{pmatrix} x_1 \\ x_2 \end{pmatrix} = \begin{pmatrix} \frac{4}{7} \\ -\frac{6}{7} \end{pmatrix}.$$

Exercises (Sect. 3.4)

1 Find the cofactor matrices of the following matrices A. Furthermore, if the matrices are regular, then find the inverse matrices using the cofactor matrices $\tilde{A}$ of A:

(1) $\begin{pmatrix} 1 & -2 & 2 \\ 4 & 1 & -1 \\ 2 & -1 & 3 \end{pmatrix}$.

(2) $\begin{pmatrix} 2 & 4 & 1 \\ 1 & -2 & 1 \\ 0 & 5 & -1 \end{pmatrix}$.

(3) $\begin{pmatrix} 2 & -1 & -2 \\ -1 & 0 & 3 \\ 3 & -2 & 5 \end{pmatrix}$.

(4) $\begin{pmatrix} a & 0 & 0 \\ d & b & 0 \\ e & f & c \end{pmatrix}$.

(5) $\begin{pmatrix} x-2 & 1 & 1 \\ 0 & 2x-1 & x-1 \\ -2 & 1 & 1 \end{pmatrix}$.

2 Solve the following linear equations by Cramer's rule:

(1) $\begin{pmatrix} 1 & -2 & 1 \\ 1 & 1 & -1 \\ 2 & -1 & 3 \end{pmatrix}\begin{pmatrix} x_1 \\ x_2 \\ x_3 \end{pmatrix} = \begin{pmatrix} 0 \\ 1 \\ 2 \end{pmatrix}$.

(2) $\begin{pmatrix} 2 & -1 & -1 \\ 3 & 1 & 5 \\ 1 & 1 & 3 \end{pmatrix}\begin{pmatrix} x_1 \\ x_2 \\ x_3 \end{pmatrix} = \begin{pmatrix} 3 \\ 5 \\ 2 \end{pmatrix}$.

(3) $\begin{pmatrix} 3 & 2 & -2 \\ -1 & 0 & 2 \\ 1 & -2 & 1 \end{pmatrix} \begin{pmatrix} x_1 \\ x_2 \\ x_3 \end{pmatrix} = \begin{pmatrix} 2 \\ 1 \\ -3 \end{pmatrix}.$

(4) $\begin{pmatrix} 1 & 2 & -3 \\ 1 & -2 & 1 \\ 2 & 1 & -1 \end{pmatrix} \begin{pmatrix} x_1 \\ x_2 \\ x_3 \end{pmatrix} = \begin{pmatrix} -4 \\ 0 \\ 1 \end{pmatrix}.$

3 Find the cofactor expansions of the following matrices with respect to the given rows or columns:

(1) $\begin{pmatrix} 2 & 1 & 3 \\ 1 & 3 & -2 \\ 0 & 6 & 5 \end{pmatrix}$ (the third row).

(2) $\begin{pmatrix} 3 & 1 & 2 \\ 1 & 0 & 2 \\ -1 & -2 & 4 \end{pmatrix}$ (the first column).

(3) $\begin{pmatrix} a & b & c \\ -2 & 1 & 0 \\ 3 & 4 & -1 \end{pmatrix}$ (the first row).

(4) $\begin{pmatrix} 1 & x & -1 \\ 3 & y & 2 \\ 2 & z & 1 \end{pmatrix}$ (the second column).

4 Find the following determinants by cofactor expansions:

(1) $\begin{pmatrix} 0 & -3 & 2 \\ 0 & 0 & 7 \\ -5 & 8 & 3 \end{pmatrix}.$

(2) $\begin{pmatrix} -7 & 0 & 1 \\ 0 & 0 & -6 \\ 3 & 2 & -9 \end{pmatrix}.$

(3) $\begin{pmatrix} 0 & 0 & 2 & -1 \\ 1 & 0 & 0 & 3 \\ 2 & -4 & -1 & 3 \\ 1 & 0 & 0 & 2 \end{pmatrix}.$

(4) $\begin{pmatrix} 3 & 0 & 0 & -5 \\ 4 & -7 & 0 & 1 \\ -1 & 0 & 0 & 2 \\ 2 & 8 & 6 & 3 \end{pmatrix}.$

5 Let A be a square matrix of degree $n \geq 2$ and $\tilde{A}$ the cofactor matrix of A. Show that $|\tilde{A}| = |A|^{n-1}$.

6 Let A be a symmetric matrix. Show that the cofactor matrix $\tilde{A}$ of A is also a symmetric matrix. Furthermore, show that if A is a regular symmetric matrix, then A^{-1} is also a symmetric matrix.

7 Let A be an alternating matrix. Is the cofactor matrix $\tilde{A}$ of A an alternating matrix?

8 Show that the determinant of an alternation matrix of odd degree is 0.

3.5   Resultants and Determinants of Special Matrices

Resultants of polynomials For non-constant polynomials

$$f(x) = a_m x^m + a_{m-1} x^{m-1} + \cdots + a_0 \quad (a_m \neq 0)$$

and

$$g(x) = b_n x^n + b_{n-1} x^{n-1} + \cdots + b_0 \quad (b_n \neq 0),$$

we consider the following square matrix:

$$A(f(x), g(x)) = \left.\left.\begin{pmatrix} a_m & a_{m-1} & \cdots & & a_0 & & & 0 \\ & a_m & a_{m-1} & \cdots & & a_0 & & \\ & 0 & \ddots & \ddots & & & \ddots & \\ & & & a_m & a_{m-1} & \cdots & a_0 \\ b_n & b_{n-1} & \cdots & \cdots & b_0 & & & 0 \\ & b_n & b_{n-1} & \cdots & \cdots & & b_0 & \\ & 0 & \ddots & \ddots & & & & \ddots \\ & & & b_n & b_{n-1} & \cdots\cdots\cdots & b_0 \end{pmatrix}\right\}n \atop \right\}m \cdot$$

Taking the determinant of $A(f(x), g(x))$, we define $R(f(x), g(x))$ by

$$R(f(x), g(x)) = \left| A(f(x), g(x)) \right|.$$

We call $R(f(x), g(x))$ the *resultant* or *Sylvester's determinant* of $f(x)$ and $g(x)$.

Theorem 3.5.1 (Resultants of polynomials) *Let $f(x)$ be a non-constant polynomial of degree m and $g(x)$ a non-constant polynomial of degree n.*

(1)   We denote all the roots of $f(x) = 0$ and $g(x) = 0$ by $\{\alpha_1, \ldots, \alpha_m\}$ and $\{\beta_1, \ldots, \beta_n\}$, respectively. Then we have

$$R(f(x), g(x)) = a_m^n b_n^m \prod_{i=1}^{m} \prod_{j=1}^{n} (\alpha_i - \beta_j).$$

(2)   The equations $f(x) = 0$ and $g(x) = 0$ have a common root if and only if $R(f(x), g(x)) = 0$.

Proof (1) We consider the following linear equation $(*)$ consisting of $m + n$ equations in $m + n$ variables $t_0, \ldots, t_{m+n-1}$:

$$(*) \begin{cases} a_m t_0 + a_{m-1} t_1 + \cdots + a_0 t_m = 0, \\ \qquad a_m t_1 + a_{m-1} t_2 + \cdots + a_0 t_{m+1} = 0, \\ \qquad \cdots \cdots = 0, \\ \qquad\qquad a_m t_{n-1} + a_{m-1} t_n + \cdots + a_0 t_{m+n-1} = 0, \\ b_n t_0 + b_{n-1} t_1 + \cdots + b_0 t_n = 0, \\ \qquad b_n t_1 + b_{n-1} t_2 + \cdots + b_0 t_{n+1} = 0, \\ \qquad \cdots \cdots = 0, \\ \qquad\qquad b_n t_{m-1} + b_{n-1} t_m + \cdots + b_0 t_{m+n-1} = 0. \end{cases}$$

We assume that $f(x) = 0$ and $g(x) = 0$ have a common root α. Then the linear equation $(*)$ has a non-trivial solution $(t_0, t_1, \ldots, t_{m+n-1})$ given by

$$t_0 = \alpha^{m+n-1}, \quad t_1 = \alpha^{m+n-2}, \quad \cdots \quad, \quad t_{m+n-2} = \alpha, \quad t_{m+n-1} = 1.$$

In fact, since $t_k = \alpha^{m+n-(1+k)}$, we have

$$\begin{aligned} & a_m t_i + a_{m-1} t_{i+1} + \cdots + a_0 t_{i+m} \\ &= a_m \alpha^{m+n-(1+i)} + a_{m-1} \alpha^{m+n-(2+i)} + \cdots + a_0 \alpha^{n-(1+i)} \\ &= \alpha^{n-(1+i)} (a_m \alpha^m + a_{m-1} \alpha^{m-1} + \cdots + a_0) = \alpha^{n-(1+i)} f(\alpha) = 0 \end{aligned}$$

for $i = 0, \ldots, n - 1$. Similarly, we have

$$b_n t_j + b_{n-1} t_{j+1} + \cdots + b_0 t_{j+n} = \alpha^{m-(1+j)} g(\alpha) = 0$$

for $j = 0, \ldots, m - 1$. Therefore the determinant $R(f(x), g(x))$ of the coefficient matrix $A(f(x), g(x))$ of the linear equation $(*)$ must be 0, namely, $R(f(x), g(x)) = 0$.

Next, we show the equality $R(f(x), g(x)) = a_m^n b_n^m \prod_{i=1}^{m} \prod_{j=1}^{n} (\alpha_i - \beta_j)$. Dividing the polynomials $f(x)$ by a_m and $g(x)$ by b_n, we let

$$f_0(x) = a_m^{-1} f(x) \quad \text{and} \quad g_0(x) = b_n^{-1} g(x).$$

Then we have

$$R(f_0(x), g_0(x)) = a_m^{-n} b_n^{-m} R(f(x), g(x)).$$

Since the roots of $f(x) = 0$ and those of $f_0(x) = 0$ are the same and the roots of $g(x) = 0$ and those of $g_0(x) = 0$ are the same, we have only to show the equality

$$R(f(x), g(x)) = \prod_{i=1}^{m} \prod_{j=1}^{n} (\alpha_i - \beta_j) \quad \text{when } a_m = b_n = 1.$$ Take variables $u_1, \ldots, u_m$
and $v_1, \ldots, v_n$. We define polynomials $\tilde{f}(x) = \tilde{f}(x ; u_1, \ldots, u_m)$ and $\tilde{g}(x) = \tilde{g}(x ; v_1, \ldots, v_n)$ by

$$\tilde{f}(x ; u_1, \ldots, u_m) = \prod_{i=1}^{m} (x - u_i) \quad \text{and} \quad \tilde{g}(x ; v_1, \ldots, v_n) = \prod_{j=1}^{n} (x - v_j).$$

Then we see that

$$f(x) = \tilde{f}(x ; \alpha_1, \ldots, \alpha_m) \quad \text{and} \quad g(x) = \tilde{g}(x ; \beta_1, \ldots, \beta_n).$$

We denote the coefficients of $\tilde{f}(x)$ by $\tilde{a}_0, \ldots, \tilde{a}_m$ and the coefficients of $\tilde{g}(x)$ by $\tilde{b}_0, \ldots, \tilde{b}_n$, respectively. More precisely

$$\tilde{f}(x) = \tilde{f}(x ; u_1, \ldots, u_m) = \prod_{i=1}^{m} (x - u_i) = \tilde{a}_m x^m + \tilde{a}_{m-1} x^{m-1} + \cdots + \tilde{a}_0$$

and

$$\tilde{g}(x) = \tilde{g}(x ; v_1, \ldots, v_n) = \prod_{j=1}^{n} (x - v_j) = \tilde{b}_n x^n + \tilde{b}_{n-1} x^{n-1} + \cdots + \tilde{b}_0.$$

Then $\tilde{a}_i$ is a polynomial of $u_1, \ldots, u_m$ for $i = 0, \ldots, m$ and $\tilde{b}_j$ is a polynomial of $v_1, \ldots, v_n$ for $j = 0, \ldots, n$. We replace a_i with $\tilde{a}_i$ for $i = 0, \ldots, m$ and b_j with $\tilde{b}_j$ for $j = 0, \ldots, n$ in the linear equation (∗) and denote the obtained linear equation by (∗∗). The resultant $R(\tilde{f}(x), \tilde{g}(x)) = R(\tilde{f}(x ; u_1, \ldots, u_m), \tilde{g}(x ; v_1, \ldots, v_n))$ is a polynomial of $\tilde{a}_0, \ldots, \tilde{a}_m$ and $\tilde{b}_0, \ldots, \tilde{b}_n$ which is obtained by replacing a_i with $\tilde{a}_i$ for $i = 0, \ldots, m$ and b_j with $\tilde{b}_j$ for $j = 0, \ldots, n$ in $R(f(x), g(x))$. Therefore $R(\tilde{f}(x), \tilde{g}(x))$ can be considered as a polynomial of $u_1, \ldots, u_m, v_1, \ldots, v_n$. We substitute v_j for u_i in the polynomial $\tilde{f}(x) = \tilde{f}(x ; u_1, \ldots, u_m)$. Since $\tilde{f}(v_j ; u_1, \ldots, v_j, \ldots, u_m) = 0$, v_j is a common root of $\tilde{f}(x ; u_1, \ldots, v_j, \ldots, u_m) = 0$ and $\tilde{g}(x ; v_1, \ldots, v_n) = 0$. Then the determinant

$$R(\tilde{f}(x ; u_1, \ldots, v_j, \ldots, u_m), \tilde{g}(x ; v_1, \ldots, v_n))$$

of the coefficient matrix of the linear equation (∗∗) replacing u_i with v_j is 0. Therefore $R(\tilde{f}(x), \tilde{g}(x)) = R(\tilde{f}(x ; u_1, \ldots, u_m), \tilde{g}(x ; v_1, \ldots, v_n))$ is divisible by

$u_i - v_j$. Since we can take any pair (i, j) for $i = 1, \ldots, m$ and $j = 1, \ldots, n$, we see that $R(\tilde{f}(x), \tilde{g}(x))$ is divisible by $\prod_{i=1}^{m} \prod_{j=1}^{n} (u_i - v_j)$.

By the relations between roots and coefficients of a polynomial, the degrees of $\tilde{a}_i$ and $\tilde{b}_j$ as polynomials of variables $u_1, \ldots, u_m$ and $v_1, \ldots, v_n$ are

$$\deg(\tilde{a}_i) = m - i \quad \text{for} \quad i = 0, \ldots, m \qquad \text{and} \qquad \deg(\tilde{b}_j) = n - j \quad \text{for} \quad j = 0, \ldots, n,$$

respectively. We let $\tilde{a}_i = 0$ if either $i \leq -1$ or $i \geq m + 1$, and $\tilde{b}_j = 0$ if either $j \leq -1$ or $j \geq n + 1$. We denote by w_{ij} the (i, j) component of the matrix $A(\tilde{f}(x), \tilde{g}(x))$, namely, $A(\tilde{f}(x), \tilde{g}(x)) = (w_{ij})$. Then for any integers p, q $(1 \leq p, q \leq m + n)$, we have

$$w_{kp} = \tilde{a}_{m+k-p} \quad \text{for} \quad k = 1, \ldots, n \qquad \text{and} \qquad w_{lq} = \tilde{b}_{l-q} \quad \text{for} \quad l = n+1, \ldots, n+m.$$

By the definition of determinants, we see that $R(\tilde{f}(x), \tilde{g}(x))$ is a summation of $\pm w_{1\sigma(1)} \cdots w_{(m+n)\sigma(m+n)}$. Here σ runs over all the permutations of elements $\{1, 2, \ldots, m + n\}$. Then $R(\tilde{f}(x), \tilde{g}(x))$ is a summation of polynomials

$$\pm \tilde{a}_{m+1-\sigma(1)} \cdots \tilde{a}_{m+n-\sigma(n)} \tilde{b}_{n+1-\sigma(n+1)} \cdots \tilde{b}_{n+m-\sigma(n+m)}$$

of $u_1, \ldots, u_m$ and $v_1, \ldots, v_n$. If

$$\tilde{a}_{m+1-\sigma(1)} \cdots \tilde{a}_{m+n-\sigma(n)} \tilde{b}_{n+1-\sigma(n+1)} \cdots \tilde{b}_{n+m-\sigma(n+m)} \neq 0,$$

then we have

$$
\begin{aligned}
&\deg(\tilde{a}_{m+1-\sigma(1)} \cdots \tilde{a}_{m+n-\sigma(n)} \tilde{b}_{n+1-\sigma(n+1)} \cdots \tilde{b}_{n+m-\sigma(n+m)}) \\
&= (m - (m + 1 - \sigma(1))) + \cdots + (m - (m + n - \sigma(n))) \\
&\quad + (n - (n + 1 - \sigma(n + 1))) + \cdots + (n - (n + m - \sigma(n + m))) \\
&= (\sigma(1) - 1) + \cdots + (\sigma(n) - n) + (\sigma(n + 1) - 1) + \cdots + (\sigma(n + m) - m) \\
&= (\sigma(1) + \cdots + \sigma(n) + \sigma(n + 1) + \cdots + \sigma(n + m)) \\
&\quad - (1 + \cdots + n) - (1 + \cdots + m) \\
&= (1 + \cdots + (n + m)) - (1 + \cdots + n) - (1 + \cdots + m) \\
&= \frac{(n + m)(n + m + 1)}{2} - \left(\frac{n(n + 1)}{2} + \frac{m(m + 1)}{2} \right) = mn.
\end{aligned}
$$

Therefore both $R(\tilde{f}(x), \tilde{g}(x))$ and $\prod_{i=1}^{m} \prod_{j=1}^{n} (u_i - v_j)$ are homogeneous polynomials of $u_1, \ldots, u_m, v_1, \ldots, v_n$ with degree mn. Since $R(\tilde{f}(x), \tilde{g}(x))$ is divisible by

$\prod_{i=1}^{m} \prod_{j=1}^{n} (u_i - v_j)$, we have

$$R(\tilde{f}(x ; u_1, \ldots, u_m), \tilde{g}(x ; v_1, \ldots, v_n)) = c \prod_{i=1}^{m} \prod_{j=1}^{n} (u_i - v_j)$$

with a constant c. Put $u_1 = \cdots = u_m = 0$ in the polynomial $\tilde{f}(x ; u_1, \ldots, u_m)$. Then we have

$$R(\tilde{f}(x ; 0, \ldots, 0), \tilde{g}(x ; v_1, \ldots, v_n)) = c \prod_{i=1}^{m} \prod_{j=1}^{n} (-v_j) = c \prod_{j=1}^{n} (-v_j)^m$$

$$= c(-1)^{mn} \prod_{j=1}^{n} v_j^m = c\tilde{b}_0^m.$$

Since $u_1 = \cdots = u_m = 0$, we have $\tilde{a}_0 = \cdots = \tilde{a}_{m-1} = 0$ and $\tilde{a}_m = 1$. Therefore the resultant of $\tilde{f}(x ; 0, \ldots, 0)$ and $\tilde{g}(x ; v_1, \ldots, v_n)$ is

$$R(\tilde{f}(x ; 0, \ldots, 0), \tilde{g}(x ; v_1, \ldots, v_n))$$

$$= \begin{vmatrix} 1 & & & & & \\ & 1 & & & & \\ 0 & & \ddots & & & 0 \\ & & & 1 & & \\ \tilde{b}_n & \tilde{b}_{n-1} & \cdots & \cdots & \tilde{b}_0 & \\ & \tilde{b}_n & \tilde{b}_{n-1} & \cdots \cdots & & \tilde{b}_0 \\ 0 & & \ddots & \ddots & & \ddots \\ & & & \tilde{b}_n & \tilde{b}_{n-1} & \cdots \cdots & \tilde{b}_0 \end{vmatrix} = \tilde{b}_0^m.$$

Comparing the two equalities, we have $c = 1$. Thus we obtain the equality

$$R(\tilde{f}(x ; u_1, \ldots, u_m), \tilde{g}(x ; v_1, \ldots, v_n)) = \prod_{i=1}^{m} \prod_{j=1}^{n} (u_i - v_j).$$

Replacing α_i with u_i for $i = 1, \ldots, m$ and β_j with v_j for $j = 1, \ldots, n$, we have

$$R(f(x), g(x)) = \prod_{i=1}^{m} \prod_{j=1}^{n} (\alpha_i - \beta_j).$$

(2) It is easy to see that $f(x) = 0$ and $g(x) = 0$ have a common root if and only if $\prod_{i=1}^{m} \prod_{j=1}^{n} (\alpha_i - \beta_j) = 0$. Since $R(f(x), g(x)) = a_m^n b_n^m \prod_{i=1}^{m} \prod_{j=1}^{n} (\alpha_i - \beta_j)$ by (1), the

necessary and sufficient condition of two equations $f(x) = 0$ and $g(x) = 0$ having a common root is $R(f(x), g(x)) = 0$. ∥

Now, we shall rewrite the resultant $R(f(x), g(x))$. Since we have

$$a_m \prod_{i=1}^{m} (\alpha_i - \beta_j) = (-1)^m a_m \prod_{i=1}^{m} (\beta_j - \alpha_i) = (-1)^m f(\beta_j)$$

and

$$b_n \prod_{j=1}^{n} (\alpha_i - \beta_j) = g(\alpha_i),$$

we have

$$(***) \qquad R(f(x), g(x)) = (-1)^{mn} b_n^m \prod_{j=1}^{n} f(\beta_j) = a_m^n \prod_{i=1}^{m} g(\alpha_i).$$

Let $f(x)$ be a non-constant polynomial. We shall find the condition that an equation $f(x) = 0$ has a multiple root. We know that the equation $f(x) = 0$ has a multiple root if and only if $f(x) = 0$ and $f'(x) = 0$ have a common root (Exercises (Sect. 3.5) 2). Therefore $f(x)=0$ has a multiple root if and only if $R(f(x), f'(x))=0$ (Theorem 3.5.1 (2)). Here $f'(x)$ is the derivative of $f(x)$. Let $\alpha_1, \ldots, \alpha_m$ be the roots of $f(x)$ and denote $f(x)$ by

$$f(x) = a_m x^m + a_{m-1} x^{m-1} + \cdots + a_0 = a_m (x - \alpha_1) \cdots (x - \alpha_m).$$

Since $f'(x) = m a_m x^{m-1} + \cdots + a_1$, the first column of $A(f(x), f'(x))$ is divisible by a_m. Then the resultant $R(f(x), f'(x))$ of $f(x)$ and $f'(x)$ is divisible by a_m. We are going to express $\dfrac{1}{a_m} R(f(x), f'(x))$ using the roots of $f(x) = 0$. As $\deg(f'(x)) = m - 1$, we have

$$\frac{1}{a_m} R(f(x), f'(x)) = a_m^{m-2} \prod_{i=1}^{m} f'(\alpha_i)$$

by $(***)$. Since we have

$$f'(x) = a_m \sum_{j=1}^{m} (x - \alpha_1) \cdots (x - \alpha_{j-1})(x - \alpha_{j+1}) \cdots (x - \alpha_m),$$

we see that

$$f'(\alpha_i) = a_m (\alpha_i - \alpha_1) \cdots (\alpha_i - \alpha_{i-1})(\alpha_i - \alpha_{i+1}) \cdots (\alpha_i - \alpha_m).$$

Therefore we have

$$\frac{1}{a_m} R(f(x), f'(x)) = a_m^{2m-2} \prod_{i=1}^{m} \prod_{\substack{j=1 \\ j \neq i}}^{m} (\alpha_i - \alpha_j)$$

$$= (-1)^{m(m-1)/2} a_m^{2m-2} \left\{ \prod_{1 \leq i < j \leq m} (\alpha_i - \alpha_j) \right\}^2$$

$$= (-1)^{m(m-1)/2} a_m^{2m-2} \Delta(\alpha_1, \ldots, \alpha_m)^2.$$

Here $\Delta(\alpha_1, \ldots, \alpha_m) = \prod_{1 \leq i < j \leq m} (\alpha_i - \alpha_j)$ is the difference product of $\alpha_1, \ldots, \alpha_m$.

Discriminant We let $D(f(x)) = a_m^{2m-2} \Delta(\alpha_1, \ldots, \alpha_m)^2$ and call it the *discriminant* of $f(x)$. Now, we have the following theorem.

Theorem 3.5.2 *Let $f(x)$ be a non-constant polynomial of degree m given by*
$$f(x) = a_m x^m + a_{m-1} x^{m-1} + \cdots + a_0 \quad (a_m \neq 0).$$
(1)   We have

$$D(f(x)) = (-1)^{m(m-1)/2} \frac{1}{a_m} R(f(x), f'(x)).$$

(2)   The equation $f(x) = 0$ has a multiple root if and only if $D(f(x)) = 0$.

Example 1   Let $f(x) = ax^2 + bx + c$ $(a \neq 0)$. Then

$$R(f(x), f'(x)) = \begin{vmatrix} a & b & c \\ 2a & b & 0 \\ 0 & 2a & b \end{vmatrix} = -ab^2 + 4a^2 c.$$

Therefore we have $D(f(x)) = -\frac{1}{a} R(f(x), f'(x)) = b^2 - 4ac$.

Example 2   Let $f(x) = a(x - \alpha)(x - \beta)$. Then we have

$$f'(x) = a(x - \beta) + a(x - \alpha) = 2a \left(x - \frac{\alpha + \beta}{2} \right).$$

Therefore by $(* * *)$, we have

$$R(f(x), f'(x)) = a(2a)^2 \left(\alpha - \frac{\alpha + \beta}{2} \right) \left(\beta - \frac{\alpha + \beta}{2} \right) = -a^3 (\alpha - \beta)^2.$$

Then $D(f(x)) = -\frac{1}{a} R(f(x), f'(x)) = a^2 (\alpha - \beta)^2 = a^2 \Delta(\alpha, \beta)^2$.

We shall calculate determinants of some special matrices.

Exercise 3.2 The Vandermonde determinant

Assume $n \geq 2$. Then show that

$$
\begin{vmatrix}
1 & 1 & \cdots & 1 \\
x_1 & x_2 & \cdots & x_n \\
x_1^2 & x_2^2 & \cdots & x_n^2 \\
\vdots & \vdots & & \vdots \\
x_1^{n-1} & x_2^{n-1} & \cdots & x_n^{n-1}
\end{vmatrix}
= \prod_{1 \leq i < j \leq n} (x_j - x_i) = (-1)^{n(n-1)/2} \prod_{1 \leq i < j \leq n} (x_i - x_j)
$$

$$
= (-1)^{n(n-1)/2} \Delta (x_1, \ldots, x_n).
$$

Answer. We use induction on n. We easily see that the equality holds for $n = 2$. We assume that the equality holds for $n - 1$. We subtract x_1 times the $(n-1)$th row from the nth row, x_1 times the $(n-2)$th row from the $(n-1)$th row and continue until we subtract x_1 times the first row from the second row. Then we get the equality

$$
\text{the left determinant} =
\begin{vmatrix}
1 & 1 & \cdots & 1 \\
0 & x_2 - x_1 & \cdots & x_n - x_1 \\
0 & x_2(x_2 - x_1) & \cdots & x_n(x_n - x_1) \\
\vdots & \vdots & & \vdots \\
0 & x_2^{n-2}(x_2 - x_1) & \cdots & x_n^{n-2}(x_n - x_1)
\end{vmatrix}
$$

$$
=
\begin{vmatrix}
x_2 - x_1 & \cdots & x_n - x_1 \\
x_2(x_2 - x_1) & \cdots & x_n(x_n - x_1) \\
\vdots & & \vdots \\
x_2^{n-2}(x_2 - x_1) & \cdots & x_n^{n-2}(x_n - x_1)
\end{vmatrix}
$$

$$
= (x_2 - x_1)(x_3 - x_1) \cdots (x_n - x_1)
\begin{vmatrix}
1 & 1 & \cdots & 1 \\
x_2 & x_3 & \cdots & x_n \\
\vdots & \vdots & & \vdots \\
x_2^{n-2} & x_3^{n-2} & \cdots & x_n^{n-2}
\end{vmatrix}
$$

$$
= (x_2 - x_1)(x_3 - x_1) \cdots (x_n - x_1) \prod_{2 \leq i < j \leq n} (x_j - x_i) \quad \text{(the induction assumption)}
$$

$$
= \prod_{1 \leq i < j \leq n} (x_j - x_i) = (-1)^{n(n-1)/2} \prod_{1 \leq i < j \leq n} (x_i - x_j)
$$

$$
= (-1)^{n(n-1)/2} \Delta (x_1, \ldots, x_n). \quad \|
$$

Exercise 3.3 Show that

$$F_n(x) = \begin{vmatrix} a_0 & -1 & 0 & \cdots & 0 \\ a_1 & x & -1 & \ddots & \vdots \\ a_2 & 0 & x & \ddots & 0 \\ \vdots & \vdots & \ddots & \ddots & -1 \\ a_n & 0 & \cdots & 0 & x \end{vmatrix} = a_0 x^n + a_1 x^{n-1} + a_2 x^{n-2} + \cdots + a_n \,(n \geq 1)$$

Answer. We use induction on n. If $n = 1$, then the equality is trivial. Assuming the equality for $n - 1$, we will show that the equality holds for n. By the cofactor expansion with respect to the first row, we see that

$$F_n(x) = a_0 \begin{vmatrix} x & -1 & 0 & \cdots & 0 \\ 0 & x & -1 & \ddots & \vdots \\ 0 & 0 & x & \ddots & 0 \\ \vdots & \vdots & \ddots & \ddots & -1 \\ 0 & 0 & \cdots & 0 & x \end{vmatrix} - (-1) \begin{vmatrix} a_1 & -1 & 0 & \cdots & 0 \\ a_2 & x & -1 & \ddots & \vdots \\ a_3 & 0 & x & \ddots & 0 \\ \vdots & \vdots & \ddots & \ddots & -1 \\ a_n & 0 & \cdots & 0 & x \end{vmatrix}$$

$$= a_0 x^n + (a_1 x^{n-1} + a_2 x^{n-2} + \cdots + a_n) \quad \text{(the induction assumption)}$$
$$= a_0 x^n + a_1 x^{n-1} + a_2 x^{n-2} + \cdots + a_n. \quad \|$$

Exercise 3.4 Show that

$$\begin{vmatrix} 1 & a & b & c+d \\ 1 & b & c & d+a \\ 1 & c & d & a+b \\ 1 & d & a & b+c \end{vmatrix} = 0.$$

Answer.
$$\begin{vmatrix} 1 & a & b & c+d \\ 1 & b & c & d+a \\ 1 & c & d & a+b \\ 1 & d & a & b+c \end{vmatrix} = \begin{vmatrix} 1 & a & b & a+b+c+d \\ 1 & b & c & b+c+d+a \\ 1 & c & d & c+d+a+b \\ 1 & d & a & d+a+b+c \end{vmatrix} \quad \text{(iv)+(ii)+(iii)}$$

((ii), (iii), and (iv) are the 2nd, the 3rd and the 4th columns, respectively)

$$= (a+b+c+d) \begin{vmatrix} 1 & a & b & 1 \\ 1 & b & c & 1 \\ 1 & c & d & 1 \\ 1 & d & a & 1 \end{vmatrix} = 0. \quad \|$$

Exercises (Sect. 3.5)
1 Find the resultant of two quadratic polynomials
$$f(x) = a_2 x^2 + a_1 x + a_0 \quad (a_2 \neq 0),$$
$$g(x) = b_2 x^2 + b_1 x + b_0 \quad (b_2 \neq 0).$$

2 Let $f(x)$ be a non-constant polynomial. Show that $f(x) = 0$ has a multiple root if and only if $f(x) = 0$ and $f'(x) = 0$ have a common root. Here $f'(x)$ is the derivative of $f(x)$.

3 Find the discriminant of the cubic polynomial $f(x) = x^3 + px + q$.

4 Find the following determinants:

(1) $\begin{vmatrix} 1 & 1 & 1 & 1 \\ 3 & 2 & 5 & 7 \\ 3^2 & 2^2 & 5^2 & 7^2 \\ 3^3 & 2^3 & 5^3 & 7^3 \end{vmatrix}.$

(2) $\begin{vmatrix} 3 & 2^2 & 1 & 1 \\ 3^2 & 2^3 & 1 & 7 \\ 3^3 & 2^4 & 1 & 7^2 \\ 3^4 & 2^5 & 1 & 7^3 \end{vmatrix}.$

(3) $\begin{vmatrix} 2^3 & 1 & 2^2 & 2 \\ -3^3 & 1 & 3^2 & -3 \\ 7^3 & 1 & 7^2 & 7 \\ 5^3 & 1 & 5^2 & 5 \end{vmatrix}.$

(4) $\begin{vmatrix} 1 & 4 & 4^3 & 4^2 \\ 2^2 & 2^3 & 2^5 & 2^4 \\ 1 & 1 & 1 & 1 \\ 2 & -2^2 & -2^4 & 2^3 \end{vmatrix}.$

5 Find the following determinants:

(1) $\begin{vmatrix} 1 & 1 & 1 \\ a & b & c \\ bc & ca & ab \end{vmatrix}.$

(2) $\begin{vmatrix} a & b & c \\ b & c & a \\ c & a & b \end{vmatrix}.$

(3) $\begin{vmatrix} a & a+b & a-b \\ a-b & a & a+b \\ a+b & a-b & a \end{vmatrix}.$

(4) $\begin{vmatrix} 1 & x & x & x \\ 1 & 2 & y & y \\ 1 & 2 & 3 & z \\ 1 & 2 & 3 & 4 \end{vmatrix}.$

(5) $\begin{vmatrix} x & 0 & 0 & 1 \\ 0 & x & 0 & 2 \\ 0 & 0 & x & 3 \\ 1 & 2 & 3 & x \end{vmatrix}.$

(6) $\begin{vmatrix} 2 & -1 & 0 & 0 \\ -3 & x & -1 & 0 \\ 4 & 0 & x & -1 \\ 5 & 0 & 0 & x \end{vmatrix}.$

6 Show the following equalities:

(1) $\begin{vmatrix} 1 & 1 & 1 & 1 \\ x & a & a & a \\ x & y & b & b \\ x & y & z & c \end{vmatrix} = -(x-a)(y-b)(z-c).$

(2) $\begin{vmatrix} a & b & b & b \\ a & b & a & a \\ a & a & b & a \\ b & b & b & a \end{vmatrix} = -(a-b)^4.$

(3) $\begin{vmatrix} 0 & a & b & c \\ -a & 0 & d & e \\ -b & -d & 0 & f \\ -c & -e & -f & 0 \end{vmatrix} = (af - be + cd)^2.$

(4)
$$\begin{vmatrix} 1+x^2 & x & 0 & \cdots & \cdots & 0 \\ x & 1+x^2 & \ddots & \ddots & & \vdots \\ 0 & \ddots & \ddots & \ddots & \ddots & \vdots \\ \vdots & & \ddots & \ddots & \ddots & 0 \\ \vdots & & \ddots & \ddots & 1+x^2 & x \\ 0 & \cdots & \cdots & 0 & x & 1+x^2 \end{vmatrix} \quad \text{(degree } n \geq 2\text{)}$$
$$= x^{2n} + x^{2n-2} + \cdots + x^2 + 1.$$

(5)
$$\begin{vmatrix} x & a & b & c \\ a & x & b & c \\ a & b & x & c \\ a & b & c & x \end{vmatrix} = (x+a+b+c)(x-a)(x-b)(x-c).$$

(6)
$$\begin{vmatrix} 0 & a^2 & b^2 & 1 \\ a^2 & 0 & c^2 & 1 \\ b^2 & c^2 & 0 & 1 \\ 1 & 1 & 1 & 0 \end{vmatrix} = -(a+b+c)(-a+b+c)(a-b+c)(a+b-c).$$

(7) For real square matrices A, B of degree n, show the equality:

$$\begin{vmatrix} A & -B \\ B & A \end{vmatrix} = |A+iB| \cdot |A-iB|$$
$$= \left(\text{abs}\left(|A+iB|\right)\right)^2.$$

Here i is the *imaginary unit* $\sqrt{-1}$, and abs(α) is the absolute value of a complex number α.

Chapter 4
Vector Spaces

4.1 Vector Spaces

Fields A non-empty set K is called a *field* when it has basic operations, addition and multiplication, satisfying the following properties. Here a, b, c are elements of K.

(1) $a + b = b + a$.
(2) $(a + b) + c = a + (b + c)$.
(3) $ab = ba$.
(4) $(ab)c = a(bc)$.
(5) $a(b + c) = ab + ac$.
(6) There exists a *zero element* 0 in K satisfying

$$a + 0 = 0 + a = a$$

for any a. The zero element 0 uniquely exists.
(7) For any a, there exists an element b satisfying

$$a + b = 0.$$

The element b uniquely exists and is denoted by $b = -a$.
(8) There exists a *unit element* 1 in K satisfying

$$a \cdot 1 = 1 \cdot a = a$$

for any a. The unit element 1 uniquely exists.
(9) If a is not 0, then there exists an element b satisfying

$$ab = ba = 1.$$

The element b uniquely exists and is called the *inverse element* of a. The inverse element b of a is denoted by $b = a^{-1}$.

© The Author(s), under exclusive license to Springer Nature Singapore Pte Ltd. 2022
T. Miyake, *Linear Algebra*, https://doi.org/10.1007/978-981-16-6994-1_4

The real number field and the complex number field The set of real numbers and the set of complex numbers satisfy the above nine conditions. Therefore they are fields. The set of real numbers is called the *real number field* and denoted by **R** as in Sect. 2.3, and the set of complex numbers is called the *complex number field* and denoted by **C**. We mainly consider the real number field **R** or the complex number field **C** as fields in this book.

To define vector spaces, we must specify the field we consider.

Vector spaces Let K be a field and V a non-empty set. We assume that there exist two operations, addition and scalar multiplication, in V like those in the set of the column vectors of degree n.

$$\text{(Addition)} \qquad \mathbf{u} + \mathbf{v} \in V \quad (\mathbf{u},\ \mathbf{v} \in V),$$
$$\text{(Scalar multiplication)} \quad c\mathbf{u} \in V \quad (\mathbf{u} \in V,\ c \in K).$$

If these two operations satisfy the eight properties below, we call V a *vector space* over K and the field K is called the *basic field* of V. Elements of V are called *vectors*.

We note that if addition and scalar multiplication are naturally defined in a set V, then the properties (1) through (8) are satisfied and V is a vector space.

Properties of Vector Spaces ($\mathbf{u},\ \mathbf{v},\ \mathbf{w} \in V,\ a,\ b \in K$)

(1) $\mathbf{u} + \mathbf{v} = \mathbf{v} + \mathbf{u}$.
(2) $(\mathbf{u} + \mathbf{v}) + \mathbf{w} = \mathbf{u} + (\mathbf{v} + \mathbf{w})$.
(3) There exists a *zero vector* $\mathbf{0}$ in V satisfying $\mathbf{u} + \mathbf{0} = \mathbf{0} + \mathbf{u} = \mathbf{u}$ for any $\mathbf{u} \in V$.
(4) For any vector $\mathbf{u} \in V$, there exists a vector $-\mathbf{u} \in V$ satisfying $\mathbf{u} + (-\mathbf{u}) = \mathbf{0}$. We call $-\mathbf{u}$ the *inverse vector* of $\mathbf{u}$.
(5) $a(b\mathbf{u}) = (ab)\mathbf{u}$.
(6) $(a + b)\mathbf{u} = a\mathbf{u} + b\mathbf{u}$.
(7) $a(\mathbf{u} + \mathbf{v}) = a\mathbf{u} + a\mathbf{v}$.
(8) $1\mathbf{u} = \mathbf{u}$ for any $\mathbf{u} \in V$.

Zero vectors The zero vector $\mathbf{0}$ uniquely exists in a vector space V. In fact, if there exist two zero vectors $\mathbf{0}_1$ and $\mathbf{0}_2$ in V, then $\mathbf{0}_1 = \mathbf{0}_1 + \mathbf{0}_2 = \mathbf{0}_2$ and the zero vector in V is unique. When we need to specify the vector space V, we denote the zero vector of V by $\mathbf{0}_V$. Let $\mathbf{u}$ be any vector in V. Adding $-0\mathbf{u}$ to both sides of $0\mathbf{u} + 0\mathbf{u} = (0 + 0)\mathbf{u} = 0\mathbf{u}$, we have $0\mathbf{u} = \mathbf{0}_V$. Since $\mathbf{u} + (-1)\mathbf{u} = (1 - 1)\mathbf{u} = 0\mathbf{u} = \mathbf{0}_V$, we also see that $(-1)\mathbf{u} = -\mathbf{u}$. Furthermore if $c(\neq 0) \in K$, then $c\mathbf{0}_V = \mathbf{0}$. In fact, for any $\mathbf{u} \in V$, we have

$$c\mathbf{0}_V + \mathbf{u} = c\left(\mathbf{0}_V + \frac{1}{2}\mathbf{u}\right) = c\left(\frac{1}{c}\mathbf{u}\right) = \mathbf{u}.$$

By the uniqueness of the zero vector, we see that $c\mathbf{0}_V = \mathbf{0}_V$.

Example 1 The set $K^n = \left\{ \mathbf{a} = \begin{pmatrix} a_1 \\ \vdots \\ a_n \end{pmatrix} \ \middle|\ a_1, \ldots, a_n \in K \right\}$ consisting of all col-

umn vectors of degree n with components in K is a vector space over K by the addition and the scalar multiplication as matrices. The zero vector of K^n is the zero column vector $\mathbf{0}_n$. That is $\mathbf{0}_n = \begin{pmatrix} 0 \\ \vdots \\ 0 \end{pmatrix} \in K^n$.

Example 2 (1) Let $\mathbf{R}[x]$ be the set of all polynomials with coefficients in $\mathbf{R}$. Then the set $\mathbf{R}[x]$ is a vector space over $\mathbf{R}$ by the addition and the scalar multiplication as polynomials. The zero vector of $\mathbf{R}[x]$ is the constant zero polynomial $f_{00}(x) \, (= 0)$. For a polynomial $f(x)$, we denote by $f'(x)$ the derivative of $f(x)$ and by $f''(x)$ the second derivative of $f(x)$.

(2) Let $\mathbf{R}[x]_n$ be the set of all polynomials with coefficients in $\mathbf{R}$ of degree n or less. Then the set $\mathbf{R}[x]_n$ is a vector space over $\mathbf{R}$ by the addition and the scalar multiplication as polynomials. The zero vector is $f_{00}(x)$.

Subspaces A non-empty subset W of a vector space V over a field K is called a *subspace* of V if W is a vector space over K by the addition and the scalar multiplication in V.

Theorem 4.1.1 *A subset W of a vector space V over a field K is a subspace of V if and only if the following three conditions are satisfied:*

(i) $\mathbf{0}_V \in W$.
(ii) *If* $\mathbf{u}, \mathbf{v} \in W$, *then* $\mathbf{u} + \mathbf{v} \in W$.
(iii) *If* $\mathbf{u} \in W$ *and* $c \in K$, *then* $c\mathbf{u} \in W$.

Proof If W is a subspace of V, then W must satisfy those conditions.

Conversely, if W satisfies three conditions, then we easily see that W satisfies conditions (1) through (8) of the properties of vector spaces. Thus W becomes a vector space, and therefore, W is a subspace of V. ∥

Example 3 Since $W = \mathbf{R}[x]_n \subset V = \mathbf{R}[x]$ satisfies three conditions in Theorem 4.1.1, $\mathbf{R}[x]_n$ is a subspace of $\mathbf{R}[x]$.

We mainly consider vector spaces over the real number field $\mathbf{R}$ or over the complex number field $\mathbf{C}$ in this book, especially in exercises and examples. So, readers may consider that K is either $\mathbf{R}$ or $\mathbf{C}$.

Matrices over K The matrices whose components are elements of a field K are called *matrices over K*. Though we did not mention the basic field in Chapters 1 through 3, the results are true for matrices over any field K.

Theorem 4.1.2 *Let A be an $m \times n$ matrix over K. Then the set*

$$W = \{\, \mathbf{x} \in K^n \mid A\mathbf{x} = \mathbf{0}_m \,\}$$

is a subspace of K^n.

Proof We have only to see that the set W satisfies conditions (i), (ii) and (iii) in Theorem 4.1.1.

(i) The condition $\mathbf{0}_n \in W$ is trivial.

(ii) Let $\mathbf{x}, \mathbf{y} \in W$. Then

$$A(\mathbf{x} + \mathbf{y}) = A\mathbf{x} + A\mathbf{y} = \mathbf{0}_m + \mathbf{0}_m = \mathbf{0}_m.$$

Therefore $\mathbf{x} + \mathbf{y} \in W$.

(iii) Let $\mathbf{x} \in W$ and $c \in K$. Then

$$A(c\mathbf{x}) = c(A\mathbf{x}) = c\mathbf{0}_m = \mathbf{0}_m.$$

Therefore $c\mathbf{x} \in W$. ∥

Solution spaces For an $m \times n$ matrix A over K, the set

$$W = \left\{ \mathbf{x} \in K^n \mid A\mathbf{x} = \mathbf{0}_m \right\}$$

of all the solutions of the homogeneous linear equation $A\mathbf{x} = \mathbf{0}_m$ is a subspace of K^n by Theorem 4.1.2. The subspace W of K^n is called the *solution space* of the homogeneous linear equation $A\mathbf{x} = \mathbf{0}_m$.

Example 4 Let $A = \begin{pmatrix} 1 & -1 & 2 \\ 2 & 1 & 3 \end{pmatrix}$ and $\mathbf{x} = \begin{pmatrix} x_1 \\ x_2 \\ x_3 \end{pmatrix}$. Then the subspace $W = \{\mathbf{x} \in$ $\mathbf{R}^3 \mid A\mathbf{x} = \mathbf{0}_2\}$ of $\mathbf{R}^3$ is a solution space of the linear equation $A\mathbf{x} = \mathbf{0}_2$.

Exercise 4.1 Find if the following sets W are subspaces of $\mathbf{R}^3$ or not:

(1) $W = \left\{ \mathbf{x} \in \mathbf{R}^3 \,\middle|\, \begin{array}{l} 2x_1 + 3x_2 - x_3 = 0, \\ x_1 - 2x_2 + 3x_3 = 0 \end{array} \right\}.$

(2) $W = \left\{ \mathbf{x} \in \mathbf{R}^3 \,\middle|\, \begin{array}{l} 2x_1 + 3x_2 - x_3 = 1, \\ x_1 - 2x_2 + 3x_3 = 2 \end{array} \right\}.$

Answers. (1) The set W is the solution space of a homogeneous linear equation. In fact, if we let $A = \begin{pmatrix} 2 & 3 & -1 \\ 1 & -2 & 3 \end{pmatrix}$, then

$$W = \left\{ \mathbf{x} \in \mathbf{R}^3 \mid A\mathbf{x} = \mathbf{0}_2 \right\}.$$

Therefore by Theorem 4.1.2, W is a subspace of $\mathbf{R}^3$.

(2) If W is a subspace of $\mathbf{R}^3$, then W must contain the zero vector $\mathbf{0}_3$ of $\mathbf{R}^3$. But $x_1 = 0$, $x_2 = 0$, $x_3 = 0$ do not satisfy the given linear equation. Thus $\mathbf{0}_3 \notin W$. Therefore W is not a subspace of $\mathbf{R}^3$. ∥

Next, we consider subspaces of $\mathbf{R}[x]_n$.

Example 5 Let $a \in \mathbf{R}$. Then the set of polynomials

$$W = \{ f(x) \in \mathbf{R}[x]_n \mid f(a) = 0 \}$$

is a subspace of $\mathbf{R}[x]_n$. To see that W is a subspace, we shall verify the three conditions of Theorem 4.1.1.

(i) Since the zero vector $f_{00}(x) \, (= 0)$ of $\mathbf{R}[x]_n$ satisfies the condition $f_{00}(a){=}0$, we see that $f_{00}(x) \in W$.

(ii) If $f(x), g(x) \in W$, then

$$(f + g)(a) = f(a) + g(a) = 0 + 0 = 0.$$

Therefore $f(x) + g(x) \in W$.

(iii) If $f(x) \in W$, $c \in \mathbf{R}$, then $(cf)(a) = cf(a) = c\,0 = 0$. Therefore

$$cf(x) \in W.$$

Since conditions (i), (ii) and (iii) of Theorem 4.1.1 are satisfied, W is a subspace of $\mathbf{R}[x]_n$.

Exercise 4.2 Find if the following sets W are subspaces of $\mathbf{R}[x]_3$ or not:

(1) $W = \{ f(x) \in \mathbf{R}[x]_3 | f(1) = 0, \ f(-1) = 0 \}$.
(2) $W = \{ f(x) \in \mathbf{R}[x]_3 | f(1) = 1, \ f(0) = 0 \}$.
(3) $W = \{ f(x) \in \mathbf{R}[x]_3 | xf'(x) = 2f(x) \}$.

Answers. We have only to verify the three conditions in Theorem 4.1.1.

(1) W is a subspace of $\mathbf{R}[x]_3$.

(i) Since the zero vector $f_{00}(x)$ of $\mathbf{R}[x]_3$ satisfies the condition $f_{00}(1){=}f_{00}(-1){=}0$, we see that $f_{00}(x) \in W$.

(ii) If $f(x), g(x) \in W$, then

$$(f + g)(1) = f(1) + g(1) = 0 + 0 = 0,$$
$$(f + g)(-1) = f(-1) + g(-1) = 0 + 0 = 0.$$

Therefore $f(x) + g(x) \in W$.

(iii) If $f(x) \in W$, $c \in \mathbf{R}$, then

$$(cf)(1) = cf(1) = c\,0 = 0, \quad (cf)(-1) = cf(-1) = c\,0 = 0.$$

Therefore $cf(x) \in W$.

By (i), (ii) and (iii), W is a subspace of $\mathbf{R}[x]_3$.

(2) W is not a subspace of $\mathbf{R}[x]_3$. If W is a subspace, then the zero vector $f_{00}(x)$ must belong to W. But $f_{00}(1) = 0 \neq 1$. Therefore the zero vector $f_{00}(x)$ does not belong to W and W is not a subspace of $\mathbf{R}[x]_3$.

(3) W is a subspace of $\mathbf{R}[x]_3$.

(i) Since both $xf'_{00}(x)$ and $2f_{00}(x)$ are equal to $f_{00}(x)$, the zero vector $f_{00}(x)$ of $\mathbf{R}[x]_3$ satisfies the condition $xf'_{00}(x) = 2f_{00}(x)$. Then $f_{00}(x) \in W$.

(ii) If $f(x), g(x) \in W$, then

$$x(f + g)'(x) = xf'(x) + xg'(x) = 2f(x) + 2g(x) = 2(f + g)(x).$$

Therefore $f(x) + g(x) \in W$.

(iii) Let $f(x) \in W$, $c \in \mathbf{R}$. Then

$$x(cf)'(x) = cxf'(x) = c(2f(x)) = 2(cf)(x).$$

Therefore $cf(x) \in W$.

By (i), (ii) and (iii), W is a subspace of $\mathbf{R}[x]_3$. $\|$

Exercises (Sect. 4.1)

1 Find if the following sets W are subspaces of the vector space $\mathbf{R}^3$ or not:

(1) $W = \left\{ \mathbf{x} \in \mathbf{R}^3 \ \middle| \ \begin{array}{l} x_1 + x_2 - x_3 = 0, \\ 3x_1 + x_2 + 2x_3 = 0 \end{array} \right\}$.

(2) $W = \left\{ \mathbf{x} \in \mathbf{R}^3 \ \middle| \ \begin{array}{l} 2x_1 - 3x_2 + x_3 \le 1, \\ 3x_1 + \ x_2 + 2x_3 \le 1 \end{array} \right\}$.

(3) $W = \left\{ \mathbf{x} \in \mathbf{R}^3 \ \middle| \ \begin{array}{l} x_3 = 2x_1 - 3x_2, \\ 3x_3 = \ x_1 + 2x_2 \end{array} \right\}$.

(4) $W = \left\{ \mathbf{x} \in \mathbf{R}^3 \ \middle| \ \begin{array}{l} x_1^2 + x_2^2 - x_3^2 = 0, \\ x_1 - x_2 + 2x_3 = 1 \end{array} \right\}$.

(5) $W = \{ \mathbf{x} \in \mathbf{R}^3 \ | \ 2x_1 + x_2 - 2x_3 = 0 \}$.

(6) $W = \left\{ \mathbf{x} \in \mathbf{R}^3 \ \middle| \ \begin{array}{l} x_1 + 3x_2 - x_3 = 0, \\ x_1 - 2x_2 + 3x_3 = 0 \end{array} \right\}$.

2 Find if the following sets W are subspaces of the vector space $\mathbf{R}[x]_3$ or not:

(1) $W = \{ f(x) \in \mathbf{R}[x]_3 | f(0) = 0, \ f(1) = 0 \}$.

(2) $W = \{ f(x) \in \mathbf{R}[x]_3 | f(0) = 1, \ f(1) = 0 \}$.

(3) $W = \{ f(x) \in \mathbf{R}[x]_3 | f(3) = 0, \ f(2) = 0 \}$.

(4) $W = \{ f(x) \in \mathbf{R}[x]_3 | f(1) \le 0, \ f(2) = 0 \}$.

(5) $W = \{ f(x) \in \mathbf{R}[x]_3 | f'(3) = 0, \ f(1) = 0 \}$.

(6) $W = \{ f(x) \in \mathbf{R}[x]_3 | f''(x) - 2xf'(x) = 0 \}$.

3 Let V be a vector space. Show that if W_1 and W_2 are subspaces of V, then $W_1 \cap W_2$ is also a subspace of V.

4 Let V be a vector space, and W_1 and W_2 subspaces of V. Assume that $W_1 \cup W_2$ is a subspace of V. Then show that either $W_1 \subset W_2$ or $W_1 \supset W_2$ holds.

4.2 Linear Independence

The notion of the linear combinations of numerical vectors of the same type defined in Sect. 1.4 is generalized to vectors in general vector spaces. Let V be a vector space over a field K.

Linear combinations Let V be a vector space over K. When a vector $\mathbf{v}$ in V is expressed as

$$\mathbf{v} = c_1\mathbf{u}_1 + c_2\mathbf{u}_2 + \cdots + c_n\mathbf{u}_n \quad (c_i \in K)$$

by n vectors $\mathbf{u}_1, \mathbf{u}_2, \ldots, \mathbf{u}_n$ in V, we say that $\mathbf{v}$ is a *linear combination* of the vectors $\mathbf{u}_1, \mathbf{u}_2, \ldots, \mathbf{u}_n$.

Unit vectors We defined the unit vectors in Sect. 2.4. We generalize the notion of unit vectors to vector spaces K^n. We let

$$\mathbf{e}_1 = \begin{pmatrix} 1 \\ 0 \\ \vdots \\ 0 \end{pmatrix}, \mathbf{e}_2 = \begin{pmatrix} 0 \\ 1 \\ \vdots \\ 0 \end{pmatrix}, \ldots, \mathbf{e}_n = \begin{pmatrix} 0 \\ \vdots \\ 0 \\ 1 \end{pmatrix},$$

and call n vectors $\mathbf{e}_i$ for $i = 1, \ldots, n$ the *unit vectors* in K^n.

Example 1 Let $\mathbf{e}_1, \mathbf{e}_2, \mathbf{e}_3$ be the unit vectors in $\mathbf{R}^3$. Then the vector $\mathbf{a} = \begin{pmatrix} 2 \\ 3 \\ -1 \end{pmatrix}$ is expressed as a linear combination of the unit vectors. In fact

$$\mathbf{a} = 2\mathbf{e}_1 + 3\mathbf{e}_2 - \mathbf{e}_3 = 2\mathbf{e}_1 + 3\mathbf{e}_2 + (-1)\mathbf{e}_3.$$

Example 2 Take vectors $1, x, x^2$ in $\mathbf{R}[x]_2$. Then the vector $5 - 2x + 4x^2$ in $\mathbf{R}[x]_2$ is expressed as a linear combination of vectors $1, x, x^2$. In fact

$$5 - 2x + 4x^2 = 5 \cdot 1 + (-2) \cdot x + 4 \cdot x^2.$$

Example 3 The vector $-2 - 2x + 7x^2$ in $\mathbf{R}[x]_2$ is expressed as a linear combination of vectors $1 + x, 2 + x - 2x^2, 2 - x + x^2$ in $\mathbf{R}[x]_2$. In fact

$$-2 - 2x + 7x^2 = 2(1 + x) - 3(2 + x - 2x^2) + (2 - x + x^2)$$
$$= 2 \cdot (1 + x) + (-3) \cdot (2 + x - 2x^2) + 1 \cdot (2 - x + x^2).$$

Linear relations Let $\mathbf{u}_1, \mathbf{u}_2, \ldots, \mathbf{u}_n$ be n vectors in V. We assume that they satisfy a relation

$$c_1\mathbf{u}_1 + c_2\mathbf{u}_2 + \cdots + c_n\mathbf{u}_n = \mathbf{0} \quad (c_i \in K).$$

This relation of $\mathbf{u}_1, \mathbf{u}_2, \ldots, \mathbf{u}_n$ is called a *linear relation* of $\mathbf{u}_1, \mathbf{u}_2, \ldots, \mathbf{u}_n$. When $c_1 = c_2 = \cdots = c_n = 0$, the linear relation is called the *trivial relation*.

Example 4 Let $\mathbf{a}_1 = \begin{pmatrix} 1 \\ 0 \\ 1 \end{pmatrix}$, $\mathbf{a}_2 = \begin{pmatrix} 0 \\ 1 \\ 1 \end{pmatrix}$, $\mathbf{a}_2 = \begin{pmatrix} -1 \\ 1 \\ 0 \end{pmatrix}$. Then there is a non-trivial linear relation $\mathbf{a}_1 - \mathbf{a}_2 + \mathbf{a}_3 = 1 \cdot \mathbf{a}_1 + (-1) \cdot \mathbf{a}_2 + 1 \cdot \mathbf{a}_3 = \mathbf{0}_3$.

Example 5 Consider three vectors $f_1(x) = 1 + 2x^2$, $f_2(x) = 1 + 2x$, $f_3(x) = 1 - 2x + 4x^2$ in $\mathbf{R}[x]_2$. Then there is a non-trivial linear relation of $f_1(x)$, $f_2(x)$, $f_3(x)$. In fact

$$2f_1(x) - f_2(x) - f_3(x) = 2f_1(x) + (-1)f_2(x) + (-1)f_3(x) = 0 = f_{00}(x).$$

Linear independence When vectors $\mathbf{u}_1, \mathbf{u}_2, \ldots, \mathbf{u}_n$ in a vector space V do not have any non-trivial linear relation, we say that these vectors are *linearly independent*. When vectors $\mathbf{u}_1, \mathbf{u}_2, \ldots, \mathbf{u}_n$ are not linearly independent, we say that these vectors are *linearly dependent*.

Example 6 The unit vectors in K^n are linearly independent. In fact, if we assume $c_1\mathbf{e}_1 + c_2\mathbf{e}_2 + \cdots + c_n\mathbf{e}_n = \mathbf{0}_n$ ($c_i \in K$), then we see that

$$\begin{pmatrix} c_1 \\ \vdots \\ c_n \end{pmatrix} = c_1\mathbf{e}_1 + \cdots + c_n\mathbf{e}_n = \mathbf{0}_n = \begin{pmatrix} 0 \\ \vdots \\ 0 \end{pmatrix}.$$

Therefore $c_1 = c_2 = \cdots = c_n = 0$ and $\mathbf{e}_1, \mathbf{e}_1, \ldots, \mathbf{e}_n$ are linearly independent.

Example 7 The $n + 1$ vectors $1, x, \ldots, x^n$ in $\mathbf{R}[x]_n$ are linearly independent. In fact, let

$$c_0 + c_1 x + c_2 x^2 + \cdots + c_n x^n = f_{00}(x) \quad (c_i \in \mathbf{R})$$

be a linear relation of vectors $1, x, \ldots, x^n$. We put

$$f(x) = c_0 + c_1 x + c_2 x^2 + \cdots + c_n x^n.$$

Substituting $x = 0$ for $f(x)$, we have $c_0 = f(0) = f_{00}(0) = 0$. Then

$$c_1 x + c_2 x^2 + \cdots + c_n x^n = x(c_1 + c_2 x + \cdots + c_n x^{n-1}) = f_{00}(x).$$

Therefore we have $c_1 + c_2 x + \cdots + c_n x^{n-1} = f_{00}(x)$. Substitute $x = 0$ for this polynomial, then we obtain $c_1 = 0$. Repeating this procedure, we obtain $c_0 = c_1 = c_2 = \cdots = c_n = 0$.

Theorem 4.2.1 *We assume that $\mathbf{u}_1, \ldots, \mathbf{u}_n \in V$ are linearly independent. If $\mathbf{v} = c_1\mathbf{u}_1 + \cdots + c_n\mathbf{u}_n$, then $c_1, \ldots, c_n$ are uniquely determined.*

Proof We assume that

$$\mathbf{v} = c_1\mathbf{u}_1 + \cdots + c_n\mathbf{u}_n = c_1'\mathbf{u}_1 + \cdots + c_n'\mathbf{u}_n.$$

Then we have

$$(c_1 - c_1')\mathbf{u}_1 + \cdots + (c_n - c_n')\mathbf{u}_n = \mathbf{0}_V.$$

Since $\mathbf{u}_1, \ldots, \mathbf{u}_n$ are linearly independent, we have $c_1 = c_1', \ldots, c_n = c_n'$. ‖

Exercise 4.3 Show that the vectors $\mathbf{a}_1 = \begin{pmatrix} 1 \\ 1 \\ -2 \end{pmatrix}$, $\mathbf{a}_2 = \begin{pmatrix} 1 \\ -1 \\ 0 \end{pmatrix}$, $\mathbf{a}_3 = \begin{pmatrix} 3 \\ 1 \\ -3 \end{pmatrix}$ in

$\mathbf{R}^3$ are linearly independent.

Answer. We let

$$c_1\mathbf{a}_1 + c_2\mathbf{a}_2 + c_3\mathbf{a}_3 = \mathbf{0}_3 (c_i \in \mathbf{R})$$

and try to find c_1, c_2, c_3. To find c_1, c_2, c_3, we only need to solve the homogeneous linear equation

$$\begin{pmatrix} 1 & 1 & 3 \\ 1 & -1 & 1 \\ -2 & 0 & -3 \end{pmatrix} \begin{pmatrix} c_1 \\ c_2 \\ c_3 \end{pmatrix} = \begin{pmatrix} 0 \\ 0 \\ 0 \end{pmatrix}.$$

By the calculation, we have a unique solution

$$c_1 = c_2 = c_3 = 0.$$

(calculation)

1	1	3	
1	−1	1	
−2	0	−3	
1	1	3	
0	−2	−2	② + ① × (−1)
0	2	3	③ + ① × 2
1	1	3	
0	1	1	② × (−1/2)
0	2	3	
1	0	2	① + ② × (−1)
0	1	1	
0	0	1	③ + ② × (−2)
1	0	0	① + ③ × (−2)
0	1	0	② + ③ × (−1)
0	0	1	

Therefore the vectors $\mathbf{a}_1, \mathbf{a}_2, \mathbf{a}_3$ are linearly independent. ‖

Exercise 4.4 Show that the three vectors

$$f_1(x) = 1 + 2x + x^2, \ f_2(x) = 1 + 3x + x^2, \ f_3(x) = -1 + x$$

in $\mathbf{R}[x]_2$ are linearly independent.
Answer. Let

$$c_1 f_1(x) + c_2 f_2(x) + c_3 f_3(x) = f_{00}(x)(c_i \in \mathbf{R})$$

be a linear relation of vectors $f_1(x)$, $f_2(x)$, $f_3(x)$. Then

$$
\begin{aligned}
& c_1 f_1(x) + c_2 f_2(x) + c_3 f_3(x) \\
&= c_1(1 + 2x + x^2) + c_2(1 + 3x + x^2) + c_3(-1 + x) \\
&= (c_1 + c_2 - c_3) + (2c_1 + 3c_2 + c_3)x + (c_1 + c_2)x^2.
\end{aligned}
$$

Since 1, x, x^2 are linearly independent by Example 7, we have the linear equation of c_1, c_2, c_3:

$$
\begin{cases}
c_1 + c_2 - c_3 = 0, \\
2c_1 + 3c_2 + c_3 = 0, \\
c_1 + c_2 \quad\ \ = 0.
\end{cases}
$$

This linear equation can be written as

$$
\begin{pmatrix} 1 & 1 & -1 \\ 2 & 3 & 1 \\ 1 & 1 & 0 \end{pmatrix}
\begin{pmatrix} c_1 \\ c_2 \\ c_3 \end{pmatrix}
= \begin{pmatrix} 0 \\ 0 \\ 0 \end{pmatrix}.
$$

The linear equation can be solved by the calculation. By the calculation, it has only a trivial solution

$$c_1 = c_2 = c_3 = 0.$$

(calculation)

$$
\begin{array}{ccc}
\hline
1 & 1 & -1 \\
2 & 3 & 1 \\
1 & 1 & 0 \\
\hline
1 & 1 & -1 \\
0 & 1 & 3 \\
0 & 0 & 1 \\
\hline
1 & 0 & -4 \\
0 & 1 & 3 \\
0 & 0 & 1 \\
\hline
1 & 0 & 0 \\
0 & 1 & 0 \\
0 & 0 & 1 \\
\hline
\end{array}
\qquad
\begin{array}{l}
\\
\\
\\
\\
② + ① \times (-2) \\
③ + ① \times (-1) \\
① + ② \times (-1) \\
\\
\\
① + ③ \times 4 \\
② + ③ \times (-3) \\
\end{array}
$$

Therefore the vectors $f_1(x)$, $f_2(x)$, $f_3(x)$ are linearly independent. ‖

Theorem 4.2.2 *Vectors $\mathbf{u}_1, \mathbf{u}_2, \ldots, \mathbf{u}_n$ in a vector space V are linearly dependent if and only if at least one vector in $\mathbf{u}_1, \mathbf{u}_2, \ldots, \mathbf{u}_n$ is expressed as a linear combination of other $n - 1$ vectors.*

Proof We assume that vectors $\mathbf{u}_1, \mathbf{u}_2, \ldots, \mathbf{u}_n$ are linearly dependent. Then there exists a linear relation

$$c_1\mathbf{u}_1 + c_2\mathbf{u}_2 + \cdots + c_n\mathbf{u}_n = \mathbf{0} \quad (c_i \in K)$$

and at least one c_i is not zero. We may assume $c_1 \neq 0$. Then

$$\mathbf{u}_1 = -\frac{c_2}{c_1}\mathbf{u}_2 - \cdots - \frac{c_n}{c_1}\mathbf{u}_n.$$

Therefore $\mathbf{u}_1$ is a linear combination of other $n - 1$ vectors.

Conversely, we assume that at least one vector of $\mathbf{u}_1, \mathbf{u}_2, \ldots, \mathbf{u}_n$ is expressed as a linear combination of other $n - 1$ vectors. We may assume that $\mathbf{u}_1$ is expressed as a linear combination of other vectors:

$$\mathbf{u}_1 = c_2\mathbf{u}_2 + \cdots + c_n\mathbf{u}_n.$$

If we let $c_1 = -1$, then we have

$$c_1\mathbf{u}_1 + c_2\mathbf{u}_2 + \cdots + c_n\mathbf{u}_n = \mathbf{0}.$$

Since $c_1 = -1 \neq 0$, we see that $\mathbf{u}_1, \mathbf{u}_2, \ldots, \mathbf{u}_n$ are linearly dependent. ‖

Theorem 4.2.3 *If vectors $\mathbf{u}_1, \mathbf{u}_2, \ldots, \mathbf{u}_n$ in a vector space V are linearly independent and $n + 1$ vectors $\mathbf{u}, \mathbf{u}_1, \mathbf{u}_2, \ldots, \mathbf{u}_n$ are linearly dependent, then the vector $\mathbf{u}$ is expressed as a linear combination of the vectors $\mathbf{u}_1, \mathbf{u}_2, \ldots, \mathbf{u}_n$.*

Proof By assumption, there exist constants $c, c_1, \ldots, c_n$ satisfying

$$c\mathbf{u} + c_1\mathbf{u}_1 + c_2\mathbf{u}_2 + \cdots + c_n\mathbf{u}_n = \mathbf{0}$$

and at least one of them is non-zero. If $c = 0$, then $c\mathbf{u} = \mathbf{0}$ and at least one of $c_1, c_2, \ldots, c_n$ must be non-zero. Therefore the vectors $\mathbf{u}_1, \mathbf{u}_2, \ldots, \mathbf{u}_n$ are linearly dependent, which is against the assumption. Thus $c \neq 0$ and

$$\mathbf{u} = -\frac{c_1}{c}\mathbf{u}_1 - \frac{c_2}{c}\mathbf{u}_2 - \cdots - \frac{c_n}{c}\mathbf{u}_n.$$

Therefore $\mathbf{u}$ is expressed as a linear combination of the vectors $\mathbf{u}_1, \mathbf{u}_2, \ldots, \mathbf{u}_n$. ‖

Expressions of linear combinations by matrices For m vectors $\mathbf{u}_1, \ldots, \mathbf{u}_m$ in V and an $m \times n$ matrix $A = \left(a_{ij} \right)$, we define

$$\left(\mathbf{u}_1 \ \cdots \ \mathbf{u}_m\right) A = \left(\mathbf{u}_1 \ \cdots \ \mathbf{u}_m\right) \begin{pmatrix} a_{11} & \cdots & a_{1n} \\ \vdots & & \vdots \\ a_{m1} & \cdots & a_{mn} \end{pmatrix}$$

$$= \left(a_{11}\mathbf{u}_1 + \cdots + a_{m1}\mathbf{u}_m \ \cdots \ a_{1n}\mathbf{u}_1 + \cdots + a_{mn}\mathbf{u}_m\right).$$

This is nothing but the product of a formal row vector $\left(\mathbf{u}_1 \ \cdots \ \mathbf{u}_m\right)$ and a matrix A.

Example 8 $\left(\mathbf{u}_1 \ \mathbf{u}_2\right)\begin{pmatrix} 3 & 2 \\ 1 & -1 \end{pmatrix} = \left(3\mathbf{u}_1 + \mathbf{u}_2 \ 2\mathbf{u}_1 - \mathbf{u}_2\right).$

Theorem 4.2.4 *Let* $\mathbf{v}_1, \mathbf{v}_2, \ldots, \mathbf{v}_n$ *and* $\mathbf{u}_1, \mathbf{u}_2, \ldots, \mathbf{u}_m$ *be vectors in a vector space* V. *We assume the following conditions:*

(i) Each vector of $\mathbf{v}_1, \mathbf{v}_2, \ldots, \mathbf{v}_n$ *is expressed as a linear combination of* $\mathbf{u}_1, \mathbf{u}_2,$
$\ldots, \mathbf{u}_m$.
(ii) $n > m$.

Then vectors $\mathbf{v}_1, \mathbf{v}_2, \ldots, \mathbf{v}_n$ *are linearly dependent.*

Proof We shall show the existence of non-trivial linear relation of $\mathbf{v}_1, \mathbf{v}_2, \ldots, \mathbf{v}_n$.
By condition (i), there exists an $m \times n$ matrix A satisfying

$$\left(\mathbf{v}_1 \ \mathbf{v}_2 \ \cdots \ \mathbf{v}_n\right) = \left(\mathbf{u}_1 \ \mathbf{u}_2 \ \cdots \ \mathbf{u}_m\right) A.$$

By condition (ii) and Theorem 2.3.3 (2), the homogeneous linear equation $A\mathbf{x} = \mathbf{0}_m$

has a non-trivial solution $\mathbf{c} = \begin{pmatrix} c_1 \\ c_2 \\ \vdots \\ c_n \end{pmatrix}$. Then

$$\begin{aligned} c_1\mathbf{v}_1 + c_2\mathbf{v}_2 + \cdots + c_n\mathbf{v}_n &= \left(\mathbf{v}_1 \ \mathbf{v}_2 \ \cdots \ \mathbf{v}_n\right)\mathbf{c} \\ &= \left(\mathbf{u}_1 \ \mathbf{u}_2 \ \cdots \ \mathbf{u}_m\right) A\mathbf{c} = \mathbf{0}_V. \end{aligned}$$

Therefore the vectors $\mathbf{v}_1, \mathbf{v}_2, \ldots, \mathbf{v}_n$ are linearly dependent. $\|$

Example 9 The vectors $\begin{pmatrix} 2 \\ 1 \end{pmatrix}, \begin{pmatrix} 4 \\ 3 \end{pmatrix}, \begin{pmatrix} 5 \\ -1 \end{pmatrix}$ in $\mathbf{R}^2$ are linearly dependent since
these three vectors can be expressed as linear combinations of two vectors $\begin{pmatrix} 1 \\ 0 \end{pmatrix}, \begin{pmatrix} 0 \\ 1 \end{pmatrix}$.

Theorem 4.2.5 *Let* $\mathbf{u}_1, \ldots, \mathbf{u}_m$ *be linearly independent vectors in a vector space* V.
Then we have the following:

(1) If the equality

$$\left(\mathbf{u}_1 \ \cdots \ \mathbf{u}_m\right) A = \left(\mathbf{0}_V \ \cdots \ \mathbf{0}_V\right)$$

holds for an $m \times n$ *matrix* A, *then* $A = O_{m,n}$.

(2) *If two $m \times n$ matrices A and B satisfy the equality*

$$\left(\mathbf{u}_1 \cdots \mathbf{u}_m \right) A = \left(\mathbf{u}_1 \cdots \mathbf{u}_m \right) B,$$

then $A = B$.

Proof (1) Let $A = \left(a_{ij} \right)$. Then we see that

$$\left(\mathbf{u}_1 \cdots \mathbf{u}_m \right) \begin{pmatrix} a_{11} & \cdots & a_{1n} \\ \vdots & & \vdots \\ a_{m1} & \cdots & a_{mn} \end{pmatrix} = \left(\mathbf{0}_V \cdots \mathbf{0}_V \right).$$

We calculate the product of the left-hand side and compare the jth columns of both sides. Then we have

$$a_{1j}\mathbf{u}_1 + \cdots + a_{mj}\mathbf{u}_m = \mathbf{0}_V.$$

Since $\mathbf{u}_1, \ldots, \mathbf{u}_m$ are linearly independent, we see that $a_{1j} = \cdots = a_{mj} = 0$. The number j can be taken for any number between 1 and n. Thus we obtain

$$A = O_{m,n}.$$

(2) We transpose the right-hand side to the left-hand side and obtain the equality

$$\left(\mathbf{u}_1 \cdots \mathbf{u}_m \right) (A - B) = \left(\mathbf{0}_V \cdots \mathbf{0}_V \right).$$

Then we see that $A - B = O_{m,n}$ by (1). Therefore we have $A = B$. ‖

Theorem 4.2.6 *Let $\mathbf{u}_1, \ldots, \mathbf{u}_m$ be linearly independent vectors in a vector space V, and the vectors $\mathbf{v}_1, \ldots, \mathbf{v}_n$ be linear combinations of $\mathbf{u}_1, \ldots, \mathbf{u}_m$ so that*

$$\left(\mathbf{v}_1 \cdots \mathbf{v}_n \right) = \left(\mathbf{u}_1 \cdots \mathbf{u}_m \right) A$$

with an $m \times n$ matrix A. We denote the column partition of A by

$$A = \left(\mathbf{a}_1 \cdots \mathbf{a}_n \right).$$

Then we have the following:

(1) *The linear relations of vectors $\mathbf{v}_1, \ldots, \mathbf{v}_n$ and the linear relations of vectors $\mathbf{a}_1, \ldots, \mathbf{a}_n$ are the same.*
(2) *Vectors $\mathbf{v}_1, \ldots, \mathbf{v}_n$ are linearly independent if and only if vectors $\mathbf{a}_1, \ldots, \mathbf{a}_n$ are linearly independent.*

Proof (1) Let $c_1\mathbf{v}_1 + \cdots + c_n\mathbf{v}_n = \mathbf{0}_V$ ($c_i \in K$) be a linear relation of $\mathbf{v}_1, \ldots, \mathbf{v}_n$. Then we have

$$\mathbf{0}_V = \begin{pmatrix} \mathbf{v}_1 \cdots \mathbf{v}_n \end{pmatrix} \begin{pmatrix} c_1 \\ \vdots \\ c_n \end{pmatrix} = \begin{pmatrix} \mathbf{u}_1 \cdots \mathbf{u}_m \end{pmatrix} A \begin{pmatrix} c_1 \\ \vdots \\ c_n \end{pmatrix}.$$

Since $\mathbf{u}_1, \ldots, \mathbf{u}_m$ are linearly independent, we have by Theorem 4.2.5 (1)

$$\mathbf{0}_m = A \begin{pmatrix} c_1 \\ \vdots \\ c_n \end{pmatrix} = c_1 \mathbf{a}_1 + \cdots + c_n \mathbf{a}_n.$$

Conversely, if $c_1 \mathbf{a}_1 + \cdots + c_n \mathbf{a}_n = \mathbf{0}_m$ is a linear relation of $\mathbf{a}_1, \ldots, \mathbf{a}_n$, then we have a linear relation of $\mathbf{v}_1, \ldots, \mathbf{v}_n$ as follows:

$$\begin{aligned} c_1 \mathbf{v}_1 + \cdots + c_n \mathbf{v}_n &= \begin{pmatrix} \mathbf{v}_1 \cdots \mathbf{v}_n \end{pmatrix} \begin{pmatrix} c_1 \\ \vdots \\ c_n \end{pmatrix} = \begin{pmatrix} \mathbf{u}_1 \cdots \mathbf{u}_m \end{pmatrix} A \begin{pmatrix} c_1 \\ \vdots \\ c_n \end{pmatrix} \\ &= \begin{pmatrix} \mathbf{u}_1 \cdots \mathbf{u}_m \end{pmatrix} \begin{pmatrix} c_1 \mathbf{a}_1 + \cdots + c_n \mathbf{a}_n \end{pmatrix} \\ &= \begin{pmatrix} \mathbf{u}_1 \cdots \mathbf{u}_m \end{pmatrix} \mathbf{0}_m = \mathbf{0}_V. \end{aligned}$$

(2) is clear by (1). ‖

Exercise 4.5 Find a non-trivial linear relation of vectors

$$f_1(x) = 1 - 3x + x^2, \ f_2(x) = 2 + x - x^2, \ f_3(x) = 4 - 5x + x^2$$

in $\mathbf{R}[x]_2$.

Answer. The vectors $f_1(x)$, $f_2(x)$, $f_3(x)$ can be expressed as a linear combination of 1, x, x^2 as

$$\begin{pmatrix} f_1(x) & f_2(x) & f_3(x) \end{pmatrix} = \begin{pmatrix} 1 & x & x^2 \end{pmatrix} \begin{pmatrix} 1 & 2 & 4 \\ -3 & 1 & -5 \\ 1 & -1 & 1 \end{pmatrix}.$$

Since the vectors 1, x, x^2 in $\mathbf{R}[x]_2$ are linearly independent by Example 7, the linear relations of vectors $f_1(x)$, $f_2(x)$, $f_3(x)$ are the same as the linear relations of vectors

$$\mathbf{a}_1 = \begin{pmatrix} 1 \\ -3 \\ 1 \end{pmatrix}, \ \mathbf{a}_2 = \begin{pmatrix} 2 \\ 1 \\ -1 \end{pmatrix}, \ \mathbf{a}_3 = \begin{pmatrix} 4 \\ -5 \\ 1 \end{pmatrix}$$ in $\mathbf{R}^3$ by Theorem 4.2.6 (1). Therefore

we have only to find a non-trivial linear relation of $\mathbf{a}_1$, $\mathbf{a}_2$, $\mathbf{a}_3$. To find a non-trivial linear relation of $\mathbf{a}_1$, $\mathbf{a}_2$, $\mathbf{a}_3$, we have only to solve the linear equation

$$\begin{pmatrix} 1 & 2 & 4 \\ -3 & 1 & -5 \\ 1 & -1 & 1 \end{pmatrix} \begin{pmatrix} x_1 \\ x_2 \\ x_3 \end{pmatrix} = \begin{pmatrix} 0 \\ 0 \\ 0 \end{pmatrix}.$$

By the calculation, the solution is

$$\left\{ c \begin{pmatrix} -2 \\ -1 \\ 1 \end{pmatrix} \,\middle|\, c \in \mathbf{R} \right\}.$$

(Calculation)

$$
\begin{array}{rrr}
1 & 2 & 4 \\
-3 & 1 & -5 \\
1 & -1 & 1 \\
\hline
1 & 2 & 4 \\
0 & 7 & 7 \\
0 & -3 & -3 \\
\hline
1 & 2 & 4 \\
0 & 1 & 1 \\
0 & -3 & -3 \\
\hline
1 & 0 & 2 \\
0 & 1 & 1 \\
0 & 0 & 0 \\
\end{array}
\qquad
\begin{array}{l}
\\
\\
\\
\\
② + ① \times 3 \\
③ + ① \times (-1) \\
\\
② \times (1/7) \\
\\
\\
① + ② \times (-2) \\
\\
③ + ② \times 3
\end{array}
$$

Then we have a non-trivial linear relation

$$-2\mathbf{a}_1 - \mathbf{a}_2 + \mathbf{a}_3 = \mathbf{0}_3$$

of $\mathbf{a}_1, \mathbf{a}_2, \mathbf{a}_3$. Therefore we have a non-trivial linear relation

$$-2f_1(x) - f_2(x) + f_3(x) = f_{00}(x) \, (= 0)$$

of $f_1(x), f_2(x), f_3(x)$. ‖

Exercise 4.6 Let $\mathbf{v}_1, \mathbf{v}_2, \mathbf{v}_3, \mathbf{v}_4$ and $\mathbf{u}_1, \mathbf{u}_2, \mathbf{u}_3, \mathbf{u}_4$ be vectors in V satisfying

$$
\begin{aligned}
\mathbf{v}_1 &= \mathbf{u}_1 - \mathbf{u}_2 + 3\mathbf{u}_3, & \mathbf{v}_2 &= 2\mathbf{u}_1 - \mathbf{u}_2 + 6\mathbf{u}_3 + \mathbf{u}_4, \\
\mathbf{v}_3 &= 2\mathbf{u}_1 - 2\mathbf{u}_2 + \mathbf{u}_3 - \mathbf{u}_4, & \mathbf{v}_4 &= \mathbf{u}_1 - \mathbf{u}_3 + 3\mathbf{u}_4.
\end{aligned}
$$

(1) Find the expression of the vectors $\mathbf{v}_1, \mathbf{v}_2, \mathbf{v}_3, \mathbf{v}_4$ as linear combinations of $\mathbf{u}_1, \mathbf{u}_2, \mathbf{u}_3, \mathbf{u}_4$ by a matrix.

(2) When $\mathbf{u}_1, \mathbf{u}_2, \mathbf{u}_3, \mathbf{u}_4$ are linearly independent, find if $\mathbf{v}_1, \mathbf{v}_2, \mathbf{v}_3, \mathbf{v}_4$ are linearly independent or not.

Answers. (1) $\begin{pmatrix} \mathbf{v}_1 & \mathbf{v}_2 & \mathbf{v}_3 & \mathbf{v}_4 \end{pmatrix} = \begin{pmatrix} \mathbf{u}_1 & \mathbf{u}_2 & \mathbf{u}_3 & \mathbf{u}_4 \end{pmatrix} \begin{pmatrix} 1 & 2 & 2 & 1 \\ -1 & -1 & -2 & 0 \\ 3 & 6 & 1 & -1 \\ 0 & 1 & -1 & 3 \end{pmatrix}.$

(2) Let

$$c_1\mathbf{v}_1 + c_2\mathbf{v}_2 + c_3\mathbf{v}_3 + c_4\mathbf{v}_4 = \mathbf{0}_V$$

be a linear relation of the vectors v_1, v_2, v_3, v_4. Then

$$0_V = c_1 v_1 + c_2 v_2 + c_3 v_3 + c_4 v_4 = \begin{pmatrix} v_1 & v_2 & v_3 & v_4 \end{pmatrix} \begin{pmatrix} c_1 \\ c_2 \\ c_3 \\ c_4 \end{pmatrix}$$

$$= \begin{pmatrix} u_1 & u_2 & u_3 & u_4 \end{pmatrix} \begin{pmatrix} 1 & 2 & 2 & 1 \\ -1 & -1 & -2 & 0 \\ 3 & 6 & 1 & -1 \\ 0 & 1 & -1 & 3 \end{pmatrix} \begin{pmatrix} c_1 \\ c_2 \\ c_3 \\ c_4 \end{pmatrix}.$$

Since the vectors u_1, u_2, u_3, u_4 are linearly independent, we see that

$$\begin{pmatrix} 1 & 2 & 2 & 1 \\ -1 & -1 & -2 & 0 \\ 3 & 6 & 1 & -1 \\ 0 & 1 & -1 & 3 \end{pmatrix} \begin{pmatrix} c_1 \\ c_2 \\ c_3 \\ c_4 \end{pmatrix} = \begin{pmatrix} 0 \\ 0 \\ 0 \\ 0 \end{pmatrix}$$

by Theorem 4.2.5 (1). We solve this linear equation on c_1, c_2, c_3, c_4. Since the coefficient matrix is a regular matrix, we have $c_1 = c_2 = c_3 = c_4 = 0$. Therefore v_1, v_2, v_3, v_4 are linearly independent. ‖

Exercises (Sect. 4.2)

1 Find if the following vectors are linearly independent or not:

(1) $\begin{pmatrix} 1 \\ 1 \\ 1 \end{pmatrix}$, $\begin{pmatrix} 0 \\ 1 \\ 1 \end{pmatrix}$, $\begin{pmatrix} 0 \\ 0 \\ 1 \end{pmatrix}$.

(2) $\begin{pmatrix} 3 \\ 2 \\ 1 \end{pmatrix}$, $\begin{pmatrix} 2 \\ 1 \\ 3 \end{pmatrix}$, $\begin{pmatrix} 5 \\ 4 \\ -3 \end{pmatrix}$.

(3) $\begin{pmatrix} 2 \\ 1 \\ 1 \end{pmatrix}$, $\begin{pmatrix} 1 \\ 1 \\ 2 \end{pmatrix}$, $\begin{pmatrix} 1 \\ 2 \\ 1 \end{pmatrix}$.

(4) $\begin{pmatrix} 2 \\ 4 \\ 1 \end{pmatrix}$, $\begin{pmatrix} 3 \\ 1 \\ 2 \end{pmatrix}$, $\begin{pmatrix} 5 \\ 1 \\ 1 \end{pmatrix}$, $\begin{pmatrix} 2 \\ 0 \\ 3 \end{pmatrix}$.

(5) $\begin{pmatrix} 2 \\ 1 \\ 1 \\ 4 \end{pmatrix}$, $\begin{pmatrix} 3 \\ 2 \\ 1 \\ 1 \end{pmatrix}$, $\begin{pmatrix} 5 \\ 4 \\ 1 \\ -5 \end{pmatrix}$.

(6) $\begin{pmatrix} 1 \\ 0 \\ 2 \\ 4 \end{pmatrix}$, $\begin{pmatrix} 1 \\ 1 \\ 0 \\ 3 \end{pmatrix}$, $\begin{pmatrix} 2 \\ 1 \\ 3 \\ 0 \end{pmatrix}$, $\begin{pmatrix} -2 \\ 0 \\ -1 \\ 1 \end{pmatrix}$.

(7) $f_1(x) = 1 + x + x^2$, $f_2(x) = 2 - x + 2x^2$, $f_3(x) = -1 + 2x + x^2$.

(8) $f_1(x) = 1 + x - x^2$, $f_2(x) = 5 + 5x - x^2$, $f_3(x) = 2 - x + 2x^2$.

2 Find non-trivial linear relations of the following vectors:

(1) $\begin{pmatrix} 1 \\ -2 \end{pmatrix}$, $\begin{pmatrix} 1 \\ 1 \end{pmatrix}$, $\begin{pmatrix} 2 \\ 5 \end{pmatrix}$.

(2) $\begin{pmatrix} 2 \\ 1 \\ 1 \end{pmatrix}$, $\begin{pmatrix} -1 \\ 1 \\ 2 \end{pmatrix}$, $\begin{pmatrix} -1 \\ 4 \\ 7 \end{pmatrix}$.

(3) $\begin{pmatrix} 1 \\ -2 \\ 1 \end{pmatrix}, \begin{pmatrix} 0 \\ -2 \\ 2 \end{pmatrix}, \begin{pmatrix} -2 \\ 2 \\ -1 \end{pmatrix}, \begin{pmatrix} -4 \\ 4 \\ -3 \end{pmatrix}.$

(4) $f_1(x) = 1 + 2x + x^2, \ f_2(x) = 1 + x^2, \ f_3(x) = 1 + 4x + x^2.$

(5) $f_1(x) = 1 + x^2, \ f_2(x) = 1 + x, \ f_3(x) = x + x^2, \ f_4(x) = 3x + x^2.$

3 Express **b** as linear combinations of $\mathbf{a}_1$, $\mathbf{a}_2$, $\mathbf{a}_3$:

(1) $\mathbf{b} = \begin{pmatrix} 1 \\ -2 \\ 2 \end{pmatrix}; \mathbf{a}_1 = \begin{pmatrix} 1 \\ 0 \\ 1 \end{pmatrix}, \ \mathbf{a}_2 = \begin{pmatrix} -1 \\ 2 \\ 0 \end{pmatrix}, \ \mathbf{a}_3 = \begin{pmatrix} 1 \\ 1 \\ 1 \end{pmatrix}.$

(2) $\mathbf{b} = \begin{pmatrix} 3 \\ 1 \\ -4 \end{pmatrix}; \mathbf{a}_1 = \begin{pmatrix} 2 \\ 1 \\ 1 \end{pmatrix}, \ \mathbf{a}_2 = \begin{pmatrix} 1 \\ 0 \\ -2 \end{pmatrix}, \ \mathbf{a}_3 = \begin{pmatrix} 3 \\ 2 \\ 1 \end{pmatrix}.$

(3) $\mathbf{b} = \begin{pmatrix} 2 \\ 1 \\ -5 \end{pmatrix}; \mathbf{a}_1 = \begin{pmatrix} 0 \\ 1 \\ 2 \end{pmatrix}, \ \mathbf{a}_2 = \begin{pmatrix} 1 \\ 1 \\ 0 \end{pmatrix}, \ \mathbf{a}_3 = \begin{pmatrix} 1 \\ 0 \\ -3 \end{pmatrix}.$

4 Express the following polynomials $g(x)$ as linear combinations of $f_1(x)$, $f_2(x)$, $f_3(x)$:

(1) $g(x) = 2 + x + 2x^2; \ f_1(x) = 1 + x - x^2, \ f_2(x) = 2 + x, \ f_3(x) = 1 + 2x^2.$

(2) $g(x) = 1 + x^2; \ f_1(x) = 1 - x + x^2, \ f_2(x) = 2 + x, \ f_3(x) = x - x^2.$

(3) $g(x) = 2 + x - 3x^2; \ f_1(x) = 1 - x^2, \ f_2(x) = 1 + x + x^2, \ f_3(x) = 1 + 2x.$

5 Find the expressions of the following linear combinations by matrices:

(1) $\mathbf{v}_1 = 2\mathbf{u}_1 + \mathbf{u}_2 - 3\mathbf{u}_3, \ \mathbf{v}_2 = \mathbf{u}_1 - \mathbf{u}_2 + \mathbf{u}_3, \ \mathbf{v}_3 = \mathbf{u}_1 + 2\mathbf{u}_2 + 4\mathbf{u}_3.$

(2) $\mathbf{v}_1 = \mathbf{u}_1 - \mathbf{u}_2 + 2\mathbf{u}_3, \ \mathbf{v}_2 = 2\mathbf{u}_1 + 2\mathbf{u}_2, \ \mathbf{v}_3 = -\mathbf{u}_1 - 7\mathbf{u}_2 + 6\mathbf{u}_3.$

(3) $\mathbf{v}_1 = -\mathbf{u}_1 - 2\mathbf{u}_2 - \mathbf{u}_3, \ \mathbf{v}_2 = \mathbf{u}_1 + \mathbf{u}_2 - 2\mathbf{u}_3, \ \mathbf{v}_3 = 3\mathbf{u}_1 + 2\mathbf{u}_2 - 3\mathbf{u}_3.$

6 Let the vectors $\mathbf{u}_1, \mathbf{u}_2, \mathbf{u}_3$ and $\mathbf{v}_1, \mathbf{v}_2, \mathbf{v}_3$ be the same as in Exercise 5 above. When the vectors $\mathbf{u}_1, \mathbf{u}_2, \mathbf{u}_3$ are linearly independent, find if the vectors $\mathbf{v}_1, \mathbf{v}_2, \mathbf{v}_3$ are linearly independent or not.

7 Let $\mathbf{u}_1, \mathbf{u}_2, \mathbf{u}_3$ and $\mathbf{v}_1, \mathbf{v}_2, \mathbf{v}_3$ be vectors. Express the following vectors **w** as linear combinations of $\mathbf{v}_1, \mathbf{v}_2, \mathbf{v}_3$:

(1) $\mathbf{w} = -2\mathbf{u}_1 + \mathbf{u}_2 + 4\mathbf{u}_3; \ \mathbf{v}_1 = \mathbf{u}_1 - \mathbf{u}_2 + \mathbf{u}_3, \ \mathbf{v}_2 = \mathbf{u}_1 - 2\mathbf{u}_3, \ \mathbf{v}_3 = -\mathbf{u}_1 + \mathbf{u}_2.$

(2) $\mathbf{w} = 3\mathbf{u}_1 + 6\mathbf{u}_2 - 4\mathbf{u}_3; \ \mathbf{v}_1 = -\mathbf{u}_1 + 2\mathbf{u}_2 + \mathbf{u}_3, \ \mathbf{v}_2 = \mathbf{u}_1 + \mathbf{u}_2, \ \mathbf{v}_3 = -2\mathbf{u}_2 + \mathbf{u}_3.$

8 Find if the statements for vectors $\mathbf{u}_1, \mathbf{u}_2, \mathbf{u}_3$ below are true or not by giving proofs or counterexamples:

(1) If $\mathbf{u}_1$ and $\mathbf{u}_2$, $\mathbf{u}_2$ and $\mathbf{u}_3$, and $\mathbf{u}_1$ and $\mathbf{u}_3$ are linearly independent, then $\mathbf{u}_1, \mathbf{u}_2, \mathbf{u}_3$ are linearly independent.

(2) If $\mathbf{u}_1$, $\mathbf{u}_1 + \mathbf{u}_2$, $\mathbf{u}_1 + \mathbf{u}_2 + \mathbf{u}_3$ are linearly independent, then $\mathbf{u}_1, \mathbf{u}_2, \mathbf{u}_3$ are linearly independent.

(3) If $\mathbf{u}_1, \mathbf{u}_2, \ldots, \mathbf{u}_n$ are linearly independent, then the r vectors $\mathbf{u}_1, \mathbf{u}_2, \ldots, \mathbf{u}_r (1 \le r \le n - 1)$ are also linearly independent.

4.3   Ranks of Sets of Vectors

Ranks of sets of vectors Let X be a non-empty subset of a vector space V. If there exist r linearly independent vectors in X and any $r + 1$ vectors in X are linearly dependent, then we call the number r the *rank* of X. The rank of X is denoted by

$$\text{rank}(X) = r.$$

If there exist r linearly independent vectors in X for any r, we say the *rank of X is infinity* and denote rank $(X) = \infty$.

Example 1 Let $V = \mathbf{R}^3$ and

$$X = \left\{ \mathbf{a}_1 = \begin{pmatrix} 1 \\ 0 \\ 0 \end{pmatrix}, \mathbf{a}_2 = \begin{pmatrix} 0 \\ 1 \\ 0 \end{pmatrix}, \mathbf{a}_3 = \begin{pmatrix} 1 \\ -3 \\ 2 \end{pmatrix}, \mathbf{a}_4 = \begin{pmatrix} 0 \\ 0 \\ 1 \end{pmatrix} \right\}.$$

Since the vectors $\mathbf{a}_1, \mathbf{a}_2, \mathbf{a}_4$ are the unit vectors in $\mathbf{R}^3$, they are linearly independent by Example 6 of Sect. 4.2. Since we have a non-trivial linear relation

$$-\mathbf{a}_1 + 3\mathbf{a}_2 + \mathbf{a}_3 - 2\mathbf{a}_4 = \mathbf{0}_3,$$

the vectors $\mathbf{a}_1, \mathbf{a}_2, \mathbf{a}_3, \mathbf{a}_4$ are linearly dependent and we have

$$\text{rank}(X) = 3.$$

Theorem 4.3.1 *Let* $\mathbf{v}_1, \ldots, \mathbf{v}_n$ *and* $\mathbf{u}_1, \ldots, \mathbf{u}_m$ *be vectors in a vector space V. If any vector in* $\{\mathbf{v}_1, \ldots, \mathbf{v}_n\}$ *is expressed as a linear combination of* $\mathbf{u}_1, \ldots, \mathbf{u}_m$, *then we have*

$$\text{rank}(\{\mathbf{v}_1, \ldots, \mathbf{v}_n\}) \leq \text{rank}(\{\mathbf{u}_1, \ldots, \mathbf{u}_m\}).$$

Proof Let $r = \text{rank}(\{\mathbf{u}_1, \ldots, \mathbf{u}_m\})$. We may assume that the vectors $\mathbf{u}_1, \ldots, \mathbf{u}_r$ are linearly independent by changing the indices of vectors if necessary. Then the vectors $\mathbf{u}_{r+1}, \ldots, \mathbf{u}_m$ are expressed as linear combinations of $\mathbf{u}_1, \ldots, \mathbf{u}_r$ by Theorem 4.2.3. Since the vectors $\mathbf{v}_1, \ldots, \mathbf{v}_n$ are expressed as linear combinations of $\mathbf{u}_1, \ldots, \mathbf{u}_m$, the vectors $\mathbf{v}_1, \ldots, \mathbf{v}_n$ are expressed as linear combinations of $\mathbf{u}_1, \ldots, \mathbf{u}_r$ similarly to the numerical vectors case in Theorem 1.4.1. By Theorem 4.2.4, any $r + 1$ vectors in $\{\mathbf{v}_1, \ldots, \mathbf{v}_n\}$ are linearly dependent. Therefore we have

$$\text{rank}(\{\mathbf{v}_1, \ldots, \mathbf{v}_n\}) \leq r = \text{rank}(\{\mathbf{u}_1, \ldots, \mathbf{u}_m\}). \|$$

Theorem 4.3.2 *Let X be a non-empty subset of a vector space V. Then the following two statements are equivalent:*

(1) rank $(X) = r$.

(2) There exist r linearly independent vectors in X and any vector in X is a linear combination of these r vectors.

Proof (1) $\Rightarrow$ (2) Take r linearly independent vectors $\mathbf{u}_1, \ldots, \mathbf{u}_r$ in X. Let $\mathbf{u}$ be a vector in X. Since rank$(X) = r$, the vectors $\mathbf{u}, \mathbf{u}_1, \ldots, \mathbf{u}_r$ are linearly dependent. Therefore $\mathbf{u}$ is expressed as a linear combination of $\mathbf{u}_1, \ldots, \mathbf{u}_r$ by Theorem 4.2.3.

(2) $\Rightarrow$ (1) We assume that there exist r linearly independent vectors $\mathbf{u}_1, \ldots, \mathbf{u}_r$ in X and any vector in X can be expressed as a linear combination of these vectors. Then by the definition of the rank, we have

$$r \leq \text{rank}(X).$$

Take any $r + 1$ vectors in X. Then these $r + 1$ vectors can be expressed as linear combinations of $\mathbf{u}_1, \ldots, \mathbf{u}_r$ by assumption. Therefore we see that those $r + 1$ vectors are linearly dependent by Theorem 4.2.4. This implies that

$$\text{rank}(X) \leq r.$$

Therefore we have

$$\text{rank}(X) = r. \|$$

The next theorem is a special case of Theorem 4.3.2. (The case where X is a finite set.)

Theorem 4.3.3 *The following two statements are equivalent:*

(1) rank$(\{\mathbf{u}_1, \ldots, \mathbf{u}_m\}) = r$.
(2) There exist r vectors in $\{\mathbf{u}_1, \ldots, \mathbf{u}_m\}$ which are linearly independent and any other $m - r$ vectors are expressed as linear combinations of these r vectors.

Exercise 4.7 We let

$$\mathbf{a}_1 = \begin{pmatrix} 1 \\ 1 \\ 3 \\ 0 \end{pmatrix}, \quad \mathbf{a}_2 = \begin{pmatrix} 1 \\ 2 \\ 0 \\ -1 \end{pmatrix}, \quad \mathbf{a}_3 = \begin{pmatrix} 1 \\ 3 \\ -3 \\ -2 \end{pmatrix}, \quad \mathbf{a}_4 = \begin{pmatrix} -2 \\ -4 \\ 1 \\ -1 \end{pmatrix}, \quad \mathbf{a}_5 = \begin{pmatrix} -1 \\ -4 \\ 7 \\ 0 \end{pmatrix}.$$

(1) Find $r = \text{rank}(\{\mathbf{a}_1, \mathbf{a}_2, \mathbf{a}_3, \mathbf{a}_4, \mathbf{a}_5\})$ and a set of the r linearly independent vectors.

(2) Express other vectors as linear combinations of these linearly independent vectors.

Answer. Let

$$A = \begin{pmatrix} \mathbf{a}_1 & \mathbf{a}_2 & \mathbf{a}_3 & \mathbf{a}_4 & \mathbf{a}_5 \end{pmatrix}.$$

Denote the reduced matrix B of A by

$$B = \begin{pmatrix} \mathbf{b}_1 & \mathbf{b}_2 & \mathbf{b}_3 & \mathbf{b}_4 & \mathbf{b}_5 \end{pmatrix}.$$

Then we have

$$c_1 \mathbf{a}_1 + \cdots + c_5 \mathbf{a}_5 = \mathbf{0}_4$$
$$\Updownarrow$$
$$c_1 \mathbf{b}_1 + \cdots + c_5 \mathbf{b}_5 = \mathbf{0}_4.$$

In other words, the vectors $\mathbf{a}_1, \ldots, \mathbf{a}_5$ and $\mathbf{b}_1, \ldots, \mathbf{b}_5$ have the same linear relations. By the calculation, $\mathbf{b}_1, \mathbf{b}_2, \mathbf{b}_4$ are linearly independent, and $\mathbf{b}_3$ and $\mathbf{b}_5$ are expressed as

$$\mathbf{b}_3 = -\mathbf{b}_1 + 2\mathbf{b}_2, \quad \mathbf{b}_5 = 2\mathbf{b}_1 - \mathbf{b}_2 + \mathbf{b}_4.$$

(Calculation)

$\mathbf{a}_1$	$\mathbf{a}_2$	$\mathbf{a}_3$	$\mathbf{a}_4$	$\mathbf{a}_5$	
1	1	1	−2	−1	
1	2	3	−4	−4	
3	0	−3	1	7	
0	−1	−2	−1	0	
1	1	1	−2	−1	
0	1	2	−2	−3	② + ① × (−1)
0	−3	−6	7	10	③ + ① × (−3)
0	−1	−2	−1	0	
1	0	−1	0	2	① + ② × (−1)
0	1	2	−2	−3	
0	0	0	1	1	③ + ② × 3
0	0	0	−3	−3	④ + ②
1	0	−1	0	2	
0	1	2	0	−1	② + ③ × 2
0	0	0	1	1	
0	0	0	0	0	④ + ③ × 3
$\mathbf{b}_1$	$\mathbf{b}_2$	$\mathbf{b}_3$	$\mathbf{b}_4$	$\mathbf{b}_5$	

Thus we obtain:

(1) $\operatorname{rank}(\{\mathbf{a}_1, \mathbf{a}_2, \mathbf{a}_3, \mathbf{a}_4, \mathbf{a}_5\}) = 3$, and the vectors $\mathbf{a}_1, \mathbf{a}_2, \mathbf{a}_4$ are linearly independent.
(2) $\mathbf{a}_3 = -\mathbf{a}_1 + 2\mathbf{a}_2, \quad \mathbf{a}_5 = 2\mathbf{a}_1 - \mathbf{a}_2 + \mathbf{a}_4.$ ‖

Let A be an $m \times n$ matrix and B the reduced matrix of A. The rows of A and the rows of B are deformed to each other by reductions. So, the rows of B can be expressed as linear combinations of the rows of A, and vice versa. Then by Theorem 4.3.1 and the definition of the rank of matrices, we have

rank({ the row vectors of A }) = rank({ the row vectors of B })
= the number of non-zero rows of B
= rank(B) = rank(A).

We note that this number is nothing but the number of primary components of B. Next, we consider column vectors of the matrices A and B. Let

$$A = \begin{pmatrix} \mathbf{a}_1 & \cdots & \mathbf{a}_n \end{pmatrix} \quad \text{and} \quad B = \begin{pmatrix} \mathbf{b}_1 & \cdots & \mathbf{b}_n \end{pmatrix}$$

be the column partitions of A and B, respectively. As we see in Exercise 4.7, we have the equivalence

$$x_1\mathbf{a}_1 + \cdots + x_n\mathbf{a}_n = \mathbf{0} \iff x_1\mathbf{b}_1 + \cdots + x_n\mathbf{b}_n = \mathbf{0}.$$

In other words, vectors $\mathbf{a}_1, \ldots, \mathbf{a}_n$ and $\mathbf{b}_1, \ldots, \mathbf{b}_n$ have the same linear relations. Therefore we have

rank({ the column vectors of A }) = rank({ the column vectors of B }).

The columns of B containing primary components can be taken as the maximal linearly independent columns of B. So, the rank of the set of column vectors of B is equal to the number of the columns containing primary components. Therefore we have the following theorem.

Theorem 4.3.4 *Let A be a matrix and B the reduced matrix of A. Then the following six numbers are equal:*

(1) rank (A).
(2) rank (B).
(3) rank ({ the row vectors of }A).
(4) rank ({ the row vectors of }B).
(5) rank ({ the column vectors of }A).
(6) rank ({ the column vectors of }B).

We obtain the following theorem, which is very important.

Theorem 4.3.5 *For a square matrix A of degree n, the following three conditions are equivalent:*

(1) *The matrix A is a regular matrix.*
(2) *The rows of A are linearly independent.*
(3) *The columns of A are linearly independent.*

Proof By Theorem 2.4.2, a square matrix A of degree n is regular if and only if rank$(A) = n$. By Theorem 4.3.4, we see that rank$(A) = n$ is equivalent to the linear independence of the rows of A, and it is also equivalent to the linear independence of the columns of A. $\parallel$

We shall prove the uniqueness of the reduced matrix stated in Theorem 2.2.1.

Theorem 4.3.6 *The reduced matrix B of a matrix A is uniquely determined.*

Proof Let A be an $m \times n$ matrix and

$$A = \begin{pmatrix} \mathbf{a}_1 & \cdots & \mathbf{a}_n \end{pmatrix} \quad \text{and} \quad B = \begin{pmatrix} \mathbf{b}_1 & \cdots & \mathbf{b}_n \end{pmatrix}$$

the column partitions of A and B, respectively. Let $r = $ rank (A) and take the r linearly independent column vectors of A from the beginning. We denote them by

$$\mathbf{a}_{k_1}, \ldots, \mathbf{a}_{k_r}.$$

By definition of the reduced matrix, the corresponding linearly independent columns $\mathbf{b}_{k_1}, \ldots, \mathbf{b}_{k_r}$ of B are r unit vectors $\mathbf{e}_1, \ldots, \mathbf{e}_r$ in $\mathbf{R}^m$, that is, $\mathbf{b}_{k_1} = \mathbf{e}_1, \ldots, \mathbf{b}_{k_r} = \mathbf{e}_r$. Since the column vectors of A and B have the same linear relations, a column vector $\mathbf{a}_i$ of A is expressed as a linear combination

$$\mathbf{a}_i = c_1 \mathbf{a}_{k_1} + \cdots + c_r \mathbf{a}_{k_r}$$

if and only if the corresponding column $\mathbf{b}_i$ of B is expressed as the linear combination

$$\mathbf{b}_i = c_1 \mathbf{b}_{k_1} + \cdots + c_r \mathbf{b}_{k_r} = c_1 \mathbf{e}_1 + \cdots + c_r \mathbf{e}_r.$$

Since $c_1, \ldots, c_r$ are uniquely determined by Theorem 4.2.1, we see that the reduced matrix B is uniquely determined. ∥

Theorem 4.3.7 *Let* $\mathbf{u}_1, \ldots, \mathbf{u}_n$ *be linearly independent vectors in a vector space* V. *When the vectors* $\mathbf{v}_1, , \ldots, \mathbf{v}_n$ *in* V *are expressed as linear combinations*

$$\left(\mathbf{v}_1 \cdots \mathbf{v}_n \right) = \left(\mathbf{u}_1 \cdots \mathbf{u}_n \right) A$$

by a square matrix A *of degree* n, *the vectors* $\mathbf{v}_1, \ldots, \mathbf{v}_n$ *are linearly independent if and only if* A *is a regular matrix.*

Proof By Theorem 4.2.6 (2), the vectors $\mathbf{v}_1, \ldots, \mathbf{v}_n$ are linearly independent if and only if the column vectors of A are linearly independent. By Theorem 4.3.5, the column vectors of A are linearly independent if and only if A is a regular matrix. Therefore

$\mathbf{v}_1, \ldots, \mathbf{v}_n$ are linearly independent if and only if A is a regular matrix. ∥

Exercise 4.8 Show that the three vectors

$$f_1(x) = 2 - x + x^2, \ f_2(x) = 1 - 3x + 2x^2, \ f_3(x) = 1 + x - x^2$$

in $\mathbf{R}[x]_2$ are linearly independent.
Answer. The three vectors $f_1(x)$, $f_2(x)$, $f_3(x)$ are expressed as linear combinations of linearly independent vectors $1, x, x^2$ as

$$\left(f_1(x) \ f_2(x) \ f_3(x) \right) = \left(1 \ x \ x^2 \right) \begin{pmatrix} 2 & 1 & 1 \\ -1 & -3 & 1 \\ 1 & 2 & -1 \end{pmatrix}.$$

Since $\begin{vmatrix} 2 & 1 & 1 \\ -1 & -3 & 1 \\ 1 & 2 & -1 \end{vmatrix} = 3 \neq 0$, the matrix $\begin{pmatrix} 2 & 1 & 1 \\ -1 & -3 & 1 \\ 1 & 2 & -1 \end{pmatrix}$ is a regular matrix. There-

fore the vectors

$$f_1(x), f_2(x), f_3(x)$$

are linearly independent by Theorem 4.3.7. ‖

Exercises (Sect. 4.3)

1 Show that the following vectors are linearly independent by using determinants of matrices:

(1) $\mathbf{a}_1 = \begin{pmatrix} 2 \\ 3 \\ 1 \end{pmatrix}$, $\mathbf{a}_2 = \begin{pmatrix} -1 \\ 1 \\ 1 \end{pmatrix}$, $\mathbf{a}_2 = \begin{pmatrix} 1 \\ 2 \\ 1 \end{pmatrix}$.

(2) $\mathbf{a}_1 = \begin{pmatrix} 2 \\ -1 \\ 1 \end{pmatrix}$, $\mathbf{a}_2 = \begin{pmatrix} 1 \\ -3 \\ 2 \end{pmatrix}$, $\mathbf{a}_3 = \begin{pmatrix} 1 \\ 1 \\ -1 \end{pmatrix}$.

(3) $f_1(x) = 1 - 2x + 3x^2$, $f_2(x) = -1 + x + 2x^2$, $f_3(x) = 1 - x + x^2$.

(4) $f_1(x) = 3 + x - x^2$, $f_2(x) = 1 + x - 2x^2$, $f_3(x) = -2x + 3x^2$.

2 Answer questions (i), (ii) and (iii) for the following sets of vectors:

(i) Find the ranks r of the sets of vectors.

(ii) Find the r linearly independent vectors from the beginning.

(iii) Express other vectors as linear combinations of the r linearly independent vectors obtained in (ii).

(1) $\mathbf{a}_1 = \begin{pmatrix} 1 \\ 0 \\ 2 \end{pmatrix}$, $\mathbf{a}_2 = \begin{pmatrix} -1 \\ 1 \\ 0 \end{pmatrix}$, $\mathbf{a}_3 = \begin{pmatrix} 1 \\ 1 \\ 4 \end{pmatrix}$, $\mathbf{a}_4 = \begin{pmatrix} -1 \\ 2 \\ 2 \end{pmatrix}$.

(2) $\mathbf{a}_1 = \begin{pmatrix} 1 \\ 0 \\ 1 \\ 1 \end{pmatrix}$, $\mathbf{a}_2 = \begin{pmatrix} 0 \\ 1 \\ 0 \\ 1 \end{pmatrix}$, $\mathbf{a}_3 = \begin{pmatrix} 0 \\ 0 \\ 1 \\ 1 \end{pmatrix}$, $\mathbf{a}_4 = \begin{pmatrix} 1 \\ -1 \\ 2 \\ 1 \end{pmatrix}$.

(3) $\mathbf{a}_1 = \begin{pmatrix} 1 \\ 0 \\ 1 \\ 0 \end{pmatrix}$, $\mathbf{a}_2 = \begin{pmatrix} 0 \\ 1 \\ 0 \\ 1 \end{pmatrix}$, $\mathbf{a}_3 = \begin{pmatrix} 1 \\ 2 \\ 1 \\ 2 \end{pmatrix}$, $\mathbf{a}_4 = \begin{pmatrix} 1 \\ 1 \\ 1 \\ 1 \end{pmatrix}$, $\mathbf{a}_5 = \begin{pmatrix} 1 \\ 1 \\ 0 \\ 1 \end{pmatrix}$.

(4) $f_1(x) = 1 + x^2$, $f_2(x) = x + x^3$, $f_3(x) = 2 + 2x^2$, $f_4(x) = x + x^2 + x^3$, $f_5(x) = -x + x^2 - x^3$.

(5) $f_1(x) = 1 + x^3$, $f_2(x) = x$, $f_3(x) = 1 + x + x^3$, $f_4(x) = 2 + x + 2x^3$, $f_5(x) = x^2 + x^3$.

3 Let A be an $l \times m$ matrix and B an $m \times n$ matrix. Show the following inequalities:

(1) rank $(AB) \leq$ rank (A).   (2) rank $(AB) \leq$ rank (B).

4 Let A be an $m \times n$ matrix, P a regular matrix of degree m and Q a regular matrix of degree n. Show the following equalities:

(1) $\operatorname{rank}(PA) = \operatorname{rank}(A)$.   (2) $\operatorname{rank}(AQ) = \operatorname{rank}(A)$.

5 Square matrices obtained by removing several rows and columns from a matrix A are called *submatrices* of A. Let $r = \operatorname{rank}(A)$. Show that there exists a regular submatrix of A of degree r.

6 For the following matrices A, find $r = \operatorname{rank}(A)$ and a regular submatrix C of A of degree r:

(1) $A = \begin{pmatrix} 1 & 2 & 4 & 3 & 1 \\ -1 & 1 & -1 & 0 & 0 \\ -2 & -1 & -5 & -3 & -1 \\ 1 & -1 & 1 & 0 & 2 \end{pmatrix}$.

(2) $A = \begin{pmatrix} -1 & 2 & 0 & 1 & 1 & 1 \\ 1 & 1 & -3 & 1 & 0 & 1 \\ 2 & 0 & -4 & -1 & 0 & 3 \\ -1 & 1 & 1 & 0 & -1 & -1 \\ 0 & 1 & -1 & 1 & 0 & 0 \end{pmatrix}$.

(3) $A = \begin{pmatrix} 1 & 3 & 1 & 1 & 0 & 2 \\ -2 & -6 & 1 & -5 & 1 & 0 \\ 1 & 3 & 1 & 1 & -1 & 1 \\ -1 & -3 & 2 & -4 & 1 & 2 \end{pmatrix}$.

4.4   Bases and Dimensions of Vector Spaces

Vector spaces generated by vectors When any vector in a vector space V can be expressed as a linear combination of vectors $\mathbf{u}_1, \mathbf{u}_2, \ldots, \mathbf{u}_n$ in V, we say that V is *generated* by the vectors $\mathbf{u}_1, \mathbf{u}_2, \ldots, \mathbf{u}_n$. The vectors $\mathbf{u}_1, \mathbf{u}_2, \ldots, \mathbf{u}_n$ are called *generators* of V.

Example 1 The unit vectors $\mathbf{e}_1, \ldots, \mathbf{e}_n$ in the vector space K^n generate K^n since

any vector $\mathbf{a} = \begin{pmatrix} a_1 \\ \vdots \\ a_n \end{pmatrix} \in K^n$ is expressed as a linear combination

$$\mathbf{a} = a_1 \mathbf{e}_1 + \cdots + a_n \mathbf{e}_n$$

of $\mathbf{e}_1, \ldots, \mathbf{e}_n$.

Bases of vector spaces A set $\{\mathbf{u}_1, \ldots, \mathbf{u}_n\}$ of vectors in a vector space V is called a *basis* of V if it satisfies the following two conditions:

(1) Vectors $\mathbf{u}_1, \ldots, \mathbf{u}_n$ are linearly independent.
(2) The vector space V is generated by vectors $\mathbf{u}_1, \ldots, \mathbf{u}_n$.

Example 2 The set $\{\mathbf{e}_1, \mathbf{e}_2, \ldots, \mathbf{e}_n\}$ of the unit vectors in K^n is a basis of K^n. In fact, the vectors are linearly independent by Example 6 in Sect. 4.2, and they generate K^n by Example 1 above. We call the basis $\{\mathbf{e}_1, \mathbf{e}_2, \ldots, \mathbf{e}_n\}$ of K^n the *standard basis* of K^n.

Though a vector space may have several bases, the numbers of the vectors constituting each basis are equal. We shall show this fact in the following theorem.

Theorem 4.4.1 *If there exist bases of a vector space V, then the numbers of vectors in each basis are equal for all bases.*

Proof Let $\{\mathbf{u}_1, \ldots, \mathbf{u}_m\}$ and $\{\mathbf{v}_1, \ldots, \mathbf{v}_n\}$ be any two bases of V. Then the vectors $\mathbf{v}_1, \ldots, \mathbf{v}_n$ can be expressed as linear combinations of the vectors $\mathbf{u}_1, \ldots, \mathbf{u}_m$. Then by Theorem 4.3.1, we see that $n \leq m$. Changing bases $\{\mathbf{u}_1, \ldots, \mathbf{u}_m\}$ and $\{\mathbf{v}_1, \ldots, \mathbf{v}_n\}$ of V, we also obtain $m \leq n$. Then we have $m = n$ and the numbers of vectors in each basis of V are equal for all bases. $\parallel$

Finite-dimensional vector spaces The vector space consisting of only the zero vector is called the *zero space*. A vector space V is called a *finite-dimensional vector space* if either V is the zero space or V has a basis consisting of finite vectors.

Dimensions of vector spaces Let V be a finite-dimensional vector space. Since the numbers of vectors in each basis of V are equal by Theorem 4.4.1, we call the number of vectors in a basis the *dimension* of V and denote it by

$$\dim (V).$$

We understand that the dimension of the zero space is 0.

Example 3 Since the standard basis $\{\mathbf{e}_1, \ldots, \mathbf{e}_n\}$ is a basis of the vector space K^n, we have
$$\dim(K^n) = n.$$

Example 4 Since $n + 1$ vectors $1, x, \ldots, x^n$ in the vector space $\mathbf{R}[x]_n$ generate the vector space $\mathbf{R}[x]_n$ and they are linearly independent by Example 7 in Sect. 4.2, the set $\{1, x, \ldots, x^n\}$ is a basis of the vector space $\mathbf{R}[x]_n$. We call the basis $\{1, x, \ldots, x^n\}$ the *standard basis* of $\mathbf{R}[x]_n$. Therefore

$$\dim (\mathbf{R}[x]_n) = n + 1.$$

Theorem 4.4.2 *A vector space V is a finite-dimensional vector space if and only if* rank (V) *is finite. Furthermore we have*

$$\dim(V) = \text{rank}(V).$$

Proof If $\dim(V) = n$, then there exists a basis $\{\mathbf{u}_1, \ldots, \mathbf{u}_n\}$. Since $\mathbf{u}_1, \ldots, \mathbf{u}_n$ are linearly independent and any vector in V is expressed as a linear combination of $\mathbf{u}_1, \ldots, \mathbf{u}_n$, we have by Theorem 4.3.2

$$\text{rank} (V) = n = \dim (V).$$

Conversely, we assume rank $(V) = n \, (< \infty)$. Take n linearly independent vectors $\mathbf{u}_1, \ldots, \mathbf{u}_n$ in V. Take an arbitrary vector $\mathbf{u}$ in V. Since $n + 1$ vectors $\mathbf{u}, \mathbf{u}_1, \ldots, \mathbf{u}_n$ are linearly dependent, the vector $\mathbf{u}$ can be expressed as a linear combination of $\mathbf{u}_1, \ldots, \mathbf{u}_n$ by Theorem 4.2.3. Therefore the set $\{\mathbf{u}_1, \ldots, \mathbf{u}_n\}$ is a basis of V, and we have

$$\dim(V) = n = \text{rank}(V). \|$$

Exercise 4.9 Find the dimension and a basis of the following solution space W of a homogeneous linear equation:

$$W = \left\{ \mathbf{x} \in \mathbf{R}^5 \;\middle|\; \begin{array}{l} x_1 \; - 2x_2 \; + \; x_3 \; + 2x_4 \; + 3x_5 \; = 0, \\ 2x_1 \; - 4x_2 \; + 3x_3 \; + 3x_4 \; + 8x_5 \; = 0 \end{array} \right\}.$$

Answer. Let

$$A = \begin{pmatrix} 1 & -2 & 1 & 2 & 3 \\ 2 & -4 & 3 & 3 & 8 \end{pmatrix}$$

be the coefficient matrix of the linear equation. By the calculation, the reduced matrix of A is

$$\begin{pmatrix} 1 & -2 & 0 & 3 & 1 \\ 0 & 0 & 1 & -1 & 2 \end{pmatrix}.$$

(Calculation)

x_1	x_2	x_3	x_4	x_5	
1	−2	1	2	3	
2	−4	3	3	8	②+① × (−2)
1	−2	1	2	3	①+② × (−1)
0	0	1	−1	2	
1	−2	0	3	1	
0	0	1	−1	2	

So, we obtain the answer for the linear equation:

$$\mathbf{x} = \begin{pmatrix} 2c_1 - 3c_2 - c_3 \\ c_1 \\ c_2 - 2c_3 \\ c_2 \\ c_3 \end{pmatrix} = c_1 \begin{pmatrix} 2 \\ 1 \\ 0 \\ 0 \\ 0 \end{pmatrix} + c_2 \begin{pmatrix} -3 \\ 0 \\ 1 \\ 1 \\ 0 \end{pmatrix} + c_3 \begin{pmatrix} -1 \\ 0 \\ -2 \\ 0 \\ 1 \end{pmatrix} \quad (c_1, c_2, c_3 \in \mathbf{R}).$$

If we let

$$\mathbf{a}_1 = \begin{pmatrix} 2 \\ 1 \\ 0 \\ 0 \\ 0 \end{pmatrix}, \quad \mathbf{a}_2 = \begin{pmatrix} -3 \\ 0 \\ 1 \\ 1 \\ 0 \end{pmatrix}, \quad \mathbf{a}_3 = \begin{pmatrix} -1 \\ 0 \\ -2 \\ 0 \\ 1 \end{pmatrix},$$

then the vectors $\mathbf{a}_1, \mathbf{a}_2, \mathbf{a}_3$ generate W. We easily see that $\mathbf{a}_1, \mathbf{a}_2, \mathbf{a}_3$ are linearly independent. Therefore the set $\{\mathbf{a}_1, \mathbf{a}_2, \mathbf{a}_3\}$ is a basis of W. Thus we have

$$\dim(W) = 3. \|$$

Basic solutions The vectors constituting a basis of the solution space of a homogeneous linear equation are called *basic solutions* of the homogeneous linear equation.

Example 5 The vectors $\mathbf{a}_1, \mathbf{a}_2, \mathbf{a}_3$ given in the answer to Exercise 4.9 are basic solutions of the homogeneous linear equation

$$\begin{cases} x_1 & - 2x_2 & + x_3 & + 2x_4 & + 3x_5 & = 0, \\ 2x_1 & - 4x_2 & + 3x_3 & + 3x_4 & + 8x_5 & = 0. \end{cases}$$

Dimensions of solution spaces We shall calculate the dimensions of the solution spaces of homogeneous linear equations. For an $m \times n$ matrix A, we put

$$W = \{\mathbf{x} \in K^n \mid A\mathbf{x} = \mathbf{0}_m\}.$$

Let B be the reduced matrix of A. By definition, the dimension of W is the number of basic solutions. This number is equal to the number of variables which take arbitrary values. It is nothing but the number of variables which do not correspond the primary components of B. Therefore

$$\dim(W) = n - \text{rank}(A).$$

Thus we obtain the following theorem.

Theorem 4.4.3 *Let A be an $m \times n$ matrix. Then the dimension of the solution space*

$$W = \{\mathbf{x} \in K^n \mid A\mathbf{x} = \mathbf{0}_m\}$$

of the homogeneous linear equation $A\mathbf{x} = \mathbf{0}_m$ is given by

$$\dim(W) = n - \mathrm{rank}(A).$$

Subspaces generated by vectors For vectors $\mathbf{u}_1, \ldots, \mathbf{u}_r$ in a vector space V, we consider the subset

$$W = \{c_1\mathbf{u}_1 + \cdots + c_r\mathbf{u}_r \mid c_i \in K \quad (1 \leq i \leq r)\}.$$

Then it is easy to see that W is a subspace of V. We call W the *subspace generated by vectors* $\mathbf{u}_1, \ldots, \mathbf{u}_r$. The subspace W is denoted by $< \mathbf{u}_1, \ldots, \mathbf{u}_r >$. Namely,

$$< \mathbf{u}_1, \ldots, \mathbf{u}_r > = \{c_1\mathbf{u}_1 + \cdots + c_r\mathbf{u}_r \mid c_i \in K \quad \text{for} \quad i = 1, \ldots, r\}.$$

The vectors $\mathbf{u}_1, \ldots, \mathbf{u}_r$ are called the *generators* of $< \mathbf{u}_1, \ldots, \mathbf{u}_r >$.

We obtain the following theorem by Theorem 4.4.2.

Theorem 4.4.4 *Let* $\mathbf{u}_1, \ldots, \mathbf{u}_r$ *be vectors of a vector space* V. *Then we have*

$$\dim(< \mathbf{u}_1, \ldots, \mathbf{u}_r >) = \mathrm{rank}(\{\mathbf{u}_1, \ldots, \mathbf{u}_r\}).$$

As for bases of vector spaces, we obtain the following theorem.

Theorem 4.4.5 *Let* $\dim(V) = n$. *For* n *vectors* $\mathbf{u}_1, \ldots, \mathbf{u}_n$ *in* V, *the following three conditions are equivalent:*

(1) The set $\{\mathbf{u}_1, \ldots, \mathbf{u}_n\}$ *is a basis of* V.
(2) The vectors $\mathbf{u}_1, \ldots, \mathbf{u}_n$ *are linearly independent.*
(3) The vectors $\mathbf{u}_1, \ldots, \mathbf{u}_n$ *generate* V.

Proof $(1) \Rightarrow (2)$ and $(1) \Rightarrow (3)$ are trivial by the definition of the basis.

$(2) \Rightarrow (3)$:Take any vector $\mathbf{u} \in V$. Since $\dim(V) = n$, the $n + 1$ vectors $\mathbf{u}, \mathbf{u}_1, \ldots, \mathbf{u}_n$ must be linearly dependent by Theorem 4.4.2. Since the vectors $\mathbf{u}_1, \ldots, \mathbf{u}_n$ are linearly independent, $\mathbf{u}$ must be expressed as a linear combination of $\mathbf{u}_1, \ldots, \mathbf{u}_n$ by Theorem 4.2.3. Therefore the vectors $\mathbf{u}_1, \ldots, \mathbf{u}_n$ generate V.

$(3) \Rightarrow (1)$:Assume that the vectors $\mathbf{u}_1, \ldots, \mathbf{u}_n$ are not linearly independent. Let $r = \mathrm{rank}(\{\mathbf{u}_1, \ldots, \mathbf{u}_n\})$. Then $r < n$ by our assumption. Take r linearly independent vectors from $\{\mathbf{u}_1, \ldots, \mathbf{u}_n\}$. We may assume that $\mathbf{u}_1, \ldots, \mathbf{u}_r$ are linearly independent by changing the indices. Then the r vectors $\mathbf{u}_1, \ldots, \mathbf{u}_r$ generate V by Theorem 4.2.3, i.e., $V = < \mathbf{u}_1, \ldots, \mathbf{u}_r >$. Since they are linearly independent, the set of vectors $\{\mathbf{u}_1, \ldots, \mathbf{u}_r\}$ is a basis of V and $\dim(V) = r < n$, which contradicts the assumption $\dim(V) = n$. $\parallel$

Example 6 Three vectors $\begin{pmatrix} 1 \\ 1 \\ 1 \end{pmatrix}$, $\begin{pmatrix} 0 \\ 1 \\ 1 \end{pmatrix}$, $\begin{pmatrix} 0 \\ 0 \\ 1 \end{pmatrix}$ in $\mathbf{R}^3$ are linearly independent.

Since $\dim(\mathbf{R}^3) = 3$, the set of vectors $\left\{ \begin{pmatrix} 1 \\ 1 \\ 1 \end{pmatrix}, \begin{pmatrix} 0 \\ 1 \\ 1 \end{pmatrix}, \begin{pmatrix} 0 \\ 0 \\ 1 \end{pmatrix} \right\}$ is a basis of $\mathbf{R}^3$ by

Theorem 4.4.5.

Theorem 4.4.6 *Let* $\{\mathbf{u}_1, \ldots, \mathbf{u}_n\}$ *be a basis of a vector space* V *of dimension* n. *Assume that vectors* $\mathbf{v}_1, \ldots, \mathbf{v}_n$ *in* V *are expressed as linear combinations of* $\mathbf{u}_1, \ldots, \mathbf{u}_n$ *by a square matrix* A *of degree* n *so that*

$$\begin{pmatrix} \mathbf{v}_1 \cdots \mathbf{v}_n \end{pmatrix} = \begin{pmatrix} \mathbf{u}_1 \cdots \mathbf{u}_n \end{pmatrix} A.$$

Then $\{\mathbf{v}_1, \ldots, \mathbf{v}_n\}$ *is a basis of* V *if and only if the matrix* A *is a regular matrix.*

Proof If A is a regular matrix, then we have

$$\begin{pmatrix} \mathbf{u}_1 \cdots \mathbf{u}_n \end{pmatrix} = \begin{pmatrix} \mathbf{v}_1 \cdots \mathbf{v}_n \end{pmatrix} A^{-1}.$$

Since V is generated by $\mathbf{u}_1, \ldots, \mathbf{u}_n$, V is also generated by $\mathbf{v}_1, \ldots, \mathbf{v}_n$. Therefore $\{\mathbf{v}_1, \ldots, \mathbf{v}_n\}$ is a basis of V by Theorem 4.4.5.

Conversely, if $\{\mathbf{v}_1, \ldots, \mathbf{v}_n\}$ is a basis of V, then there exists a square matrix B satisfying $\begin{pmatrix} \mathbf{u}_1 \cdots \mathbf{u}_n \end{pmatrix} = \begin{pmatrix} \mathbf{v}_1 \cdots \mathbf{v}_n \end{pmatrix} B$. Therefore we see that

$$\begin{aligned} \begin{pmatrix} \mathbf{u}_1 \cdots \mathbf{u}_n \end{pmatrix} AB &= \begin{pmatrix} \mathbf{v}_1 \cdots \mathbf{v}_n \end{pmatrix} B \\ &= \begin{pmatrix} \mathbf{u}_1 \cdots \mathbf{u}_n \end{pmatrix} = \begin{pmatrix} \mathbf{u}_1 \cdots \mathbf{u}_n \end{pmatrix} E_n. \end{aligned}$$

Since $\mathbf{u}_1, \ldots, \mathbf{u}_n$ are linearly independent, we have $AB = E_n$ by Theorem 4.2.5 (2). Then A is a regular matrix by Theorem 2.4.1. $\|$

Example 7 The set of vectors

$$\{f_1(x) = 1 + x - x^2, \ f_2(x) = 1 + 2x^2, \ f_3(x) = -1 - x + 3x^2\}$$

is a basis of $\mathbf{R}[x]_2$. In fact, since $\dim(\mathbf{R}[x]_2) = 3$ and

$$\begin{pmatrix} f_1(x) \ f_2(x) \ f_3(x) \end{pmatrix} = \begin{pmatrix} 1 & x & x^2 \end{pmatrix} \begin{pmatrix} 1 & 1 & -1 \\ 1 & 0 & -1 \\ -1 & 2 & 3 \end{pmatrix},$$

we have only to see that the matrix $A = \begin{pmatrix} 1 & 1 & -1 \\ 1 & 0 & -1 \\ -1 & 2 & 3 \end{pmatrix}$ is a regular matrix. Since

$|A| = -2 \neq 0$, A is a regular matrix and $\{f_1(x), \ f_2(x), \ f_3(x)\}$ is a basis of $\mathbf{R}[x]_2$.

Exercises (Sect. 4.4)

1 Show that the following sets $\{a_1, \ldots, a_n\}$ of vectors are bases of V:

(1) $\left\{ a_1 = \begin{pmatrix} -3 \\ 1 \\ -2 \end{pmatrix}, \quad a_2 = \begin{pmatrix} 1 \\ -2 \\ 1 \end{pmatrix}, \quad a_3 = \begin{pmatrix} 1 \\ -1 \\ 2 \end{pmatrix} \right\}, V = \mathbf{R}^3.$

(2) $\left\{ a_1 = \begin{pmatrix} 2 \\ -1 \\ 1 \end{pmatrix}, \quad a_2 = \begin{pmatrix} -1 \\ 0 \\ -1 \end{pmatrix}, \quad a_3 = \begin{pmatrix} 1 \\ 3 \\ 1 \end{pmatrix} \right\}, V = \mathbf{R}^3.$

(3) $\left\{ a_1 = \begin{pmatrix} 1 \\ -1 \\ 2 \\ 0 \end{pmatrix}, \quad a_2 = \begin{pmatrix} 0 \\ 1 \\ -1 \\ 2 \end{pmatrix}, \quad a_3 = \begin{pmatrix} 2 \\ -1 \\ 1 \\ 1 \end{pmatrix}, \quad a_4 = \begin{pmatrix} 1 \\ -1 \\ 2 \\ -1 \end{pmatrix} \right\}, V = \mathbf{R}^4.$

2 Show that the following set $\{ f_1(x), f_2(x), f_3(x) \}$ of vectors are bases of $\mathbf{R}[x]_2$:

(1) $\{ f_1(x) = 1 - x + x^2, \ f_2(x) = -1 + 2x + 2x^2, \ f_3(x) = 1 - 2x - x^2 \}$.
(2) $\{ f_1(x) = 1 + x + x^2, \ f_2(x) = x + 2x^2, \ f_3(x) = -2 - x^2 \}$.
(3) $\{ f_1(x) = 3 + x - x^2, \ f_2(x) = -1 - x + 2x^2, \ f_3(x) = 1 + 3x^2 \}$.

3 Find the dimensions and bases of the following vector spaces W:

(1) $W = \left\{ x \in \mathbf{R}^5 \ \middle| \ \begin{pmatrix} 1 & 1 & 1 & 1 & 1 \\ 1 & -1 & 1 & 0 & 2 \\ 2 & 1 & 2 & -1 & 5 \end{pmatrix} x = 0 \right\}.$

(2) $W = \left\{ x \in \mathbf{R}^5 \ \middle| \ \begin{pmatrix} 2 & 0 & -1 & 3 & 4 \\ 1 & 2 & 3 & 1 & -5 \\ 3 & 1 & 4 & -7 & 10 \end{pmatrix} x = 0 \right\}.$

(3) $W = \left\{ x \in \mathbf{R}^3 \ \middle| \ \begin{array}{l} x_1 + 2x_2 - x_3 = 0, \\ 3x_1 - 3x_2 + 2x_3 = 0 \end{array} \right\}.$

(4) $W = \left\{ x \in \mathbf{R}^4 \ \middle| \ \begin{array}{l} x_1 + x_2 - x_3 + x_4 = 0, \\ 3x_1 + x_2 + 2x_3 - x_4 = 0 \end{array} \right\}.$

(5) $W = \{ f(x) \in \mathbf{R}[x]_3 | f(1) = f(-1) = 0 \}.$

(6) $W = \{ f(x) \in \mathbf{R}[x]_3 | f(1) = f'(1) = 0 \}.$

4 Let $\{u_1, u_2, u_3\}$ be a basis of a vector space V. Find if the following sets $\{v_1, v_2, v_3\}$ of vectors in V are bases of V or not:

(1) $v_1 = 2u_1 + u_2 - u_3, v_2 = u_1 + 2u_2 + u_3, v_3 = u_1 + u_2 + u_3.$
(2) $v_1 = u_1 - u_2 + u_3, v_2 = -u_1 + 3u_2 - u_3, v_3 = u_1 + u_3.$
(3) $v_1 = 2u_1 + u_2 - u_3, v_2 = u_1 + 2u_2 - u_3, v_3 = -u_1 + u_2 - u_3.$

5 Let V be a vector space of dimension n. Show that if the vectors $v_1, \ldots, v_r (0 < r < n)$ in V are linearly independent, then there exists a basis of V containing $v_1, \ldots, v_r$.

6 Find a basis of $V = \mathbf{R}^n$ containing the following linearly independent vectors a_1 and a_2:

(1) $V = \mathbf{R}^3, \mathbf{a}_1 = \begin{pmatrix} 1 \\ 2 \\ 1 \end{pmatrix}, \quad \mathbf{a}_2 = \begin{pmatrix} 0 \\ 2 \\ 1 \end{pmatrix}.$

(2) $V = \mathbf{R}^4, \mathbf{a}_1 = \begin{pmatrix} 1 \\ 1 \\ 2 \\ -1 \end{pmatrix}, \quad \mathbf{a}_2 = \begin{pmatrix} 0 \\ 2 \\ 4 \\ -2 \end{pmatrix}.$

7 Let W be a subspace of V. Show that if $\dim(W) = \dim(V)$, then $W = V$.

Chapter 5
Linear Mappings

5.1 Linear Mappings and Isomorphisms

One of the simplest functions of real numbers is a proportional function $f(x) = ax$ $(a, x \in \mathbf{R})$. Linear mappings can be considered as a generalization of proportional functions to higher-dimensional spaces.

Linear mappings Let U and V be vector spaces over K. A mapping T of U into V is called a *linear mapping* (over K) when T satisfies the following conditions:

(1) $T(\mathbf{u} + \mathbf{v}) = T(\mathbf{u}) + T(\mathbf{v})$ $(\mathbf{u}, \mathbf{v} \in U)$,
(2) $T(c\mathbf{u}) = cT(\mathbf{u})$ $(\mathbf{u} \in U, c \in K)$.

When T is a linear mapping of U into V, we denote it by

$$T : U \longrightarrow V.$$

Any linear mapping T of U into V maps the zero vector $\mathbf{0}_U$ of U to the zero vector $\mathbf{0}_V$ of V. In fact, we see that

$$T(\mathbf{0}_U) = T(0 \cdot \mathbf{0}_U) = 0 \cdot T(\mathbf{0}_U) = \mathbf{0}_V.$$

The linear mapping of U into V which maps all vectors in U to the zero vector $\mathbf{0}_V$ of V is called the *zero mapping*. We denote the zero mapping of U into V by $O_{U,V}$ or simply by O.

Example 1 Let A be an $m \times n$ matrix and T_A the mapping of K^n into K^m defined by

$$T_A(\mathbf{x}) = A\mathbf{x} \quad (\mathbf{x} \in K^n).$$

Then T_A is a linear mapping of K^n into K^m. In fact, T_A satisfies the two conditions of the definition of linear mapping:

© The Author(s), under exclusive license to Springer Nature Singapore Pte Ltd. 2022   139
T. Miyake, *Linear Algebra*, https://doi.org/10.1007/978-981-16-6994-1_5

$$T_A(\mathbf{x} + \mathbf{y}) = A(\mathbf{x} + \mathbf{y}) = A\mathbf{x} + A\mathbf{y}$$
$$= T_A(\mathbf{x}) + T_A(\mathbf{y}) \qquad (\mathbf{x}, \mathbf{y} \in K^n),$$
$$T_A(c\mathbf{x}) = A(c\mathbf{x}) = cA\mathbf{x} = cT_A(\mathbf{x}) \qquad (\mathbf{x} \in K^n, \quad c \in K).$$

Linear mappings associated with matrices We call the linear mapping T_A in Example 1 the *linear mapping associated with a matrix A*.

Images and kernels of linear mappings Let T be a linear mapping of a vector space U into a vector space V. The subset of V defined by

$$\mathrm{Im}\,(T) = \{\, T(\mathbf{u}) \mid \mathbf{u} \in U \,\}$$

is called the *image* of T. The image of T is also expressed as $T(U)$. Further, the subset of U defined by

$$\mathrm{Ker}\,(T) = \{\, \mathbf{u} \in U \mid T(\mathbf{u}) = \mathbf{0}_V \,\}$$

is called the *kernel* of T.

Theorem 5.1.1 *Let U, V be vector spaces and T a linear mapping of U into V. Then we have the following assertions:*

(1) The image $\mathrm{Im}\,(T)$ of T is a subspace of V.
(2) The kernel $\mathrm{Ker}\,(T)$ of T is a subspace of U.

Proof (1) We note $\mathbf{0}_V = T(\mathbf{0}_U) \in \mathrm{Im}(T)$. Let $\mathbf{v}_1, \mathbf{v}_2 \in \mathrm{Im}\,(T)$. Then there exist $\mathbf{u}_1, \mathbf{u}_2 \in U$ so that $T(\mathbf{u}_1) = \mathbf{v}_1$ and $T(\mathbf{u}_2) = \mathbf{v}_2$. Then we see that

$$\mathbf{v}_1 + \mathbf{v}_2 = T(\mathbf{u}_1) + T(\mathbf{u}_2) = T(\mathbf{u}_1 + \mathbf{u}_2) \in \mathrm{Im}\,(T),$$

and

$$c\mathbf{v}_1 = cT(\mathbf{u}_1) = T(c\mathbf{u}_1) \in \mathrm{Im}\,(T)$$

for a constant c. Therefore by Theorem 4.1.1, $\mathrm{Im}\,(T)$ is a subspace of V.
(2) Since $T(\mathbf{0}_U) = \mathbf{0}_V$, we see that $\mathbf{0}_U \in \mathrm{Ker}\,(T)$. Let $\mathbf{u}_1, \mathbf{u}_2 \in \mathrm{Ker}\,(T)$. We see that

$$T(\mathbf{u}_1 + \mathbf{u}_2) = T(\mathbf{u}_1) + T(\mathbf{u}_2) = \mathbf{0}_V + \mathbf{0}_V = \mathbf{0}_V.$$

Then we have $\mathbf{u}_1 + \mathbf{u}_2 \in \mathrm{Ker}\,(T)$. We also see that

$$T(c\mathbf{u}_1) = cT(\mathbf{u}_1) = c\mathbf{0}_V = \mathbf{0}_V.$$

Thus we have $c\mathbf{u}_1 \in \mathrm{Ker}\,(T)$. Therefore by Theorem 4.1.1, $\mathrm{Ker}\,(T)$ is a subspace of U. $\parallel$

From now on, we consider only finite-dimensional vector spaces.

Ranks and nullities of linear mappings For a linear mapping T of U into V, we define the *rank* and the *nullity* of T by

$$\text{rank}\,(T) = \dim\,(\text{Im}(T)) \quad \text{and} \quad \text{null}\,(T) = \dim\,(\text{Ker}(T)),$$

respectively.

Kernels and images of matrices For an $m \times n$ matrix A, we define Ker (A) by the solution space of the homogeneous linear equation $A\mathbf{x} = \mathbf{0}_m$, namely,

$$\text{Ker}\,(A) = \{\mathbf{x} \in K^n \mid A\mathbf{x} = \mathbf{0}_m\},$$

and call Ker (A) the *kernel* of A. We let

$$\text{null}\,(A) = \dim\,(\text{Ker}\,(A))$$

and call it the *nullity* of A. Let T_A be a linear mapping of K^n into K^m associated with the matrix A. Then we see that

$$\text{Ker}\,(A) = \text{Ker}\,(T_A) \quad \text{and} \quad \text{null}\,(A) = \text{null}\,(T_A).$$

We let $A = \begin{pmatrix} \mathbf{a}_1 \ldots \mathbf{a}_n \end{pmatrix}$ be the column partition of A. We define the subspace Im (A) of K^m by the subspace generated by the column vectors $\mathbf{a}_1, \ldots, \mathbf{a}_n$, namely,

$$\text{Im}\,(A) = <\mathbf{a}_1, \ldots, \mathbf{a}_n>$$
$$= \{c_1\mathbf{a}_1 + \cdots + c_n\mathbf{a}_n \mid c_1, \ldots, c_n \in K\}$$

and call Im (A) the *image* of A. By the definition of rank (A), Theorems 4.3.4 and 4.4.4, we see that

$$\dim\,(\text{Im}\,(A)) = \text{rank}\,(A).$$

Furthermore, we see that $T_A(\mathbf{e}_1) = \mathbf{a}_1, \ldots, T_A(\mathbf{e}_n) = \mathbf{a}_n$. Here $\{\mathbf{e}_1, \ldots, \mathbf{e}_n\}$ is the standard basis of K^n. Then we see that

$$\text{Im}\,(A) = \text{Im}\,(T_A) = \{A\mathbf{x} \mid \mathbf{x} \in K^n\}.$$

Therefore we have

$$\text{rank}\,(T_A) = \dim\,(\text{Im}\,(T_A)) = \dim\,(\text{Im}\,(A)) = \text{rank}\,(A).$$

Since rank $(A) + \text{null}\,(A) = n$ by Theorem 4.4.3, we have

$$\text{rank}\,(T_A) + \text{null}\,(T_A) = n.$$

Example 2 Let

$$A = \begin{pmatrix} 1 & -1 & 0 & 1 \\ -2 & 2 & 1 & -1 \end{pmatrix}.$$

We solve the linear equation $Ax = 0_2$. Then by the calculation

$$
\begin{array}{rrrr}
1 & -1 & 0 & 1 \\
-2 & 2 & 1 & -1 \\
\hline
1 & -1 & 0 & 1 \\
0 & 0 & 1 & 1 \\
\end{array}
\quad ②+①\times 2,
$$

we have

$$\mathrm{Ker}\,(A) = \left\{ a \begin{pmatrix} 1 \\ 1 \\ 0 \\ 0 \end{pmatrix} + b \begin{pmatrix} -1 \\ 0 \\ -1 \\ 1 \end{pmatrix} \;\middle|\; a, b \in \mathbf{R} \right\}$$

and $\mathrm{null}\,(A) = \dim\,(\mathrm{Ker}\,(A)) = 2$. Let $A = \begin{pmatrix} \mathbf{a}_1 & \mathbf{a}_2 & \mathbf{a}_3 & \mathbf{a}_4 \end{pmatrix}$ be the column partition of A. Then we have

$$\mathrm{Im}\,(A) = \{ c_1 \mathbf{a}_1 + c_2 \mathbf{a}_2 + c_3 \mathbf{a}_3 + c_4 \mathbf{a}_4 | c_1, c_2, c_3, c_4 \in \mathbf{R} \}.$$

Looking at relations of column vectors of the reduced matrix $B = \begin{pmatrix} 1 & -1 & 0 & 1 \\ 0 & 0 & 1 & 1 \end{pmatrix}$ of A, we see that $\mathbf{a}_1$ and $\mathbf{a}_3$ are linearly independent and $\mathbf{a}_2 = -\mathbf{a}_1$, $\mathbf{a}_4 = \mathbf{a}_1 + \mathbf{a}_3$. Thus we have

$$\mathrm{Im}\,(A) = \{ c_1 \mathbf{a}_1 + c_3 \mathbf{a}_3 | c_1, c_3 \in \mathbf{R} \}.$$

Therefore

$$\mathrm{rank}\,(A) = \dim\,(\mathrm{Im}\,(A)) = 2$$

and

$$\mathrm{rank}\,(A) + \mathrm{null}\,(A) = 4.$$

The following theorem is the generalization of the above-mentioned result to linear mappings of vector spaces.

Theorem 5.1.2 *Let U, V are vector spaces and T a linear mapping of U into V. Then we have*

$$\mathrm{rank}\,(T) + \mathrm{null}\,(T) = \dim\,(U).$$

Proof We let $r = \text{null}\,(T)$ and $\{\mathbf{u}_1, \ldots, \mathbf{u}_r\}$ a basis of Ker (T). We also let $s = \text{rank}\,(T)$ and $\{\mathbf{v}_1, \ldots, \mathbf{v}_s\}$ a basis of Im (T). Take s vectors $\mathbf{u}_{r+1}, \ldots, \mathbf{u}_{r+s}$ in U so that

$$T(\mathbf{u}_{r+1}) = \mathbf{v}_1, \quad T(\mathbf{u}_{r+2}) = \mathbf{v}_2, \quad \ldots, \quad T(\mathbf{u}_{r+s}) = \mathbf{v}_s.$$

If we can show that the set of vectors

$$\{\mathbf{u}_1, \quad \ldots, \quad \mathbf{u}_r, \quad \mathbf{u}_{r+1}, \quad \ldots, \quad \mathbf{u}_{r+s}\}$$

is a basis of U, then the theorem is proved.

(1) We shall show that those vectors generate U. Let $\mathbf{u}$ be a vector in U. Since $T(\mathbf{u}) \in \text{Im}\,(T)$, we have $T(\mathbf{u}) = b_1\mathbf{v}_1 + \cdots + b_s\mathbf{v}_s$ $(b_j \in K)$ and

$$\begin{aligned}
&T(\mathbf{u} - b_1\mathbf{u}_{r+1} - \cdots - b_s\mathbf{u}_{r+s}) \\
&= b_1\mathbf{v}_1 + \cdots + b_s\mathbf{v}_s - (b_1\mathbf{v}_1 + \cdots + b_s\mathbf{v}_s) \\
&= \mathbf{0}_V.
\end{aligned}$$

Thus we see that $\mathbf{u} - b_1\mathbf{u}_{r+1} - \cdots - b_s\mathbf{u}_{r+s} \in \text{Ker}\,(T)$. Since $\{\mathbf{u}_1, \ldots, \mathbf{u}_r\}$ is a basis of Ker (T), we can express

$$\mathbf{u} - b_1\mathbf{u}_{r+1} - \cdots - b_s\mathbf{u}_{r+s} = a_1\mathbf{u}_1 + \cdots + a_r\mathbf{u}_r \quad (a_i \in K).$$

Therefore $\mathbf{u} = a_1\mathbf{u}_1 + \cdots + a_r\mathbf{u}_r + b_1\mathbf{u}_{r+1} + \cdots + b_s\mathbf{u}_{r+s}$.

(2) We shall show that vectors

$$\mathbf{u}_1, \quad \ldots, \quad \mathbf{u}_r, \quad \mathbf{u}_{r+1}, \quad \ldots, \quad \mathbf{u}_{r+s}$$

are linearly independent. Assume

$$a_1\mathbf{u}_1 + \cdots + a_r\mathbf{u}_r + b_1\mathbf{u}_{r+1} + \cdots + b_s\mathbf{u}_{r+s} = \mathbf{0}_U \quad (a_i, b_j \in K).$$

We apply T to both sides. Since $T(\mathbf{u}_i) = \mathbf{0}_V$ for $i = 1, \ldots, r$, we have

$$\begin{aligned}
\mathbf{0}_V = T(\mathbf{0}_U) &= T(a_1\mathbf{u}_1 + \cdots + a_r\mathbf{u}_r + b_1\mathbf{u}_{r+1} + \cdots + b_s\mathbf{u}_{r+s}) \\
&= \mathbf{0}_V + b_1\mathbf{v}_1 + \cdots + b_s\mathbf{v}_s \\
&= b_1\mathbf{v}_1 + \cdots + b_s\mathbf{v}_s.
\end{aligned}$$

Since $\mathbf{v}_1, \ldots, \mathbf{v}_s$ are linearly independent, we see that

$$b_1 = \cdots = b_s = 0.$$

Then we have

$$a_1\mathbf{u}_1 + \cdots + a_r\mathbf{u}_r = \mathbf{0}_U.$$

As $\mathbf{u}_1, \ldots, \mathbf{u}_r$ are linearly independent, we see that $a_1 = \cdots = a_r = 0$. Therefore the vectors $\mathbf{u}_1, \quad \ldots, \quad \mathbf{u}_r, \quad \mathbf{u}_{r+1}, \quad \ldots, \quad \mathbf{u}_{r+s}$ are linearly independent.   ‖

Exercise 5.1 Let $A = \begin{pmatrix} 1 & 0 & 1 & 2 & 1 \\ 2 & -1 & 1 & 5 & 0 \\ 1 & 3 & 4 & -1 & 7 \end{pmatrix}$.

(1) Find null (A) and a basis of Ker (A).
(2) Find rank (A) and a basis of Im(A).

Answers. (1) The kernel of A is the solution space of $A\mathbf{x} = \mathbf{0}_3$. By the calculation

$\mathbf{a}_1$	$\mathbf{a}_2$	$\mathbf{a}_3$	$\mathbf{a}_4$	$\mathbf{a}_5$	
1	0	1	2	1	
2	−1	1	5	0	
1	3	4	−1	7	
1	0	1	2	1	
0	−1	−1	1	−2	②+①×(−2)
0	3	3	−3	6	③+①×(−1)
1	0	1	2	1	
0	−1	−1	1	−2	
0	0	0	0	0	③+②×3
1	0	1	2	1	
0	1	1	−1	2	②×(−1),
0	0	0	0	0	
$\mathbf{b}_1$	$\mathbf{b}_2$	$\mathbf{b}_3$	$\mathbf{b}_4$	$\mathbf{b}_5$	

the reduced matrix of the matrix A is

$$B = \begin{pmatrix} 1 & 0 & 1 & 2 & 1 \\ 0 & 1 & 1 & -1 & 2 \\ 0 & 0 & 0 & 0 & 0 \end{pmatrix}.$$

Solving $A\mathbf{x} = \mathbf{0}_3$, we obtain

$$\left\{ \begin{pmatrix} -1 \\ -1 \\ 1 \\ 0 \\ 0 \end{pmatrix}, \begin{pmatrix} -2 \\ 1 \\ 0 \\ 1 \\ 0 \end{pmatrix}, \begin{pmatrix} -1 \\ -2 \\ 0 \\ 0 \\ 1 \end{pmatrix} \right\}$$

as a basis of Ker(A). Therefore null $(A) = \dim (\mathrm{Ker}\,(A)) = 3$.

(2) Let $A = \begin{pmatrix} \mathbf{a}_1 & \ldots & \mathbf{a}_5 \end{pmatrix}$ and $B = \begin{pmatrix} \mathbf{b}_1 & \ldots & \mathbf{b}_5 \end{pmatrix}$ be the column partitions of the matrices A and B, respectively. Since Im (A) is generated by $\mathbf{a}_1, \ldots, \mathbf{a}_5$, we shall find the linearly independent vectors among $\mathbf{a}_1, \ldots, \mathbf{a}_5$. By the calculation above, we see that the column vectors $\mathbf{b}_1$ and $\mathbf{b}_2$ are linearly independent and $\mathbf{b}_3, \mathbf{b}_4, \mathbf{b}_5$ are

linear combinations of them. Then $\mathbf{a}_1$ and $\mathbf{a}_2$ are linearly independent, and $\mathbf{a}_3, \mathbf{a}_4, \mathbf{a}_5$ are linear combinations of $\mathbf{a}_1$ and $\mathbf{a}_2$. Therefore, we have

$$\mathrm{rank}\,(A) = \dim\,(\mathrm{Im}\,(A)) = 2$$

and

$$\left\{ \mathbf{a}_1 = \begin{pmatrix} 1 \\ 2 \\ 1 \end{pmatrix}, \quad \mathbf{a}_2 = \begin{pmatrix} 0 \\ -1 \\ 3 \end{pmatrix} \right\} \text{ is a basis of Im } (A). \;\|$$

Addition, scalar multiplication and product of linear mappings We define the addition, scalar multiplication and product of linear mappings as follows:

(Addition) For linear mappings T_1 and T_2 of U into V, we define the linear mapping $T_1 + T_2$ of U into V by

$$(T_1 + T_2)(\mathbf{u}) = T_1(\mathbf{u}) + T_2(\mathbf{u}) \quad (\mathbf{u} \in U).$$

(Scalar multiplication) For a linear mapping T of U into V and $c \in K$, we define the linear mapping cT of U into V by

$$(cT)(\mathbf{u}) = c\,T(\mathbf{u}) \quad (\mathbf{u} \in U,\, c \in K).$$

(Product) Let T be a linear mapping of a vector space U into a vector space V and S a linear mapping of V into a vector space W. We define the mapping ST of V into W by

$$ST(\mathbf{u}) = S(T(\mathbf{u})) \quad (\mathbf{u} \in U).$$

It is easy to see that ST is a linear mapping of U into W. We call ST the *product of linear mappings* S and T.

Linear transformations Linear mappings of a vector space U into U itself are called *linear transformations* of a vector space U. The linear transformation of U which does not change any vector is called the *identity transformation* of U and denoted by I_U, and the zero mapping of U into U is called the *zero transformation* of U and denoted by O_U.

Isomorphisms and automorphisms Let U and V be vector spaces and T a linear mapping of U into V. We denote by I_U and I_V the identity transformations of U and V, respectively. If there exists a linear mapping S of V into U satisfying

$$ST = I_U \quad \text{and} \quad TS = I_V,$$

then the linear mapping T is called an *isomorphism* of U into V and we write

$$T : U \cong V.$$

The linear mapping S is called the *inverse linear mapping* or the *inverse isomorphism* of T denoted by T^{-1}. When there exists an isomorphism T of U into V, the vector spaces U and V are called *isomorphic*. If a linear transformation T of U is an isomorphism, we call T an *isomorphic transformation* of U or an *automorphism* of U and T^{-1} the *inverse transformation* or the *inverse automorphism* of T.

One-to-one mappings For two sets X and Y, a mapping T of X into Y is called a *one-to-one mapping* when the following holds:

$$\text{if } T(a) = T(b), \quad \text{then } \quad a = b \text{ for any elements } a, b \in X.$$

This is equivalent to that if $a \neq b$, then $T(a) \neq T(b)$.

Theorem 5.1.3 *Let T be a linear mapping of U into V. Then conditions (1) and (2) are equivalent.*

(1) The linear mapping T is a one-to-one mapping.
(2) $\operatorname{Ker}(T) = \{0_U\}$.

Proof $(1) \Rightarrow (2)$ Since $T(0_U) = 0_V$ and T is a one-to-one linear mapping, we have $\operatorname{Ker}(T) = \{0_U\}$.

$(2) \Rightarrow (1)$ Assume that $T(\mathbf{u}_1) = T(\mathbf{u}_2)$ $(\mathbf{u}_1, \mathbf{u}_2 \in U)$. Then by the linearity of T, we have $T(\mathbf{u}_1 - \mathbf{u}_2) = 0_V$. Thus we obtain that $\mathbf{u}_1 - \mathbf{u}_2 \in \operatorname{Ker}(T) = \{0_U\}$. Therefore we have

$$\mathbf{u}_1 - \mathbf{u}_2 = 0_U, \quad \text{namely,} \quad \mathbf{u}_1 = \mathbf{u}_2.$$

This implies that T is a one-to-one mapping. $\|$

Surjective mappings For a mapping T of a set X into a set Y, we let

$$T(X) = \operatorname{Im}(T) = \{T(a) \mid a \in X\} \subset Y.$$

We call T a *surjective mapping* if it satisfies $T(X) = Y$.

Bijective mappings A mapping T of a set X into a set Y is called a *bijective mapping* if T is a one-to-one and surjective mapping.

Theorem 5.1.4 *Assume that U and V are vector spaces.*

(1) A linear mapping T of U into V is an isomorphism if and only if T is a bijective mapping.
(2) If U and V are isomorphic, then $\dim(U) = \dim(V)$.
(3) Assume that $\dim(U) = \dim(V)$. Then a linear mapping T of U into V is an isomorphism if and only if T is either a one-to-one mapping or a surjective mapping.

Proof (1) Let T be an isomorphism of U into V. We assume that for vectors $\mathbf{u}_1$ and $\mathbf{u}_2$ in U, $T(\mathbf{u}_1) = T(\mathbf{u}_2)$ holds. Then we have

$$\mathbf{u}_1 = T^{-1}(T(\mathbf{u}_1)) = T^{-1}(T(\mathbf{u}_2)) = \mathbf{u}_2.$$

Therefore T is a one-to-one mapping. Furthermore, for any vector $\mathbf{v} \in V$, we see that $\mathbf{v} = T(T^{-1}(\mathbf{v}))$. Then T is a surjective mapping.

Conversely, assume that T is a one-to-one and surjective linear mapping of U into V. Let $\mathbf{v} \in V$. Since T is surjective, there exists a vector $\mathbf{u} \in U$ satisfying $T(\mathbf{u}) = \mathbf{v}$. Since T is a one-to-one mapping, $\mathbf{u}$ uniquely exists. If we define a mapping S of V into U by $S(\mathbf{v}) = \mathbf{u}$, then it is easy to see that S is the inverse linear mapping of T. Therefore T is an isomorphism.

(2) Let T be an isomorphism of U into V. Assume that $\mathbf{u}_1, \dots, \mathbf{u}_n$ are vectors in U. Then by applying T and T^{-1}, we have

$$c_1\mathbf{u}_1 + \cdots + c_n\mathbf{u}_n = \mathbf{0}_U \iff c_1 T(\mathbf{u}_1) + \cdots + c_n T(\mathbf{u}_n) = \mathbf{0}_V.$$

Therefore we have $\dim(U) = \dim(V)$.

(3) Since $\mathrm{rank}(T) + \mathrm{null}(T) = \dim(U)$ by Theorem 5.1.2, we have:

$$
\begin{aligned}
T \text{ is one-to-one} &\iff \mathrm{Ker}\,(T) = \{\mathbf{0}_U\} \text{ (Theorem 5.1.3)} \\
&\iff \mathrm{null}\,(T) = 0 \\
&\iff \dim\,(\mathrm{Im}\,(T)) = \mathrm{rank}\,(T) = \dim\,(V) \\
&\iff T \text{ is surjective.}
\end{aligned}
$$

Therefore, a linear mapping T is one-to-one if and only if T is surjective. Thus (3) is proved by (1). ‖

Theorem 5.1.5

(1) All vector spaces of dimension n are isomorphic each other.

(2) Let U be a vector space of dimension n and $\{\mathbf{u}_1, \dots, \mathbf{u}_n\}$ be a basis of U. Let $\mathbf{u} = a_1\mathbf{u}_1 + \cdots + a_n\mathbf{u}_n \in U$. Let S be the linear mapping of U into K^n defined by $S(\mathbf{u}) = \begin{pmatrix} a_1 \\ \vdots \\ a_n \end{pmatrix}$. Then S is an isomorphism of U into K^n. We call S the isomorphism of U into K^n with respect to the basis $\{\mathbf{u}_1, \dots, \mathbf{u}_n\}$.

Proof (1) Let U and V be vector spaces of dimension n. Take bases $\{\mathbf{u}_1, \dots, \mathbf{u}_n\}$ and $\{\mathbf{v}_1, \dots, \mathbf{v}_n\}$ of U and V, respectively. For $\mathbf{u} = a_1\mathbf{u}_1 + \cdots + a_n\mathbf{u}_n$, we define the linear mapping S of U into V by

$$S(\mathbf{u}) = a_1\mathbf{v}_1 + \cdots + a_n\mathbf{v}_n.$$

Since $\{\mathbf{v}_1, \dots, \mathbf{v}_n\}$ is a basis of V, the linear mapping S is a surjective mapping. Therefore S is an isomorphism by Theorem 5.1.4 (3).

(2) is the special case of (1) by taking the standard basis $\{e_1, \ldots, e_n\}$ of K^n.   ‖

Exercise 5.2 Let T be a linear mapping of $\mathbf{R}[x]_2$ into $\mathbf{R}^3$ defined by

$$T(f(x)) = \begin{pmatrix} f(0) \\ f(1) \\ f(-1) \end{pmatrix} \in \mathbf{R}^3.$$

Show that the linear mapping T is an isomorphism of $\mathbf{R}[x]_2$ into $\mathbf{R}^3$.
  Answer.   Let $f(x) = a_0 + a_1 x + a_2 x^2$. Then we have

$$T(f(x)) = \begin{pmatrix} a_0 \\ a_0 + a_1 + a_2 \\ a_0 - a_1 + a_2 \end{pmatrix}$$
$$= a_0 e_1 + (a_0 + a_1 + a_2)e_2 + (a_0 - a_1 + a_2)e_3.$$

Here $\{e_1, e_2, e_3\}$ is the standard basis of $\mathbf{R}^3$. We assume that $T(f(x)) = 0_3$. Since e_1, e_2, e_3 are linearly independent, we have a linear equation

$$\begin{pmatrix} 1 & 0 & 0 \\ 1 & 1 & 1 \\ 1 & -1 & 1 \end{pmatrix} \begin{pmatrix} a_0 \\ a_1 \\ a_2 \end{pmatrix} = \begin{pmatrix} 0 \\ 0 \\ 0 \end{pmatrix}.$$

Since the matrix $\begin{pmatrix} 1 & 0 & 0 \\ 1 & 1 & 1 \\ 1 & -1 & 1 \end{pmatrix}$ is a regular matrix, the linear equation has only the trivial
solution by Theorem 2.4.2. Therefore $f(x) = 0 = f_{00}(x)$ and the linear mapping T
is a one-to-one mapping. Since dim $(\mathbf{R}[x]_2) = $ dim $(\mathbf{R}^3) = 3$, T is an isomorphism
by Theorem 5.1.4 (3).   ‖

Exercises (Sect. 5.1)
1.   Find if the following mappings T are linear mappings or not:

(1) $T\left(\begin{pmatrix} x_1 \\ x_2 \\ x_3 \end{pmatrix}\right) = \begin{pmatrix} 3x_1 - x_2 + 2x_3 \\ x_1 + 3x_2 - x_3 \end{pmatrix}$: $\mathbf{R}^3 \to \mathbf{R}^2$.

(2) $T\left(\begin{pmatrix} x_1 \\ x_2 \end{pmatrix}\right) = \begin{pmatrix} x_1 + x_2 + 2 \\ 2x_1 + 3x_2 - 1 \end{pmatrix}$: $\mathbf{R}^2 \to \mathbf{R}^2$.

(3) $T(f(x)) = 2f'(x) + f(2x)$: $\mathbf{R}[x]_3 \to \mathbf{R}[x]_3$.

(4) $T(f(x)) = \begin{pmatrix} f(0) \\ f(1) \end{pmatrix}$: $\mathbf{R}[x]_3 \to \mathbf{R}^2$.

(5) $T(f(x)) = f'(x) + a$: $\mathbf{R}[x]_3 \to \mathbf{R}[x]_3$.

2.   Find (i) and (ii) for the following matrices A:

(i) null (A) and a basis of Ker (A).
(ii) rank (A) and a basis of Im (A).

(1) $A = \begin{pmatrix} 2 & 4 & 3 & 1 \\ 0 & 0 & 1 & 1 \\ 1 & 2 & 1 & 0 \end{pmatrix}$.

(2) $A = \begin{pmatrix} 1 & 1 & 0 & 1 \\ 0 & 2 & -4 & -6 \\ 2 & 4 & -4 & -4 \end{pmatrix}$.

(3) $A = \begin{pmatrix} 1 & 3 & 1 & 0 \\ 0 & 1 & 1 & 0 \\ -1 & -2 & 0 & 1 \end{pmatrix}$.

(4) $A = \begin{pmatrix} 1 & -2 & 1 & 0 \\ 1 & -2 & 1 & 0 \\ -2 & 4 & -2 & 0 \\ 1 & -1 & 2 & 1 \end{pmatrix}$.

(5) $A = \begin{pmatrix} 0 & 1 & 1 & 1 & 3 \\ -1 & -2 & -5 & -1 & -4 \\ 1 & 1 & 4 & 0 & 1 \\ 1 & -1 & 2 & -2 & -5 \end{pmatrix}$.

3. Show that any linear mapping T of K^n into K^m can be expressed as $T = T_A$ with a suitable $m \times n$ matrix A. Furthermore, show that the matrix A is uniquely determined.

4. Find if the following linear mappings are one-to-one mappings or not. Also find if they are surjective mappings or not:

(1) $T : \mathbf{R}^2 \to \mathbf{R}$, $\quad T\left(\begin{pmatrix} x_1 \\ x_2 \end{pmatrix}\right) = x_1 - 2x_2$.

(2) $T : \mathbf{R}^2 \to \mathbf{R}^3$, $\quad T\left(\begin{pmatrix} x_1 \\ x_2 \end{pmatrix}\right) = \begin{pmatrix} 2x_1 + x_2 \\ x_1 + x_2 \\ x_1 - 3x_2 \end{pmatrix}$.

5. Show that the following linear mappings T are isomorphisms:

(1) $T : U = \mathbf{R}^3 \to V = \mathbf{R}^3$, $\quad T\left(\begin{pmatrix} x_1 \\ x_2 \\ x_3 \end{pmatrix}\right) = \begin{pmatrix} -x_1 + x_2 - 2x_3 \\ 2x_1 + x_2 - x_3 \\ x_1 \quad\quad + x_3 \end{pmatrix}$.

(2) $T : U = \mathbf{R}[x]_2 \to V = \mathbf{R}^3$, $\quad T(f(x)) = \begin{pmatrix} f(0) + f(1) \\ f'(1) + f'(-1) \\ f''(0) \end{pmatrix}$.

6. Let A be an $l \times m$ matrix and B an $m \times n$ matrix. Show that

$$T_A T_B = T_{AB}.$$

7. Find the products ST of the following linear mappings S and T:

(1) $S = T_A : \mathbf{R}^3 \to \mathbf{R}^3$, $\quad A = \begin{pmatrix} 1 & 0 & -1 \\ -1 & 2 & 1 \\ 0 & -1 & 2 \end{pmatrix}$.

$T = T_B : \mathbf{R}^2 \to \mathbf{R}^3$, $\quad B = \begin{pmatrix} -3 & 2 \\ 1 & -1 \\ 0 & 3 \end{pmatrix}$.

(2) $S = T_A : \mathbf{R}^2 \to \mathbf{R}^2$, $\quad A = \begin{pmatrix} 2 & 1 \\ 3 & 1 \end{pmatrix}$.

$\quad T : \mathbf{R}[x]_2 \to \mathbf{R}^2$, $\quad T(f(x)) = \begin{pmatrix} f(1) \\ f(-1) \end{pmatrix}$.

(3) $S : \mathbf{R}[x]_2 \to \mathbf{R}^2$, $\quad S(f(x)) = \begin{pmatrix} f(0) \\ f(1) \end{pmatrix}$.

$\quad T : \mathbf{R}[x]_2 \to \mathbf{R}[x]_2$, $\quad T(f(x)) = f(x) - f'(x)$.

5.2   Matrix Representations of Linear Mappings

Matrix representation of linear mappings Let T be a linear mapping of a vector space U into a vector space V. We take a basis $\{\mathbf{u}_1, \ldots, \mathbf{u}_n\}$ of U and a basis $\{\mathbf{v}_1, \ldots, \mathbf{v}_m\}$ of V. Since the vectors $T(\mathbf{u}_1), \ldots, T(\mathbf{u}_n)$ are vectors in V, we can express them as linear combinations of $\mathbf{v}_1, \ldots, \mathbf{v}_m$. Therefore there exists an $m \times n$ matrix A satisfying

$$\left(T(\mathbf{u}_1) \ldots T(\mathbf{u}_n) \right) = \left(\mathbf{v}_1 \ldots \mathbf{v}_m \right) A.$$

We call this equality the *matrix representation* of the linear mapping T with respect to the bases $\{\mathbf{u}_i\}$ of U and $\{\mathbf{v}_j\}$ of V. The matrix A is called the *representation matrix* of T with respect to the bases $\{\mathbf{u}_i\}$ of U and $\{\mathbf{v}_j\}$ of V.

Example 1 Let T_A be a linear mapping of $\mathbf{R}^3$ into $\mathbf{R}^2$ associated with the matrix

$$A = \begin{pmatrix} 3 & -2 & 1 \\ -1 & 1 & 5 \end{pmatrix}.$$

For the standard basis $\left\{ \mathbf{e}_1 = \begin{pmatrix} 1 \\ 0 \\ 0 \end{pmatrix}, \mathbf{e}_2 = \begin{pmatrix} 0 \\ 1 \\ 0 \end{pmatrix}, \mathbf{e}_3 = \begin{pmatrix} 0 \\ 0 \\ 1 \end{pmatrix} \right\}$ of $\mathbf{R}^3$ and the stan-

dard basis $\left\{ \mathbf{e}_1' = \begin{pmatrix} 1 \\ 0 \end{pmatrix}, \mathbf{e}_2' = \begin{pmatrix} 0 \\ 1 \end{pmatrix} \right\}$ of $\mathbf{R}^2$, we see that

$$T_A(\mathbf{e}_1) = \begin{pmatrix} 3 \\ -1 \end{pmatrix} = 3\mathbf{e}_1' - \mathbf{e}_2', \quad T_A(\mathbf{e}_2) = \begin{pmatrix} -2 \\ 1 \end{pmatrix} = -2\mathbf{e}_1' + \mathbf{e}_2',$$

$$T_A(\mathbf{e}_3) = \begin{pmatrix} 1 \\ 5 \end{pmatrix} = \mathbf{e}_1' + 5\mathbf{e}_2'.$$

Then we have

$$\left(T_A(\mathbf{e}_1) \ T_A(\mathbf{e}_2) \ T_A(\mathbf{e}_3) \right) = \left(\mathbf{e}_1' \ \mathbf{e}_2' \right) \begin{pmatrix} 3 & -2 & 1 \\ -1 & 1 & 5 \end{pmatrix}.$$

Therefore the representation matrix of T_A with respect to the standard bases of $\mathbf{R}^3$ and $\mathbf{R}^2$ is the matrix

$$A = \begin{pmatrix} 3 & -2 & 1 \\ -1 & 1 & 5 \end{pmatrix}.$$

Example 2 Let T be a linear mapping T of $\mathbf{R}[x]_2$ into $\mathbf{R}[x]_1$ defined by $T(f(x)) = f'(x) - f(0)x$. We shall consider the matrix representation of T with respect to the standard basis $\{1, x, x^2\}$ of $\mathbf{R}[x]_2$ and the standard basis $\{1, x\}$ of $\mathbf{R}[x]_1$. We see that

$$\left(T(1) \ T(x) \ T(x^2) \right) = \left(-x \ 1 \ 2x \right) = \left(1 \ x \right) \begin{pmatrix} 0 & 1 & 0 \\ -1 & 0 & 2 \end{pmatrix}.$$

Therefore the representation matrix of T with respect to the standard basis $\{1, x, x^2\}$ of $\mathbf{R}[x]_2$ and the standard basis $\{1, x\}$ of $\mathbf{R}[x]_1$ is

$$\begin{pmatrix} 0 & 1 & 0 \\ -1 & 0 & 2 \end{pmatrix}.$$

We generalize Example 1 to any linear mapping T_A associated with a matrix A.

Theorem 5.2.1 *Let A be an $m \times n$ matrix and T_A the linear mapping of K^n into K^m associated with A. Then the representation matrix of T_A with respect to the standard bases of K^n and K^m is the matrix A itself.*

Proof Let $A = \left(\mathbf{a}_1 \ \dots \ \mathbf{a}_n \right)$ be the column partition of the matrix A. We denote by $\{\mathbf{e}_1, \dots, \mathbf{e}_n\}$ the standard basis of K^n and by $\{\mathbf{e}_1', \dots, \mathbf{e}_m'\}$ the standard basis of K^m. Then we have

$$T_A(\mathbf{e}_i) = A\mathbf{e}_i = \left(\mathbf{a}_1 \ \dots \ \mathbf{a}_i \ \dots \ \mathbf{a}_n \right) \begin{pmatrix} 0 \\ \vdots \\ 1 \\ \vdots \\ 0 \end{pmatrix} = \mathbf{a}_i$$

for $i = 1, \dots, n$. Therefore we have

$$\left(T_A(\mathbf{e}_1) \ T_A(\mathbf{e}_2) \ \dots \ T_A(\mathbf{e}_n) \right) = \left(\mathbf{a}_1 \ \mathbf{a}_2 \ \dots \ \mathbf{a}_n \right)$$
$$= A = E_m A$$
$$= \left(\mathbf{e}_1' \ \mathbf{e}_2' \ \dots \ \mathbf{e}_m' \right) A.$$

Thus the representation matrix of T_A with respect to the standard bases is the matrix A. $\parallel$

Theorem 5.2.2 *Assume that* $\dim(U) = \dim(V) = n$. *For a linear mapping* T *of* U *into* V, *the following are equivalent:*

(1) A linear mapping T *is an isomorphism.*
(2) The representation matrix A *of* T *with respect to a basis* $\{\mathbf{u}_1, \ldots, \mathbf{u}_n\}$ *of* U *and a basis* $\{\mathbf{v}_1, \ldots, \mathbf{v}_n\}$ *of* V *is a regular matrix.*

The representation matrix of the inverse isomorphism T^{-1} *of* T *with respect to the same bases is the inverse matrix* A^{-1} *of* A.

Proof For a vector $\mathbf{u} = x_1\mathbf{u}_1 + \cdots + x_n\mathbf{u}_n \in U$ and a vector $\mathbf{v} = y_1\mathbf{v}_1 + \cdots + y_n\mathbf{v}_n \in V$, we let $\mathbf{x} = \begin{pmatrix} x_1 \\ \vdots \\ x_n \end{pmatrix}$ and $\mathbf{y} = \begin{pmatrix} y_1 \\ \vdots \\ y_n \end{pmatrix}$, respectively. Since

$$\big(T(\mathbf{u}_1) \ldots T(\mathbf{u}_n)\big) = \big(\mathbf{v}_1 \ldots \mathbf{v}_n\big)A,$$

we have

$$T(\mathbf{u}) = \big(T(\mathbf{u}_1) \ldots T(\mathbf{u}_n)\big)\mathbf{x} = \big(\mathbf{v}_1 \ldots \mathbf{v}_n\big)A\mathbf{x}.$$

Therefore $T(\mathbf{u}) = \mathbf{0}_V$ if and only if $A\mathbf{x} = \mathbf{0}_n$ by Theorem 4.2.5 (1). Since $\dim(U) = \dim(V)$, we have the following equivalences:

$$\begin{aligned} T \text{ is an isomorphism} &\Leftrightarrow \text{Ker}(T) = \{\mathbf{0}_U\} \quad \text{(Theorems 5.1.3 and 5.1.4 (3))} \\ &\Leftrightarrow \text{Ker}(A) = \{\mathbf{0}_n\} \\ &\Leftrightarrow A \text{ is a regular matrix} \quad \text{(Theorem 2.4.2).} \end{aligned}$$

Furthermore, we assume that A is a regular matrix. We define a linear mapping S of V into U by

$$S(\mathbf{v}) = \big(\mathbf{u}_1 \ldots \mathbf{u}_n\big)A^{-1}\mathbf{y}.$$

Then $\big(S(\mathbf{u}_1) \ldots S(\mathbf{u}_n)\big) = \big(\mathbf{u}_1 \ldots \mathbf{u}_n\big)A^{-1}$. Therefore for $\mathbf{u}_1 = x_1\mathbf{u}_1 + \ldots + x_n\mathbf{u}_n$, we have

$$\begin{aligned} ST(\mathbf{u}) = S\big((\mathbf{u}_1 \ldots \mathbf{u}_n)A\mathbf{x}\big) &= \big(\mathbf{u}_1 \ldots \mathbf{u}_n\big)A^{-1}A\mathbf{x} \\ &= \big(\mathbf{u}_1 \ldots \mathbf{u}_n\big)\mathbf{x} = \mathbf{u}. \end{aligned}$$

Thus we obtain $ST = I_U$. Similarly we have $TS = I_V$. Therefore S is the inverse isomorphism T^{-1} of T and the representation matrix of T^{-1} with respect to the bases $\{\mathbf{v}_j\}$ of V and $\{\mathbf{u}_i\}$ of U is A^{-1}. ∥

The following theorem shows the meaning of the representation matrices of linear mappings.

Theorem 5.2.3 *Let* U *and* V *be vector spaces of dimension* n *and* m, *respectively. For a linear mapping* T *of* U *into* V, *let* A *be the representation matrix of* T *with respect to the bases* $\{\mathbf{u}_1, \ldots, \mathbf{u}_n\}$ *and* $\{\mathbf{v}_1, \ldots, \mathbf{v}_m\}$ *of* U *and* V, *respectively. Then we have*

$$T_A S_U = S_V T.$$

This equality means that the following diagram is commutative:

$$
\begin{array}{ccc}
& S_U & \\
U & \longrightarrow & K^n \\
T \downarrow & & \downarrow T_A \\
V & \longrightarrow & K^m. \\
& S_V &
\end{array}
$$

Here S_U and S_V are the isomorphism of U into K^n with respect to the basis $\{\mathbf{u}_1, \ldots, \mathbf{u}_n\}$ and the isomorphism of V into K^m with respect to the basis $\{\mathbf{v}_1, \ldots, \mathbf{v}_m\}$, respectively, defined in Theorem 5.1.5(2).

Proof Let $\mathbf{u} = a_1 \mathbf{u}_1 + \cdots + a_n \mathbf{u}_n$. Then

$$
T_A(S_U(\mathbf{u})) = T_A\left(\begin{pmatrix} a_1 \\ \vdots \\ a_n \end{pmatrix}\right) = A \begin{pmatrix} a_1 \\ \vdots \\ a_n \end{pmatrix}.
$$

On the other hand, we have

$$
S_V(T(\mathbf{u})) = S_V\left(\begin{pmatrix} \mathbf{v}_1 \ldots \mathbf{v}_m \end{pmatrix} A \begin{pmatrix} a_1 \\ \vdots \\ a_n \end{pmatrix}\right)
$$

$$
= A \begin{pmatrix} a_1 \\ \vdots \\ a_n \end{pmatrix}.
$$

Therefore we have

$$
T_A(S_U(\mathbf{u})) = S_V(T(\mathbf{u})). \quad \|
$$

Theorem 5.2.4 *Let T be a linear mapping of vector spaces U into V. We let A be the representation matrix of T with respect to bases $\{\mathbf{u}_1, \ldots, \mathbf{u}_n\}$ of U and $\{\mathbf{v}_1, \ldots, \mathbf{v}_m\}$ of V. Then we have:*

(1) $S_U(\mathrm{Ker}\,(T)) = \mathrm{Ker}\,(T_A) = \mathrm{Ker}\,(A)$ and null $(T) = $ null (A).
(2) $S_V(\mathrm{Im}\,(T)) = \mathrm{Im}\,(T_A) = \mathrm{Im}\,(A)$ and rank $(T) = $ rank (A).

Here S_U is the isomorphism of U into K^n with respect to the basis $\{\mathbf{u}_1, \ldots, \mathbf{u}_n\}$ and S_V is the isomorphism of V into K^m with respect to the basis $\{\mathbf{v}_1, \ldots, \mathbf{v}_m\}$ as in Theorem 5.2.3.

Proof By Theorem 5.2.3, we have

$$S_V(T(\mathbf{u})) = T_A(S_U(\mathbf{u})) \quad \text{for} \quad \mathbf{u} \in U.$$

(1) Since $T(\mathbf{u}) = \mathbf{0}_V$ if and only if $T_A(S_U(\mathbf{u})) = S_V(T(\mathbf{u})) = \mathbf{0}_m$, we have

$$S_U(\text{Ker}\,(T)) = \text{Ker}\,(T_A) = \text{Ker}\,(A).$$

Then we have

$$\text{null}\,(T) = \text{null}\,(A).$$

(2) Since S_U is an isomorphism, we have $S_U(U) = K^n$. Therefore

$$S_V(\text{Im}(T)) = S_V(T(U)) = T_A(S_U(U)) = T_A(K^n) = \text{Im}(T_A) = \text{Im}(A).$$

Since S_V is an isomorphism, we have

$$\text{rank}\,(T) = \text{rank}\,(T_A) = \text{rank}\,(A). \quad \|$$

Exercise 5.3 Let T be the linear mapping of $\mathbf{R}[x]_2$ into $\mathbf{R}^3$ defined by

$$T(f(x)) = \begin{pmatrix} f(0) \\ f(1) \\ f(0) - f(1) \end{pmatrix} \quad (f(x) \in \mathbf{R}[x]_2).$$

Find the following:

(i) null(T) and Ker(T).   (ii)   rank(T) and Im(T).

Answer. (i) Taking the standard basis $\{1, x, x^2\}$ of $\mathbf{R}[x]_2$ and the standard basis $\{\mathbf{e}_1, \mathbf{e}_2, \mathbf{e}_3\}$ of $\mathbf{R}^3$, we have

$$\begin{pmatrix} T(1) \ T(x) \ T(x^2) \end{pmatrix} = \begin{pmatrix} \mathbf{e}_1 \ \mathbf{e}_2 \ \mathbf{e}_3 \end{pmatrix} \begin{pmatrix} 1 & 0 & 0 \\ 1 & 1 & 1 \\ 0 & -1 & -1 \end{pmatrix}.$$

Therefore the representation matrix of T with respect to these bases is

$$A = \begin{pmatrix} 1 & 0 & 0 \\ 1 & 1 & 1 \\ 0 & -1 & -1 \end{pmatrix}.$$

To obtain Ker (T) and Im (T), we have only to calculate Ker (A) and Im (A). The reduced matrix of A is $B = \begin{pmatrix} 1 & 0 & 0 \\ 0 & 1 & 1 \\ 0 & 0 & 0 \end{pmatrix}$ by the calculation

$$
\begin{array}{ccc}
1 & 0 & 0 \\
1 & 1 & 1 \\
0 & -1 & -1 \\
\hline
1 & 0 & 0 \\
0 & 1 & 1 \\
0 & -1 & -1 \\
\hline
1 & 0 & 0 \\
0 & 1 & 1 \\
0 & 0 & 0 \\
\hline
\end{array}
\qquad
\begin{array}{l}
\\
\\
\\
\\
②+①\times(-1) \\
\\
\\
\\
③+②.
\end{array}
$$

Then the solutions of $A\mathbf{x} = \mathbf{0}_3$ are $\mathbf{x} = a \begin{pmatrix} 0 \\ -1 \\ 1 \end{pmatrix}$ $(a \in \mathbf{R})$. By Theorem 5.2.4, we have

$$\text{null}\,(T) = 1$$

and

$$\text{Ker}\,(T) = \{\, a(-x + x^2) \mid a \in \mathbf{R} \,\}.$$

(ii) We easily see that the first and the second column vectors of B are linearly independent and the third column vector of B is a linear combination of these columns. Therefore the corresponding column vectors $\begin{pmatrix} 1 \\ 1 \\ 0 \end{pmatrix}$ and $\begin{pmatrix} 0 \\ 1 \\ -1 \end{pmatrix}$ of A are linearly independent and $\begin{pmatrix} 0 \\ 1 \\ -1 \end{pmatrix}$ is a linear combination of them. By Theorem 5.2.4, we have

$$\text{rank}\,(T) = 2$$

and

$$\text{Im}\,(T) = \left\{\, b \begin{pmatrix} 1 \\ 1 \\ 0 \end{pmatrix} + c \begin{pmatrix} 0 \\ 1 \\ -1 \end{pmatrix} \;\middle|\; b, c \in \mathbf{R} \,\right\}. \quad \|$$

Matrix of changing bases Let $\{\mathbf{u}_1, \ldots, \mathbf{u}_n\}$ and $\{\mathbf{u}'_1, \ldots, \mathbf{u}'_n\}$ be bases of a vector space U. Then there is a regular matrix P satisfying

$$\begin{pmatrix} \mathbf{u}'_1 & \cdots & \mathbf{u}'_n \end{pmatrix} = \begin{pmatrix} \mathbf{u}_1 & \cdots & \mathbf{u}_n \end{pmatrix} P.$$

We call the regular matrix P the *matrix of changing bases* from $\{\mathbf{u}_1, \ldots, \mathbf{u}_n\}$ to $\{\mathbf{u}'_1, \ldots, \mathbf{u}'_n\}$.

Let T be a linear mapping of U into V. We denote by A the representation matrix of T with respect to a basis $\{\mathbf{u}_1, \ldots, \mathbf{u}_n\}$ of U and a basis $\{\mathbf{v}_1, \ldots, \mathbf{v}_m\}$ of V. We also let B be another representation matrix of T with respect to another basis $\{\mathbf{u}'_1, \ldots, \mathbf{u}'_n\}$

of U and another basis $\{\mathbf{v}_1', \ldots, \mathbf{v}_m'\}$ of V. The following theorem gives the relation of A and B.

Theorem 5.2.5 *Let T be a linear mapping of U into V. Let A be the representation matrix of T with respect to bases $\{\mathbf{u}_1, \ldots, \mathbf{u}_n\}$ of U and $\{\mathbf{v}_1, \ldots, \mathbf{v}_m\}$ of V and B the representation matrix of T with respect to bases $\{\mathbf{u}_1', \ldots, \mathbf{u}_n'\}$ of U and $\{\mathbf{v}_1', \ldots, \mathbf{v}_m'\}$ of V. We denote by P the matrix of changing bases from $\{\mathbf{u}_1, \ldots, \mathbf{u}_n\}$ to $\{\mathbf{u}_1', \ldots, \mathbf{u}_n'\}$ and by Q that from $\{\mathbf{v}_1, \ldots, \mathbf{v}_m\}$ to $\{\mathbf{v}_1', \ldots, \mathbf{v}_m'\}$, respectively. Then we have*

$$B = Q^{-1}AP.$$

Proof Let $P = (p_{ij})$. By definition, we see that

$$\left(\mathbf{u}_1' \ldots \mathbf{u}_n'\right) = \left(\mathbf{u}_1 \ldots \mathbf{u}_n\right)\begin{pmatrix} p_{11} & \cdots & p_{1n} \\ \vdots & & \vdots \\ p_{n1} & \cdots & p_{nn} \end{pmatrix}.$$

We also see that

$$\left(\mathbf{v}_1' \ldots \mathbf{v}_m'\right) = \left(\mathbf{v}_1 \ldots \mathbf{v}_m\right) Q.$$

By the linearity of T and the definition of the representation matrix A, we have

$$
\begin{aligned}
&\left(T(\mathbf{u}_1') \ldots T(\mathbf{u}_n')\right) \\
&= \left(T(p_{11}\mathbf{u}_1 + \cdots + p_{n1}\mathbf{u}_n) \ldots T(p_{1n}\mathbf{u}_1 + \cdots + p_{nn}\mathbf{u}_n)\right) \\
&= \left(p_{11}T(\mathbf{u}_1) + \cdots + p_{n1}T(\mathbf{u}_n) \ldots p_{1n}T(\mathbf{u}_1) + \cdots + p_{nn}T(\mathbf{u}_n)\right) \\
&= \left(T(\mathbf{u}_1) \ldots T(\mathbf{u}_n)\right) P \\
&= \left(\mathbf{v}_1 \ldots \mathbf{v}_m\right) AP.
\end{aligned}
$$

On the other hand, by the definition of the representation matrix B and the matrix of changing bases Q, we have

$$\left(T(\mathbf{u}_1') \ldots T(\mathbf{u}_n')\right) = \left(\mathbf{v}_1' \ldots \mathbf{v}_m'\right) B = \left(\mathbf{v}_1 \ldots \mathbf{v}_m\right) QB.$$

Thus we have

$$\left(\mathbf{v}_1 \quad \cdots \quad \mathbf{v}_m\right) AP = \left(\mathbf{v}_1 \quad \cdots \quad \mathbf{v}_m\right) QB.$$

Since $\mathbf{v}_1, \ldots, \mathbf{v}_m$ are linearly independent, we have $AP = QB$ by Theorem 4.2.5 (2). Therefore we have $B = Q^{-1}AP$. ∥

Exercise 5.4 Let T_A be the linear mapping of $\mathbf{R}^3$ into $\mathbf{R}^2$ associated with the matrix $A = \begin{pmatrix} 2 & 4 & 1 \\ 1 & -1 & 0 \end{pmatrix}$. Find the representation matrix B of T_A with respect to the following bases of $\mathbf{R}^3$ and $\mathbf{R}^2$:

$$\text{a basis of } \mathbf{R}^3 : \left\{ \mathbf{a}_1 = \begin{pmatrix} 2 \\ 0 \\ 3 \end{pmatrix}, \mathbf{a}_2 = \begin{pmatrix} 0 \\ 1 \\ 1 \end{pmatrix}, \mathbf{a}_3 = \begin{pmatrix} 1 \\ 0 \\ 1 \end{pmatrix} \right\},$$

$$\text{a basis of } \mathbf{R}^2 : \left\{ \mathbf{b}_1 = \begin{pmatrix} 1 \\ 1 \end{pmatrix}, \mathbf{b}_2 = \begin{pmatrix} 2 \\ 3 \end{pmatrix} \right\}.$$

Answer. By Theorem 5.2.1, the representation matrix of T_A with respect to the standard bases is A. We change bases of $\mathbf{R}^3$ and $\mathbf{R}^2$. Let the standard basis of $\mathbf{R}^3$ and the standard basis of $\mathbf{R}^2$ be $\{\mathbf{e}_1, \mathbf{e}_2, \mathbf{e}_3\}$ and $\{\mathbf{e}_1', \mathbf{e}_2'\}$, respectively. We denote the matrix of changing bases from the standard basis $\{\mathbf{e}_1, \mathbf{e}_2, \mathbf{e}_3\}$ to the basis $\{\mathbf{a}_1, \mathbf{a}_2, \mathbf{a}_3\}$ by P:

$$\begin{pmatrix} \mathbf{a}_1 & \mathbf{a}_2 & \mathbf{a}_3 \end{pmatrix} = \begin{pmatrix} \mathbf{e}_1 & \mathbf{e}_2 & \mathbf{e}_3 \end{pmatrix} P.$$

Then $P = \begin{pmatrix} 2 & 0 & 1 \\ 0 & 1 & 0 \\ 3 & 1 & 1 \end{pmatrix}$. We also denote the matrix of changing bases from the standard basis $\{\mathbf{e}_1', \mathbf{e}_2'\}$ to the basis $\{\mathbf{b}_1, \mathbf{b}_2\}$ by Q:

$$\begin{pmatrix} \mathbf{b}_1 & \mathbf{b}_2 \end{pmatrix} = \begin{pmatrix} \mathbf{e}_1' & \mathbf{e}_2' \end{pmatrix} Q.$$

Then $Q = \begin{pmatrix} 1 & 2 \\ 1 & 3 \end{pmatrix}$. By Theorem 5.2.5, we have

$$B = Q^{-1}AP = \begin{pmatrix} 3 & -2 \\ -1 & 1 \end{pmatrix} \begin{pmatrix} 2 & 4 & 1 \\ 1 & -1 & 0 \end{pmatrix} \begin{pmatrix} 2 & 0 & 1 \\ 0 & 1 & 0 \\ 3 & 1 & 1 \end{pmatrix}$$

$$= \begin{pmatrix} 17 & 17 & 7 \\ -5 & -6 & -2 \end{pmatrix}. \quad \|$$

Linear transformations and representation matrices Let T be a linear transformation of a vector space U. For a linear transformation T of U and a basis $\{\mathbf{u}_1, \ldots, \mathbf{u}_n\}$ of U, the matrix A defined by

$$\begin{pmatrix} T(\mathbf{u}_1) & \ldots & T(\mathbf{u}_n) \end{pmatrix} = \begin{pmatrix} \mathbf{u}_1 & \ldots & \mathbf{u}_n \end{pmatrix} A$$

is called the *representation matrix of the linear transformation T* with respect to the basis $\{\mathbf{u}_1, \ldots, \mathbf{u}_n\}$ of U. The representation matrix of the identity transformation I_U (resp. the zero transformation O_U) with respect to any basis of U is the unit matrix E_n (resp. the zero matrix O_n).

Exercise 5.5 Let T be the linear transformation of $\mathbf{R}[x]_2$ defined by

$$T(f(x)) = (x+1)f''(x) - 2f'(x) + f(0)x.$$

Find the following:

    (i) Ker (T) and null (T).   (ii) Im (T) and rank (T).

Answers.   By Theorem 5.2.4, we have only to find them for the representation matrix.

    (i) Since

$$\left(T(1)\ T(x)\ T(x^2) \right) = \left(1\ x\ x^2 \right) \begin{pmatrix} 0 & -2 & 2 \\ 1 & 0 & -2 \\ 0 & 0 & 0 \end{pmatrix},$$

the representation matrix of T with respect to the standard basis $\{1, x, x^2\}$ of $\mathbf{R}[x]_2$ is

$$A = \begin{pmatrix} 0 & -2 & 2 \\ 1 & 0 & -2 \\ 0 & 0 & 0 \end{pmatrix} = \left(\mathbf{a}_1\ \mathbf{a}_2\ \mathbf{a}_3 \right).$$

The reduced matrix of A is

$$B = \begin{pmatrix} 1 & 0 & -2 \\ 0 & 1 & -1 \\ 0 & 0 & 0 \end{pmatrix} = \left(\mathbf{b}_1\ \mathbf{b}_2\ \mathbf{b}_3 \right)$$

by the calculation

$$
\begin{array}{ccc}
\mathbf{a}_1 & \mathbf{a}_2 & \mathbf{a}_3 \\
\hline
0 & -2 & 2 \\
1 & 0 & -2 \\
0 & 0 & 0 \\
\hline
1 & 0 & -2 \\
0 & -2 & 2 \\
0 & 0 & 0 \\
\hline
1 & 0 & -2 \\
0 & 1 & -1 \\
0 & 0 & 0 \\
\hline
\mathbf{b}_1 & \mathbf{b}_2 & \mathbf{b}_3
\end{array}
\qquad
\begin{array}{l}
\\
\\
\\
① \leftrightarrow ② \\
\\
\\
② \times (-1/2).
\end{array}
$$

Then the solution of $A\mathbf{x} = \mathbf{0}_3$ is $\mathbf{x} = a \begin{pmatrix} 2 \\ 1 \\ 1 \end{pmatrix}$ $(a \in \mathbf{R})$.

    Considering the corresponding vectors in $\mathbf{R}[x]_2$, we have

$$\text{Ker } (T) = \{a(2 + x + x^2) \,|\, a \in \mathbf{R}\} \quad \text{and} \quad \text{null } (T) = 1.$$

    (ii) Since the vectors $\mathbf{b}_1$ and $\mathbf{b}_2$ are linearly independent and $\mathbf{b}_3 = -2\mathbf{b}_1 - \mathbf{b}_2$, the vectors $\mathbf{a}_1$ and $\mathbf{a}_2$ are linearly independent and $\mathbf{a}_3 = -2\mathbf{a}_1 - \mathbf{a}_2$. So, the corresponding vectors x and -2 in $\mathbf{R}[x]_2$ are linearly independent and the vector $2 - 2x$ is a linear combination of x and -2. Therefore we have

$$\text{Im } (T) = \{bx + c \mid b, c \in \mathbf{R}\} \quad \text{and} \quad \text{rank } (T) = 2. \quad \|$$

The following theorem is the restatement of Theorem 5.2.5 for linear transformations.

Theorem 5.2.6 *Let* $\{\mathbf{u}_1, \ldots, \mathbf{u}_n\}$ *and* $\{\mathbf{u}_1', \ldots, \mathbf{u}_n'\}$ *be two bases of a vector space* U *of dimension* n. *Let* A *be the representation matrix of a linear transformation* T *of* U *with respect to the basis* $\{\mathbf{u}_1, \ldots, \mathbf{u}_n\}$ *and* B *the representation matrix of* T *with respect to the basis* $\{\mathbf{u}_1', \ldots, \mathbf{u}_n'\}$. *If* P *is the matrix of changing bases from* $\{\mathbf{u}_1, \ldots, \mathbf{u}_n\}$ *to* $\{\mathbf{u}_1', \ldots, \mathbf{u}_n'\}$, *then we have*

$$B = P^{-1}AP.$$

Linear transformations associated with square matrices For a square matrix A of degree n, we call the linear mapping T_A of K^n into itself associated with the matrix A the *linear transformation associated with* A.

Exercise 5.6 Let $A = \begin{pmatrix} 2 & 3 \\ -1 & -1 \end{pmatrix}$ and T_A be the linear transformation of $\mathbf{R}^2$ associated with the matrix A.

(1) Find the representation matrix of T_A with respect to the standard basis of $\mathbf{R}^2$.
(2) Find the representation matrix B of T_A with respect to the basis

$$\left\{ \mathbf{b}_1 = \begin{pmatrix} 1 \\ 1 \end{pmatrix}, \mathbf{b}_2 = \begin{pmatrix} 1 \\ 2 \end{pmatrix} \right\}.$$

Answers.   (1) By Theorem 5.2.1, the representation matrix of T_A with respect to the standard basis $\{\mathbf{e}_1, \mathbf{e}_2\}$ of $\mathbf{R}^2$ is $A = \begin{pmatrix} 2 & 3 \\ -1 & -1 \end{pmatrix}$.

(2) Since

$$\begin{pmatrix} \mathbf{b}_1 & \mathbf{b}_2 \end{pmatrix} = \begin{pmatrix} \mathbf{e}_1 & \mathbf{e}_2 \end{pmatrix} \begin{pmatrix} 1 & 1 \\ 1 & 2 \end{pmatrix},$$

the matrix of changing bases from the standard basis $\{\mathbf{e}_1, \mathbf{e}_2\}$ to the basis $\{\mathbf{b}_1, \mathbf{b}_2\}$ is $P = \begin{pmatrix} 1 & 1 \\ 1 & 2 \end{pmatrix}$. Since $P^{-1} = \begin{pmatrix} 2 & -1 \\ -1 & 1 \end{pmatrix}$, we obtain by Theorem 5.2.6

$$B = P^{-1}AP = \begin{pmatrix} 12 & 19 \\ -7 & -11 \end{pmatrix}. \quad \|$$

Similarity of square matrices We say that square matrices A and B of the same degree are *similar* if there exists a regular matrix P satisfying

$$B = P^{-1}AP.$$

Example 3 Let

$$A = \begin{pmatrix} 2 & 3 \\ -1 & -1 \end{pmatrix} \quad \text{and} \quad B = \begin{pmatrix} 12 & 19 \\ -7 & -11 \end{pmatrix}.$$

Then A and B are similar by Exercise 5.6.

Exercises (Sect. 5.2)

1   Find the representation matrices of the following linear mappings T with respect to the given bases:

(1) $T(\mathbf{x}) = \begin{pmatrix} 1 & -2 \\ 1 & 1 \end{pmatrix} \mathbf{x} : U = \mathbf{R}^2 \to V = \mathbf{R}^2;\ \left\{ \begin{pmatrix} 1 \\ 0 \end{pmatrix}, \begin{pmatrix} 2 \\ 1 \end{pmatrix} \right\}$: a basis of U and $\left\{ \begin{pmatrix} 2 \\ 1 \end{pmatrix}, \begin{pmatrix} 3 \\ 1 \end{pmatrix} \right\}$: a basis of V.

(2) $T(\mathbf{x}) = \begin{pmatrix} 2 & 4 & 1 \\ 1 & 5 & 3 \end{pmatrix} \mathbf{x} : \mathbf{R}^3 \to \mathbf{R}^2;\ \left\{ \begin{pmatrix} 1 \\ 0 \\ 1 \end{pmatrix}, \begin{pmatrix} 1 \\ 2 \\ 2 \end{pmatrix}, \begin{pmatrix} 0 \\ 1 \\ 1 \end{pmatrix} \right\}$: a basis of $\mathbf{R}^3$ and $\left\{ \begin{pmatrix} 1 \\ 2 \end{pmatrix}, \begin{pmatrix} 2 \\ 3 \end{pmatrix} \right\}$: a basis of $\mathbf{R}^2$.

(3) $T(f(x)) = f(0)x + 2f'(x) : \mathbf{R}[x]_2 \to \mathbf{R}[x]_1;\ \{1,\ 1 + x,\ x + x^2\}$: a basis of $\mathbf{R}[x]_2$ and $\{1 - x,\ x\}$: a basis of $\mathbf{R}[x]_1$.

(4) $T(f(x)) = \begin{pmatrix} f(0) + f(-1) \\ f(1) \\ f(0) - f(-1) \end{pmatrix}$: $\mathbf{R}[x]_2 \to \mathbf{R}^3;\ \{x^2,\ x,\ 1\}$: a basis of $\mathbf{R}[x]_2$ and $\left\{ \begin{pmatrix} 1 \\ 1 \\ 0 \end{pmatrix}, \begin{pmatrix} 1 \\ 0 \\ 1 \end{pmatrix}, \begin{pmatrix} 0 \\ 1 \\ 0 \end{pmatrix} \right\}$: a basis of $\mathbf{R}^3$.

2   Find the representation matrices of the following linear transformations T with respect to the given bases:

(1) $T(\mathbf{x}) = \begin{pmatrix} 0 & 3 & -1 \\ -1 & 2 & 1 \\ 2 & 1 & 1 \end{pmatrix} \mathbf{x} : \mathbf{R}^3 \to \mathbf{R}^3;\ \left\{ \begin{pmatrix} 1 \\ 1 \\ 1 \end{pmatrix}, \begin{pmatrix} 1 \\ 0 \\ 1 \end{pmatrix}, \begin{pmatrix} 1 \\ 1 \\ 0 \end{pmatrix} \right\}$: a basis of $\mathbf{R}^3$.

(2) $T(\mathbf{x}) = \begin{pmatrix} 3 & 1 & 2 \\ -2 & 0 & -1 \\ 1 & 2 & -2 \end{pmatrix} \mathbf{x} : \mathbf{R}^3 \to \mathbf{R}^3;\ \left\{ \begin{pmatrix} 1 \\ 0 \\ -1 \end{pmatrix}, \begin{pmatrix} 0 \\ 2 \\ 1 \end{pmatrix}, \begin{pmatrix} 1 \\ 1 \\ -1 \end{pmatrix} \right\}$: a basis of $\mathbf{R}^3$.

(3) $T(f(x)) = f(x) + f(1)x + f(0)x^2 : \mathbf{R}[x]_2 \to \mathbf{R}[x]_2;$ $\{1 + x,\ 2 + x,\ x - x^2\}$: a basis of $\mathbf{R}[x]_2$.

(4) $T(f(x)) = f(-1)(1 + x^2) + xf'(x) : \mathbf{R}[x]_2 \to \mathbf{R}[x]_2;$ $\{1 + x^2,\ x + x^2,\ 1 + 2x^2\}$: a basis of $\mathbf{R}[x]_2$.

3   For the following linear mappings or linear transformations T, find (i) and (ii) below:

(i) null(T) and a basis of Ker(T).

(ii) rank(T) and a basis of Im(T).

(1) The linear transformation T of $\mathbf{R}[x]_2$ defined by

$$T(f(x)) = 2f(x) - xf'(x).$$

(2) The linear transformation T of $\mathbf{R}[x]_2$ defined by

$$T(f(x)) = f(0)x + f(x) - xf'(x).$$

(3) The linear mapping $T : \mathbf{R}[x]_3 \to \mathbf{R}^3$ defined by

$$T(f(x)) = \begin{pmatrix} f(-1) - f'(0) \\ f'(0) - f(1) \\ f(1) - f(-1) \end{pmatrix}.$$

(4) The linear mapping $T : \mathbf{R}^3 \to \mathbf{R}[x]_2$ defined by

$$T\left(\begin{pmatrix} a_1 \\ a_2 \\ a_3 \end{pmatrix}\right) = (a_1 - a_2) + (a_2 - a_3)x + (a_3 - a_1)x^2.$$

4 Let T be a linear transformation of an n-dimensional vector space V satisfying $T^n = O_V$ and $T^{n-1} \neq O_V$.

(1) Take a vector $\mathbf{u}$ in V satisfying $T^{n-1}(\mathbf{u}) \neq \mathbf{0}_V$. Show that the set $\{T^{n-1}(\mathbf{u}), \ldots, T(\mathbf{u}), \mathbf{u}\}$ of vectors is a basis of V.
(2) Find the representation matrix of T with respect to the basis $\{T^{n-1}(\mathbf{u}), \ldots, T(\mathbf{u}), \mathbf{u}\}$ of V.

5 Show that the following linear mappings T are isomorphisms:

(1) $T : \mathbf{R}[x]_2 \to \mathbf{R}[x]_2$, $T(f(x)) = 3f(x) - xf'(x) + f(1)$.

(2) $T = T_A : \mathbf{R}^3 \to \mathbf{R}^3$, $A = \begin{pmatrix} 1 & 1 & -1 \\ 2 & -1 & 1 \\ -1 & 2 & 0 \end{pmatrix}$.

(3) $T : \mathbf{R}[x]_2 \to \mathbf{R}^3$, $T(f(x)) = \begin{pmatrix} f(1) \\ f(0) \\ f(-1) \end{pmatrix}$.

5.3 Eigenvalues and Eigenvectors

Eigenvalues and eigenvectors Let T be a linear transformation of a vector space U. If a non-zero vector $\mathbf{u} \in U$ satisfies the condition

$$T(\mathbf{u}) = \lambda \mathbf{u} \qquad (\lambda \in K),$$

then we call the vector $\mathbf{u}$ an *eigenvector* of T with the *eigenvalue* λ.

Eigenvalues and eigenvectors of square matrices Let A be a square matrix of degree n. If non-zero vector $\mathbf{a} \in K^n$ satisfies

$$A\mathbf{a} = \lambda \mathbf{a} \qquad (\lambda \in K),$$

then we call the vector $\mathbf{a}$ an *eigenvector* of A with the *eigenvalue* λ. We see that an eigenvector $\mathbf{a}$ of A is an eigenvector of the linear transformation T_A associated with A.

Example 1 Let $A = \begin{pmatrix} 5 & 0 \\ 0 & -2 \end{pmatrix}$. Let $\mathbf{e}_1$ and $\mathbf{e}_2$ be the unit vectors of $\mathbf{R}^2$. Then we have

$$A\mathbf{e}_1 = \begin{pmatrix} 5 & 0 \\ 0 & -2 \end{pmatrix}\begin{pmatrix} 1 \\ 0 \end{pmatrix} = \begin{pmatrix} 5 \\ 0 \end{pmatrix} = 5\mathbf{e}_1,$$

$$A\mathbf{e}_2 = \begin{pmatrix} 5 & 0 \\ 0 & -2 \end{pmatrix}\begin{pmatrix} 0 \\ 1 \end{pmatrix} = \begin{pmatrix} 0 \\ -2 \end{pmatrix} = -2\mathbf{e}_2.$$

Therefore 5 and -2 are eigenvalues of A. The unit vector $\mathbf{e}_1$ is an eigenvector of A with the eigenvalue 5, and the unit vector $\mathbf{e}_2$ is an eigenvector of A with the eigenvalue -2.

Example 2 Let $A = \begin{pmatrix} 7 & -6 \\ 3 & -2 \end{pmatrix}$. We take $\mathbf{a}_1 = \begin{pmatrix} 1 \\ 1 \end{pmatrix}$ and $\mathbf{a}_2 = \begin{pmatrix} 2 \\ 1 \end{pmatrix}$. Then we have

$$A\mathbf{a}_1 = \begin{pmatrix} 7 & -6 \\ 3 & -2 \end{pmatrix}\begin{pmatrix} 1 \\ 1 \end{pmatrix} = \begin{pmatrix} 1 \\ 1 \end{pmatrix} = \mathbf{a}_1,$$

$$A\mathbf{a}_2 = \begin{pmatrix} 7 & -6 \\ 3 & -2 \end{pmatrix}\begin{pmatrix} 2 \\ 1 \end{pmatrix} = \begin{pmatrix} 8 \\ 4 \end{pmatrix} = 4\mathbf{a}_2.$$

Therefore 1 and 4 are eigenvalues of A. The vector $\mathbf{a}_1$ is an eigenvector of A with the eigenvalue 1 and $\mathbf{a}_2$ is an eigenvector of A with the eigenvalue 4.

Eigenspaces Let U be a vector space and T a linear transformation of U. For an eigenvalue λ of T, we let

$$\begin{aligned} W(\lambda; T) &= \{\mathbf{u} \in U | T(\mathbf{u}) = \lambda \mathbf{u}\} \\ &= \mathrm{Ker}(T - \lambda I_U). \end{aligned}$$

Here I_U is the identity transformation of U. Therefore the subset $W(\lambda; T)$ of U is a non-zero subspace of U (Theorem 5.1.1 (2)). We call the subspace $W(\lambda; T)$ the *eigenspace* of T with the eigenvalue λ. Non-zero vectors in $W(\lambda; T)$ are nothing but eigenvectors of T with the eigenvalue λ.

Eigenspaces of square matrices Let T_A be the linear transformation of K^n associated with a square matrix A of degree n. Then for an eigenvalue λ of A, we define the *eigenspace* $W(\lambda; A)$ with the eigenvalue λ of A by

$$W(\lambda; A) = \{\mathbf{x} \in K^n \mid A\mathbf{x} = \lambda \mathbf{x}\}$$
$$= \text{the solution space of the linear equation } (\lambda E_n - A)\mathbf{x} = \mathbf{0}_n.$$

We note that eigenvalues of A are eigenvalues of T_A. It is easy to see that

$$W(A; \lambda) = W(\lambda; T_A).$$

Characteristic polynomials of matrices For a square matrix A of degree n, we define the *characteristic polynomial* of A by

$$g_A(t) = \left| t E_n - A \right|.$$

Characteristic roots of matrices We call the roots of the characteristic polynomial $g_A(t)$ the *characteristic roots* of A. We note that even if A is a real matrix, characteristic roots are not necessarily real numbers and non-real complex numbers can appear as characteristic roots of A.

Example 3 Let $A = \begin{pmatrix} 7 & -6 \\ 3 & -2 \end{pmatrix}$. Then the characteristic polynomial of the matrix A is

$$g_A(t) = \left| t E_2 - A \right| = \begin{vmatrix} t-7 & 6 \\ -3 & t+2 \end{vmatrix}$$
$$= t^2 - 5t + 4 = (t-1)(t-4).$$

Therefore the characteristic roots of A are 1 and 4.

Theorem 5.3.1

(1) Let A be a square matrix of degree n. Then we have
λ is an eigenvalue of A
$\iff \lambda$ is a characteristic root of A belonging to K.
(2) Let A be a real square matrix. Then when we consider A as a matrix over $\mathbf{R}$, eigenvalues of A are real characteristic roots.

Proof (1) ($\Rightarrow$) Let $\mathbf{a} \in K^n$ be an eigenvector of A with the eigenvalue $\lambda \in K$. Then $\mathbf{a} \neq \mathbf{0}_n$ and we have

$$A\mathbf{a} = \lambda \mathbf{a} = \lambda E_n \mathbf{a}.$$

Since the linear equation $(\lambda E_n - A)\mathbf{x} = \mathbf{0}_n$ has a non-trivial solution $\mathbf{a}$, we have $g_A(\lambda) = \left| \lambda E_n - A \right| = 0$ by Theorems 2.4.2 and 3.4.2.

($\Leftarrow$) Let $\lambda \in K$ is a characteristic root of A. Since $\left| \lambda E_n - A \right| = g_A(\lambda) = 0$, there exists a non-zero vector $\mathbf{a} \in K^n$ satisfying $(\lambda E_n - A)\mathbf{a} = \mathbf{0}_n$, namely,

$$Aa = \lambda E_n \mathbf{a} = \lambda \mathbf{a}.$$

Therefore $\mathbf{a}$ is an eigenvector of A with the eigenvalue λ.

(2) This is the special case of (1) for $K = \mathbf{R}$. $\|$

Example 4 Let $A = \begin{pmatrix} 2 & 0 & 0 \\ 0 & 0 & 1 \\ 0 & -1 & 0 \end{pmatrix}$. Then the characteristic polynomial of A is

$$g_A(t) = \left| tE_3 - A \right| = \begin{vmatrix} t-2 & 0 & 0 \\ 0 & t & -1 \\ 0 & 1 & t \end{vmatrix}$$

$$= (t-2)(t^2 + 1).$$

Since the characteristic roots are the roots of $g_A(t) = (t-2)(t^2 + 1) = 0$, they are $2, i$ and $-i$. If we consider A as a matrix over $\mathbf{R}$, then the eigenvalue of A is only 2. But if we consider A as a matrix over $\mathbf{C}$, then the eigenvalues of A are $2, i$ and $-i$. Here i is the imaginary unit $\sqrt{-1}$.

We shall find eigenvectors of the matrix $\begin{pmatrix} 7 & -6 \\ 3 & -2 \end{pmatrix}$ considered in Example 2.

Example 5 Let $A = \begin{pmatrix} 7 & -6 \\ 3 & -2 \end{pmatrix}$. Then the characteristic polynomial of A is

$$g_A(t) = t^2 - 5t + 4 = (t-1)(t-4)$$

and the characteristic roots of A are 1, 4 as we obtained in Example 3. Since they are real numbers, they are eigenvalues of A by Theorem 5.3.1 (2). We shall find the eigenvectors of A.

To find the eigenvectors of A with the eigenvalue 1, we shall solve the linear equation $A\mathbf{x} = \mathbf{x}$, or $(E_2 - A)\mathbf{x} = \mathbf{0}_2$.

The reduced matrix of

$$E_2 - A = \begin{pmatrix} -6 & 6 \\ -3 & 3 \end{pmatrix}$$

is $\begin{pmatrix} 1 & -1 \\ 0 & 0 \end{pmatrix}$ by the calculation

$$\begin{array}{rr} -6 & 6 \\ \hline -3 & 3 \\ \hline 1 & -1 \\ \hline -3 & 3 \\ \hline 1 & -1 \\ \hline 0 & 0 \end{array} \quad \begin{array}{l} \\ \\ ① \times (-1/6) \\ \\ \\ ②+①\times 3. \end{array}$$

Therefore the eigenvectors with the eigenvalue $\lambda = 1$ are

$$c \begin{pmatrix} 1 \\ 1 \end{pmatrix} \in \mathbf{R}^2 \quad (c \neq 0).$$

To find the eigenvectors of A with the eigenvalue 4, we shall solve the linear equation $A\mathbf{x} = 4\mathbf{x}$, or $(4E_2 - A)\mathbf{x} = \mathbf{0}_2$. The reduced matrix of

$$4E_2 - A = \begin{pmatrix} -3 & 6 \\ -3 & 6 \end{pmatrix}$$

is $\begin{pmatrix} 1 & -2 \\ 0 & 0 \end{pmatrix}$ by the calculation

$$\begin{array}{rr} -3 & 6 \\ \hline -3 & 6 \\ \hline -3 & 6 \\ \hline 0 & 0 \\ \hline 1 & -2 \\ \hline 0 & 0 \end{array} \quad \begin{array}{l} \\ \\ \\ ②+①\times (-1) \\ ①\times (-1/3). \end{array}$$

Then the eigenvectors with the eigenvalue 4 are

$$c \begin{pmatrix} 2 \\ 1 \end{pmatrix} \in \mathbf{R}^2 \quad (c \neq 0).$$

Exercise 5.7 For a real matrix $A = \begin{pmatrix} 2 & 1 & -1 \\ 1 & 2 & -1 \\ 1 & 1 & 0 \end{pmatrix}$, answer the following questions:

(1) Calculate the characteristic polynomial $g_A(t)$ of A.
(2) Find the eigenvalues of A.
(3) Find the eigenspace $W(\lambda ; A)$ for each eigenvalue λ.

Answers. (1) By definition, we have

$$g_A(t) = |tE_3 - A| = \begin{vmatrix} t-2 & -1 & 1 \\ -1 & t-2 & 1 \\ -1 & -1 & t \end{vmatrix}$$
$$= t(t-2)^2 + 1 + 1 + (t-2) - t + (t-2)$$
$$= (t-1)^2(t-2).$$

(2) Since the characteristic polynomial of A is $g_A(t) = (t-1)^2(t-2)$, the characteristic roots are $1, 2$. As they are real numbers, the eigenvalues of A are $1, 2$ by Theorem 5.3.1 (2).

(3) We note that the eigenspace of A with the eigenvalue λ is the solution space of $(\lambda E_3 - A)\mathbf{x} = \mathbf{0}_3$.

(i) Let $\lambda = 1$. We shall solve the linear equation

$$(\lambda E_3 - A)\mathbf{x} = (E_3 - A)\mathbf{x} = \mathbf{0}_3.$$

The reduced matrix of

$$E_3 - A = \begin{pmatrix} -1 & -1 & 1 \\ -1 & -1 & 1 \\ -1 & -1 & 1 \end{pmatrix}$$

is $\begin{pmatrix} 1 & 1 & -1 \\ 0 & 0 & 0 \\ 0 & 0 & 0 \end{pmatrix}$ by the calculation

-1	-1	1	
-1	-1	1	
-1	-1	1	
1	1	-1	①× (−1)
-1	-1	1	
-1	-1	1	
1	1	-1	
0	0	0	②+ ①
0	0	0	③+ ①.

We solve the linear equation

$$\begin{pmatrix} 1 & 1 & -1 \\ 0 & 0 & 0 \\ 0 & 0 & 0 \end{pmatrix} \mathbf{x} = \mathbf{0}_3,$$

and obtain

$$W(1; A) = \left\{ c_1 \begin{pmatrix} -1 \\ 1 \\ 0 \end{pmatrix} + c_2 \begin{pmatrix} 1 \\ 0 \\ 1 \end{pmatrix} \ \middle|\ c_1, c_2 \in \mathbf{R} \right\}.$$

(ii) Let $\lambda = 2$. We shall solve the linear equation

$$(\lambda E_3 - A)\mathbf{x} = (2E_3 - A)\mathbf{x} = \mathbf{0}_3.$$

The reduced matrix of

$$2E_3 - A = \begin{pmatrix} 0 & -1 & 1 \\ -1 & 0 & 1 \\ -1 & -1 & 2 \end{pmatrix}$$

is $\begin{pmatrix} 1 & 0 & -1 \\ 0 & 1 & -1 \\ 0 & 0 & 0 \end{pmatrix}$ by the calculation

$$
\begin{array}{ccc}
0 & -1 & 1 \\
-1 & 0 & 1 \\
-1 & -1 & 2 \\
\hline
-1 & 0 & 1 \\
0 & -1 & 1 \\
-1 & -1 & 2 \\
\hline
1 & 0 & -1 \\
0 & 1 & -1 \\
-1 & -1 & 2 \\
\hline
1 & 0 & -1 \\
0 & 1 & -1 \\
0 & -1 & 1 \\
\hline
1 & 0 & -1 \\
0 & 1 & -1 \\
0 & 0 & 0 \\
\end{array}
\qquad
\begin{array}{l}
\\
\\
\\
① \leftrightarrow ② \\
\\
\\
① \times (-1) \\
② \times (-1) \\
\\
\\
\\
③ + ① \\
\\
\\
③ + ②.
\end{array}
$$

Solving the linear equation

$$\begin{pmatrix} 1 & 0 & -1 \\ 0 & 1 & -1 \\ 0 & 0 & 0 \end{pmatrix}\mathbf{x} = \mathbf{0}_3,$$

we obtain

$$W(2\,;A) = \left\{ c_3 \begin{pmatrix} 1 \\ 1 \\ 1 \end{pmatrix} \,\middle|\, c_3 \in \mathbf{R} \right\}. \quad \|$$

Polynomials of a square matrix Let $f(t) = a_m t^m + a_{m-1}t^{m-1} + \cdots + a_1 t + a_0$ be a polynomial of degree m. Substituting a square matrix A of degree n for t formally, we define the matrix $f(A)$ of degree n by

$$f(A) = a_m A^m + a_{m-1}A^{m-1} + \cdots + a_1 A + a_0 E_n.$$

For the characteristic polynomial $g_A(t)$ of A, we have the following theorem.

Theorem 5.3.2 (Cayley–Hamilton) *Let $g_A(t)$ be the characteristic polynomial of a square matrix A of degree n. Then we have*

$$g_A(A) = O_n.$$

Proof Let $B(t) = tE_n - A$ and $\tilde{B}(t)$ the cofactor matrix of $B(t)$. Since $g_A(t) = |tE_n - A| = |B(t)|$, we see by Theorem 3.4.1 that

$(*)\qquad B(t)\tilde{B}(t) = g_A(t)E_n.$

Since each component of $\tilde{B}(t)$ is a determinant of a matrix of degree $n - 1$ obtained by removing a column and a row from $B(t)$, it is a polynomial of t with degree at most $n - 1$. Therefore $\tilde{B}(t)$ is expressed as

$$\tilde{B}(t) = t^{n-1}B_{n-1} + t^{n-2}B_{n-2} + \cdots + tB_1 + B_0.$$

Here B_k is a square matrix of degree n for $k = 0, \ldots, n - 1$. Putting this equality into $(*)$, we obtain

$(**)\qquad (tE_n - A)(t^{n-1}B_{n-1} + t^{n-2}B_{n-2} + \cdots + tB_1 + B_0) = g_A(t)E_n.$

We want to substitute A for t. But it is not so simple because of non-commutativity of matrices. Expanding the left-hand side of $(**)$, we have

$$t^n B_{n-1} + t^{n-1}(B_{n-2} - AB_{n-1}) + \cdots + t(B_0 - AB_1) + (-AB_0) = g_A(t)E_n.$$

We let $g_A(t) = t^n + b_{n-1}t^{n-1} + \cdots + b_1t + b_0$ and compare both sides. Then

$(***)\qquad B_{n-1} = E_n,\ B_{n-2} - AB_{n-1} = b_{n-1}E_n,\ \ldots,\ -AB_0 = b_0E_n.$

Therefore by $(***)$ we have

$$\begin{aligned}
O_n &= (A - A)(A^{n-1}B_{n-1} + A^{n-2}B_{n-2} + \cdots + AB_1 + B_0)\\
&= A^n B_{n-1} + A^{n-1}(B_{n-2} - AB_{n-1}) + \cdots + \cdots + (-AB_0)\\
&= A^n + b_{n-1}A^{n-1} + \cdots + b_1A + b_0E_n = g_A(A). \quad \|
\end{aligned}$$

Example 6 The characteristic polynomial of $A = \begin{pmatrix} -1 & 1 \\ -6 & 4 \end{pmatrix}$ is

$$\begin{aligned}
g_A(t) = |tE_2 - A| &= \begin{vmatrix} t+1 & -1 \\ 6 & t-4 \end{vmatrix}\\
&= (t+1)(t-4) + 6 = t^2 - 3t + 2.
\end{aligned}$$

Therefore $g_A(A) = A^2 - 3A + 2E_2 = O_2$. In fact, we have

$$
\begin{aligned}
g_A(A) &= A^2 - 3A + 2E_2 \\
&= \begin{pmatrix} -1 & 1 \\ -6 & 4 \end{pmatrix}^2 - 3 \begin{pmatrix} -1 & 1 \\ -6 & 4 \end{pmatrix} + 2 \begin{pmatrix} 1 & 0 \\ 0 & 1 \end{pmatrix} \\
&= \begin{pmatrix} -5 & 3 \\ -18 & 10 \end{pmatrix} - \begin{pmatrix} -3 & 3 \\ -18 & 12 \end{pmatrix} + \begin{pmatrix} 2 & 0 \\ 0 & 2 \end{pmatrix} = O_2.
\end{aligned}
$$

Now, we shall define characteristic polynomials for linear transformations of vector spaces. For this purpose, we need the following theorem.

Theorem 5.3.3 *Let A be a square matrix of degree n and P a regular matrix of degree n. Then we have*

$$
g_{P^{-1}AP}(t) = g_A(t).
$$

Proof By the definition of the characteristic polynomial, we have

$$
\begin{aligned}
g_{P^{-1}AP}(t) &= \left| tE_n - P^{-1}AP \right| = \left| tP^{-1}P - P^{-1}AP \right| \\
&= \left| P^{-1}(tE_n)P - P^{-1}AP \right| = \left| P^{-1}(tE_n - A)P \right| \\
&= \left| P^{-1} \right| \cdot \left| tE_n - A \right| \cdot \left| P \right| = \left| tE_n - A \right| \\
&= g_A(t). \quad \|
\end{aligned}
$$

Characteristic polynomials and roots of linear transformations Let T be a linear transformation of a vector space V of dimension n and A the representation matrix of T with respect to a basis of V. We define the *characteristic polynomial* of T by

$$
g_T(t) = g_A(t) = \left| tE_n - A \right|.
$$

If B is another representation matrix of T with respect to another basis of V, then B is similar to A by Theorem 5.2.6 and $g_B(t) = g_A(t)$ by Theorem 5.3.3. Therefore the characteristic polynomial of T is uniquely determined. We call roots of $g_T(t) = 0$ *characteristic roots* of T.

Theorem 5.3.4 *Let T be a linear transformation of a vector space V of dimension n and A a representation matrix of T. Then the following statements are equivalent:*

(1) λ is an eigenvalue of T.
(2) λ is a characteristic root of T and $\lambda \in K$.
(3) λ is an eigenvalue of A.
(4) λ is a characteristic root of A and $\lambda \in K$.

Proof (3) $\Leftrightarrow$ (4) is nothing but Theorem 5.3.1 (1). Since $g_T(t) = g_A(t)$, (2) $\Leftrightarrow$ (4) is clear. Therefore we have only to see that (1) $\Leftrightarrow$ (3) holds.

(1) $\Leftrightarrow$ (3)   Let A be the representation matrix of T with respect to a basis $\{\mathbf{u}_1, \ldots, \mathbf{u}_n\}$ of V. Any vector $\mathbf{u}$ in V is expressed as a linear combination

$$\mathbf{u} = c_1 \mathbf{u}_1 + \cdots + c_n \mathbf{u}_n$$

of $\mathbf{u}_1, \ldots, \mathbf{u}_n$. Let $\mathbf{c} = \begin{pmatrix} c_1 \\ \vdots \\ c_n \end{pmatrix} \in K^n$. Then

$$T(\mathbf{u}) = T(c_1 \mathbf{u}_1 + \cdots + c_n \mathbf{u}_n) = c_1 T(\mathbf{u}_1) + \cdots + c_n T(\mathbf{u}_n)$$

$$= \left(T(\mathbf{u}_1) \ldots T(\mathbf{u}_n) \right) \begin{pmatrix} c_1 \\ \vdots \\ c_n \end{pmatrix}$$

$$= \left(\mathbf{u}_1 \ldots \mathbf{u}_n \right) A\mathbf{c}.$$

Now, let λ be an eigenvalue of T with an eigenvector $\mathbf{u}$. Then we have

$$T(\mathbf{u}) = \lambda \mathbf{u} = \lambda (c_1 \mathbf{u}_1 + \cdots + c_n \mathbf{u}_n) = \left(\mathbf{u}_1 \ldots \mathbf{u}_n \right) \lambda \mathbf{c}.$$

Since vectors $\mathbf{u}_1, \ldots, \mathbf{u}_n$ are linearly independent, we have

$$A\mathbf{c} = \lambda \mathbf{c}$$

by Theorem 4.2.5 (2). Since $\mathbf{u} \neq \mathbf{0}_V$, we have $\mathbf{c} \neq \mathbf{0}_n$. Then λ is an eigenvalue of A with an eigenvector $\mathbf{c}$.

Conversely, let λ be an eigenvalue of A with an eigenvector $\mathbf{c} = \begin{pmatrix} c_1 \\ \vdots \\ c_n \end{pmatrix}$. As $A\mathbf{c} = \lambda \mathbf{c}$, we see that $T(\mathbf{u}) = \lambda \mathbf{u}$ by the above consideration. Since $\mathbf{c} \neq \mathbf{0}_n$, we have $\mathbf{u} \neq \mathbf{0}_V$. Therefore λ is an eigenvalue of T with an eigenvector $\mathbf{u}$. $\|$

Exercise 5.8   Let T be a linear transformation of a vector space $V = \mathbf{R}[x]_2$ defined by

$$T(f(x)) = f(1 + 2x).$$

For the linear transformation T, answer the following questions:

(1)   Find the characteristic polynomial $g_T(t)$ of T.
(2)   Find the eigenvalues of T.
(3)   For each eigenvalue λ of T, find the eigenspace $W(\lambda \, ; T)$.

Answers. (1) Let $\{1, x, x^2\}$ be the standard basis of V. Since we have

$$\left(T(1)\ T(x)\ T(x^2) \right) = \left(1\ \ 1 + 2x\ \ 1 + 4x + 4x^2 \right)$$
$$= \left(1\ x\ x^2 \right) \begin{pmatrix} 1\ 1\ 1 \\ 0\ 2\ 4 \\ 0\ 0\ 4 \end{pmatrix},$$

the representation matrix with respect to the basis is

$$A = \begin{pmatrix} 1\ 1\ 1 \\ 0\ 2\ 4 \\ 0\ 0\ 4 \end{pmatrix}.$$

Therefore the characteristic polynomial of T is

$$g_T(t) = g_A(t) = \begin{vmatrix} t-1 & -1 & -1 \\ 0 & t-2 & -4 \\ 0 & 0 & t-4 \end{vmatrix}$$
$$= (t-1)(t-2)(t-4).$$

(2) We solve $g_T(t) = 0$ and get the characteristic roots $1, 2, 4$ of T. Since they are real numbers, they are the eigenvalues of T by Theorem 5.3.4.

(3) For each eigenvalue λ, we shall find the eigenspace $W(\lambda; T)$. Assume that $f(x) = a_0 + a_1 x + a_2 x^2$ is an eigenvector of T with the eigenvalue λ. If we let

$$\mathbf{a} = \begin{pmatrix} a_0 \\ a_1 \\ a_2 \end{pmatrix}, \text{ then } f(x) = \left(1\ x\ x^2 \right) \mathbf{a}. \text{ Since we have}$$

$$T(f(x)) = T\left(\left(1\ x\ x^2 \right) \mathbf{a} \right)$$
$$= \left(T(1)\ T(x)\ T(x^2) \right) \mathbf{a} = \left(1\ x\ x^2 \right) A\mathbf{a}$$

and

$$\lambda f(x) = \lambda \left(1\ x\ x^2 \right) \mathbf{a} = \left(1\ x\ x^2 \right) \lambda \mathbf{a},$$

we see that $T(f(x)) = \lambda f(x)$ is equivalent to that the column vector $\mathbf{a} = \begin{pmatrix} a_0 \\ a_1 \\ a_2 \end{pmatrix}$ is a solution of the linear equation $A\mathbf{x} = \lambda\mathbf{x}$, or a solution of the linear equation $(\lambda E_3 - A)\mathbf{x} = \mathbf{0}_3$. Therefore we have only to find the eigenspace $W(\lambda; A)$ of A.

(i) Let $\lambda = 4$. The reduced matrix of $4E_3 - A$ is $\begin{pmatrix} 1\ 0\ -1 \\ 0\ 1\ -2 \\ 0\ 0\ 0 \end{pmatrix}$ by the calculation

$$\begin{array}{ccc} 3 & -1 & -1 \\ 0 & 2 & -4 \\ 0 & 0 & 0 \end{array}$$

$$\begin{array}{ccc} 3 & -1 & -1 \\ 0 & 1 & -2 \\ 0 & 0 & 0 \end{array} \quad \text{②} \times (1/2)$$

$$\begin{array}{ccc} 3 & 0 & -3 \\ 0 & 1 & -2 \\ 0 & 0 & 0 \end{array} \quad \text{①} + \text{②}$$

$$\begin{array}{ccc} 1 & 0 & -1 \\ 0 & 1 & -2 \\ 0 & 0 & 0 \end{array} \quad \text{①} \times (1/3).$$

The solutions of $(4E_3 - A)\mathbf{x} = \mathbf{0}_3$ are $\mathbf{a} = c\begin{pmatrix} 1 \\ 2 \\ 1 \end{pmatrix}$ $(c \in \mathbf{R})$. Therefore

$$f(x) = \begin{pmatrix} 1 & x & x^2 \end{pmatrix}\mathbf{a} = c(1 + 2x + x^2)$$

and

$$W(4\,;T) = \{c(1 + 2x + x^2)\,|\,c \in \mathbf{R}\}.$$

(ii) Let $\lambda = 2$. The reduced matrix of $2E_3 - A$ is $\begin{pmatrix} 1 & -1 & 0 \\ 0 & 0 & 1 \\ 0 & 0 & 0 \end{pmatrix}$ by the calculation

$$\begin{array}{ccc} 1 & -1 & -1 \\ 0 & 0 & -4 \\ 0 & 0 & -2 \end{array}$$

$$\begin{array}{ccc} 1 & -1 & -1 \\ 0 & 0 & 1 \\ 0 & 0 & -2 \end{array} \quad \text{②} \times (-1/4)$$

$$\begin{array}{ccc} 1 & -1 & 0 \\ 0 & 0 & 1 \\ 0 & 0 & 0 \end{array} \quad \begin{array}{l} \text{①} + \text{②} \\ \\ \text{③} + \text{②} \times 2. \end{array}$$

The solutions of $(2E_3 - A)\mathbf{x} = \mathbf{0}_3$ are $\mathbf{a} = c\begin{pmatrix} 1 \\ 1 \\ 0 \end{pmatrix}$ $(c \in \mathbf{R})$. Therefore

$$f(x) = \begin{pmatrix} 1 & x & x^2 \end{pmatrix}\mathbf{a} = c(1 + x)$$

and

$$W(2\,;T) = \{c\,(1+x)\,|\,c \in \mathbf{R}\}.$$

(iii) Let $\lambda = 1$. The reduced matrix of $E_3 - A$ is $\begin{pmatrix} 0 & 1 & 0 \\ 0 & 0 & 1 \\ 0 & 0 & 0 \end{pmatrix}$ by the calculation

$$
\begin{array}{ccc}
0 & -1 & -1 \\
0 & -1 & -4 \\
0 & 0 & -3 \\
\hline
0 & -1 & -1 \\
0 & 0 & -3 \\
0 & 0 & -3 \\
\hline
0 & -1 & 0 \\
0 & 0 & -3 \\
0 & 0 & 0 \\
\hline
0 & 1 & 0 \\
0 & 0 & 1 \\
0 & 0 & 0 \\
\end{array}
\qquad
\begin{array}{l}
\\
\\
\\
\\
②+①\times(-1) \\
\\
①+②\times(-1/3) \\
\\
③+②\times(-1) \\
\hline
①\times(-1) \\
②\times(-1/3).
\end{array}
$$

The solutions of $(E_3 - A)\mathbf{x} = \mathbf{0}_3$ are $\mathbf{a} = c \begin{pmatrix} 1 \\ 0 \\ 0 \end{pmatrix}$ $(c \in \mathbf{R})$. Therefore

$$f(x) = \begin{pmatrix} 1 & x & x^2 \end{pmatrix} \mathbf{a} = c$$

and

$$W(1\,;T) = \{c\,|\,c \in \mathbf{R}\}. \quad \|$$

The vector spaces over the complex number field We mainly consider the real number field $\mathbf{R}$ and the complex number field $\mathbf{C}$ as the basic field K of vector spaces. But there are differences between the field $\mathbf{R}$ and the field $\mathbf{C}$. One of the differences is that any algebraic equation $f(x) = 0$ of degree n has n roots in $\mathbf{C}$ admitting multiple roots (*the fundamental theorem of algebra*). Therefore any linear transformation of V of dimension n ($\neq 0$) over $\mathbf{C}$ has always n eigenvalues admitting overlap. But this is not true for the real number field $\mathbf{R}$. Then the following four conditions are equivalent for a linear transformation T of a vector space V over $\mathbf{C}$ and a representation matrix A of the linear transformation T:

(1) λ is an eigenvalue of T. (2) λ is a characteristic root of T.

(3) λ is an eigenvalue of A. (4) λ is a characteristic root of A.

Commutative linear transformations We say that linear transformations T_1 and T_2 of a vector space V are *commutative* if $T_1 T_2 = T_2 T_1$ is satisfied.

For commutative linear transformations of a vector space over $\mathbf{C}$, we have the following theorem.

Theorem 5.3.5

(1) Let T_1 and T_2 be linear transformations of a vector space V over $\mathbf{C}$. If T_1 and T_2 are commutative, then T_1 and T_2 have a common eigenvector in V.

(2) Let A and B be complex square matrices of degree n. If A and B are commutative, then A and B have a common eigenvector in $\mathbf{C}^n$.

Proof (1) Let λ_1 be an eigenvalue of T_1 and $W(\lambda_1 ; T_1)$ the eigenspace of T_1 with the eigenvalue λ_1. If $\mathbf{v} \in W(\lambda_1 ; T_1)$, then

$$T_1(T_2(\mathbf{v})) = T_2(T_1(\mathbf{v})) = T_2(\lambda_1 \mathbf{v}) = \lambda_1 T_2(\mathbf{v}).$$

Therefore $T_2(\mathbf{v}) \in W(\lambda_1 ; T_1)$. Then T_2 induces a linear transformation T_2' of $W(\lambda_1 ; T_1)$. Let λ_2 be an eigenvalue of T_2' and $\mathbf{u} \in W(\lambda_1 ; T_1)$ an eigenvector of T_2' with the eigenvalue λ_2. Then we have

$$T_1(\mathbf{u}) = \lambda_1 \mathbf{u}, \quad \text{and} \quad T_2(\mathbf{u}) = T_2'(\mathbf{u}) = \lambda_2 \mathbf{u}.$$

Therefore $\mathbf{u}$ is a common eigenvector of T_1 and T_2.

(2) Since A and B are commutative, the linear transformations T_A and T_B associated with A and B, respectively are also commutative. Then (2) is obtained from (1) by taking $T_1 = T_A$ and $T_2 = T_B$.   ‖

Exercises (Sect. 5.3)

1   For the following polynomials $f(t)$ and matrices A, calculate $f(A)$:

(1) $A = \begin{pmatrix} 2 & 1 \\ 4 & -3 \end{pmatrix}$, $f(t) = 2t^2 + t - 1$.

(2) $A = \begin{pmatrix} 2 & 2 \\ 1 & -3 \end{pmatrix}$, $f(t) = t^2 + t - 8$.

2   For the following real matrices A, find (i), (ii) and (iii):

(i) Characteristic polynomials $g_A(t)$.
(ii) Eigenvalues λ of A.
(iii) Eigenspaces $W(\lambda ; A)$ for each eigenvalue λ of A.

(1) $A = \begin{pmatrix} 5 & -3 \\ 2 & 0 \end{pmatrix}$.   (2) $A = \begin{pmatrix} 2 & 2 \\ 3 & 1 \end{pmatrix}$.

(3) $A = \begin{pmatrix} 4 & -1 & 5 \\ 1 & 2 & 3 \\ -1 & 1 & 0 \end{pmatrix}$.   (4) $A = \begin{pmatrix} -1 & 0 & -2 \\ 3 & 2 & 2 \\ 1 & -1 & 3 \end{pmatrix}$.

$$
(5)\ A = \begin{pmatrix} 2 & 0 & 0 & 0 \\ 3 & 2 & 3 & 0 \\ -3 & 0 & -1 & 0 \\ 3 & 3 & 3 & -1 \end{pmatrix}. \qquad (6)\ A = \begin{pmatrix} -1 & -2 & 3 & -3 \\ 0 & 1 & 0 & 0 \\ 3 & 3 & -1 & 3 \\ 3 & 3 & -3 & 5 \end{pmatrix}.
$$

3 For the following linear transformations T of $\mathbf{R}[x]_2$, find (i), (ii) and (iii):

(i) Characteristic polynomials $g_T(t)$ of T.
(ii) Eigenvalues λ of T.
(iii) Eigenspaces $W(\lambda; T)$ for each eigenvalue λ of T.

(1) $T(f(x)) = f(1 - x)$. \qquad (2) $T(f(x)) = f(2x) + f'(x)$.

4 Let $A = \begin{pmatrix} 2 & 1 \\ -7 & -3 \end{pmatrix}$. For the following polynomials $f(t)$, calculate $f(A)$ by the Cayley–Hamilton theorem:

(1) $f(t) = t^{20}$. \qquad (2) $f(t) = t^{11} + t^7 - 2$.

5.4 Direct Sums of Vector Spaces and Minimal Polynomials

Sums of subspaces Let $W_1, \ldots, W_r$ be subspaces of a vector space V. By

$$
W_1 + \cdots + W_r = \{\, \mathbf{w}_1 + \cdots + \mathbf{w}_r \mid \mathbf{w}_i \in W_i \ (1 \le i \le r)\,\},
$$

we denote the subspace of V generated by vectors $\mathbf{w}_i$ in W_i for $i = 1, \ldots, r$. The subspace $W_1 + \cdots + W_r$ is called the *sum of the subspaces* $W_1, \ldots, W_r$.

Direct sums of subspaces Let V be a vector space and $W_1, \ldots, W_r$ non-zero subspaces of V. We say that V is the *direct sum of subspaces* $W_1, \ldots, W_r$ if any vector $\mathbf{v}$ in V is uniquely expressed as

$$
\mathbf{v} = \mathbf{w}_1 + \cdots + \mathbf{w}_r \qquad (\mathbf{w}_i \in W_i).
$$

When V is the direct sum of subspaces $W_1, \ldots, W_r$, we write

$$
V = W_1 \oplus \cdots \oplus W_r.
$$

We call $V = W_1 \oplus \cdots \oplus W_r$ the *direct sum decomposition* of V and the subspace W_i for $i = 1, \ldots, r$ the *direct factor* of the direct sum decomposition. If $V = W_1 \oplus \cdots \oplus W_r$, then we easily see that

$$
\dim(V) = \sum_{i=1}^{r} \dim(W_i).
$$

Example 1 Let $V = \mathbf{R}^2$. We take two subspaces W_1 and W_2 of $\mathbf{R}^2$ defined by

$$W_1 = \left\{ \begin{pmatrix} a \\ 0 \end{pmatrix} \middle| a \in \mathbf{R} \right\} \quad \text{and} \quad W_2 = \left\{ \begin{pmatrix} 0 \\ b \end{pmatrix} \middle| b \in \mathbf{R} \right\},$$

respectively. Then we have $V = W_1 \oplus W_2$. In fact, any vector $\begin{pmatrix} a_1 \\ a_2 \end{pmatrix}$ in V is uniquely expressed as a sum of vectors in W_1 and W_2, namely,

$$\begin{pmatrix} a_1 \\ a_2 \end{pmatrix} = \begin{pmatrix} a_1 \\ 0 \end{pmatrix} + \begin{pmatrix} 0 \\ a_2 \end{pmatrix}.$$

Example 2 Let $V = \mathbf{R}^2$. Take two subspaces of V defined by

$$W_1 = \left\{ a \begin{pmatrix} 1 \\ 1 \end{pmatrix} \middle| a \in \mathbf{R} \right\} \quad \text{and} \quad W_2 = \left\{ b \begin{pmatrix} 1 \\ 2 \end{pmatrix} \middle| b \in \mathbf{R} \right\}.$$

Then $V = W_1 \oplus W_2$. In fact, any vector $\mathbf{a} = \begin{pmatrix} a_1 \\ a_2 \end{pmatrix}$ in V is uniquely expressed as a linear combination of $\begin{pmatrix} 1 \\ 1 \end{pmatrix}$ and $\begin{pmatrix} 1 \\ 2 \end{pmatrix}$:

$$\mathbf{a} = (2a_1 - a_2) \begin{pmatrix} 1 \\ 1 \end{pmatrix} + (-a_1 + a_2) \begin{pmatrix} 1 \\ 2 \end{pmatrix}.$$

Theorem 5.4.1 *Let $W_1, \ldots, W_r$ be subspaces of a vector space V. When $V = W_1 + \cdots + W_r$, the following two conditions are equivalent:*

(1) $V = W_1 \oplus \cdots \oplus W_r$.
(2) $(W_1 + \cdots + W_{k-1}) \cap W_k = \{\mathbf{0}_V\}$ for $k = 2, \ldots, r$.

Proof $(1) \Rightarrow (2)$ Assume that there exists k $(2 \leq k \leq r)$ satisfying

$$(W_1 + \cdots + W_{k-1}) \cap W_k \neq \{\mathbf{0}_V\}.$$

Then there exists a non-zero vector $\mathbf{w}_k \in (W_1 + \cdots + W_{k-1}) \cap W_k$. Since $\mathbf{w}_k$ belongs to $W_1 + \cdots + W_{k-1}$, there exist vectors $\mathbf{w}_i \in W_i$ for $i = 1, \ldots, k-1$ satisfying $\mathbf{w}_k = \mathbf{w}_1 + \cdots + \mathbf{w}_{k-1}$. Therefore we have

$$\mathbf{0}_V = \mathbf{w}_1 + \cdots + \mathbf{w}_{k-1} - \mathbf{w}_k \quad \text{and} \quad \mathbf{w}_k \neq \mathbf{0}_V.$$

Since $\mathbf{0}_V = \mathbf{0}_V + \cdots + \mathbf{0}_V$, the zero vector $\mathbf{0}_V$ has at least two different expressions as sums of vectors in W_i for $i = 1, \ldots, k$. This is a contradiction.

$(2) \Rightarrow (1)$ Assume that a vector $\mathbf{v} \in V$ can be expressed as sums of vectors in two ways:

$$\mathbf{v} = \mathbf{w}_1 + \cdots + \mathbf{w}_{r-1} + \mathbf{w}_r = \mathbf{w}_1' + \cdots + \mathbf{w}_{r-1}' + \mathbf{w}_r'.$$

Here $\mathbf{w}_i$, $\mathbf{w}_i' \in W_i$ for $i = 1, \ldots, r$. Then we have

$$(\mathbf{w}_1 - \mathbf{w}_1') + \cdots + (\mathbf{w}_{r-1} - \mathbf{w}_{r-1}') = -(\mathbf{w}_r - \mathbf{w}_r'),$$

which is an element in $(W_1 + \cdots + W_{r-1}) \cap W_r$. Since $(W_1 + \cdots + W_{r-1}) \cap W_r = \{\mathbf{0}_V\}$, we have $\mathbf{w}_r - \mathbf{w}_r' = \mathbf{0}_V$, or $\mathbf{w}_r = \mathbf{w}_r'$. Thus we have

$$\mathbf{w}_1 + \cdots + \mathbf{w}_{r-1} = \mathbf{w}_1' + \cdots + \mathbf{w}_{r-1}'.$$

Repeating this argument, we have

$$\mathbf{w}_{r-1} = \mathbf{w}_{r-1}', \quad \ldots, \quad \mathbf{w}_1 = \mathbf{w}_1'.$$

Then the vector $\mathbf{v}$ is uniquely expressed as the sum of vectors in $W_1, \ldots, W_r$. Therefore $V = W_1 \oplus \cdots \oplus W_r$. ∥

Theorem 5.4.2 *Let V be a vector space and I_V the identity transformation of V. Let I_i $(1 \leq i \leq r)$ be linear transformations of V satisfying the following three conditions:*

(i) $I_V = I_1 + \cdots + I_r$,
(ii) $I_i I_j = O_V$ if $i \neq j$ for $i, j = 1, \ldots, r$,
(iii) $I_i I_i = I_i$ for $i = 1, \ldots, r$.

Then V is a direct sum of the subspaces $I_1(V), \ldots, I_r(V)$, namely,

$$V = I_1(V) \oplus \cdots \oplus I_r(V).$$

Here O_V is the zero transformation of V and $I_i I_j$ is the product of the linear transformations I_i and I_j.

Proof By condition (i), any vector $\mathbf{v}$ in V is expressed as

$$\mathbf{v} = I_V(\mathbf{v}) = I_1(\mathbf{v}) + \cdots + I_r(\mathbf{v}).$$

Thus we have $V = I_1(V) + \cdots + I_r(V)$. Therefore, by Theorem 5.4.1, we have only to show

$$(I_1(V) + \cdots + I_{k-1}(V)) \cap I_k(V) = \{\mathbf{0}_V\} \quad \text{for} \quad k = 2, \ldots, r.$$

We assume that

$$\mathbf{w} \in (I_1(V) + \cdots + I_{k-1}(V)) \cap I_k(V).$$

Then there exist vectors $\mathbf{v}_i \in V$ for $i = 1, \ldots, k$ satisfying

$$\mathbf{w} = I_1(\mathbf{v}_1) + \cdots + I_{k-1}(\mathbf{v}_{k-1}) = I_k(\mathbf{v}_k).$$

By condition (ii), we have

$$I_k(\mathbf{w}) = I_k I_1(\mathbf{v}_1) + \cdots + I_k I_{k-1}(\mathbf{v}_{k-1})$$
$$= \mathbf{0}_V + \cdots + \mathbf{0}_V = \mathbf{0}_V.$$

On the other hand, since $\mathbf{w} = I_k(\mathbf{v}_k)$, by condition (iii), we have

$$I_k(\mathbf{w}) = I_k I_k(\mathbf{v}_k) = I_k(\mathbf{v}_k) = \mathbf{w}.$$

Thus we see that $\mathbf{w} = \mathbf{0}_V$ and $(I_1(V) + \cdots + I_{k-1}(V)) \cap I_k(V) = \{\mathbf{0}_V\}$ for $k = 2, \ldots, r$. Therefore

$$V = I_1(V) \oplus \cdots \oplus I_r(V). \quad \|$$

Direct sums of linear transformations Let V be a vector space. Assume that $V = W_1 \oplus \cdots \oplus W_r$ is a direct sum of subspaces W_i of V for $i = 1, \ldots, r$. Then any vector $\mathbf{v}$ in V is uniquely expressed as a sum of vectors of W_i. For linear transformations T_i of W_i, we define the linear transformation T of V by

$$T(\mathbf{v}) = T_1(\mathbf{w}_1) + \cdots + T_r(\mathbf{w}_r) \quad \text{for} \quad \mathbf{v} = \mathbf{w}_1 + \cdots + \mathbf{w}_r \quad (\mathbf{w}_i \in W_i).$$

We call T the *direct sum of linear transformations* $T_1, \ldots, T_r$ and write

$$T = T_1 \oplus \cdots \oplus T_r.$$

Direct sums of square matrices Let A_i be a square matrix of degree n_i for $i = 1, \ldots, r$. We consider the square matrix

$$A = \begin{pmatrix} A_1 & & 0 \\ & \ddots & \\ 0 & & A_r \end{pmatrix}$$

of degree $n = n_1 + \cdots + n_r$. The matrix A is called the *direct sum of square matrices* $A_1, \ldots, A_r$ and denoted by

$$A = A_1 \oplus \cdots \oplus A_r.$$

Direct sums of square matrices are considered as a special case of partitions of matrices.

Example 3
$$\begin{pmatrix} -3 & 0 & 0 & 0 & 0 \\ 0 & 2 & -1 & 0 & 0 \\ 0 & -3 & 2 & 0 & 0 \\ 0 & 0 & 0 & 3 & 4 \\ 0 & 0 & 0 & -1 & 3 \end{pmatrix}$$
$$= (-3) \oplus \begin{pmatrix} 2 & -1 \\ -3 & 2 \end{pmatrix} \oplus \begin{pmatrix} 3 & 4 \\ -1 & 3 \end{pmatrix}.$$

Example 4 $\begin{pmatrix} 2 & 0 & 0 \\ 0 & 4 & 0 \\ 0 & 0 & -3 \end{pmatrix} = (2) \oplus (4) \oplus (-3).$

Now, we have the following theorem.

Theorem 5.4.3 *Let V be a vector space and $W_1, \ldots, W_r$ subspaces of V satisfying $V = W_1 \oplus \cdots \oplus W_r$. Let T_i be a linear transformation of W_i and A_i the representation matrix of T_i with respect to the basis $\{\mathbf{w}_{i1}, \ldots, \mathbf{w}_{in_i}\}$ of W_i for $i = 1, \ldots, r$.*

(1) The representation matrix A of the direct sum $T = T_1 \oplus \cdots \oplus T_r$ with respect to the basis $\{\mathbf{w}_{11}, \ldots, \mathbf{w}_{1n_1}, \ldots, \mathbf{w}_{r1}, \ldots, \mathbf{w}_{rn_r}\}$ of V is

$$A = A_1 \oplus \cdots \oplus A_r.$$

(2) Let $g_T(t)$ and $g_{T_i}(t)$ be the characteristic polynomials of T and T_i for $i = 1, \ldots, r$. Then we have

$$g_T(t) = \prod_{i=1}^{r} g_{T_i}(t) = g_{T_1}(t) \ldots g_{T_r}(t).$$

Proof We have only to prove the theorem when $r = 2$.

(1) Let $\{\mathbf{w}_{11}, \ldots, \mathbf{w}_{1n_1}\}$ be a basis of W_1 and $\{\mathbf{w}_{21}, \ldots, \mathbf{w}_{2n_2}\}$ a basis of W_2. Since

$$\left(T(\mathbf{w}_{11}) \ldots T(\mathbf{w}_{1n_1}) \ T(\mathbf{w}_{21}) \ldots T(\mathbf{w}_{2n_2}) \right)$$

$$= \left(T_1(\mathbf{w}_{11}) \ldots T_1(\mathbf{w}_{1n_1}) \ T_2(\mathbf{w}_{21}) \ldots T_2(\mathbf{w}_{2n_2}) \right)$$

$$= \left(\mathbf{w}_{11} \ldots \mathbf{w}_{1n_1} \ \mathbf{w}_{21} \ldots \mathbf{w}_{2n_2} \right) \begin{pmatrix} A_1 & O_{n_1, n_2} \\ O_{n_2, n_1} & A_2 \end{pmatrix},$$

we have $A = A_1 \oplus A_2$.

(2) Put $n = n_1 + n_2$. Since $E_n = E_{n_1} \oplus E_{n_2}$ and $A = A_1 \oplus A_2$ by (1), we have

$$\begin{aligned} g_T(t) &= \left| tE_n - A \right| = \left| t(E_{n_1} \oplus E_{n_2}) - (A_1 \oplus A_2) \right| \\ &= \left| (tE_{n_1} - A_1) \oplus (tE_{n_2} - A_2) \right| \\ &= \left| tE_{n_1} - A_1 \right| \cdot \left| tE_{n_2} - A_2 \right| = g_{T_1}(t) \, g_{T_2}(t). \quad \| \end{aligned}$$

Ideals of polynomials We let $K[t]$ be the set of polynomials with coefficients in K. A non-empty subset J of $K[t]$ is called an *ideal* of $K[t]$ if for any polynomials $f(t), g(t) \in K[t]$ and $p(t), q(t) \in J$,

$$f(t)p(t) + g(t)q(t) \in J$$

holds.

Theorem 5.4.4 *Any ideal J of $K[t]$ is generated by one polynomial. More precisely, there exists a polynomial $p(t) \in J$ satisfying*

$$J = \{ f(t)p(t) \mid f(t) \in K[t] \} = p(t)K[t].$$

Proof Let $f_{00}(t)$ be the constant zero polynomial. If $J = \{ f_{00}(t) \}$, then we can take $p(t) = f_{00}(t)$. We assume $J \neq \{ f_{00}(t) \}$. Let $p(t)$ be a non-zero polynomial in J whose degree is the minimum among the degrees of non-zero polynomials in J. Let $q(t)$ be a polynomial in J. Divide $q(t)$ by $p(t)$. Then we have

$$q(t) = f(t)p(t) + r(t) \quad (f(t), r(t) \in K[t], \ \deg(r(t)) < \deg(p(t))).$$

Assume that $r(t) \neq f_{00}(t)$. Then we see that

$$r(t) = q(t) - f(t)p(t).$$

Since both $p(t)$ and $q(t)$ are elements of J, the polynomial $r(t)$ belongs to J. Since the degree of $p(t)$ is the minimum among the degrees of non-zero polynomials in J, we see the existence of the non-zero polynomial $p(x)$ in J satisfying

$$\deg(r(t)) < \deg(p(t)),$$

which contradicts the choice of $p(t)$. Then we have $r(t) = f_{00}(t)$ and $q(t)$ is divisible by $p(t)$. Therefore $J = p(t)K[t]$. $\|$

Greatest common divisors of polynomials Let $n \geq 2$. We say that a polynomial $g(t)$ is a *common divisor* of non-zero polynomials $f_1(t), \ldots, f_n(t)$ if every polynomial $f_i(t)$ is divisible by $g(t)$ for $i = 1, \ldots, n$. If $\deg(g(t))$ is the largest among the degrees of common divisors of $f_1(t), \ldots, f_n(t)$, we call $g(t)$ the *greatest common divisor* of $f_1(t), \ldots, f_n(t)$. The greatest common divisor of $f_1(t), \ldots, f_n(t)$ is determined uniquely up to constant multiples. We say that polynomials $f_1(t), \ldots, f_n(t)$ have *no common divisor* when their common divisors are only constants.

Generators of ideals We call the polynomial $p(t)$ in Theorem 5.4.4 the *generator* of the ideal J, which is the greatest common divisor of the non-zero polynomials in J. The generator of an ideal is determined uniquely up to constant multiples.

Exercise 5.9

(1) Show that

$$J = \{ f(t)(t^3 + t^2 - 2t) + g(t)(t^2 - 1) \mid f(t), \, g(t) \in \mathbf{R}[\,t\,]\}$$

is an ideal of $\mathbf{R}[\,t\,]$.

(2) Find a generator of the ideal J.

Answers. (1) Let $f_1(t)(t^3 + t^2 - 2t) + g_1(t)(t^2 - 1)$ and $f_2(t)(t^3 + t^2 - 2t) + g_2(t)(t^2 - 1)$ be polynomials in J. Then for any polynomials $f(t)$ and $g(t)$ in $\mathbf{R}[\,t\,]$, we have

$$\begin{aligned}
& f(t)(f_1(t)(t^3 + t^2 - 2t) + g_1(t)(t^2 - 1)) \\
& \quad + g(t)(f_2(t)(t^3 + t^2 - 2t) + g_2(t)(t^2 - 1)) \\
& = (f(t)f_1(t) + g(t)f_2(t))(t^3 + t^2 - 2t) \\
& \quad + (f(t)g_1(t) + g(t)g_2(t))(t^2 - 1) \in J.
\end{aligned}$$

Therefore J is an ideal of $\mathbf{R}[\,t\,]$.

(2) Let $p_1(t) = t^3 + t^2 - 2t$ and $p_2(t) = t^2 - 1$. We divide $p_1(t)$ by $p_2(t)$ and obtain

$$p_1(t) = (t + 1)p_2(t) + (-t + 1).$$

Let $p_3(t) = t - 1$. Since

$$p_3(t) = -p_1(t) + (t + 1)p_2(t),$$

$p_3(t)$ belongs to J. Dividing $p_2(t)$ by $p_3(t) = t - 1$ we see that

$$p_2(t) = (t + 1)p_3(t).$$

We also have

$$\begin{aligned}
p_1(t) &= (t + 1)p_2(t) - p_3(t) \\
&= ((t + 1)^2 - 1)p_3(t) \\
&= (t^2 + 2t)p_3(t).
\end{aligned}$$

Since both $p_1(t)$ and $p_2(t)$ are divisible by $p_3(t) = t - 1$, we see that $J = p_3(t)\mathbf{R}[\,t\,] = (t - 1)\mathbf{R}[\,t\,]$. Therefore $p_3(t) = t - 1$ is the generator of the ideal J. ‖

By Theorem 5.4.4, we obtain the following theorem.

Theorem 5.4.5 *Assume that $n \geq 2$. If non-zero polynomials $p_1(t), \ldots, p_n(t)$ in $K[\,t\,]$ have no common divisor, then there exist polynomials $f_1(t), \ldots, f_n(t)$ in $K[\,t\,]$ satisfying*

$$f_1(t)p_1(t) + \cdots + f_n(t)p_n(t) = 1.$$

Proof We define a subset J of $K[t]$ by

$$J = \{ g_1(t)p_1(t) + \cdots + g_n(t)p_n(t) \mid g_1(t), \ldots, g_n(t) \in K[t] \}.$$

Then it is easy to see that J is an ideal of $K[t]$. Therefore the ideal J has the generator $p(t) \in J$ by Theorem 5.4.4, i.e.,

$$J = p(t)K[t].$$

Since $p(t)$ belongs to J, there exist polynomials $p_1(t), \ldots, p_n(t) \in K[t]$ satisfying

$$p(t) = g_1(t)p_1(t) + \cdots + g_n(t)p_n(t).$$

Since we have $J = p(t)K[t]$ and $p_i(t) \in J$, $p_i(t)$ are divisible by $p(t)$. By the assumption, $p_1(t), \ldots, p_n(t)$ have no common divisor. Therefore $p(t)$ must be a non-zero constant c. Let

$$f_i(t) = c^{-1}g_i(t) \quad \text{for} \quad i = 1, \ldots, n.$$

Then we have the equality

$$1 = f_1(t)p_1(t) + \cdots + f_n(t)p_n(t).$$

Thus we have the polynomials $f_1(t), \ldots, f_n(t) \in K[t]$ which we want.   ‖

Minimal polynomials of square matrices For a non-zero square matrix A of degree n, a non-zero polynomial $p_A(t)$ is called the *minimal polynomial* of A if $p_A(t)$ satisfies the following conditions:

(i) $p_A(A) = O_n$ and the degree of $p_A(t)$ is the minimum among the degrees of non-zero polynomials $f(t)$ satisfying $f(A) = O_n$.
(ii) The leading coefficient of $p_A(t)$ is 1.

For minimal polynomials of square matrices, we have the following theorem.

Theorem 5.4.6 *Let A be a non-zero square matrix of degree n.*

(1) There uniquely exists the minimal polynomial $p_A(t)$ of A.
(2) A polynomial $f(t)$ satisfies $f(A) = O_n$ if and only if $f(t)$ is divisible by $p_A(t)$. Especially, the characteristic polynomial $g_A(t)$ is divisible by $p_A(t)$.
(3) Any characteristic root of A is a root of the minimal polynomial $p_A(t)$.

Proof (1) Let $J = \{ f(t) \in K[t] \mid f(A) = O_n \}$. We shall show that the set J is a non-zero ideal of $K[t]$. Since $g_A(A) = O_n$ by Theorem 5.3.2 (Cayley–Hamilton), we have $g_A(t) \in J$ and $J \neq \{ f_{00}(t) \}$. For polynomials $f_1(t), f_2(t) \in J$ and $g_1(t), g_2(t) \in K[t]$, we have

$$g_1(A)f_1(A) + g_2(A)f_2(A) = g_1(A)O_n + g_2(A)O_n = O_n.$$

Thus J is a non-zero ideal of $K[t]$. Therefore there exists a non-zero polynomial $p_0(t)$ satisfying $J = p_0(t)K[t]$ by Theorem 5.4.4. We let $p_A(t)$ be the polynomial obtained by dividing $p_0(t)$ by the leading coefficient of $p_0(t)$. Since the leading coefficient of $p_A(t)$ is 1, $p_A(t)$ is the minimal polynomial of A. As the generators of the ideal J are uniquely determined up to constant multiples, the minimal polynomial $p_A(t)$ uniquely exists.

(2) Let $f(t) \in K[t]$. Since

$$f(A) = O_n \iff f(t) \in J$$
$$\iff f(t) \text{ is divisible by } p_A(t),$$

we have (2).

(3) By a similar argument in the proof of Theorem 5.3.2 (Cayley–Hamilton), there exists a polynomial $f(t)$ with matrix coefficients of degree n and a square matrix C of degree n satisfying

$$p_A(t)E_n = p_A(tE_n) = (tE_n - A)f(t) + C.$$

Substitute A for t. Since $p_A(A) = O_n$ and $(A - A)f(A) = O_n$, we have $C = O_n$. Thus we obtain

$$p_A(t)E_n = (tE_n - A)f(t).$$

Taking the determinants of both sides, we have $p_A(t)^n = g_A(t)\,|\,f(t)\,|$. Here $|\,f(t)\,|$ is the determinant of the matrix $f(t)$. Therefore $p_A(t)^n$ is divisible by the characteristic polynomial $g_A(t)$ of A and any characteristic root of A is a root of $p_A(t)$. ∥

Exercise 5.10 Find the minimal polynomial $p_A(t)$ of $A = \begin{pmatrix} 2 & 1 & 0 \\ 0 & 2 & 0 \\ 0 & 0 & 2 \end{pmatrix}$.

Answer.   Let $g_A(t)$ be the characteristic polynomial of A. Then we have $g_A(t) = |\,tE_3 - A\,| = (t - 2)^3$. Since $g_A(t)$ is divisible by $p_A(t)$, $p_A(t)$ is either

$$t - 2, \quad (t - 2)^2 \quad \text{or} \quad (t - 2)^3.$$

We calculate $A - 2E_3$ and $(A - 2E_3)^2$:

$$A - 2E_3 = \begin{pmatrix} 2 & 1 & 0 \\ 0 & 2 & 0 \\ 0 & 0 & 2 \end{pmatrix} - 2E_3 = \begin{pmatrix} 0 & 1 & 0 \\ 0 & 0 & 0 \\ 0 & 0 & 0 \end{pmatrix} \neq O_3,$$
$$(A - 2E_3)^2 = O_3.$$

Therefore the minimal polynomial of A is $p_A(t) = (t - 2)^2$. ∥

Polynomials of linear transformations Let T be a linear transformation of a vector space V. For a polynomial $f(t) = a_m t^m + \cdots + a_1 t + a_0$, we define the linear transformation $f(T)$ by

$$f(T) = a_m T^m + \cdots + a_1 T + a_0 I_V.$$

Here $T^i = \underbrace{T \ldots T}_{i}$ for $i \geq 1$ and T^i is called the ith product of T. We understand $T^0 = I_V$ (the identity transformation of V). If T is the linear transformation T_A associated with a square matrix A, then it is easy to see that $f(T_A)$ is the linear transformation associated with $f(A)$.

Minimal polynomials of linear transformations For a non-zero transformation T of a vector space V, we consider the non-zero polynomial $p_T(t)$ satisfies the following conditions:

(i) $p_T(T) = O_V$ and the degree of $p_T(t)$ is the minimum among the degrees of non-zero polynomials $f(t)$ satisfying $f(T) = O_V$.
(ii) The leading coefficient of $p_T(t)$ is 1.

Here O_V is the zero transformation of V. The polynomial $p_T(t)$ is called the *minimal polynomial* of T.

Theorem 5.4.7 *Let T be a linear transformation of V and A a representation matrix of T with respect to a basis $\{\mathbf{u}_1, \ldots, \mathbf{u}_n\}$ of V. Then we have the following:*

(1) For a polynomial $f(t)$, the representation matrix of $f(T)$ with respect to the same basis $\{\mathbf{u}_1, \ldots, \mathbf{u}_n\}$ is $f(A)$.
(2) For the characteristic polynomial $g_T(t)$ of T, we have

$$g_T(T) = O_V.$$

(3) The minimal polynomial $p_T(t)$ of T uniquely exists and is equal to the minimal polynomial $p_A(t)$ of A, namely,

$$p_T(t) = p_A(t).$$

(4) A polynomial $f(t)$ satisfies $f(T) = O_V$ if and only if $f(t)$ is divisible by $p_T(t)$. In particular, the characteristic polynomial $g_T(t)$ of T is divisible by $p_T(t)$.
(5) Any characteristic root of T is a root of the minimal polynomial $p_T(t)$.

Proof (1) Let $A = \begin{pmatrix} \mathbf{a}_1 & \ldots & \mathbf{a}_n \end{pmatrix}$ be the column partition of A. Since

$$\begin{pmatrix} T(\mathbf{u}_1) & \ldots & T(\mathbf{u}_n) \end{pmatrix} = \begin{pmatrix} \mathbf{u}_1 & \ldots & \mathbf{u}_n \end{pmatrix} A,$$

we have

$$T(\mathbf{u}_j) = \left(\mathbf{u}_1 \ \ldots \ \mathbf{u}_n \right) \mathbf{a}_j.$$

By the linearity of T, we see that

$$T^2(\mathbf{u}_j) = \left(T(\mathbf{u}_1) \ \ldots \ T(\mathbf{u}_n) \right) \mathbf{a}_j = \left(\mathbf{u}_1 \ \ldots \ \mathbf{u}_n \right) A \mathbf{a}_j,$$

$$\ldots,$$

$$T^m(\mathbf{u}_j) = \left(T(\mathbf{u}_1) \ \ldots \ T(\mathbf{u}_n) \right) A^{m-2} \mathbf{a}_j = \left(\mathbf{u}_1 \ \ldots \ \mathbf{u}_n \right) A^{m-1} \mathbf{a}_j,$$

$$\ldots.$$

Therefore for any positive integer m, we have

$$(T^m(\mathbf{u}_1) \ \ldots \ T^m(\mathbf{u}_n)) = (\mathbf{u}_1 \ldots \mathbf{u}_n) A^{m-1}(\mathbf{a}_1 \ldots \mathbf{a}_n) = (\mathbf{u}_1 \ldots \mathbf{u}_n) A^m$$

and for any polynomial $f(t)$, we see

$$(f(T)(\mathbf{u}_1) \ \ldots \ f(T)(\mathbf{u}_n)) = (\mathbf{u}_1 \ldots \mathbf{u}_n) f(A).$$

Therefore $f(A)$ is the representation matrix of $f(T)$.

(2) Let $g_A(t)$ be the characteristic polynomial of A. Then we have $g_T(t) = g_A(t)$ by definition. Since $g_A(A) = O_n$ by Theorem 5.3.2 (Cayley–Hamilton) and $g_A(A)$ is the representation matrix of $g_A(T)$ by (1), we have $g_A(T) = O_V$. Therefore we have

$$g_T(T) = g_A(T) = O_V.$$

(3) By (1), for any polynomial $f(t)$, we see that $f(T) = O_V$ if and only if $f(A) = O_n$. Therefore $p_T(t) = p_A(t)$ and the existence and the uniqueness of $p_T(t)$ follow from Theorem 5.4.6 (1).

(4) Follows from Theorem 5.4.6 (2).

(5) Follows from $p_T(t) = p_A(t)$, $g_T(t) = g_A(t)$ and Theorem 5.4.6 (3).   ‖

Least common multiples of polynomials Let $f_1(t), \ldots, f_r(t)$ be non-zero polynomials in $K[t]$. Then we easily see that the set

$$J = \{ f(t) \in K[t] \mid f(t) \text{ is divisible by } f_i(t) \text{ for } i = 1, \ldots, r \}$$

is an ideal of $K[t]$. In fact, if $f(t), g(t) \in J$, then $f(t)$ and $g(t)$ are divisible by $f_i(t)$ for $i = 1, \ldots, f_r(t)$. Then for any polynomials $p(t)$ and $q(t)$, the polynomial $p(t)f(t) + q(t)g(t)$ is divisible by $f_i(t)$ for $i = 1, \ldots, r$. Therefore J is an ideal of $K[t]$. Then there exists a non-zero polynomial $f_0(t)$ satisfying

$$J = f_0(t) K[t]$$

by Theorem 5.4.4. We call the generator $f_0(t)$ of J the *least common multiple of polynomials* $f_1(t), \ldots, f_r(t)$.

Theorem 5.4.8 *Let V be a vector space and*

$$V = W_1 \oplus \cdots \oplus W_r$$

a direct sum of subspaces $W_1, \ldots, W_r$ of V. We assume that the linear transformation T of V maps W_i into itself. Let T_i be the linear transformation of W_i induced by T and $p_{T_i}(t)$ the minimal polynomial of T_i for $i = 1, \ldots, r$, and $f_0(t)$ the least common multiple of the minimal polynomials $p_{T_1}(t), \ldots, p_{T_r}(t)$ with the leading coefficient 1. We denote the minimal polynomial of T by $p_T(t)$. Then we have

$$p_T(t) = f_0(t).$$

Proof Since $T = T_1 \oplus \cdots \oplus T_r$, we have

$$f(T) = f(T_1) \oplus \cdots \oplus f(T_r)$$

for any polynomial $f(t) \in K[t]$. Since the least common multiple $f_0(t)$ of $p_{T_1}(t)$, $\ldots, p_{T_r}(t)$ is divisible by $p_{T_i}(t)$ and $p_{T_i}(T_i) = O_{W_i}$ by definition, we have $f_0(T_i) = O_{W_i}$. Then we see that $f_0(T) = f_0(T_1) \oplus \cdots \oplus f_0(T_r) = O_{W_1} \oplus \cdots \oplus O_{W_r} = O_V$. Therefore $f_0(t)$ is divisible by $p_T(t)$ by Theorem 5.4.7 (4).

Conversely, since $p_T(T_i)$ is the restriction of $p_T(T)$ on W_i, we have $p_T(T_i) = O_{W_i}$ for $i = 1, \ldots, r$. Therefore $p_T(t)$ is divisible by $p_{T_i}(t)$ for $i = 1, \ldots, r$. Then

$$p_T(t) \in J_0 = \{ f(t) \mid f(t) \text{ is divisible by } p_{T_i}(t) \text{ for } i = 1, \ldots, r \}.$$

Therefore $p_T(t)$ is divisible by the generator $f_0(t)$ of J_0. Since both leading coefficients of $p_T(t)$ and $f_0(t)$ are 1, we have $p_T(t) = f_0(t)$.   ‖

We shall restate Theorem 5.4.8 in terms of matrices.

Theorem 5.4.9 *Let $A_1, \ldots, A_r$ be square matrices of degree $n_1, \ldots, n_r$, respectively. Let $n = n_1 + \cdots + n_r$ and*

$$A = A_1 \oplus \cdots \oplus A_r.$$

Let $f_0(t)$ be the least common multiple of the minimal polynomials $p_{A_1}(t), \ldots, p_{A_r}(t)$ of $A_1, \ldots, A_r$, respectively, with the leading coefficient 1. Then the minimal polynomial $p_A(t)$ of A is equal to $f_0(t)$.

Proof Since $f_0(t)$ is the least common multiple of $p_{A_1}(t), \ldots, p_{A_r}(t)$, we have

$$f_0(A) = f_0(A_1) \oplus \cdots \oplus f_0(A_r) = O_{n_1} \oplus \cdots \oplus O_{n_r} = O_n.$$

Then $f_0(t)$ is divisible by $p_A(t)$ by Theorem 5.4.6 (2).

Conversely, since $O_n = p_A(A) = p_A(A_1) \oplus \cdots \oplus p_A(A_r)$, we have

$$p_A(A_i) = O_{n_i} \quad \text{for} \quad i = 1, \ldots, r.$$

Then $p_A(t)$ is divisible by $p_{A_i}(t)$ by Theorem 5.4.6(2) for $i = 1, \ldots, r$ and

$$p_A(t) \in J_0 = \{ f(t) \mid f(t) \text{ is divisible by } p_{A_i}(t) \text{ for } i = 1, \ldots, r \}.$$

Therefore we have that $p_A(t)$ is divisible by the generator $f_0(t)$ of J_0. Since both leading coefficients of $p_T(t)$ and $f_0(t)$ are 1, we have $p_A(t) = f_0(t)$. ∥

Exercises (Sect. 5.4)

1 For the following subspaces W_1 and W_2 of V, show $V = W_1 \oplus W_2$:

(1) $V = \mathbf{R}^2, \quad W_1 = \left\{ c \begin{pmatrix} 1 \\ 1 \end{pmatrix} \;\middle|\; c \in \mathbf{R} \right\}, \quad W_2 = \left\{ c \begin{pmatrix} 1 \\ -1 \end{pmatrix} \;\middle|\; c \in \mathbf{R} \right\}.$

(2) $V = \mathbf{R}^3, W_1 = \left\{ c_1 \begin{pmatrix} 1 \\ 0 \\ -1 \end{pmatrix} + c_2 \begin{pmatrix} 0 \\ 1 \\ 1 \end{pmatrix} \;\middle|\; c_1, c_2 \in \mathbf{R} \right\},$

$\qquad W_2 = \left\{ c \begin{pmatrix} 1 \\ 2 \\ 0 \end{pmatrix} \;\middle|\; c \in \mathbf{R} \right\}.$

(3) $V = \mathbf{R}^4, W_1 = \left\{ \begin{pmatrix} c_1 \\ c_2 \\ c_3 \\ c_4 \end{pmatrix} \in \mathbf{R}^4 \;\middle|\; \begin{array}{l} c_1 + c_2 + c_3 + c_4 = 0 \\ c_1 - c_2 + c_3 + c_4 = 0 \end{array} \right\},$

$\qquad W_2 = \left\{ \begin{pmatrix} c_1 \\ c_2 \\ c_3 \\ c_4 \end{pmatrix} \in \mathbf{R}^4 \;\middle|\; \begin{array}{l} c_1 + c_2 - c_3 - c_4 = 0 \\ c_1 + c_2 - c_3 + c_4 = 0 \end{array} \right\}.$

2 For the following subspaces W_1 and W_2 of V, show that $V = W_1 \oplus W_2$:

(1) $V = \mathbf{R}[x]_2, \quad W_1 = \{ f(x) \in \mathbf{R}[x]_2 \mid f(1) = 0 \},$
$\qquad\qquad\qquad W_2 = \{ f(x) \in \mathbf{R}[x]_2 \mid f(0) = f(-1) = 0 \}.$

(2) $V = \mathbf{R}[x]_3, \quad W_1 = \{ f(x) \in \mathbf{R}[x]_3 \mid f(-1) = f(0) = 0 \},$
$\qquad\qquad\qquad W_2 = \{ f(x) \in \mathbf{R}[x]_3 \mid f(1) = f(2) = 0 \}.$

3 Find the minimal polynomials of the following matrices:

$$(1) \ A = \begin{pmatrix} 2 & 1 & 0 & 0 & 0 \\ 0 & 2 & 1 & 0 & 0 \\ 0 & 0 & 2 & 0 & 0 \\ 0 & 0 & 0 & 2 & 0 \\ 0 & 0 & 0 & 0 & 5 \end{pmatrix}. \qquad (2) \ A = \begin{pmatrix} 3 & 1 & 0 & 0 & 0 \\ 0 & 3 & 1 & 0 & 0 \\ 0 & 0 & 3 & 0 & 0 \\ 0 & 0 & 0 & 3 & 0 \\ 0 & 0 & 0 & 0 & 3 \end{pmatrix}.$$

4. Let $V = W_1 \oplus \cdots \oplus W_r$ be a direct sum of subspaces $W_1, \ldots, W_r$ of a vector space V. For a vector $\mathbf{v} = \mathbf{w}_1 + \cdots + \mathbf{w}_r$ $(\mathbf{w}_i \in W_i)$ in V, let $I_i(\mathbf{v}) = \mathbf{w}_i$ for $i = 1, \ldots, r$. Then show that I_i is a linear transformation of V and $I_1, \ldots, I_r$ satisfy conditions (i), (ii) and (iii) of Theorem 5.4.2.

5.5   Diagonalization

Diagonalization of linear transformations Let T be a linear transformation of a vector space V. We say that T is *diagonalizable* if there exists a basis of V consisting of eigenvectors of T. The *diagonalization* of T is finding such a basis of V.

Diagonalization of square matrices We say that a square matrix A is *diagonalizable* if there exists a regular matrix P so that the matrix $B = P^{-1}AP$ is a diagonal matrix. The *diagonalization* of a square matrix A is finding such matrices P and B.

Semisimple matrices We call diagonalizable matrices *semisimple matrices*.

We emphasize that *not all linear transformations nor all square matrices are diagonalizable*.

We begin with an application of the diagonalization.

Exercise 5.11 Let $A = \begin{pmatrix} 8 & -10 \\ 5 & -7 \end{pmatrix}$, $P = \begin{pmatrix} 1 & 2 \\ 1 & 1 \end{pmatrix}$.

(1) Show the equality $P^{-1}AP = \begin{pmatrix} -2 & 0 \\ 0 & 3 \end{pmatrix}$.

(2) Calculate A^n for a positive integer n by this equality.

Answers. (1) We let $B = P^{-1}AP$. Since $P^{-1} = \begin{pmatrix} -1 & 2 \\ 1 & -1 \end{pmatrix}$, we see that

$$B = P^{-1}AP = \begin{pmatrix} -1 & 2 \\ 1 & -1 \end{pmatrix} \begin{pmatrix} 8 & -10 \\ 5 & -7 \end{pmatrix} \begin{pmatrix} 1 & 2 \\ 1 & 1 \end{pmatrix} = \begin{pmatrix} -2 & 0 \\ 0 & 3 \end{pmatrix}.$$

(2) Calculations of diagonal matrices are easy. We easily see

$$B^n = \begin{pmatrix} (-2)^n & 0 \\ 0 & 3^n \end{pmatrix}.$$

Since $A = PBP^{-1}$, we have

$$\begin{aligned}
A^n &= (PBP^{-1})(PBP^{-1})\ldots(PBP^{-1}) = PB^nP^{-1} \\
&= \begin{pmatrix} 1 & 2 \\ 1 & 1 \end{pmatrix}\begin{pmatrix} (-2)^n & 0 \\ 0 & 3^n \end{pmatrix}\begin{pmatrix} -1 & 2 \\ 1 & -1 \end{pmatrix} \\
&= \begin{pmatrix} -(-2)^n + 2\cdot 3^n & 2\{(-2)^n - 3^n\} \\ -(-2)^n + 3^n & 2(-2)^n - 3^n \end{pmatrix}. \quad \|
\end{aligned}$$

Theorem 5.5.1 *Let A be a square matrix of degree n and T_A the linear transformation of K^n associated with A. Then the following two statements are equivalent:*

(1) The linear transformation T_A is diagonalizable.
(2) The matrix A is diagonalizable.

Proof (1) $\Rightarrow$ (2) Since T_A is diagonalizable, there exists a basis $\{\mathbf{p}_1, \ldots, \mathbf{p}_n\}$ of K^n consisting of eigenvectors of T_A. Let λ_i the eigenvalue of T_A with the eigenvector $\mathbf{p}_i$ for $i = 1, \ldots, n$. Let

$$P = \begin{pmatrix} \mathbf{p}_1 & \cdots & \mathbf{p}_n \end{pmatrix} \quad \text{and} \quad \Lambda = \begin{pmatrix} \lambda_1 & & 0 \\ & \ddots & \\ 0 & & \lambda_n \end{pmatrix}.$$

Then P is a regular matrix. Since

$$A\mathbf{p}_i = T_A(\mathbf{p}_i) = \lambda_i\mathbf{p}_i \quad \text{for} \quad i = 1, \ldots, n,$$

we have

$$\begin{aligned}
AP &= \begin{pmatrix} A\mathbf{p}_1 & \cdots & A\mathbf{p}_n \end{pmatrix} = \begin{pmatrix} \lambda_1\mathbf{p}_1 & \cdots & \lambda_n\mathbf{p}_n \end{pmatrix} \\
&= \begin{pmatrix} \mathbf{p}_1 & \cdots & \mathbf{p}_n \end{pmatrix}\Lambda = P\Lambda.
\end{aligned}$$

Therefore we have $AP = P\Lambda$. Since $\Lambda = P^{-1}AP$ is a diagonal matrix, the matrix A is diagonalizable.

(2) $\Rightarrow$ (1) Since the matrix A is diagonalizable, there exists a regular matrix P such that $\Lambda = P^{-1}AP$ is a diagonal matrix. Then $AP = P\Lambda$. Let

$$P = \begin{pmatrix} \mathbf{p}_1 \cdots \mathbf{p}_n \end{pmatrix} \quad \text{and} \quad \Lambda = \begin{pmatrix} \lambda_1 & & 0 \\ & \ddots & \\ 0 & & \lambda_n \end{pmatrix}.$$

Then we have

$$\left(A\mathbf{p}_1 \ \ldots \ A\mathbf{p}_n \right) = AP = P\Lambda = \left(\lambda_1\mathbf{p}_1 \ \ldots \ \lambda_n\mathbf{p}_n \right).$$

Therefore we have $T_A(\mathbf{p}_i) = A\mathbf{p}_i = \lambda_i\mathbf{p}_i$ for $i = 1, \ldots, n$. Namely, $\mathbf{p}_i$ is an eigenvector of T_A. Since the matrix P is a regular matrix, the vectors $\mathbf{p}_1, \ldots, \mathbf{p}_n$ are linearly independent and the set $\{ \mathbf{p}_1, \ldots, \mathbf{p}_n \}$ is a basis of K^n. Thus we see that the vector space K^n has a basis consisting of eigenvectors of T_A. $\quad \|$

Theorem 5.5.2 *Let T be a linear transformation of a vector space V. We denote by $\lambda_1, \ldots, \lambda_r$ all the different eigenvalues of T. Put $W = W(\lambda_1 ; T) + \cdots + W(\lambda_r ; T)$. Then W is a direct sum of subspaces $W(\lambda_1 ; T), \ldots, W(\lambda_r ; T)$, namely,*

$$W = W(\lambda_1 ; T) \oplus \cdots \oplus W(\lambda_r ; T).$$

Proof Assume that $\mathbf{w} \in W$ is expressed as

$$\mathbf{w} = \mathbf{w}_1 + \cdots + \mathbf{w}_r = \mathbf{w}_1' + \cdots + \mathbf{w}_r'.$$

Here $\mathbf{w}_i, \mathbf{w}_i' \in W(\lambda_i ; T)$ for $i = 1, \ldots, r$. Then

$$\mathbf{0}_V = (\mathbf{w}_1 - \mathbf{w}_1) + \cdots + (\mathbf{w}_r - \mathbf{w}_r').$$

Therefore we have only to show that the zero vector is expressed uniquely as a sum of vectors in $W(\lambda_i ; T)$ for $i = 1, \ldots, r$. We assume

$$\mathbf{0}_V = \mathbf{w}_1 + \cdots + \mathbf{w}_r \quad (\mathbf{w}_i \in W(\lambda_i ; T)).$$

Applying T to both sides of this equality repeatedly, we obtain

$$\mathbf{0}_V = \lambda_1\mathbf{w}_1 + \cdots + \lambda_r\mathbf{w}_r,$$

$$\ldots\ldots\ldots,$$

$$\mathbf{0}_V = \lambda_1^{r-1}\mathbf{w}_1 + \cdots + \lambda_r^{r-1}\mathbf{w}_r.$$

These equalities are expressed as

$$\left(\mathbf{0}_V \ \ldots \ \mathbf{0}_V \right) = \left(\mathbf{w}_1 \ \ldots \ \mathbf{w}_r \right) P$$

with the matrix $P = \begin{pmatrix} 1 & \lambda_1 & \ldots & \lambda_1^{r-1} \\ \vdots & \vdots & & \vdots \\ 1 & \lambda_r & \ldots & \lambda_r^{r-1} \end{pmatrix}$. The determinant of ${}^tP$ is the Vandermonde determinant considered in Sect. 3.5, Exercise 3.2. Since the λ_i's are all different, we have $|P| \neq 0$. Therefore P is a regular matrix. Multiplying both sides of the equality by P^{-1} from the right, we obtain

$$\mathbf{w}_1 = \cdots = \mathbf{w}_r = \mathbf{0}_V. \quad \|$$

Theorem 5.5.3 *Let T be a linear transformation of a vector space V and $\lambda_1, \ldots, \lambda_r$ all the different eigenvalues of T. Then the following three statements are equivalent:*

(1) The linear transformation T is diagonalizable.
(2) $V = W(\lambda_1 ; T) \oplus \cdots \oplus W(\lambda_r ; T)$.

(3) The minimal polynomial of T is $p_T(t) = \prod_{i=1}^{r}(t - \lambda_i)$, namely, $p_T(t)$ has no multiple roots.

Proof (1) $\Leftrightarrow$ (2)   Assume that T is diagonalizable. Then there exists a basis of V consisting of eigenvectors of T. Changing the order of vectors in the basis, we may assume that the basis is

$$\{\mathbf{w}_{11}, \ldots, \mathbf{w}_{1n_1}, \ldots, \mathbf{w}_{r1}, \ldots, \mathbf{w}_{rn_r}\},$$

and the vectors $\mathbf{w}_{i1}, \ldots, \mathbf{w}_{in_i}$ are in $W(\lambda_i ; T)$ for $i = 1, \ldots, r$. Since $V = W(\lambda_1 ; T) + \cdots + W(\lambda_r ; T)$, we have

$$V = W(\lambda_1 ; T) \oplus \cdots \oplus W(\lambda_r ; T)$$

by Theorem 5.5.2. Conversely, we assume that $V = W(\lambda_1 ; T) \oplus \cdots \oplus W(\lambda_r ; T)$. Take a basis $\{\mathbf{w}_{i1}, \ldots, \mathbf{w}_{in_i}\}$ of $W(\lambda_i ; T)$ for $i = 1, \ldots, r$. Then the set of vectors $\{\mathbf{w}_{11}, \ldots, \mathbf{w}_{1n_1}, \ldots, \mathbf{w}_{r1}, \ldots, \mathbf{w}_{rn_r}\}$ consisting of eigenvectors of T is a basis of V. Therefore the linear transformation T is diagonalizable.

(2) $\Leftrightarrow$ (3)   We easily see that the minimal polynomial $p_i(t)$ of the restriction T_i of T to the subspace $W(\lambda_i ; T)$ is $t - \lambda_i$ for $i = 1, \ldots, r$. Since V is the direct sum of $W(\lambda_i ; T)$ and the least common multiple of $p_i(t) = t - \lambda_i$ for $i = 1, \ldots, r$ is $\prod_{i=1}^{r}(t - \lambda_i)$, we have $p_T(t) = \prod_{i=1}^{r}(t - \lambda_i)$ by Theorem 5.4.8.

Conversely, we assume $p_T(t) = \prod_{i=1}^{r}(t - \lambda_i)$. Let $f_i(t) = \prod_{j \neq i}(t - \lambda_j)$ for $i = 1, \ldots, r$. Since polynomials $f_i(t)$ for $i = 1, \ldots, r$ have no common divisor, there exist polynomials $g_1(t), \ldots, g_r(t)$ satisfying

$$g_1(t)f_1(t) + \cdots + g_r(t)f_r(t) = 1$$

by Theorem 5.4.5. Then any vector $\mathbf{u} \in V$ is expressed as

$$\mathbf{u} = I_V(\mathbf{u}) = g_1(T)f_1(T)(\mathbf{u}) + \cdots + g_r(T)f_r(T)(\mathbf{u}).$$

Put $\mathbf{u}_i = g_i(T)f_i(T)(\mathbf{u})$ for $i = 1, \ldots, r$. Since

$$(T - \lambda_i I_V)(\mathbf{u}_i) = g_i(T)f_i(T)(T - \lambda_i I_V)(\mathbf{u}) = g_i(T)p_T(T)(\mathbf{u}) = \mathbf{0}_V,$$

we have $\mathbf{u}_i \in W(\lambda_i ; T)$ for $i = 1, \ldots, r$. Then $V = W(\lambda_1 ; T) + \cdots + W(\lambda_r ; T)$. Therefore $V = W(\lambda_1 ; T) \oplus \cdots \oplus W(\lambda_r ; T)$ by Theorem 5.5.2.   ‖

We restate Theorem 5.5.3 in terms of matrices.

Theorem 5.5.4 *Let A be a square matrix of degree n and* $\lambda_1, \ldots, \lambda_r$ *all the different eigenvalues of A. Then the following three statements are equivalent:*

(1) The square matrix A is diagonalizable.
(2) $K^n = W(\lambda_1 ; A) \oplus \cdots \oplus W(\lambda_r ; A)$.
(3) The minimal polynomial of A is $p_A(t) = \prod_{i=1}^{r}(t - \lambda_i)$, namely, $p_A(t)$ has no multiple roots.

Exercise 5.12 Find if the real matrix $A = \begin{pmatrix} 5 & 6 & 0 \\ -1 & 0 & 0 \\ 1 & 2 & 2 \end{pmatrix}$ is diagonalizable or not. Furthermore, if the matrix A is diagonalizable, then diagonalize it.

Answer.   First, we calculate the characteristic polynomial of the matrix A. The characteristic polynomial is

$$g_A(t) = \left| tE_3 - A \right| = \begin{vmatrix} t-5 & -6 & 0 \\ 1 & t & 0 \\ -1 & -2 & t-2 \end{vmatrix} = (t-2)^2(t-3).$$

Therefore the eigenvalues of A are 2, 3 by Theorem 5.3.1. We shall find the eigenspace $W(\lambda ; A)$ for each eigenvalue λ.

(i) Let $\lambda = 2$. Solve the linear equation $\begin{pmatrix} -3 & -6 & 0 \\ 1 & 2 & 0 \\ -1 & -2 & 0 \end{pmatrix} \mathbf{x} = \mathbf{0}$. By the calculation

$$\begin{array}{ccc} -3 & -6 & 0 \\ 1 & 2 & 0 \\ -1 & -2 & 0 \\ \hline 1 & 2 & 0 \\ -3 & -6 & 0 \\ -1 & -2 & 0 \\ \hline 1 & 2 & 0 \\ 0 & 0 & 0 \\ 0 & 0 & 0 \end{array} \quad \begin{array}{l} \\ \\ \\ ① \leftrightarrow ② \\ \\ \\ \\ ② + ① \times 3 \\ ③ + ①, \end{array}$$

we have

$$W(2\,;A) = \left\{ c_1 \begin{pmatrix} -2 \\ 1 \\ 0 \end{pmatrix} + c_2 \begin{pmatrix} 0 \\ 0 \\ 1 \end{pmatrix} \middle| c_1, c_2 \in \mathbf{R} \right\}.$$

(ii) Let $\lambda = 3$. Solve the linear equation $\begin{pmatrix} -2 & -6 & 0 \\ 1 & 3 & 0 \\ -1 & -2 & 1 \end{pmatrix} \mathbf{x} = \mathbf{0}$. By the calcula-

tion

$$
\begin{array}{ccc}
-2 & -6 & 0 \\
1 & 3 & 0 \\
-1 & -2 & 1 \\
\hline
1 & 3 & 0 \\
-1 & -2 & 1 \\
-2 & -6 & 0 \\
\hline
1 & 3 & 0 \\
0 & 1 & 1 \\
0 & 0 & 0 \\
\hline
1 & 0 & -3 \\
0 & 1 & 1 \\
0 & 0 & 0
\end{array}
\qquad
\begin{array}{l}
① \leftarrow ② \\
② \leftarrow ③ \\
③ \leftarrow ① \\[6pt]
② + ① \\
③ + ① \times 2 \\
① + ② \times (-3),
\end{array}
$$

we have

$$W(3\,;A) = \left\{ c \begin{pmatrix} 3 \\ -1 \\ 1 \end{pmatrix} \middle| c \in \mathbf{R} \right\}.$$

Since

$$\dim (W(2\,;A) \oplus W(3\,;A))$$
$$= \dim (W(2\,;A)) + \dim (W(3\,;A))$$
$$= 2 + 1 = 3 = \dim(\mathbf{R}^3),$$

we have $\mathbf{R}^3 = W(2\,;A) \oplus W(3\,;A)$ and the matrix A is diagonalizable by Theorem 5.5.4.

Let $P = \begin{pmatrix} -2 & 0 & 3 \\ 1 & 0 & -1 \\ 0 & 1 & 1 \end{pmatrix}$. Then we have $B = P^{-1}AP = \begin{pmatrix} 2 & 0 & 0 \\ 0 & 2 & 0 \\ 0 & 0 & 3 \end{pmatrix}$. ‖

We note that the choice of the pair of matrices P and B is not unique. In fact, if we take

$$P = \begin{pmatrix} -2 & 3 & 0 \\ 1 & -1 & 0 \\ 0 & 1 & 1 \end{pmatrix}, \quad \text{then} \quad B = P^{-1}AP = \begin{pmatrix} 2 & 0 & 0 \\ 0 & 3 & 0 \\ 0 & 0 & 2 \end{pmatrix}.$$

This pair of matrices P and B are different from the pair of matrices obtained in the answer to Exercise 5.11. But the diagonal components of the diagonal matrices B are unchanged except for the order of the components.

Exercise 5.13 Find if the matrix $A = \begin{pmatrix} 1 & 3 & 2 \\ 0 & -1 & 0 \\ 1 & 2 & 0 \end{pmatrix}$ is diagonalizable or not.

Answer. The characteristic polynomial of the matrix A is

$$g_A(t) = \left| tE_3 - A \right| = \begin{vmatrix} t-1 & -3 & -2 \\ 0 & t+1 & 0 \\ -1 & -2 & t \end{vmatrix} = (t+1)^2(t-2).$$

Therefore the eigenvalues of A are -1, 2 by Theorem 5.3.1. We shall calculate the eigenspaces with eigenvalues $\lambda = -1$ and $\lambda = 2$.

(i) Let $\lambda = -1$. To obtain the eigenspace $W(-1; A)$, we solve the linear equation $\begin{pmatrix} -2 & -3 & -2 \\ 0 & 0 & 0 \\ -1 & -2 & -1 \end{pmatrix} \mathbf{x} = \mathbf{0}_3$. The reduced matrix of $\begin{pmatrix} -2 & -3 & -2 \\ 0 & 0 & 0 \\ -1 & -2 & -1 \end{pmatrix}$ is $\begin{pmatrix} 1 & 0 & 1 \\ 0 & 1 & 0 \\ 0 & 0 & 0 \end{pmatrix}$.

Then we have

$$W(-1;\ A) = \left\{ c \begin{pmatrix} -1 \\ 0 \\ 1 \end{pmatrix} \middle| c \in \mathbf{R} \right\}.$$

(ii) Let $\lambda = 2$. To obtain the eigenspace $W(2; A)$, we solve the linear equation $\begin{pmatrix} 1 & -3 & -2 \\ 0 & 3 & 0 \\ -1 & -2 & 2 \end{pmatrix} \mathbf{x} = \mathbf{0}_3$. The reduced matrix of $\begin{pmatrix} 1 & -3 & -2 \\ 0 & 3 & 0 \\ -1 & -2 & 2 \end{pmatrix}$ is $\begin{pmatrix} 1 & 0 & -2 \\ 0 & 1 & 0 \\ 0 & 0 & 0 \end{pmatrix}$.

Then we have

$$W(2;\ A) = \left\{ c \begin{pmatrix} 2 \\ 0 \\ 1 \end{pmatrix} \middle| c \in \mathbf{R} \right\}.$$

Since

$$\dim\left(W(-1;\ A) \oplus W(2;\ A)\right) = \dim\left(W(-1;\ A)\right) + \dim\left(W(2;\ A)\right)$$
$$= 1 + 1 \neq 3 = \dim(\mathbf{R}^3),$$

we have $W(-1;\ A) \oplus W(2;\ A) \neq \mathbf{R}^3$. Therefore the matrix A is not diagonalizable by Theorem 5.5.4. ∥

Exercise 5.14 Find if the linear transformation T of $\mathbf{R}[x]_2$ defined by

$$T(f(x)) = (1+x)(f(1) + f'(x))$$

is diagonalizable or not. If T is diagonalizable, then diagonalize it.

Answer. Since

$$\left(T(1)\ T(x)\ T(x^2)\right) = \left(1\ x\ x^2\right) \begin{pmatrix} 1 & 2 & 1 \\ 1 & 2 & 3 \\ 0 & 0 & 2 \end{pmatrix},$$

the representation matrix of T is $A = \begin{pmatrix} 1 & 2 & 1 \\ 1 & 2 & 3 \\ 0 & 0 & 2 \end{pmatrix}$ and

$$g_T(t) = \begin{vmatrix} t-1 & -2 & -1 \\ -1 & t-2 & -3 \\ 0 & 0 & t-2 \end{vmatrix} = t(t-2)(t-3).$$

Therefore the eigenvalues of T are 0, 2 and 3 by Theorem 5.3.1.

(i) Solving the linear equation $A\mathbf{x} = 0\mathbf{x}$, we have $\mathbf{x} = c \begin{pmatrix} -2 \\ 1 \\ 0 \end{pmatrix}$ $(c \in \mathbf{R})$.

(ii) Solving the linear equation $A\mathbf{x} = 2\mathbf{x}$, we have $\mathbf{x} = c \begin{pmatrix} -3 \\ -2 \\ 1 \end{pmatrix}$ $(c \in \mathbf{R})$.

(iii) Solving the linear equation $A\mathbf{x} = 3\mathbf{x}$, we have $\mathbf{x} = c \begin{pmatrix} 1 \\ 1 \\ 0 \end{pmatrix}$ $(c \in \mathbf{R})$.

Thus we obtain

$$W(0; T) = \{c(-2+x)|c \in \mathbf{R}\}, \quad W(2; T) = \{c(-3-2x+x^2)|c \in \mathbf{R}\},$$

$$W(3; T) = \{c(1+x)|c \in \mathbf{R}\}.$$

Since dim ($W(0; T)$) + dim ($W(2; T)$) + dim ($W(3; T)$) = 3 = dim ($\mathbf{R}[x]_2$), the linear transformation T is diagonalizable by Theorem 5.5.3. The basis of $\mathbf{R}[x]_2$ consisting of eigenvectors of T is

$$\{-2+x, \ -3-2x+x^2, \ 1+x\}. \ \|$$

To prove Theorem 5.5.6 below, we need the following theorem.

Theorem 5.5.5 *Let T be a linear transformation of a vector space V and $W_1, \ldots, W_r$ be subspaces of V. We assume that*

$$V = W_1 \oplus \cdots \oplus W_r \quad and \quad T(W_i) \subset W_i \quad for\ i = 1, \ldots, r.$$

Let $\mathbf{v}$ be an eigenvector of T with an eigenvalue λ. If

$$\mathbf{v} = \mathbf{w}_1 + \cdots + \mathbf{w}_r \quad (\mathbf{w}_i \in W_i),$$

then we have $T(\mathbf{w}_i) = \lambda \mathbf{w}_i$ *for* $i = 1, \ldots, r$.

Proof We have only to prove the case for $r = 2$. Let $\mathbf{v}$ be an eigenvector of T with an eigenvalue λ. We assume that $\mathbf{v} = \mathbf{w}_1 + \mathbf{w}_2$ $(\mathbf{w}_1 \in W_1, \mathbf{w}_2 \in W_2)$. Then we have

$$T(\mathbf{w}_1) + T(\mathbf{w}_2) = T(\mathbf{w}_1 + \mathbf{w}_2) = \lambda(\mathbf{w}_1 + \mathbf{w}_2) = \lambda \mathbf{w}_1 + \lambda \mathbf{w}_2.$$

Therefore

$$T(\mathbf{w}_1) - \lambda \mathbf{w}_1 = -(T(\mathbf{w}_2) - \lambda \mathbf{w}_2).$$

Since $(T(\mathbf{w}_1) - \lambda \mathbf{w}_1) \in W_1$ and $(T(\mathbf{w}_2) - \lambda \mathbf{w}_2) \in W_2$, they must be the zero vector of V by Theorem 5.4.1. Therefore

$$T(\mathbf{w}_1) = \lambda \mathbf{w}_1 \quad \text{and} \quad T(\mathbf{w}_2) = \lambda \mathbf{w}_2. \quad \|$$

For commutative linear transformations, we have the following theorem.

Theorem 5.5.6

(1) Let S and T be linear transformations of a vector space V. Let $n = \dim(V)$. Assume that both S and T are diagonalizable. If S and T are commutative, then there exists a basis of V consisting of eigenvectors of both S and T.

(2) Let A and B be square matrices of degree n. Assume that both A and B are diagonalizable. If A and B are commutative, then there exists a regular matrix P so that

$$P^{-1}AP \text{ and } P^{-1}BP \text{ are diagonal matrices.}$$

Proof (1) Let $\lambda_1, \ldots, \lambda_r$ be all the different eigenvalues of T. Since T is diagonalizable, V is the direct sum

$$(*) \qquad V = W(\lambda_1 ; T) \oplus \cdots \oplus W(\lambda_r ; T)$$

of eigenspaces $W(\lambda_j ; T)$ for $j = 1, \ldots, r$. Since T and S are commutative, S induces a linear transformation of $W(\lambda_j ; T)$ for $j = 1, \ldots, r$. In fact, if $\mathbf{w}_j \in W(\lambda_j ; T)$, then

$$T(S(\mathbf{w}_j)) = S(T(\mathbf{w}_j)) = S(\lambda_j \mathbf{w}_j) = \lambda_j S(\mathbf{w}_j).$$

Therefore we have $S(\mathbf{w}_j) \in W(\lambda_j ; T)$. Since S is diagonalizable, there exists a basis $\{\mathbf{v}_1, \ldots, \mathbf{v}_n\}$ of V consisting of eigenvectors of S. By $(*)$, $\mathbf{v}_i$ is written as a sum of vectors $\mathbf{w}_{ij} \in W(\lambda_j ; T)$:

$$\mathbf{v}_i = \mathbf{w}_{i1} + \cdots + \mathbf{w}_{ir}$$

for $i = 1, \ldots, n$. By Theorem 5.5.5, non-zero vectors in the set $\{\mathbf{w}_{11}, \ldots, \mathbf{w}_{1r}, \ldots, \mathbf{w}_{n1}, \ldots, \mathbf{w}_{nr}\}$ are eigenvectors of S. The set of vectors $\{\mathbf{v}_1, \ldots, \mathbf{v}_n\}$ is a basis of V and each vector $\mathbf{v}_i$ is a linear combination of the vectors in $\{\mathbf{w}_{11}, \ldots, \mathbf{w}_{1r}, \ldots, \mathbf{w}_{n1}, \ldots, \mathbf{w}_{nr}\}$. Then we can find a basis of V by taking vectors from $\{\mathbf{w}_{11}, \ldots, \mathbf{w}_{1r}, \ldots, \mathbf{w}_{n1}, \ldots, \mathbf{w}_{nr}\}$. Since $\mathbf{w}_{ij} \in W(\lambda_j ; T)$, the basis is consisting of eigenvectors of both S and T.

(2) Though (2) is just the restatement of (1) in terms of matrices, we give a brief proof here. Let T_A and T_B be the linear transformations associated with A and B, respectively. Since the matrices A and B are diagonalizable, both T_A and T_B are also diagonalizable. Since A and B are commutative, T_A and T_B are commutative. Then by (1), there exists a basis $\{\mathbf{p}_1, \ldots, \mathbf{p}_n\}$ of K^n consisting of eigenvectors of both T_A and T_B. Let

$$P = (\mathbf{p}_1 \; \cdots \; \mathbf{p}_n).$$

We let λ_i be the eigenvalue of T_A and μ_i the eigenvalue of T_B for the vector $\mathbf{p}_i$ for $i = 1, \ldots, n$. In other words,

$$A\mathbf{p}_i = \lambda_i \mathbf{p}_i \quad \text{and} \quad B\mathbf{p}_i = \mu_i \mathbf{p}_i$$

for $i = 1, \ldots, n$. Then we have

$$P^{-1}AP = \begin{pmatrix} \lambda_1 & & 0 \\ & \ddots & \\ 0 & & \lambda_n \end{pmatrix} \quad \text{and} \quad P^{-1}BP = \begin{pmatrix} \mu_1 & & 0 \\ & \ddots & \\ 0 & & \mu_n \end{pmatrix}. \quad \|$$

Example 1 Let $A = \begin{pmatrix} 3 & 2 \\ -1 & 0 \end{pmatrix}$ and $B = \begin{pmatrix} -5 & -8 \\ 4 & 7 \end{pmatrix}$. We have

$$g_A(t) = (t-1)(t-2) \quad \text{and} \quad g_B(t) = (t-3)(t+1).$$

Since $p_A(t) = g_A(t) = (t-1)(t-2)$ and $p_B(t) = g_B(t) = (t-3)(t+1)$, both A and B are diagonalizable by Theorem 5.5.4. It is easy to see that A and B are commutative. Then the matrices A and B can be diagonalized by the same regular matrix by Theorem 5.5.6. In fact, solving a linear equation $(E_2 - A)\mathbf{x} = \mathbf{0}_2$, we have $\mathbf{x} = a \begin{pmatrix} -1 \\ 1 \end{pmatrix}$ $(a \in \mathbf{R})$. Solving a linear equation $(2E_2 - A)\mathbf{x} = \mathbf{0}_2$, we have $\mathbf{x} = b \begin{pmatrix} -2 \\ 1 \end{pmatrix}$ $(b \in \mathbf{R})$. Let $P = \begin{pmatrix} -1 & -2 \\ 1 & 1 \end{pmatrix}$. Then we see that

$$P^{-1}AP = \begin{pmatrix} 1 & 0 \\ 0 & 2 \end{pmatrix} \quad \text{and} \quad P^{-1}BP = \begin{pmatrix} 3 & 0 \\ 0 & -1 \end{pmatrix}.$$

Nilpotent transformations A linear transformation T of a vector space V is called a *nilpotent transformation* if there exists a positive integer m satisfying $T^m = O_V$.

Theorem 5.5.7 *Any non-zero nilpotent transformation is not diagonalizable.*

Proof Let T be a non-zero nilpotent transformation of a vector space V of dimension n satisfying $T^m = O_V$ $(m \geq 2)$. We assume that T is diagonalizable. Then there exists a basis $\{\mathbf{u}_1, \ldots, \mathbf{u}_n\}$ of V consisting of eigenvectors of T. We let

$$T(\mathbf{u}_i) = \lambda_i \mathbf{u}_i \quad \text{for} \quad i = 1, \ldots, n.$$

Then we have

$$T^m(\mathbf{u}_i) = \lambda_i^m \mathbf{u}_i \quad \text{for} \quad i = 1, \ldots, n.$$

Since $T^m = O_V$, we have $\lambda_i = 0$ for $i = 1, \ldots, n$. Therefore $T = O_V$ and this contradicts the assumption that $T \neq O_V$. ‖

Exercises (Sect. 5.5)

1 Diagonalize the linear transformations T_A of $\mathbf{R}^3$ which are associated with the following matrices A:

(1) $A = \begin{pmatrix} 1 & 0 & 2 \\ -3 & -2 & -2 \\ 0 & 0 & 3 \end{pmatrix}$.

(2) $A = \begin{pmatrix} 2 & 0 & 1 \\ 1 & 1 & 1 \\ 0 & 0 & 1 \end{pmatrix}$.

2 Find if the following real matrices are diagonalizable or not. If the matrices are diagonalizable, then diagonalize them:

(1) $\begin{pmatrix} 7 & -6 \\ 3 & -2 \end{pmatrix}$.

(2) $\begin{pmatrix} 13 & -30 \\ 5 & -12 \end{pmatrix}$.

(3) $\begin{pmatrix} 2 & -3 \\ -1 & 2 \end{pmatrix}$.

(4) $\begin{pmatrix} -3 & -2 & -2 \\ 4 & 3 & 2 \\ 8 & 4 & 5 \end{pmatrix}$.

(5) $\begin{pmatrix} 2 & -1 & 2 \\ 1 & 0 & 2 \\ -2 & 2 & -1 \end{pmatrix}$.

(6) $\begin{pmatrix} 2 & -2 & -2 \\ 0 & 1 & -1 \\ 0 & 0 & 2 \end{pmatrix}$.

(7) $\begin{pmatrix} 2 & -1 & 4 \\ 0 & 1 & 4 \\ -3 & 3 & -1 \end{pmatrix}$.

(8) $\begin{pmatrix} 2 & 0 & 2 & 0 \\ 1 & 1 & 2 & 0 \\ 0 & 0 & 1 & 0 \\ -1 & -3 & -2 & -2 \end{pmatrix}$.

(9) $\begin{pmatrix} -1 & -6 & 0 & 0 \\ 0 & 2 & 0 & 0 \\ 0 & -3 & -1 & 0 \\ -4 & -6 & 0 & 3 \end{pmatrix}$.

3. Find if the following linear transformations T of $\mathbf{R}[x]_2$ are diagonalizable or not. If the linear transformations are diagonalizable, then diagonalize them:

(1) $T(f(x)) = f(x) + f(1)(x + x^2)$.
(2) $T(f(x)) = f(0)x + f(1)x^2 + f(x)$.
(3) $T(f(x)) = xf'(x) + f(0)x + f(1)x^2$.

4 Calculate the n-th power of the following matrices A:

(1) $A = \begin{pmatrix} 7 & -6 \\ 3 & -2 \end{pmatrix}$.

(2) $A = \begin{pmatrix} 13 & -30 \\ 5 & -12 \end{pmatrix}$.

5 Show that all non-zero nilpotent matrices are not diagonalizable.

6 Answer questions (i), (ii) and (iii) for the following matrices A and B:

(i) Show that the matrices A and B are commutative.
(ii) Show that the matrices A and B are diagonalizable.
(iii) Find a regular matrix P such that both $P^{-1}AP$ and $P^{-1}BP$ are diagonal matrices.

(1) $A = \begin{pmatrix} 0 & 2 & 1 \\ 1 & 1 & 1 \\ 2 & -2 & 1 \end{pmatrix}$, $B = \begin{pmatrix} 2 & 1 & 1 \\ 1 & 2 & 1 \\ 1 & -1 & 2 \end{pmatrix}$.

(2) $A = \begin{pmatrix} 2 & 0 & 0 \\ 3 & -1 & 0 \\ 3 & -3 & 2 \end{pmatrix}$, $B = \begin{pmatrix} 5 & -2 & 2 \\ 4 & -1 & 2 \\ 0 & 0 & 1 \end{pmatrix}$.

7 For a square matrix $A = \begin{pmatrix} a_{11} & \cdots & a_{1n} \\ \vdots & & \vdots \\ a_{n1} & \cdots & a_{nn} \end{pmatrix}$, we define the *trace* of A by

$$\mathrm{tr}(A) = a_{11} + a_{22} + \cdots + a_{nn}.$$

Let

$$g_A(t) = t^n + a_{n-1}t^{n-1} + \cdots + a_1 t + a_0$$

be the characteristic polynomial of A. Then show that

$$a_{n-1} = -\mathrm{tr}(A) \quad \text{and} \quad a_0 = (-1)^n \, |A|.$$

8 Let A be an $m \times n$ matrix and B an $n \times m$ matrix. Show the equality

$$\mathrm{tr}(AB) = \mathrm{tr}(BA).$$

9. Let A and B be similar square matrices of degree n. Show the equality

$$\mathrm{tr}(A) = \mathrm{tr}(B).$$

10 Show that a square matrix A is a regular matrix if and only if every characteristic root of A is not zero.

11. Let T be a linear transformation of a vector space V and A a representation matrix of T. Show that T is diagonalizable if and only if A is diagonalizable.

5.6   Spaces of Matrices and Equivalence Relations

Spaces of matrices Let $M_{m \times n}(K)$ be the set of all $m \times n$ matrices over a field K, or

$$M_{m \times n}(K) = \{A \mid A \text{ is an } m \times n \text{ matrix over} K \}.$$

The set $M_{m \times n}(K)$ is a vector space over K (Theorem 5.6.1 below). We call $M_{m \times n}(K)$ the *space of $m \times n$ matrices* over K. We note that the zero vector in the vector space $M_{m \times n}(K)$ is the zero matrix $O_{m,n}$.

Matrix units For a pair (i, j) $(1 \le i \le m,\ 1 \le j \le n)$, let E_{ij} be the $m \times n$ matrix whose (i, j) component is 1 and other components are zeros. The matrix E_{ij} is called the (i, j)-*matrix unit*:

$$
E_{ij} =
\begin{array}{c}
\text{the } j\text{th column} \\
\downarrow \\
\begin{pmatrix}
0 & \cdots & 0 & \cdots & 0 \\
\vdots & & \vdots & & \vdots \\
0 & \cdots & 1 & \cdots & 0 \\
\vdots & & \vdots & & \vdots \\
0 & \cdots & 0 & \cdots & 0
\end{pmatrix}
\end{array}
\leftarrow \text{the } i\text{th row.}
$$

By Theorem 5.6.1 below, $M_{m \times n}(K)$ is a vector space and $\{E_{11}, E_{12}, \ldots, E_{mn}\}$ is a basis of the vector space $M_{m \times n}(K)$. We call the basis the *standard basis* of $M_{m \times n}(K)$.

Theorem 5.6.1 *The set $M_{m \times n}(K)$ is a vector space over K by the addition and the scalar multiplication of matrices. Furthermore, $\dim(M_{m \times n}(K)) = mn$.*

Proof It is easy to see that $M_{m \times n}(K)$ is a vector space over K. Since the (i, j)-matrix units E_{ij} $(i = 1, \ldots, m;\ j = 1, \ldots, n)$ are linearly independent and generate the space $M_{m \times n}(K)$, the set

$$\{E_{11}, E_{12}, \ldots, E_{mn}\}$$

is a basis of $M_{m \times n}(K)$. Therefore we have

$$\dim(M_{m \times n}(K)) = mn. \quad \|$$

Spaces of linear mappings Let U and V be vector spaces over K. We denote by $\mathrm{Hom}(U, V)$ or $\mathrm{Hom}_K(U, V)$ the set of all the linear mappings of U into V. By the addition and the scalar multiplication of linear mappings defined in Sect. 5.1, the set $\mathrm{Hom}(U, V)$ is a vector space over K (Theorem 5.6.2 below). We call $\mathrm{Hom}(U, V)$ the *space of linear mappings* of U into V. We note that the zero vector in $\mathrm{Hom}(U, V)$ is the zero mapping $O_{U,V}$ of U into V.

Theorem 5.6.2 *Let U and V be vector spaces. The set* $\mathrm{Hom}(U, V)$ *is a vector space by the addition and the scalar multiplication of linear mappings defined in Sect. 5.1.*

Proof Since the addition and the scalar multiplication of $\mathrm{Hom}(U, V)$ satisfy the eight properties of vector spaces, $\mathrm{Hom}(U, V)$ is a vector space over K and the zero vector in $\mathrm{Hom}(U, V)$ is the zero mapping $O_{U,V}$. ‖

Let $\{\mathbf{u}_1, \ldots, \mathbf{u}_n\}$ be a basis of a vector space U and $\{\mathbf{v}_1, \ldots, \mathbf{v}_m\}$ a basis of a vector space V. For a linear mapping $T \in \mathrm{Hom}(U, V)$, we denote by A_T the representation matrix of T with respect to the bases $\{\mathbf{u}_1, \ldots, \mathbf{u}_n\}$ and $\{\mathbf{v}_1, \ldots, \mathbf{v}_m\}$. We define the mapping Φ of $\mathrm{Hom}(U, V)$ into $\mathrm{M}_{m \times n}(K)$ by

$$\Phi(T) = A_T.$$

Theorem 5.6.3 *Let U be a vector space of dimension n and V a vector space of dimension m.*

(1) The mapping Φ is an isomorphism of $\mathrm{Hom}(U, V)$ *into* $\mathrm{M}_{m \times n}(K)$.
(2) $\dim(\mathrm{Hom}(U, V)) = mn$.

Proof (1) Let T, T_1, T_2 be vectors in $\mathrm{Hom}(U, V)$. Then it is easy to see $A_{T_1+T_2} = A_{T_1} + A_{T_2}$. Thus we have

$$\Phi(T_1 + T_2) = A_{T_1+T_2} = A_{T_1} + A_{T_2} = \Phi(T_1) + \Phi(T_2).$$

We also see $A_{cT} = cA_T$ for any $c \in K$. Then we have

$$\Phi(cT) = A_{cT} = cA_T = c\Phi(T).$$

Therefore the mapping Φ is a linear mapping of $\mathrm{Hom}(U, V)$ into $\mathrm{M}_{m \times n}(K)$.

We shall show that the linear mapping Φ is an isomorphism. To see that Φ is a one-to-one mapping, we show that $\mathrm{Ker}(\Phi) = \{O_{U,V}\}$. Here $O_{U,V}$ is the zero mapping of U into V. We assume that $T \in \mathrm{Ker}(\Phi)$. Then $A_T = \Phi(T) = O_{m,n}$. Then $T(\mathbf{u}_1) = \cdots = T(\mathbf{u}_n) = \mathbf{0}_V$ and $T = O_{U,V}$.

Next, we shall show that Φ is surjective. Take any matrix $A \in \mathrm{M}_{m \times n}(K)$. We want to show the existence of a linear mapping $T \in \mathrm{Hom}(U, V)$ satisfying $\Phi(T) = A$. Since a linear mapping T is determined by $\{T(\mathbf{u}_1), \ldots, T(\mathbf{u}_n)\}$, we define the vectors $T(\mathbf{u}_1), \ldots, T(\mathbf{u}_n)$ in V by

$$\left(T(\mathbf{u}_1) \ldots T(\mathbf{u}_n) \right) = \left(\mathbf{v}_1 \ldots \mathbf{v}_m \right) A$$

with the matrix A. Then we have $\Phi(T) = A$. Since Φ is one-to-one and surjective, it is an isomorphism of $\mathrm{Hom}(U, V)$ into $\mathrm{M}_{m \times n}(K)$.

(2) By Theorems 5.1.4 and 5.6.1, we have

$$\dim(\mathrm{Hom}(U, V)) = \dim(\mathrm{M}_{m \times n}(K)) = mn. \;\|$$

Linear functionals and dual spaces Let U be a vector space over K. We call linear mappings of U into K *linear functionals* of U. We denote by U^* the set of linear functionals of U:

$$U^* = \mathrm{Hom}(U, K) = \text{the set of linear functionals of } U.$$

Therefore U^* is a vector space over K. The vector space U^* is called the *dual space* of U over K. Since $\dim(K) = 1$, we have

$$\dim(U^*) = \dim(U)$$

by Theorem 5.6.3 (2). The addition and the scalar multiplication of U^* are the addition and the scalar multiplication as linear mappings of U into K:

(Addition) $\qquad\qquad (f + g)(\mathbf{u}) = f(\mathbf{u}) + g(\mathbf{u}),$
(Scalar multiplication) $\qquad (cf)(\mathbf{u}) = cf(\mathbf{u}).$

Here $f, g \in U^*$, $\mathbf{u} \in U$ and $c \in K$.

Dual bases Let $\dim(U) = n$. For a basis $\{\mathbf{u}_1, \ldots, \mathbf{u}_n\}$ of U, we take vectors $f_1, \ldots, f_n$ (i.e. linear functionals) in U^* defined by

$$f_i(a_1\mathbf{u}_1 + \cdots + a_n\mathbf{u}_n) = a_i \quad \text{for} \quad i = 1, \ldots, n.$$

We note that they satisfy $f_i(\mathbf{u}_j) = \delta_{ij}$ (the Kronecker delta) for $i, j = 1, \ldots, n$. The vectors $f_1, \ldots, f_n$ in U^* are linearly independent. In fact, assume that they satisfy

$$a_1 f_1 + \cdots + a_n f_n = \mathbf{0}_{U^*} \quad (a_i \in K).$$

Then for any $\mathbf{u}_i (1 \leq i \leq n)$, we have

$$0 = (a_1 f_1 + \cdots + a_n f_n)(\mathbf{u}_i) = a_i.$$

Therefore they are linearly independent. Since

$$\dim(U^*) = \dim(U) = n,$$

we see that the set $\{f_1, \ldots, f_n\}$ is a basis of U^* by Theorem 4.4.5. We call this basis $\{f_1, \ldots, f_n\}$ of U^* the *dual basis* of the basis $\{\mathbf{u}_1, \ldots, \mathbf{u}_n\}$ of U.

Example 1 Let $V = \mathbf{R}^2$ and f_1, f_2 be vectors in V^* defined by

$$f_1\left(\begin{pmatrix} a_1 \\ a_2 \end{pmatrix}\right) = a_1, \quad f_2\left(\begin{pmatrix} a_1 \\ a_2 \end{pmatrix}\right) = a_2$$

for $\begin{pmatrix} a_1 \\ a_2 \end{pmatrix} \in V$. Then $\{f_1, f_2\}$ is the dual basis of the standard basis $\{e_1, e_2\}$ of V since

$$f_i(e_j) = \delta_{ij}$$

holds.

Theorem 5.6.4 *Let $\{u_1, \ldots, u_n\}$ and $\{v_1, \ldots, v_n\}$ be bases of a vector space U of the dimension n, and let $\{f_1, \ldots, f_n\}$ and $\{g_1, \ldots, g_n\}$ be dual bases of the bases $\{u_1, \ldots, u_n\}$ and $\{v_1, \ldots, v_n\}$, respectively. If*

$$\begin{pmatrix} v_1 & \ldots, & v_n \end{pmatrix} = \begin{pmatrix} u_1 & \ldots, & u_n \end{pmatrix} P,$$

then we have

$$\begin{pmatrix} g_1 & \ldots, & g_n \end{pmatrix} = \begin{pmatrix} f_1 & \ldots, & f_n \end{pmatrix} {}^t P^{-1}.$$

Proof Let Q be a square matrix of degree n satisfying

$$\begin{pmatrix} g_1 & \ldots, & g_n \end{pmatrix} = \begin{pmatrix} f_1 & \ldots, & f_n \end{pmatrix} Q.$$

Then we have

$$
\begin{aligned}
E_n &= \begin{pmatrix} g_1(v_1) & \ldots & g_1(v_n) \\ \vdots & & \vdots \\ g_n(v_1) & \ldots & g_n(v_n) \end{pmatrix} = \begin{pmatrix} g_1 \\ \vdots \\ g_n \end{pmatrix} \begin{pmatrix} v_1 & \ldots & v_n \end{pmatrix} \\
&= {}^t Q \begin{pmatrix} f_1 \\ \vdots \\ f_n \end{pmatrix} \begin{pmatrix} v_1 & \ldots & v_n \end{pmatrix} = {}^t Q \begin{pmatrix} f_1 \\ \vdots \\ f_n \end{pmatrix} \begin{pmatrix} u_1 & \ldots & u_n \end{pmatrix} P \\
&= {}^t Q \begin{pmatrix} f_1(u_1) & \ldots & f_1(u_n) \\ \vdots & & \vdots \\ f_n(u_1) & \ldots & f_n(u_n) \end{pmatrix} P = {}^t Q E_n P = {}^t Q P.
\end{aligned}
$$

Thus we have

$$Q = {}^t P^{-1}. \quad \|$$

Example 2 Let $\{e_1, e_2\}$ be the standard basis of $\mathbf{R}^2$ and $\{f_1, f_2\}$ the dual basis of $\{e_1, e_2\}$. We shall find the dual basis $\{g_1, g_2\}$ of the basis $\left\{ \begin{pmatrix} 1 \\ -2 \end{pmatrix}, \begin{pmatrix} 1 \\ -1 \end{pmatrix} \right\}$ of $\mathbf{R}^2$. Since

$$\left(\begin{pmatrix} 1 \\ -2 \end{pmatrix} \begin{pmatrix} 1 \\ -1 \end{pmatrix} \right) = \begin{pmatrix} e_1 & e_2 \end{pmatrix} \begin{pmatrix} 1 & 1 \\ -2 & -1 \end{pmatrix},$$

we have

$$\left(g_1 \; g_2 \right) = \left(f_1 \; f_2 \right) \; {}^{t}\left(\begin{matrix} 1 & 1 \\ -2 & -1 \end{matrix} \right)^{-1} = \left(f_1 \; f_2 \right) \left(\begin{matrix} -1 & 2 \\ -1 & 1 \end{matrix} \right).$$

Therefore

$$g_1 = -f_1 - f_2, \quad \text{and} \quad g_2 = 2f_1 + f_2.$$

Dual mappings and dual transformations For a linear mapping $T : U \to V$, we define $T^* : V^* \to U^*$ by

$$T^*(g)(\mathbf{u}) = g(T(\mathbf{u})) \quad (g \in V^*, \; \mathbf{u} \in U).$$

It is easy to see that T^* is a linear mapping of V^* into U^*. The linear mapping T^* is called the *dual mapping* of T. When $V = U$ and T is a linear transformation of U, the dual mapping T^* is called the *dual transformation* of T.

Example 3 Let $U = V = \mathbf{R}^2$ and T the transformation mapping defined by

$$T\left(\left(\begin{matrix} x_1 \\ x_2 \end{matrix} \right) \right) = \left(\begin{matrix} 2x_1 + 3x_2 \\ -x_1 + 5x_2 \end{matrix} \right).$$

Let $\{f_1, f_2\}$ be the dual basis of the standard basis $\{\mathbf{e}_1, \mathbf{e}_2\}$ of V (see Example 1 above). Then we have

$$T^*(f_1)\left(\left(\begin{matrix} x_1 \\ x_2 \end{matrix} \right) \right) = f_1\left(T\left(\left(\begin{matrix} x_1 \\ x_2 \end{matrix} \right) \right) \right) = 2x_1 + 3x_2,$$

$$T^*(f_2)\left(\left(\begin{matrix} x_1 \\ x_2 \end{matrix} \right) \right) = f_2\left(T\left(\left(\begin{matrix} x_1 \\ x_2 \end{matrix} \right) \right) \right) = -x_1 + 5x_2.$$

Theorem 5.6.5 *Assume* $\dim(U) = n$ *and* $\dim(V) = m$. *Let* $T : U \to V$ *be a linear mapping and* $T^* : V^* \to U^*$ *the dual mapping of* T. *Take a basis* $\{\mathbf{u}_1, \ldots, \mathbf{u}_n\}$ *of* U *and a basis* $\{\mathbf{v}_1, \ldots, \mathbf{v}_m\}$ *of* V. *We denote the dual basis of* $\{\mathbf{u}_1, \ldots, \mathbf{u}_n\}$ *by* $\{f_1, \ldots, f_n\}$ *and the dual basis of* $\{\mathbf{v}_1, \ldots, \mathbf{v}_m\}$ *by* $\{g_1, \ldots, g_m\}$. *Let* A *be the representation matrix of* T *with respect to the bases* $\{\mathbf{u}_1, \ldots, \mathbf{u}_n\}$ *and* $\{\mathbf{v}_1, \ldots, \mathbf{v}_m\}$. *Then the representation matrix of* T^* *with respect to the dual bases* $\{g_1, \ldots, g_m\}$ *and* $\{f_1, \ldots, f_n\}$ *is the transposed matrix* ${}^{t}A$ *of* A. *More precisely,*

$$\left(T^*(g_1) \; \ldots \; T^*(g_m) \right) = \left(f_1 \; \ldots \; f_n \right)^{t} A.$$

Proof We denote by B the representation matrix of $T^* : V^* \to U^*$ with respect to the bases $\{g_1, \ldots, g_m\}$ of V^* and $\{f_1, \ldots, f_n\}$ of U^*. Then

$(*)$ $\quad \left(T^*(g_1) \; \ldots \; T^*(g_m) \right) = \left(f_1 \; \ldots \; f_n \right) B.$

Since $T^*(g)(\mathbf{u}) = g(T(\mathbf{u}))(g \in V^*, \mathbf{u} \in U)$, we have

$$(**) \quad \begin{pmatrix} T^*(g_1) \\ \vdots \\ T^*(g_m) \end{pmatrix} (\mathbf{u}_1 \ \ldots \ \mathbf{u}_n) = \begin{pmatrix} g_1 \\ \vdots \\ g_m \end{pmatrix} (T(\mathbf{u}_1) \ \ldots \ T(\mathbf{u}_n)).$$

We calculate both sides of this equality. By the relation $(*)$ and $f_i(\mathbf{u}_j) = \delta_{ij}$ for $i, j = 1, \ldots, n$, we obtain

$$\text{the left-hand side of } (**) = {}^t B \begin{pmatrix} f_1 \\ \vdots \\ f_n \end{pmatrix} (\mathbf{u}_1 \ \ldots \ \mathbf{u}_n)$$

$$= {}^t B E_n = {}^t B.$$

Next, by the relation

$$(T(\mathbf{u}_1) \ \ldots \ T(\mathbf{u}_n)) = (\mathbf{v}_1 \ \ldots \ \mathbf{v}_m) A$$

and $g_k(\mathbf{v}_l) = \delta_{kl}$ for $k, l = 1, \ldots, m$, we obtain

$$\text{the right-hand side of } (**) = \begin{pmatrix} g_1 \\ \vdots \\ g_m \end{pmatrix} (\mathbf{v}_1 \ \ldots \ \mathbf{v}_m) A$$

$$= E_m A = A.$$

Therefore we have ${}^t B = A$, or $B = {}^t A$. $\|$

Exercise 5.15 (1) Let T be a linear mapping of $M_{2 \times 2}(\mathbf{R})$ into $\mathbf{R}^2$ defined by

$$T\left(\begin{pmatrix} a_{11} & a_{12} \\ a_{21} & a_{22} \end{pmatrix}\right) = \begin{pmatrix} a_{11} + 2a_{22} \\ -a_{12} + 3a_{21} \end{pmatrix}.$$

Find the representation matrix A of the linear mapping T with respect to the standard basis $\{E_{11}, E_{12}, E_{21}, E_{22}\}$ of $M_{2 \times 2}(\mathbf{R})$ and the standard basis $\{\mathbf{e}_1, \mathbf{e}_2\}$ of $\mathbf{R}^2$.

(2) Let $\{f_1, f_2, f_3, f_4\}$ be the dual basis of $\{E_{11}, E_{12}, E_{21}, E_{22}\}$ and $\{g_1, g_2\}$ the dual basis of $\{\mathbf{e}_1, \mathbf{e}_2\}$. Find the representation matrix B of the dual mapping T^* of V^* into U^* with respect to these dual bases.

Answers. (1) By definition, we have

$$T(E_{11}) = T\left(\begin{pmatrix} 1 & 0 \\ 0 & 0 \end{pmatrix}\right) = \begin{pmatrix} 1 \\ 0 \end{pmatrix} = (\mathbf{e}_1 \ \mathbf{e}_2) \begin{pmatrix} 1 \\ 0 \end{pmatrix},$$

$$T(E_{12}) = T\left(\begin{pmatrix} 0 & 1 \\ 0 & 0 \end{pmatrix}\right) = \begin{pmatrix} 0 \\ -1 \end{pmatrix} = (\mathbf{e}_1 \ \mathbf{e}_2) \begin{pmatrix} 0 \\ -1 \end{pmatrix},$$

$$T(E_{21}) = T\left(\begin{pmatrix} 0 & 0 \\ 1 & 0 \end{pmatrix}\right) = \begin{pmatrix} 0 \\ 3 \end{pmatrix} = (\mathbf{e}_1 \ \mathbf{e}_2) \begin{pmatrix} 0 \\ 3 \end{pmatrix},$$

$$T(E_{22}) = T\left(\begin{pmatrix} 0 & 0 \\ 0 & 1 \end{pmatrix}\right) = \begin{pmatrix} 2 \\ 0 \end{pmatrix} = (\mathbf{e}_1 \ \mathbf{e}_2) \begin{pmatrix} 2 \\ 0 \end{pmatrix}.$$

Then we see that

$$\left(T(E_{11})\ T(E_{12})\ T(E_{21})\ T(E_{22}) \right) = \left(\mathbf{e}_1\ \mathbf{e}_2 \right) \begin{pmatrix} 1 & 0 & 0 & 2 \\ 0 & -1 & 3 & 0 \end{pmatrix}.$$

Thus we obtain the representation matrix

$$A = \begin{pmatrix} 1 & 0 & 0 & 2 \\ 0 & -1 & 3 & 0 \end{pmatrix}.$$

(2) By Theorem 5.6.5, the representation matrix B is the transposed matrix $^t A$ of A. Therefore we have

$$B = ^t A = \begin{pmatrix} 1 & 0 \\ 0 & -1 \\ 0 & 3 \\ 2 & 0 \end{pmatrix}. \quad \|$$

Equivalence relations Let X be a set. A relation $\sim$ among elements of X is called an *equivalence relation* if it satisfies the following three conditions ($a, b, c \in X$):

(1) $a \sim a$,   (2) $a \sim b \ \Rightarrow \ b \sim a$,   (3) $a \sim b, \ b \sim c \ \Rightarrow \ a \sim c$.

Example 4 Let $X = M_{n \times n}(K)$. Then similarity of matrices is an equivalence relation of the set X. In fact, we see:

(1) Let A be a square matrix of degree n. Then $A = E_n^{-1} A E_n$.
(2) Let A and B be square matrices of degree n. If there exists a regular matrix P satisfying $B = P^{-1} A P$, then $A = P B P^{-1} = (P^{-1})^{-1} B (P^{-1})$.
(3) Let A, B and C be square matrices of degree n. If there exist regular matrices P and Q satisfying $B = P^{-1} A P$ and $C = Q^{-1} B Q$, then $C = Q^{-1} P^{-1} A P Q = (PQ)^{-1} A (PQ)$.

The symbol $\equiv$ is sometimes used as a symbol of equivalence relation.

Example 5 We denote by $\mathbf{Z}$ the set of integers and let N be a positive integer. We define a relation $\equiv$ for $m, n \in \mathbf{Z}$ by

$$m \equiv n \pmod{N} \ \Leftrightarrow \ m - n \text{ is divisible by } N.$$

Then the relation $\equiv$ is an equivalence relation of $\mathbf{Z}$. We call the equivalence relation $\equiv$ the *congruence relation* modulo N, and when $m \equiv n \pmod{N}$, we say that m and n are *congruent* modulo N.

Equivalence classes Let X be a set in which an equivalence relation $\sim$ is defined. For an element $a \in X$, the subset

$$\mathrm{cl}(a) = \{ b \in X \mid b \sim a \}$$

of X is called the *equivalence class* containing a. Equivalence classes are often called *cosets*. We call a a *representative* of the equivalence class $\mathrm{cl}(a)$. We denote by $X/\sim$ the set of equivalence classes of X with respect to the equivalence relation $\sim$ and call it the *quotient set* of X by the equivalence relation $\sim$.

Equivalence relations of vector spaces defined by subspaces Let V be a vector space and W a subspace of V. We define an equivalence relation $\sim$ of V by

$$\mathbf{u} \sim \mathbf{v} \qquad \Leftrightarrow \qquad \mathbf{u} - \mathbf{v} \in W.$$

It is easy to see that the relation $\sim$ satisfies the three conditions of the equivalence relation. Therefore it is an equivalence relation. This equivalence relation of V is called the *equivalence relation defined by a subspace W*. We denote the quotient set $V/\sim$ by V/W.

Quotient spaces We can define addition and scalar multiplication in V/W. For $\mathbf{u}, \mathbf{v} \in V$, we define

$$\text{(Addition)} \qquad \mathrm{cl}(\mathbf{u}) + \mathrm{cl}(\mathbf{v}) = \mathrm{cl}(\mathbf{u} + \mathbf{v}).$$

The sum of $\mathrm{cl}(\mathbf{u})$ and $\mathrm{cl}(\mathbf{v})$ is independent of the choice of the representatives of $\mathrm{cl}(\mathbf{u})$ and $\mathrm{cl}(\mathbf{v})$. In fact, assume that $\mathrm{cl}(\mathbf{u}_1) = \mathrm{cl}(\mathbf{u}_2)$ and $\mathrm{cl}(\mathbf{v}_1) = \mathrm{cl}(\mathbf{v}_2)$. Then $\mathbf{u}_2 = \mathbf{u}_1 + \mathbf{w}_1$ and $\mathbf{v}_2 = \mathbf{v}_1 + \mathbf{w}_2$ ($\mathbf{w}_1, \mathbf{w}_2 \in W$) and we have

$$\mathbf{u}_2 + \mathbf{v}_2 = \mathbf{u}_1 + \mathbf{v}_1 + (\mathbf{w}_1 + \mathbf{w}_2).$$

Since $\mathbf{w}_1 + \mathbf{w}_2 \in W$, we have

$$\mathrm{cl}(\mathbf{u}_2 + \mathbf{v}_2) = \mathrm{cl}(\mathbf{u}_1 + \mathbf{v}_1).$$

Next, we define scalar multiplication in V/W. For an element $k \in K$ and an element $\mathrm{cl}(\mathbf{u}) \in V/W$, we define

$$\text{(Scalar multiplication)} \qquad k \cdot \mathrm{cl}(\mathbf{u}) = \mathrm{cl}(k\mathbf{u}).$$

The scalar multiplication $\mathrm{cl}(k\mathbf{u})$ is also independent of the choice of the representatives of $\mathrm{cl}(\mathbf{u})$. In fact, if $\mathrm{cl}(\mathbf{u}_1) = \mathrm{cl}(\mathbf{u}_2)$, then $\mathbf{u}_2 = \mathbf{u}_1 + \mathbf{w}$ ($\mathbf{w} \in W$). Therefore we see that $k\mathbf{u}_2 = k\mathbf{u}_1 + k\mathbf{w}$. Since $k\mathbf{w} \in W$, we have

$$\mathrm{cl}(k\mathbf{u}_1) = \mathrm{cl}(k\mathbf{u}_2).$$

The quotient set V/W is a vector space (Theorem 5.6.6 below). We call the vector space V/W the *quotient space* of V by W. The quotient space is also called the *factor space*.

Theorem 5.6.6 *Let V be a vector space and W a subspace of V. Then the quotient set V/W is a vector space by the addition and the scalar multiplication defined above.*

Proof We easily see that the addition and the scalar multiplication of V/W defined above satisfy the eight properties of vector spaces. Therefore V/W is a vector space. We note that the zero vector in V/W is $\mathbf{0}_{V/W} = \mathrm{cl}(\mathbf{0}_V)$. ‖

Theorem 5.6.7 *Let V be a vector space and W a subspace of V. Then*

$$\dim(V/W) = \dim(V) - \dim(W).$$

Proof Let T be the linear mapping of V into V/W defined by

$$T(\mathbf{v}) = \mathrm{cl}(\mathbf{v}).$$

Then we have

$$\mathrm{Ker}(T) = W \quad \text{and} \quad \mathrm{Im}(T) = V/W.$$

Therefore by Theorem 5.1.2, we have

$$\dim(W) + \dim(V/W) = \mathrm{null}(T) + \mathrm{rank}(T) = \dim(V). \quad ‖$$

Theorem 5.6.8 *Let U and V be vector spaces and T a linear mapping of U into V. Then T induces an isomorphism T_0 of $U/\mathrm{Ker}(T)$ into $T(U)$, namely,*

$$T_0 : U/\mathrm{Ker}(T) \cong T(U).$$

Proof For a vector $\mathrm{cl}(\mathbf{u}) \in U/\mathrm{Ker}(T)$, we define a mapping T_0 of $U/\mathrm{Ker}(T)$ into $T(U)$ by

$$T_0(\mathrm{cl}(\mathbf{u})) = T(\mathbf{u}).$$

If $\mathrm{cl}(\mathbf{u}_1) = \mathrm{cl}(\mathbf{u}_2)$ $(\mathbf{u}_1, \mathbf{u}_2 \in U)$, then $\mathbf{u}_2 = \mathbf{u}_1 + \mathbf{w}$ $(\mathbf{w} \in \mathrm{Ker}(T))$. Therefore $T(\mathbf{u}_2) = T(\mathbf{u}_1) + T(\mathbf{w}) = T(\mathbf{u}_1)$ and the mapping T_0 is well defined, namely, the image $T_0(\mathrm{cl}(\mathbf{u}))$ is independent of the choice of the representatives of the equivalence class $\mathrm{cl}(\mathbf{u})$. Since we have by the linearity of T

$$\begin{aligned} T_0(\mathrm{cl}(\mathbf{u}) + \mathrm{cl}(\mathbf{v})) &= T_0(\mathrm{cl}(\mathbf{u} + \mathbf{v})) = T(\mathbf{u} + \mathbf{v}) = T(\mathbf{u}) + T(\mathbf{v}) \\ &= T_0(\mathrm{cl}(\mathbf{u})) + T_0(\mathrm{cl}(\mathbf{v})), \\ T_0(k \cdot \mathrm{cl}(\mathbf{u})) &= T_0(\mathrm{cl}(k\mathbf{u})) = T(k\mathbf{u}) = kT(\mathbf{u}) \\ &= kT_0(\mathrm{cl}(\mathbf{u})), \end{aligned}$$

the mapping T_0 is a linear mapping. Here $\mathbf{u}, \mathbf{v} \in U$ and $k \in K$.

Furthermore, since $T_0(U/\mathrm{Ker}(T)) = T(U)$, T_0 is a surjective linear mapping. By definition, we see that $\mathrm{cl}(\mathbf{u}) \in \mathrm{Ker}(T_0)$ if and only if $\mathbf{u} \in \mathrm{Ker}(T)$. Therefore we have

$$\text{Ker}(T_0) = \{\text{cl}(\mathbf{u}) \mid \mathbf{u} \in \text{Ker}(T)\} = \{\mathbf{0}_{U/\text{Ker}(T)}\}.$$

Thus we see that T_0 is a one-to-one and surjective linear mapping of $U/\text{Ker}(T)$ into $T(U)$. Therefore the linear mapping T_0 is an isomorphism. ‖

Example 6 Let $U = M_{2\times 2}(\mathbf{R})$ and $V = \mathbf{R}^2$. We denote by T the linear mapping of U into V defined by

$$T\left(\begin{pmatrix} a_{11} & a_{12} \\ a_{21} & a_{22} \end{pmatrix}\right) = \begin{pmatrix} a_{11} + a_{12} \\ a_{21} - a_{22} \end{pmatrix}.$$

For a vector $\begin{pmatrix} a_1 \\ a_2 \end{pmatrix} \in V = \mathbf{R}^2$, we see that

$$T\left(\begin{pmatrix} a_1 & 0 \\ a_2 & 0 \end{pmatrix}\right) = \begin{pmatrix} a_1 \\ a_2 \end{pmatrix}.$$

Therefore T is a surjective mapping, or $T(U) = V$. We shall find Ker (T). If

$$T\left(\begin{pmatrix} a_{11} & a_{12} \\ a_{21} & a_{22} \end{pmatrix}\right) = \begin{pmatrix} a_{11} + a_{12} \\ a_{21} - a_{22} \end{pmatrix} = \begin{pmatrix} 0 \\ 0 \end{pmatrix},$$

then $a_{11} + a_{12} = 0$, $a_{21} - a_{22} = 0$. Therefore we have

$$\text{Ker}\,(T) = \left\{\begin{pmatrix} a & -a \\ b & b \end{pmatrix} \middle| a, b \in \mathbf{R}\right\}.$$

Thus we obtain the isomorphism $T_0 : M_{2\times 2}(\mathbf{R})/\text{Ker}(T) \cong \mathbf{R}^2$ induced by T.

Exercises (Sect. 5.6)

1 Express the dual bases of the following bases by the dual bases of the standard bases:

(1) A basis $\left\{\mathbf{a}_1 = \begin{pmatrix} 1 \\ -1 \end{pmatrix}, \mathbf{a}_2 = \begin{pmatrix} 1 \\ 0 \end{pmatrix}\right\}$ of $\mathbf{R}^2$.

(2) A basis $\left\{\mathbf{a}_1 = \begin{pmatrix} 2 \\ 1 \\ 0 \end{pmatrix}, \mathbf{a}_2 = \begin{pmatrix} 0 \\ 1 \\ -1 \end{pmatrix}, \mathbf{a}_3 = \begin{pmatrix} 1 \\ 0 \\ 1 \end{pmatrix}\right\}$ of $\mathbf{R}^3$.

2 Let $\left\{\mathbf{a}_1 = \begin{pmatrix} 2 \\ -1 \end{pmatrix}, \mathbf{a}_2 = \begin{pmatrix} -1 \\ 1 \end{pmatrix}\right\}$ be a basis of $V = \mathbf{R}^2$.

(1) We define a linear transformation T of V by

$$T\left(\begin{pmatrix} x_1 \\ x_2 \end{pmatrix}\right) = \begin{pmatrix} 3x_1 - x_2 \\ 2x_1 + x_2 \end{pmatrix} = \begin{pmatrix} 3 & -1 \\ 2 & 1 \end{pmatrix}\begin{pmatrix} x_1 \\ x_2 \end{pmatrix}.$$

Find the representation matrix A of T with respect to the basis $\{\mathbf{a}_1, \mathbf{a}_2\}$.

(2) Let $\{g_1, g_2\}$ be the dual basis of the basis $\{\mathbf{a}_1, \mathbf{a}_2\}$. Find the representation matrix of the dual transformation T^* of T with respect to the basis $\{g_1, g_2\}$ of V^*.

3 We define a linear transformation T of $\mathbf{R}[x]_2$ by

$$T(f(x)) = xf'(x) + 2f(x) + f(1)(x + x^2).$$

(1) Find the representation matrix A of T with respect to the standard basis of $\mathbf{R}[x]_2$.
(2) Let $f_1(x) = 1 + 2x$, $f_2(x) = x + x^2$, $f_3(x) = -1 + x^2$. Show that $\{f_1(x), f_2(x), f_3(x)\}$ is a basis of $\mathbf{R}[x]_2$.
(3) Find the representation matrix B of T with respect to the basis $\{f_1(x), f_2(x), f_3(x)\}$ of $\mathbf{R}[x]_2$.
(4) Find the representation matrix of the dual transformation T^* of T with respect to the dual basis of $\{f_1(x), f_2(x), f_3(x)\}$.

4 Let W be the set of upper triangular matrices of degree n. Show that W is a subspace of $M_{n \times n}(\mathbf{R})$ and find the dimension of W.

5 Let A be a real $m \times n$ matrix of rank r and

$$W = \{X \in M_{n \times n}(\mathbf{R}) \mid AX = O_{m,n} \}.$$

Show that W is a subspace of $M_{n \times n}(\mathbf{R})$ and find the dimension of W.

6 Let W be the set of real symmetric matrices of degree n. Show that W is a subspace of $M_{n \times n}(\mathbf{R})$ and find the dimension of W.

7 Let W be the set of alternating matrices of degree n. Show that W is a subspace of $M_{n \times n}(\mathbf{R})$ and find the dimension of W.

8 Show that $\mathbf{C}^n$ $(n \geq 1)$ is a vector space over $\mathbf{R}$ and find the dimension of $\mathbf{C}^n$ as a vector space over $\mathbf{R}$.

9 Let W_1 and W_2 be subspaces of a vector space V and put $U = W_1 \cap W_2$. Show that if $V = W_1 + W_2$, then $V/U = W_1/U \oplus W_2/U$.

Chapter 6
Inner Product Spaces

6.1 Inner Products

In this chapter, we define and explain the inner products on vector spaces over the real number field $\mathbf{R}$. The matrices we consider are real matrices unless otherwise stated. The inner products on vector spaces over the complex number field $\mathbf{C}$ are called the Hermitian inner products and will be discussed in Chap. 7.

Inner products Let V be a vector space over $\mathbf{R}$. To two vectors $\mathbf{u}, \mathbf{v}$ in V, we take a real number $(\mathbf{u}, \mathbf{v})$. The mapping $(\mathbf{u}, \mathbf{v})$ of $V \times V$ into $\mathbf{R}$ is called an *inner product* of V if it satisfies the following four conditions:

(1) $(\mathbf{u}_1 + \mathbf{u}_2, \mathbf{v}) = (\mathbf{u}_1, \mathbf{v}) + (\mathbf{u}_2, \mathbf{v})$.
(2) $(c\mathbf{u}, \mathbf{v}) = c(\mathbf{u}, \mathbf{v})$.
(3) $(\mathbf{u}, \mathbf{v}) = (\mathbf{v}, \mathbf{u})$.
(4) For any vector $\mathbf{u} \in V$, $(\mathbf{u}, \mathbf{u}) \geq 0$. Furthermore, if $\mathbf{u} \neq \mathbf{0}_V$, then $(\mathbf{u}, \mathbf{u}) > 0$.

Here, $\mathbf{u}, \mathbf{u}_1, \mathbf{u}_2, \mathbf{v} \in V$ and $c \in \mathbf{R}$. The real number $(\mathbf{u}, \mathbf{v})$ is called the *inner product* of vectors $\mathbf{u}$ and $\mathbf{v}$ in V.

By (1) and (3), we have

$$(\mathbf{u}, \mathbf{v}_1 + \mathbf{v}_2) = (\mathbf{v}_1 + \mathbf{v}_2, \mathbf{u}) = (\mathbf{v}_1, \mathbf{u}) + (\mathbf{v}_2, \mathbf{u}) = (\mathbf{u}, \mathbf{v}_1) + (\mathbf{u}, \mathbf{v}_2).$$

By (2) and (3), we have

$$(\mathbf{u}, c\mathbf{v}) = (c\mathbf{v}, \mathbf{u}) = c(\mathbf{v}, \mathbf{u}) = c(\mathbf{u}, \mathbf{v}).$$

We also have $(\mathbf{0}_V, \mathbf{u}) = (\mathbf{u}, \mathbf{0}_V) = 0$ for any $\mathbf{u} \in V$. In fact

$$(\mathbf{0}_V, \mathbf{u}) = (0\,\mathbf{0}_V, \mathbf{u}) = 0(\mathbf{0}_V, \mathbf{u}) = 0$$

© The Author(s), under exclusive license to Springer Nature Singapore Pte Ltd. 2022
T. Miyake, *Linear Algebra*, https://doi.org/10.1007/978-981-16-6994-1_6

and

$$(\mathbf{u}, \mathbf{0}_V) = (\mathbf{0}_V, \mathbf{u}) = 0.$$

Inner product spaces Vector spaces with inner products are called *inner product spaces*.

Example 1 Let $\mathbf{a} = \begin{pmatrix} a_1 \\ \vdots \\ a_n \end{pmatrix}$ and $\mathbf{b} = \begin{pmatrix} b_1 \\ \vdots \\ b_n \end{pmatrix}$ be vectors in $\mathbf{R}^n$. We define an inner product of $\mathbf{R}^n$ by

$$(\mathbf{a}, \mathbf{b}) = {}^t\mathbf{ab} = a_1 b_1 + \cdots + a_n b_n.$$

Then this is an inner product (see Exercise 6.1 below). We call this inner product the *standard inner product* of $\mathbf{R}^n$. The inner product space $\mathbf{R}^n$ with the standard inner product is called the *Euclidean space* of dimension n.

Exercise 6.1 Prove that the standard inner product of $\mathbf{R}^n$ satisfies conditions (1) through (4) of the definition of the inner product.

Answer. Let $\mathbf{a}, \mathbf{a}_1, \mathbf{a}_2, \mathbf{b} \in \mathbf{R}^n$ and $c \in \mathbf{R}$. Then we see that:

(1) $(\mathbf{a}_1 + \mathbf{a}_2, \mathbf{b}) = {}^t(\mathbf{a}_1 + \mathbf{a}_2)\mathbf{b} = ({}^t\mathbf{a}_1 + {}^t\mathbf{a}_2)\mathbf{b}$
$= {}^t\mathbf{a}_1\mathbf{b} + {}^t\mathbf{a}_2\mathbf{b} = (\mathbf{a}_1, \mathbf{b}) + (\mathbf{a}_2, \mathbf{b}).$

(2) $(c\mathbf{a}, \mathbf{b}) = {}^t(c\mathbf{a})\mathbf{b} = c\,{}^t\mathbf{ab} = c(\mathbf{a}, \mathbf{b}).$

(3) $(\mathbf{a}, \mathbf{b}) = {}^t\mathbf{ab} = {}^t({}^t\mathbf{ab}) = {}^t\mathbf{b}^t({}^t\mathbf{a}) = {}^t\mathbf{ba} = (\mathbf{b}, \mathbf{a}).$
   (Since ${}^t\mathbf{ab}$ is a scalar, we see that ${}^t\mathbf{ab} = {}^t({}^t\mathbf{ab})$.)

(4) Let $\mathbf{a} = \begin{pmatrix} a_1 \\ \vdots \\ a_n \end{pmatrix} \in \mathbf{R}^n$, then we have

$$(\mathbf{a}, \mathbf{a}) = {}^t\mathbf{aa} = a_1^2 + \cdots + a_n^2 \geq 0.$$

Furthermore, if $\mathbf{a} \neq \mathbf{0}_n$, then there exists i satisfying $a_i \neq 0$. Therefore

$$(\mathbf{a}, \mathbf{a}) = {}^t\mathbf{aa} = a_1^2 + \cdots + a_n^2 \geq a_i^2 > 0. \quad \|$$

Example 2 For vectors $\mathbf{a} = \begin{pmatrix} a_1 \\ a_2 \end{pmatrix}, \mathbf{b} = \begin{pmatrix} b_1 \\ b_2 \end{pmatrix} \in \mathbf{R}^2$, we let

$$(\mathbf{a}, \mathbf{b}) = 2a_1 b_1 + 3a_2 b_2.$$

Then $(\ ,\ )$ is an inner product of $\mathbf{R}^2$. But this inner product is different from the standard inner product.

Norms of vectors Let V be an inner product space. Since we see $(\mathbf{u}, \mathbf{u}) \geq 0$ for any vector $\mathbf{u} \in V$, we can take a square root of $(\mathbf{u}, \mathbf{u})$. We let

$$\|\mathbf{u}\| = \sqrt{(\mathbf{u}, \mathbf{u})}$$

and call it the *norm* or the *length* of a vector $\mathbf{u}$.

Example 3 Let $V = \mathbf{R}^2$, $\mathbf{a} = \begin{pmatrix} 3 \\ -2 \end{pmatrix}$. Then with respect to the standard inner product, we have

$$\|\mathbf{a}\| = \sqrt{(\mathbf{a}, \mathbf{a})} = \sqrt{3^2 + (-2)^2} = \sqrt{13}.$$

Standard inner product of $\mathbf{R}[x]_n$ Let $n \geq 1$. For $f(x), g(x) \in \mathbf{R}[x]_n$, we define

$$(*) \qquad (f(x), g(x)) = \int_{-1}^{1} f(x)g(x)dx.$$

Then this is an inner product of $\mathbf{R}[x]_n$ (see Exercise 6.2 below). We call this inner product the *standard inner product* of $\mathbf{R}[x]_n$ in this book.

Exercise 6.2

1. Show that $(f(x), g(x))$ defined by $(*)$ is an inner product of $\mathbf{R}[x]_n$.
2. Calculate $\|x\|$ with respect to this inner product.

Answer. 1. We shall verify conditions (1) through (4) of the definition of the inner product. Here $f(x), f_1(x), f_2(x), g(x) \in \mathbf{R}[x]_n$ and $c \in \mathbf{R}$.

(1) $(f_1(x) + f_2(x), g(x)) = \int_{-1}^{1} (f_1(x) + f_2(x))g(x)dx$

$\qquad = \int_{-1}^{1} (f_1(x)g(x) + f_2(x)g(x)) \, dx$

$\qquad = \int_{-1}^{1} f_1(x)g(x)dx + \int_{-1}^{1} f_2(x)g(x)dx$

$\qquad = (f_1(x), g(x)) + (f_2(x), g(x)).$

(2) $(cf(x), g(x)) = \int_{-1}^{1} (cf(x))g(x)dx = c \int_{-1}^{1} f(x)g(x)dx = c(f(x), g(x)).$

(3) $(f(x), g(x)) = \int_{-1}^{1} f(x)g(x)dx = \int_{-1}^{1} g(x)f(x)dx = (g(x), f(x)).$

(4) Assume that $f(x) \neq f_{00}(x)$. Then there exists a point $x = a$ ($|a| < 1$) satisfying $f(a)^2 > 0$. Since $f(x)^2$ is a continuous function, there exists $\varepsilon > 0$ ($|a \pm \varepsilon| < 1$) satisfying $f(x)^2 > f(a)^2/2$ ($x \in [a - \varepsilon, a + \varepsilon]$). Then

$$(f(x), f(x)) = \int_{-1}^{1} f(x)^2 dx \geq \int_{a-\varepsilon}^{a+\varepsilon} f(x)^2 dx \geq f(a)^2 \varepsilon > 0.$$

2. Since $(x, x) = \int_{-1}^{1} x^2 dx = \left[\dfrac{x^3}{3}\right]_{-1}^{1} = \dfrac{2}{3}$, we have $\|x\| = \dfrac{\sqrt{2}}{\sqrt{3}}.$ ‖

Theorem 6.1.1 *Norms of an inner product space V have the following properties. Here* $\mathbf{u}, \mathbf{v} \in V$ *and* $c \in \mathbf{R}$.

(1) $\|c\mathbf{u}\| = |c| \cdot \|\mathbf{u}\|$.
(2) $|(\mathbf{u}, \mathbf{v})| \leq \|\mathbf{u}\| \cdot \|\mathbf{v}\|$ *(the Schwarz inequality)*.
(3) $\|\mathbf{u} + \mathbf{v}\| \leq \|\mathbf{u}\| + \|\mathbf{v}\|$ *(the triangular inequality)*.

Proof   (1) We see that

$$\|c\mathbf{u}\|^2 = (c\mathbf{u}, c\mathbf{u}) = c^2(\mathbf{u}, \mathbf{u}) = c^2\|\mathbf{u}\|^2.$$

Then by taking the non-negative square roots of both sides, we obtain (1).

(2) If $\mathbf{u} = \mathbf{0}_V$, then both sides of the Schwarz inequality are zeros. So, the Schwarz inequality holds. For $\mathbf{u} \neq \mathbf{0}_V$, we let $f(t) = \|t\mathbf{u} + \mathbf{v}\|^2$ (≥ 0) for $t \in \mathbf{R}$. Then

$$f(t) = (t\mathbf{u} + \mathbf{v}, t\mathbf{u} + \mathbf{v}) = t^2\|\mathbf{u}\|^2 + 2t(\mathbf{u}, \mathbf{v}) + \|\mathbf{v}\|^2.$$

We note $\|\mathbf{u}\|^2 > 0$. Since $f(t) \geq 0$, the discriminant D of the quadratic polynomial $f(t)$ must be 0 or negative, or $D/4 = (\mathbf{u}, \mathbf{v})^2 - \|\mathbf{u}\|^2\|\mathbf{v}\|^2 \leq 0$. Therefore $(\mathbf{u}, \mathbf{v})^2 \leq \|\mathbf{u}\|^2\|\mathbf{v}\|^2$. Taking the non-negative square roots of both sides, we have the Schwarz inequality.

(3) $\|\mathbf{u} + \mathbf{v}\|^2 = (\mathbf{u} + \mathbf{v}, \mathbf{u} + \mathbf{v}) = \|\mathbf{u}\|^2 + 2(\mathbf{u}, \mathbf{v}) + \|\mathbf{v}\|^2$

$$\leq \|\mathbf{u}\|^2 + 2\|\mathbf{u}\| \cdot \|\mathbf{v}\| + \|\mathbf{v}\|^2 = (\|\mathbf{u}\| + \|\mathbf{v}\|)^2.$$

We used the Schwarz inequality in the above calculation. Taking the non-negative square roots of both sides, we have the triangular inequality.   ‖

Orthogonality of vectors If vectors $\mathbf{u}$ and $\mathbf{v}$ in an inner product space V satisfy $(\mathbf{u}, \mathbf{v}) = 0$, we call the vectors $\mathbf{u}$ and $\mathbf{v}$ *orthogonal*. The orthogonality of the vectors $\mathbf{u}$ and $\mathbf{v}$ is denoted by $\mathbf{u} \perp \mathbf{v}$.

Example 4 The vectors $\mathbf{a} = \begin{pmatrix} 1 \\ -2 \\ 3 \end{pmatrix}$ and $\mathbf{b} = \begin{pmatrix} 7 \\ 2 \\ -1 \end{pmatrix}$ in the Euclidean space $\mathbf{R}^3$ are orthogonal. In fact

$$(\mathbf{a}, \mathbf{b}) = 1 \cdot 7 + (-2) \cdot 2 + 3 \cdot (-1) = 0.$$

Example 5 The vectors 1 and x in the vector space $\mathbf{R}[x]_n$ are orthogonal with respect to the standard inner product of $\mathbf{R}[x]_n$. In fact

$$(1, x) = \int_{-1}^{1} 1 \cdot x \, dx = \left[\frac{x^2}{2}\right]_{-1}^{1} = \frac{1}{2} - \frac{1}{2} = 0.$$

Theorem 6.1.2 *If non-zero vectors* $u_1, \ldots, u_r$ *in an inner product space* V *are orthogonal to each other, then they are linearly independent.*

Proof By assumption, we have $(u_i, u_j) = 0$ if $i \neq j$. Let $c_1 u_1 + \cdots + c_r u_r = 0_V$ be a linear relation of vectors $u_1, \ldots, u_r$. We have

$$0 = (0_V, u_i) = (c_1 u_1 + \cdots + c_r u_r, u_i) = c_i(u_i, u_i) \quad \text{for} \quad i = 1, \ldots, r.$$

Since $u_i \neq 0_V$, we have $(u_i, u_i) \neq 0$. Thus we obtain $c_i = 0$ for any $i = 1, \ldots, r$. Then the vectors $u_1, \ldots, u_r$ are linearly independent. $\|$

Example 6 The vectors $a_1 = \begin{pmatrix} 1 \\ 0 \\ 1 \end{pmatrix}$, $a_2 = \begin{pmatrix} 1 \\ 1 \\ -1 \end{pmatrix}$, $a_3 = \begin{pmatrix} -1 \\ 2 \\ 1 \end{pmatrix}$ in the Euclidean space $\mathbf{R}^3$ are linearly independent. In fact, they are non-zero vectors and satisfy $(a_1, a_2) = (a_2, a_3) = (a_1, a_3) = 0$. Therefore they are linearly independent by Theorem 6.1.2.

Orthogonality of subspaces Let V be an inner product space. We say that subspaces W_1 and W_2 of V are *orthogonal* if $(w_1, w_2) = 0$ for any $w_1 \in W_1$ and $w_2 \in W_2$. The orthogonality of W_1 and W_2 is denoted by $W_1 \perp W_2$.

Exercises (Sect. 6.1)

1 Let $\mathbf{R}^3$ be the Euclidean space of dimension 3 and $\mathbf{R}[x]_2$ the inner product space with the standard inner product. Find the following inner products of vectors:

(1) $\left(\begin{pmatrix} 2 \\ 4 \\ 2 \end{pmatrix}, \begin{pmatrix} 3 \\ -2 \\ 4 \end{pmatrix} \right).$ (2) $\left(\begin{pmatrix} -1 \\ 3 \\ -2 \end{pmatrix}, \begin{pmatrix} 5 \\ -2 \\ -7 \end{pmatrix} \right).$

(3) $(f(x), g(x)); \quad f(x) = 2 + x - x^2, \quad g(x) = 1 - x + x^2.$

2 Find the norms of the following vectors with respect to the inner products in Exercise 1:

(1) $u = \begin{pmatrix} 1 \\ -4 \\ -1 \end{pmatrix}.$ (2) $u = \begin{pmatrix} -1 \\ 2 \\ 3 \end{pmatrix}.$ (3) $f(x) = 2 + x - x^2.$

3 Find the number a so that it satisfies the following conditions for vectors in the Euclidean space $\mathbf{R}^3$:

(1) The vectors $\begin{pmatrix} 1 \\ a \\ -1 \end{pmatrix}$ and $\begin{pmatrix} 3 \\ -2 \\ a \end{pmatrix}$ are orthogonal.

(2)   The vectors $\begin{pmatrix} a-1 \\ 2 \\ 1 \end{pmatrix}$ and $\begin{pmatrix} 2 \\ 1 \\ a+1 \end{pmatrix}$ are orthogonal.

4 (1) Find the vector **u** in the Euclidean space $\mathbf{R}^3$ which is of norm 1 and orthogonal to the vectors $\begin{pmatrix} 1 \\ 1 \\ -1 \end{pmatrix}$ and $\begin{pmatrix} 2 \\ -2 \\ 1 \end{pmatrix}$.

(2) Let $\mathbf{R}[x]_2$ be the inner product space with the standard inner product. Find the vector $g(x) \in \mathbf{R}[x]_2$ of norm 1 which is orthogonal to the vectors $f_1(x) = 1$ and $f_2(x) = x$.

5 For an inner product space V, show the following for vectors $\mathbf{u}, \mathbf{v} \in V$:

(1) $\|\mathbf{u} + \mathbf{v}\|^2 + \|\mathbf{u} - \mathbf{v}\|^2 = 2(\|\mathbf{u}\|^2 + \|\mathbf{v}\|^2)$.
(2) Vectors $\mathbf{u}$ and $\mathbf{v}$ are orthogonal $\Leftrightarrow$ $\|\mathbf{u} + \mathbf{v}\|^2 = \|\mathbf{u}\|^2 + \|\mathbf{v}\|^2$.
(3) Vectors $\mathbf{u} + \mathbf{v}$ and $\mathbf{u} - \mathbf{v}$ are orthogonal $\Leftrightarrow$ $\|\mathbf{u}\| = \|\mathbf{v}\|$.
(4) $(\mathbf{u}, \mathbf{v}) = \dfrac{1}{2}(\|\mathbf{u} + \mathbf{v}\|^2 - \|\mathbf{u}\|^2 - \|\mathbf{v}\|^2)$.

6 For a subspace W of an inner product space V, we let

$$W^\perp = \{\mathbf{u} \in V \mid (\mathbf{u}, \mathbf{v}) = 0 \quad \text{for any} \quad \mathbf{v} \in W\}.$$

Then show that $W^\perp$ is a subspace of V. The subspace $W^\perp$ is called the *orthogonal complement* of W in V.

7 Find the dimensions and bases of the subspaces $W^\perp$ for the following subspaces W of the Euclidean space $\mathbf{R}^3$:

(1) $W = \{\mathbf{x} \in \mathbf{R}^3 \mid x_1 + x_2 + x_3 = 0\}$.

(2) $W = \left\{\mathbf{x} \in \mathbf{R}^3 \left| \begin{array}{l} x_1 + 2x_2 + x_3 = 0, \\ 2x_1 + 3x_2 - x_3 = 0 \end{array} \right.\right\}$.

(3) $W = \left\{\mathbf{x} \in \mathbf{R}^3 \left| \begin{array}{l} 3x_1 - 5x_2 + x_3 = 0, \\ -x_1 + 2x_2 - x_3 = 0 \end{array} \right.\right\}$.

8 Show that the following subspaces W_1 and W_2 of the Euclidean space $\mathbf{R}^4$ are orthogonal:

(1) $W_1 = \{\mathbf{x} \in \mathbf{R}^4 \mid x_1 + x_2 + x_3 + x_4 = 0\}$.

$W_2 = \left\{\mathbf{x} \in \mathbf{R}^4 \left| \begin{array}{l} 2x_1 + 3x_2 - x_3 - 4x_4 = 0, \\ x_1 + 2x_2 \qquad - 3x_4 = 0, \\ x_1 + 2x_2 + x_3 - 4x_4 = 0 \end{array} \right.\right\}$.

(2) $W_1 = \left\{ \mathbf{x} \in \mathbf{R}^4 \left| \begin{array}{l} x_1 \quad\;\;\; + x_3 + x_4 = 0, \\ x_1 + x_2 + x_3 - x_4 = 0 \end{array} \right. \right\}$.

$\; W_2 = \left\{ \mathbf{x} \in \mathbf{R}^4 \left| \begin{array}{l} x_1 - 2x_2 \quad\;\;\; - x_4 = 0, \\ 2x_1 - 2x_2 - x_3 - x_4 = 0 \end{array} \right. \right\}$.

(3) $W_1 = \left\{ \mathbf{x} \in \mathbf{R}^4 \left| \begin{array}{l} x_1 \quad\;\;\; - 2x_3 - x_4 = 0, \\ x_1 + x_2 - \;\; x_3 \quad\quad\; = 0 \end{array} \right. \right\}$.

$\; W_2 = \left\{ \mathbf{x} \in \mathbf{R}^4 \left| \begin{array}{l} x_1 \quad\;\;\; + x_3 - x_4 = 0, \\ x_1 - x_2 \quad\quad\; + x_4 = 0 \end{array} \right. \right\}$.

6.2 Orthonormal Bases and Orthogonal Matrices

In the rest of this chapter, we only consider inner product spaces and the Euclidean spaces.

Orthonormal bases A basis $\{\mathbf{u}_1, \ldots, \mathbf{u}_n\}$ of an inner product space V of dimension n is called an *orthonormal basis* if it satisfies the condition

$\quad(*)\qquad\qquad (\mathbf{u}_i, \mathbf{u}_j) = \delta_{ij} \qquad (\delta_{ij}$: the Kronecker delta$)$

for $i = 1, \ldots, n$ and $j = 1, \ldots, n$.

We note that if $\dim (V) = n$ and n vectors $\mathbf{u}_1, \ldots, \mathbf{u}_n$ satisfy the orthonormal condition $(*)$, they are linearly independent by Theorem 6.1.2 and $\{\mathbf{u}_1, \ldots, \mathbf{u}_n\}$ is an orthonormal basis of V by Theorem 4.4.5.

Example 1 The standard basis $\{\mathbf{e}_1, \ldots, \mathbf{e}_n\}$ of $\mathbf{R}^n$ is an orthonormal basis of the Euclidean space $\mathbf{R}^n$.

Example 2 Take two vectors $\mathbf{a}_1 = \begin{pmatrix} 1/\sqrt{2} \\ 1/\sqrt{2} \end{pmatrix}, \mathbf{a}_2 = \begin{pmatrix} 1/\sqrt{2} \\ -1/\sqrt{2} \end{pmatrix}$ in $\mathbf{R}^2$. Since we see that

$$\|\mathbf{a}_1\| = \sqrt{\left(1/\sqrt{2}\right)^2 + \left(1/\sqrt{2}\right)^2} = 1, \qquad \|\mathbf{a}_2\| = \sqrt{\left(1/\sqrt{2}\right)^2 + \left(-1/\sqrt{2}\right)^2} = 1$$

and

$$(\mathbf{a}_1, \mathbf{a}_2) = \left(1/\sqrt{2}\right)\left(1/\sqrt{2}\right) + \left(1/\sqrt{2}\right)\left(-1/\sqrt{2}\right) = 0,$$

$\{\mathbf{a}_1, \mathbf{a}_2\}$ is an orthonormal basis of the Euclidean space $\mathbf{R}^2$.

Example 3 We consider the standard inner product as the inner product of $\mathbf{R}[x]_1$. By Example 4 in Sect. 4.4, we have $\dim(\mathbf{R}[x]_1) = 2$. Let $f_1(x) = \dfrac{1}{\sqrt{2}}$ and $f_2(x) = \dfrac{\sqrt{3}}{\sqrt{2}}x$. Since we see that

$$\|f_1(x)\|^2 = \int_{-1}^{1} \frac{1}{2}dx = 1, \qquad \|f_2(x)\|^2 = \frac{3}{2}\int_{-1}^{1} x^2 dx = \frac{3}{2}\left[\frac{x^3}{3}\right]_{-1}^{1} = 1$$

and

$$(f_1(x), f_2(x)) = \frac{\sqrt{3}}{2}\int_{-1}^{1} x\,dx = \frac{\sqrt{3}}{2}\left[\frac{x^2}{2}\right]_{-1}^{1} = 0,$$

$\{f_1(x), f_2(x)\}$ is an orthonormal basis of $\mathbf{R}[x]_1$.

To obtain orthonormal bases, the following theorem is quite useful.

Theorem 6.2.1 (Schmidt orthonormalization) *Let* $\{\mathbf{v}_1, \dots, \mathbf{v}_n\}$ *be a basis of an inner product space* V. *Then there exists an orthonormal basis* $\{\mathbf{u}_1, \dots, \mathbf{u}_n\}$ *of* V *satisfying*

$$< \mathbf{u}_1, \dots, \mathbf{u}_r > = < \mathbf{v}_1, \dots, \mathbf{v}_r > \quad \text{for} \quad r = 1, \dots, n.$$

In particular, any inner product space has an orthonormal basis. Finding the orthonormal basis $\{\mathbf{u}_1, \dots, \mathbf{u}_n\}$ *from the basis* $\{\mathbf{v}_1, \dots, \mathbf{v}_n\}$ *is called the orthonormalization of the basis* $\{\mathbf{v}_1, \dots, \mathbf{v}_n\}$.

Proof We are going to prove the theorem by induction on r.

If $r = 1$, we let

$$\mathbf{u}_1 = \frac{1}{\|\mathbf{v}_1\|}\mathbf{v}_1.$$

Then we see that $\|\mathbf{u}_1\| = 1$ and $< \mathbf{u}_1 > = < \mathbf{v}_1 >$.

Let r be an integer $1 \le r < n$. We assume that there exist vectors $\mathbf{u}_1, \dots, \mathbf{u}_r$ satisfying

$$< \mathbf{u}_1, \dots, \mathbf{u}_r > = < \mathbf{v}_1, \dots, \mathbf{v}_r > \quad \text{and} \quad (\mathbf{u}_i, \mathbf{u}_j) = \delta_{ij} \quad \text{for} \quad i, j = 1, \dots, r.$$

We put

$$\tilde{\mathbf{v}}_{r+1} = \mathbf{v}_{r+1} - \sum_{i=1}^{r}(\mathbf{v}_{r+1}, \mathbf{u}_i)\mathbf{u}_i \quad \text{and} \quad \mathbf{u}_{r+1} = \frac{1}{\|\tilde{\mathbf{v}}_{r+1}\|}\tilde{\mathbf{v}}_{r+1}.$$

Since $\mathbf{v}_1, \dots, \mathbf{v}_{r+1}$ are linearly independent, $\mathbf{v}_{r+1} \notin < \mathbf{v}_1, \dots, \mathbf{v}_r > = < \mathbf{u}_1, \dots, \mathbf{u}_r >$. Therefore we see that $\tilde{\mathbf{v}}_{r+1} \ne \mathbf{0}_V$ and $\|\mathbf{u}_{r+1}\| = 1$. For $j = 1, \dots, r$, we have

$$(\mathbf{u}_j, \tilde{\mathbf{v}}_{r+1}) = \left(\mathbf{u}_j, \mathbf{v}_{r+1} - \sum_{i=1}^{r}(\mathbf{v}_{r+1}, \mathbf{u}_i)\mathbf{u}_i\right)$$
$$= (\mathbf{u}_j, \mathbf{v}_{r+1}) - (\mathbf{v}_{r+1}, \mathbf{u}_j) = 0.$$

Therefore $(\mathbf{u}_j, \mathbf{u}_{r+1}) = 0$ for $j = 1, \ldots, r$. We also have

$$< \mathbf{u}_1, \ldots, \mathbf{u}_r, \mathbf{u}_{r+1} > \; = \; < \mathbf{v}_1, \ldots, \mathbf{v}_r, \tilde{\mathbf{v}}_{r+1} >$$
$$= \; < \mathbf{v}_1, \ldots, \mathbf{v}_r, \mathbf{v}_{r+1} > .$$

We continue these procedures until $r = n - 1$. When $r = n - 1$, we obtain the orthonormal basis $\{\mathbf{u}_1, \ldots, \mathbf{u}_n\}$ we want. $\|$

Exercise 6.3 Orthonormalize the following basis of the Euclidean space $\mathbf{R}^3$ by the Schmidt orthonormalization:

$$\left\{ \mathbf{v}_1 = \begin{pmatrix} 1 \\ 1 \\ 0 \end{pmatrix}, \mathbf{v}_2 = \begin{pmatrix} 1 \\ 3 \\ 1 \end{pmatrix}, \mathbf{v}_3 = \begin{pmatrix} 2 \\ -1 \\ 1 \end{pmatrix} \right\}.$$

Answer. We shall find an orthonormal basis $\{\mathbf{u}_1, \mathbf{u}_2, \mathbf{u}_3\}$ step by step from $\{\mathbf{v}_1, \mathbf{v}_2, \mathbf{v}_3\}$ by the Schmidt orthonormalization.

$$\mathbf{u}_1 = \frac{1}{\|\mathbf{v}_1\|}\mathbf{v}_1 = \frac{1}{\sqrt{2}} \begin{pmatrix} 1 \\ 1 \\ 0 \end{pmatrix} = \begin{pmatrix} 1/\sqrt{2} \\ 1/\sqrt{2} \\ 0 \end{pmatrix}.$$

$$\tilde{\mathbf{v}}_2 = \mathbf{v}_2 - (\mathbf{v}_2, \mathbf{u}_1)\mathbf{u}_1 = \begin{pmatrix} 1 \\ 3 \\ 1 \end{pmatrix} - \frac{4}{\sqrt{2}} \cdot \frac{1}{\sqrt{2}} \begin{pmatrix} 1 \\ 1 \\ 0 \end{pmatrix} = \begin{pmatrix} -1 \\ 1 \\ 1 \end{pmatrix},$$

$$\mathbf{u}_2 = \frac{1}{\|\tilde{\mathbf{v}}_2\|}\tilde{\mathbf{v}}_2 = \frac{1}{\sqrt{3}} \begin{pmatrix} -1 \\ 1 \\ 1 \end{pmatrix} = \begin{pmatrix} -1/\sqrt{3} \\ 1/\sqrt{3} \\ 1/\sqrt{3} \end{pmatrix}.$$

$$\tilde{\mathbf{v}}_3 = \mathbf{v}_3 - (\mathbf{v}_3, \mathbf{u}_1)\mathbf{u}_1 - (\mathbf{v}_3, \mathbf{u}_2)\mathbf{u}_2$$
$$= \begin{pmatrix} 2 \\ -1 \\ 1 \end{pmatrix} - \frac{1}{2} \begin{pmatrix} 1 \\ 1 \\ 0 \end{pmatrix} - \frac{-2}{3} \begin{pmatrix} -1 \\ 1 \\ 1 \end{pmatrix} = \frac{5}{6} \begin{pmatrix} 1 \\ -1 \\ 2 \end{pmatrix},$$

$$\mathbf{u}_3 = \frac{1}{\|\tilde{\mathbf{v}}_3\|}\tilde{\mathbf{v}}_3 = \frac{1}{\sqrt{6}} \begin{pmatrix} 1 \\ -1 \\ 2 \end{pmatrix} = \begin{pmatrix} 1/\sqrt{6} \\ -1/\sqrt{6} \\ 2/\sqrt{6} \end{pmatrix}.$$

Thus we have an orthonormal basis

$$\left\{ \begin{pmatrix} 1/\sqrt{2} \\ 1/\sqrt{2} \\ 0 \end{pmatrix}, \begin{pmatrix} -1/\sqrt{3} \\ 1/\sqrt{3} \\ 1/\sqrt{3} \end{pmatrix}, \begin{pmatrix} 1/\sqrt{6} \\ -1/\sqrt{6} \\ 2/\sqrt{6} \end{pmatrix} \right\}$$

of the Euclidean space $\mathbf{R}^3$. ‖

Exercise 6.4 We consider the standard inner product of $\mathbf{R}[x]_2$ defined in Sect. 6.1. Orthonormalize the standard basis $\{1, x, x^2\}$ of $\mathbf{R}[x]_2$ by the Schmidt orthonormalization.

Answer. Let $f_0(x) = 1$, $f_1(x) = x$ and $f_2(x) = x^2$. We shall obtain an orthonormal basis $\{g_0(x), g_1(x), g_2(x)\}$ from $\{f_0(x), f_1(x), f_2(x)\}$ following the Schmidt orthonormalization. First we see that

$$\| f_0(x) \|^2 = \int_{-1}^{1} dx = 2, \qquad \| f_0(x) \| = \sqrt{2}.$$

We let $g_0(x) = \dfrac{1}{\| f_0(x) \|} f_0(x) = \dfrac{1}{\sqrt{2}}$ and

$$\tilde{g}_1(x) = f_1(x) - (f_1(x), g_0(x))g_0(x) = x - \int_{-1}^{1} \frac{x}{\sqrt{2}} dx \cdot \frac{1}{\sqrt{2}} = x.$$

Then we see that

$$\| \tilde{g}_1(x) \|^2 = \int_{-1}^{1} x^2 dx = \frac{2}{3}, \qquad \| \tilde{g}_1(x) \| = \frac{\sqrt{2}}{\sqrt{3}}.$$

We let $g_1(x) = \dfrac{1}{\| \tilde{g}_1(x) \|} \tilde{g}_1(x) = \dfrac{\sqrt{3}}{\sqrt{2}} x$. Next, let

$$\tilde{g}_2(x) = f_2(x) - (f_2(x), g_0(x))g_0(x) - (f_2(x), g_1(x))g_1(x)$$
$$= x^2 - \int_{-1}^{1} \frac{x^2}{\sqrt{2}} dx \cdot \frac{1}{\sqrt{2}} - \int_{-1}^{1} \frac{\sqrt{3}}{\sqrt{2}} x^3 dx \cdot \frac{\sqrt{3}}{\sqrt{2}} x$$
$$= x^2 - \frac{1}{3}.$$

Then we see that

$$\| \tilde{g}_2(x) \|^2 = \int_{-1}^{1} \left(x^2 - \frac{1}{3} \right)^2 dx = \frac{8}{45}, \qquad \| \tilde{g}_2(x) \| = \frac{2\sqrt{2}}{3\sqrt{5}}.$$

We let $g_2(x) = \dfrac{1}{\| \tilde{g}_2(x) \|} \tilde{g}_2(x) = \dfrac{3\sqrt{5}}{2\sqrt{2}} \left(x^2 - \frac{1}{3} \right)$. Then

$$\left\{ \frac{1}{\sqrt{2}}, \frac{\sqrt{3}}{\sqrt{2}} x, \frac{3\sqrt{5}}{2\sqrt{2}} \left(x^2 - \frac{1}{3} \right) \right\}$$

is the orthonormal basis of $\mathbf{R}[x]_2$ we want. ‖

Theorem 6.2.2 *Let V be an inner product space of dimension n and $\{\mathbf{u}_1, \ldots, \mathbf{u}_n\}$ an orthonormal basis of V. For vectors $\mathbf{a} = \begin{pmatrix} a_1 \\ \vdots \\ a_n \end{pmatrix}$ and $\mathbf{b} = \begin{pmatrix} b_1 \\ \vdots \\ b_n \end{pmatrix}$ in $\mathbf{R}^n$, we let vectors $\mathbf{u}$ and $\mathbf{v}$ in V be $\mathbf{u} = a_1\mathbf{u}_1 + \cdots + a_n\mathbf{u}_n = \begin{pmatrix} \mathbf{u}_1 & \ldots & \mathbf{u}_n \end{pmatrix}\mathbf{a}$ and $\mathbf{v} = b_1\mathbf{u}_1 + \cdots + b_n\mathbf{u}_n = \begin{pmatrix} \mathbf{u}_1 & \ldots & \mathbf{u}_n \end{pmatrix}\mathbf{b}$, respectively.*

(1) We have

$$(\mathbf{u}, \mathbf{v}) = a_1 b_1 + \cdots + a_n b_n = {}^t\mathbf{ab}.$$

(2) Let T be a linear transformation of V and A the representation matrix of T with respect to the orthonormal basis $\{\mathbf{u}_1, \ldots, \mathbf{u}_n\}$. Then we have

$$(T(\mathbf{u}), \mathbf{v}) = {}^t\mathbf{a}\,{}^t A\mathbf{b}.$$

Proof (1)

$$(\mathbf{u}, \mathbf{v}) = (a_1\mathbf{u}_1 + \cdots + a_n\mathbf{u}_n, \ b_1\mathbf{u}_1 + \cdots + b_n\mathbf{u}_n)$$

$$= \sum_{j=1}^{n}\sum_{i=1}^{n} a_i b_j (\mathbf{u}_i, \mathbf{u}_j) = a_1 b_1 + \cdots + a_n b_n = {}^t\mathbf{ab}.$$

(2) Since $T(\mathbf{u}) = \begin{pmatrix} \mathbf{u}_1 & \ldots & \mathbf{u}_n \end{pmatrix} A\mathbf{a}$, we have

$$(T(\mathbf{u}), \mathbf{v}) = {}^t(A\mathbf{a})\mathbf{b} = {}^t\mathbf{a}\,{}^t A\mathbf{b}. \quad \|$$

Orthogonal transformations A linear transformation T of an inner product space V is called an *orthogonal transformation* if T satisfies

$$(T(\mathbf{u}), T(\mathbf{v})) = (\mathbf{u}, \mathbf{v}) \quad (\mathbf{u}, \ \mathbf{v} \in V).$$

Theorem 6.2.3 *Let $\{\mathbf{u}_1, \ldots, \mathbf{u}_n\}$ be an orthonormal basis of an inner product space V and T a linear transformation of V. Then T is an orthogonal transformation of V if and only if $\{T(\mathbf{u}_1), \ldots, T(\mathbf{u}_n)\}$ is an orthonormal basis of V.*

Proof ("Only if" part.) Since T is an orthogonal transformation, we have

$$(T(\mathbf{u}_i), T(\mathbf{u}_j)) = (\mathbf{u}_i, \mathbf{u}_j) = \delta_{ij} \quad (1 \le i, j \le n).$$

Therefore by Theorem 6.1.2, the vectors $T(\mathbf{u}_1), \ldots, T(\mathbf{u}_n)$ are linearly independent. Since $\dim(V) = n$, $\{T(\mathbf{u}_1), \ldots, T(\mathbf{u}_n)\}$ is an orthonormal basis by Theorem 4.4.5.

("If" part.) Let

$$\mathbf{u} = a_1\mathbf{u}_1 + \cdots + a_n\mathbf{u}_n \quad \text{and} \quad \mathbf{v} = b_1\mathbf{u}_1 + \cdots + b_n\mathbf{u}_n$$

be vectors in V. Then we see that

$$T(\mathbf{u}) = a_1 T(\mathbf{u}_1) + \cdots + a_n T(\mathbf{u}_n)$$

and

$$T(\mathbf{v}) = b_1 T(\mathbf{u}_1) + \cdots + b_n T(\mathbf{u}_n).$$

Since both $\{\mathbf{u}_1, \ldots, \mathbf{u}_n\}$ and $\{T(\mathbf{u}_1), \ldots, T(\mathbf{u}_n)\}$ are orthonormal bases of V, we have

$$(T(\mathbf{u}), T(\mathbf{v})) = a_1 b_1 + \cdots + a_n b_n = (\mathbf{u}, \mathbf{v})$$

by Theorem 6.2.2. Therefore T is an orthogonal transformation. $\parallel$

Orthogonal matrices A square matrix P of degree n is called an *orthogonal matrix* if it satisfies

$$^t P P = E_n.$$

If P is an orthogonal matrix, then we have $\left| P \right|^2 = 1$ by taking the determinants of both sides. Thus $\left| P \right| = \pm 1$. Therefore P is a regular matrix. We easily see that an orthogonal matrix P satisfies $P^{-1} = {}^t P$.

Example 4 The matrix $P = \begin{pmatrix} \cos\theta & -\sin\theta \\ \sin\theta & \cos\theta \end{pmatrix}$ is an orthogonal matrix of degree 2. In fact, we have

$$
\begin{aligned}
^t P P &= \begin{pmatrix} \cos\theta & \sin\theta \\ -\sin\theta & \cos\theta \end{pmatrix} \begin{pmatrix} \cos\theta & -\sin\theta \\ \sin\theta & \cos\theta \end{pmatrix} \\
&= \begin{pmatrix} \cos^2\theta + \sin^2\theta & -\cos\theta\sin\theta + \sin\theta\cos\theta \\ -\sin\theta\cos\theta + \cos\theta\sin\theta & \sin^2\theta + \cos^2\theta \end{pmatrix} \\
&= E_2.
\end{aligned}
$$

Therefore P is an orthogonal matrix.

Theorem 6.2.4 *Let* $A = \begin{pmatrix} \mathbf{a}_1 & \ldots & \mathbf{a}_n \end{pmatrix}$ *be the column partition of a square matrix A of degree n. Then the following three conditions are equivalent:*

(1) The matrix A is an orthogonal matrix.
(2) $\{\mathbf{a}_1, \ldots, \mathbf{a}_n\}$ is an orthonormal basis of the Euclidean space $\mathbf{R}^n$.
(3) The linear transformation T_A of $\mathbf{R}^n$ associated with A is an orthogonal transformation.

Proof $(1) \Leftrightarrow (2)$ Since the (i, j) component of $^t A A$ is $^t \mathbf{a}_i \mathbf{a}_j$, we have

$$^t A A = E_n \quad \Longleftrightarrow \quad (\mathbf{a}_i, \mathbf{a}_j) = {}^t \mathbf{a}_i \mathbf{a}_j = \delta_{ij} \text{ for } i, j = 1, \ldots, n.$$

$(2) \Leftrightarrow (3)$ For the standard basis $\{\mathbf{e}_1, \ldots, \mathbf{e}_n\}$ of $\mathbf{R}^n$, we have

$$T_A(\mathbf{e}_1) = A\mathbf{e}_1 = \mathbf{a}_1, \quad \ldots \quad , \quad T_A(\mathbf{e}_n) = A\mathbf{e}_n = \mathbf{a}_n.$$

Since $\{\mathbf{e}_1, \ldots, \mathbf{e}_n\}$ is an orthonormal basis of $\mathbf{R}^n$, we see by Theorem 6.2.3

$$T_A \text{ is an orthogonal transformation}$$
$$\Longleftrightarrow \{\mathbf{a}_1, \ldots, \mathbf{a}_n\} \text{ is an orthonormal basis of } \mathbf{R}^n. \quad \|$$

Example 5 Let P be a matrix given by

$$P = \begin{pmatrix} 1/\sqrt{2} & -1/\sqrt{3} & 1/\sqrt{6} \\ 1/\sqrt{2} & 1/\sqrt{3} & -1/\sqrt{6} \\ 0 & 1/\sqrt{3} & 2/\sqrt{6} \end{pmatrix}.$$

Then P is an orthogonal matrix. To see this, let

$$\mathbf{p}_1 = \begin{pmatrix} 1/\sqrt{2} \\ 1/\sqrt{2} \\ 0 \end{pmatrix}, \quad \mathbf{p}_2 = \begin{pmatrix} -1/\sqrt{3} \\ 1/\sqrt{3} \\ 1/\sqrt{3} \end{pmatrix}, \quad \mathbf{p}_3 = \begin{pmatrix} 1/\sqrt{6} \\ -1/\sqrt{6} \\ 2/\sqrt{6} \end{pmatrix}.$$

Since $(\mathbf{p}_i, \mathbf{p}_j) = \delta_{ij}$ for $i, j = 1, 2, 3$, $\{\mathbf{p}_1, \mathbf{p}_2, \mathbf{p}_3\}$ is an orthonormal basis of the Euclidean space $\mathbf{R}^3$. Therefore the matrix $P = (\mathbf{p}_1 \ \mathbf{p}_2 \ \mathbf{p}_3)$ is an orthogonal matrix by Theorem 6.2.4.

Exercises (Sect. 6.2)

1 Orthonormalize the following bases of the Euclidean spaces $\mathbf{R}^3$ or $\mathbf{R}^4$ by the Schmidt orthonormalization:

(1) $\left\{ \begin{pmatrix} 1 \\ 1 \\ 0 \end{pmatrix}, \begin{pmatrix} 1 \\ 1 \\ 1 \end{pmatrix}, \begin{pmatrix} 1 \\ 0 \\ 0 \end{pmatrix} \right\}.$

(3) $\left\{ \begin{pmatrix} 1 \\ 1 \\ 1 \end{pmatrix}, \begin{pmatrix} 1 \\ 0 \\ 1 \end{pmatrix}, \begin{pmatrix} 1 \\ 2 \\ 0 \end{pmatrix} \right\}.$

(2) $\left\{ \begin{pmatrix} 2 \\ 1 \\ 1 \end{pmatrix}, \begin{pmatrix} 1 \\ 0 \\ 1 \end{pmatrix}, \begin{pmatrix} 1 \\ 2 \\ 1 \end{pmatrix} \right\}.$

(4) $\left\{ \begin{pmatrix} 1 \\ 1 \\ 0 \\ 0 \end{pmatrix}, \begin{pmatrix} 0 \\ 1 \\ 1 \\ 0 \end{pmatrix}, \begin{pmatrix} 1 \\ 0 \\ 0 \\ 1 \end{pmatrix}, \begin{pmatrix} 0 \\ 1 \\ 0 \\ 0 \end{pmatrix} \right\}.$

2 Orthonormalize the following bases by the Schmidt orthonormalization with respect to the standard inner product of $\mathbf{R}[x]_2$:

(1) $\{x^2, x, 1\}.$

(2) $\{1 + x, x + x^2, 1\}.$

3 Show that the following matrices are orthogonal matrices:

(1) $\begin{pmatrix} 1/\sqrt{3} & 0 & 2/\sqrt{6} \\ 1/\sqrt{3} & 1/\sqrt{2} & -1/\sqrt{6} \\ -1/\sqrt{3} & 1/\sqrt{2} & 1/\sqrt{6} \end{pmatrix}.$

(2) $\begin{pmatrix} \cos\phi & -\sin\phi & 0 \\ \cos\theta\sin\phi & \cos\theta\cos\phi & -\sin\theta \\ \sin\theta\sin\phi & \sin\theta\cos\phi & \cos\theta \end{pmatrix}.$

4 Find the numbers a, b, c so that the following matrices are orthogonal matrices:

(1) $\begin{pmatrix} a & -b & -c \\ a & b & -c \\ a & 0 & 2c \end{pmatrix}.$

(2) $\begin{pmatrix} a & 2a & a \\ b & 0 & -b \\ c & -c & c \end{pmatrix}.$

5 Show the following statements for subspaces W, W_1, W_2 of an inner product space V:

(1) $W \cap W^\perp = \{\mathbf{0}_V\}.$
(2) $V = W \oplus W^\perp.$
(3) $(W^\perp)^\perp = W.$
(4) $W_1 \subset W_2 \Leftrightarrow W_1^\perp \supset W_2^\perp.$
(5) $\dim(W) + \dim(W^\perp) = \dim(V)$

6 Show the following statements for orthogonal matrices:

(1) If P is an orthogonal matrix of degree n, then ${}^tP$ and P^{-1} are also orthogonal matrices of degree n.
(2) If both P and Q are orthogonal matrices of degree n, then the product PQ is also an orthogonal matrix of degree n.

7 Show that a linear transformation T of an inner product space V is an orthogonal transformation if and only if $\|T(\mathbf{u})\| = \|\mathbf{u}\|$ for any vector $\mathbf{u} \in V$.

6.3   Diagonalization of Symmetric Matrices

The arguments in Sect. 5.5 on diagonalization are applicable to matrices over any field. In this section, we *only* consider the inner product spaces and real matrices unless otherwise stated.

Adjoint transformation Let V be an inner product space and $(\mathbf{u}, \mathbf{v})$ the inner product of V. For a linear transformation T of V, the linear transformation T^* of V satisfying

$$(T(\mathbf{u}), \mathbf{v}) = (\mathbf{u}, T^*(\mathbf{v})) \quad \text{for any vectors} \quad \mathbf{u}, \mathbf{v} \in V$$

is called the *adjoint transformation* of T.

Theorem 6.3.1

(1) There uniquely exists the adjoint transformation T^* for a linear transformation T of V.

(2) Let $\{\mathbf{u}_1, \ldots, \mathbf{u}_n\}$ be an orthonormal basis of V. If A is the representation matrix of T with respect to the basis, then the representation matrix of T^* with respect to the basis is ${}^tA$.

Proof (1) Let $A = \left(a_{ij} \right)$ be the representation matrix of T with respect to the basis $\{\mathbf{u}_1 \ldots, \mathbf{u}_n\}$. Let $\mathbf{u} = a_1\mathbf{u}_1 + \cdots + a_n\mathbf{u}_n$ and $\mathbf{a} = \begin{pmatrix} a_1 \\ \vdots \\ a_n \end{pmatrix}$. We define T^* by

$$T^*(\mathbf{u}) = \left(\mathbf{u}_1 \ldots \mathbf{u}_n \right) {}^t A\mathbf{a}.$$

Let $\mathbf{v} = b_1\mathbf{u}_1 + \cdots + b_n\mathbf{u}_n \in V$ and $\mathbf{b} = \begin{pmatrix} b_1 \\ \vdots \\ b_n \end{pmatrix}$. Then $\mathbf{v} = \left(\mathbf{u}_1 \ldots \mathbf{u}_n \right) \mathbf{b}$. We have by Theorem 6.2.2 (2) that

$$\begin{aligned} (T(\mathbf{u}), \mathbf{v}) = {}^t(A\mathbf{a})\mathbf{b} &= ({}^t\mathbf{a}\,{}^tA)\mathbf{b} \\ &= {}^t\mathbf{a}\,({}^tA\mathbf{b}) = (\mathbf{u}, T^*(\mathbf{v})). \end{aligned}$$

Thus T^* is an adjoint transformation of T. When an orthonormal basis is fixed, the linear transformation T and its representation matrix correspond bijectively by Theorem 5.6.3. Therefore the adjoint transformation of T uniquely exists.

(2) is clear by the proof of (1). ∥

Linear functionals associated with vectors For a vector $\mathbf{u} \in V$, we define the linear functional $p_{\mathbf{u}}$ of V by

$$p_{\mathbf{u}}(\mathbf{v}) = (\mathbf{v}, \mathbf{u}) \quad \text{for} \quad \mathbf{v} \in V.$$

We call $p_{\mathbf{u}}$ the *linear functional associated with a vector* $\mathbf{u}$. The linear mapping

$$p : V \ni \mathbf{u} \longmapsto p(\mathbf{u}) = p_{\mathbf{u}} \in V^*$$

is an isomorphism of V into V^*. In fact, if $p(\mathbf{u}) = p_{\mathbf{u}} = \mathbf{0}_{V^*}$, then $(\mathbf{v}, \mathbf{u}) = p_{\mathbf{u}}(\mathbf{v}) = {}^t0$ for any $\mathbf{v} \in V$. In particular, $(\mathbf{u}, \mathbf{u}) = 0$. Then $\mathbf{u} = \mathbf{0}_V$ by condition (4) of the inner product. Therefore p is a one-to-one mapping. Since $\dim (V) = \dim (V^*)$, p is an isomorphism by Theorem 5.1.4.

As for adjoint transformations and dual transformations, we have the following theorem.

Theorem 6.3.2 *For a linear transformation T of V, we have*

$$T^*p = pT^*.$$

Here the transformation T^* in the left-hand side is the dual transformation of T and the transformation T^* in the right-hand side is the adjoint transformation of T. This equality means that the following diagram is commutative:

$$\begin{array}{ccc}
 & p & \\
V & \longrightarrow & V^* \\
T^* \downarrow & & \downarrow T^* \\
V & \longrightarrow & V^*. \\
 & p &
\end{array}$$

Proof We shall show that $T^* p = p\, T^*$. Let **u** and **v** be vectors in V. By the definition of the dual transformation T^*, we have

$$(T^* p)(\mathbf{u})(\mathbf{v}) = T^*(p(\mathbf{u}))(\mathbf{v}) = p(\mathbf{u})(T(\mathbf{v}))$$
$$= p_{\mathbf{u}}(T(\mathbf{v})) = (T(\mathbf{v}), \mathbf{u}).$$

By the definition of the adjoint transformation, we see that

$$(p T^*)(\mathbf{u})(\mathbf{v}) = p\,(T^*(\mathbf{u}))(\mathbf{v}) = p_{T^*(\mathbf{u})}(\mathbf{v})$$
$$= (\mathbf{v}, T^*(\mathbf{u})) = (T(\mathbf{v}), \mathbf{u}).$$

Therefore $T^* p(\mathbf{u}) = p\, T^*(\mathbf{u})$, namely, $T^* p = p\, T^*$. ‖

Self-adjoint transformations Let T be a transformation of an inner product space V and T^* the adjoint transformation of T. The transformation T is called a *self-adjoint transformation* if $T^* = T$ is satisfied.

Theorem 6.3.3 *Let T be a transformation of V and A the representation matrix of T with respect to an orthonormal basis of V. Then T is a self-adjoint transformation if and only if A is a symmetric matrix.*

Proof By Theorem 6.3.1 (2), the representation matrices of T and T^* with respect to the orthonormal basis are A and $^t A$, respectively. Therefore we see that

$$T = T^* \quad \text{if and only if} \quad A = {}^t A. \ \ \|$$

Complex conjugation Though we consider only real numbers in this chapter, we need some knowledge of complex numbers here. Let $i = \sqrt{-1}$ be the imaginary unit. For a complex number $\alpha = a + bi$, we let $\overline{\alpha} = a - bi$ and call it the *complex conjugate* of α. A complex number α is a real number if and only if $\overline{\alpha} = \alpha$. Complex conjugation preserves addition, subtraction, multiplication and division of complex numbers:

$$\overline{\alpha \pm \beta} = \overline{\alpha} \pm \overline{\beta}, \quad \overline{\alpha\beta} = \overline{\alpha} \cdot \overline{\beta}, \quad \overline{\left(\dfrac{\alpha}{\beta}\right)} = \dfrac{\overline{\alpha}}{\overline{\beta}} \ \ (\beta \neq 0).$$

We define the *complex conjugate* of a complex matrix $A = (\alpha_{ij})$ by $\overline{A} = (\overline{\alpha}_{ij})$. Let A and B be two square complex matrices of the same degree. Since complex conjugation preserves addition, subtraction and multiplication, we have

$$\overline{A \pm B} = \overline{A} \pm \overline{B}, \qquad \overline{AB} = \overline{A} \cdot \overline{B},$$
$$\overline{\alpha A} = \overline{\alpha}\,\overline{A} \ (\alpha \in \mathbf{C}), \qquad \overline{{}^t A} = {}^t \overline{A}.$$

Theorem 6.3.4 *All the characteristic roots of a real symmetric matrix are real numbers. Therefore any characteristic roots of real symmetric matrices are eigenvalues of them.*

Proof Let A be a real symmetric matrix of degree n and $\lambda \in \mathbf{C}$ be a characteristic root of A, i.e., $g_A(\lambda) = |\lambda E_n - A| = 0$. Since $\lambda E_n - A$ is not regular, there exists a non-zero column vector $\mathbf{x} \in \mathbf{C}^n$ satisfying $A\mathbf{x} = \lambda \mathbf{x}$. Since A is a real matrix, we see $\overline{A} = A$. Then $A\overline{\mathbf{x}} = \overline{A}\overline{\mathbf{x}} = \overline{A\mathbf{x}} = \overline{\lambda\mathbf{x}} = \overline{\lambda}\overline{\mathbf{x}}$. Therefore we obtain

$$(*) \qquad \overline{\lambda}\,{}^t\overline{\mathbf{x}}\mathbf{x} = {}^t(\overline{\lambda}\overline{\mathbf{x}})\mathbf{x} = {}^t(A\overline{\mathbf{x}})\mathbf{x} = {}^t\overline{\mathbf{x}}\,{}^t A\mathbf{x}$$
$$= {}^t\overline{\mathbf{x}}A\mathbf{x} = {}^t\overline{\mathbf{x}}(\lambda\mathbf{x}) = \lambda\,{}^t\overline{\mathbf{x}}\mathbf{x}.$$

Write $\mathbf{x} = \begin{pmatrix} x_1 \\ \vdots \\ x_n \end{pmatrix}$. Since

$${}^t\overline{\mathbf{x}}\mathbf{x} = \overline{x_1}x_1 + \cdots + \overline{x_n}x_n = |x_1|^2 + \cdots + |x_n|^2 \neq 0,$$

we can divide both sides of $(*)$ by ${}^t\overline{\mathbf{x}}\mathbf{x}$. Then we have $\overline{\lambda} = \lambda$. Therefore λ is a real number. $\parallel$

Upper triangularization of square matrices Let A be a square matrix. We say that the matrix A is *upper triangularizable* if there exists a regular matrix P so that the matrix $B = P^{-1}AP$ is an upper triangular matrix. Finding a regular matrix P and an upper triangular matrix $B = P^{-1}AP$ is called the *upper triangularization* of A. We note that the upper triangularization of A does not uniquely exist. We can similarly define *lower triangularizable* and the *lower triangularization*.

Example 1 Let $A = \begin{pmatrix} 4 & -1 & 0 \\ 10 & -4 & 5 \\ 2 & -1 & 2 \end{pmatrix}$ and $P = \begin{pmatrix} 1 & 0 & -1 \\ 3 & -1 & -5 \\ 1 & 0 & -2 \end{pmatrix}$. Then we see that

$$B = P^{-1}AP = \begin{pmatrix} 1 & 1 & 3 \\ 0 & -1 & -1 \\ 0 & 0 & 2 \end{pmatrix}.$$

Then A is upper triangularizable and B is the upper triangularization of A.

Theorem 6.3.5 *Let A be a real square matrix of degree n. If all characteristic roots of A are real, then A is upper triangularizable by an orthogonal matrix. More precisely, there exists an orthogonal matrix P satisfying*

$$P^{-1}AP = \begin{pmatrix} \lambda_1 & & * \\ & \ddots & \\ 0 & & \lambda_n \end{pmatrix}.$$

The orthogonal matrix P can be taken so that $\left| P \right| = 1$.

Proof We are going to prove the theorem by induction on n. The theorem is trivial for $n = 1$. Let $n > 1$. We assume that the theorem holds for matrices whose degrees are less than n. Let λ_1 be a characteristic root of A, then it is a real number, and thus it is an eigenvalue of A. Let $\mathbf{q}_1$ be an eigenvector of A with the eigenvalue λ_1 of norm 1 in the Euclidean space $\mathbf{R}^n$. Take an orthonormal basis $\{\mathbf{q}_1, \ldots, \mathbf{q}_n\}$ of the Euclidean space $\mathbf{R}^n$ which contains the vector $\mathbf{q}_1$, and define a square matrix $Q = \begin{pmatrix} \mathbf{q}_1 \cdots \mathbf{q}_n \end{pmatrix}$ by column partition. By Theorem 6.2.4, the matrix Q is an orthogonal matrix. Since $Q^{-1} = {}^tQ$ and $\mathbf{q}_1$ is an eigenvector of A, or $A\mathbf{q}_1 = \lambda_1\mathbf{q}_1$, we have

$$Q^{-1}AQ = \begin{pmatrix} {}^t\mathbf{q}_1 \\ {}^t\mathbf{q}_2 \\ \vdots \\ {}^t\mathbf{q}_n \end{pmatrix} \begin{pmatrix} \lambda_1\mathbf{q}_1 & A\mathbf{q}_2 & \cdots & A\mathbf{q}_n \end{pmatrix} = \begin{pmatrix} \lambda_1 & * & \cdots & * \\ \hline 0 & & B & \end{pmatrix}.$$

Here B is a real square matrix of degree $n - 1$. We calculate the characteristic polynomial $g_A(t)$. Since the characteristic roots of A are all real numbers and $g_A(t) = g_{Q^{-1}AQ}(t) = (t - \lambda_1)g_B(t)$, all characteristic roots of B are also real numbers. Therefore by the induction assumption, there exists an orthogonal matrix R of degree $n - 1$ satisfying

$$R^{-1}BR = \begin{pmatrix} \lambda_2 & & * \\ & \ddots & \\ 0 & & \lambda_n \end{pmatrix}.$$

We define a matrix P by
$$P = Q\begin{pmatrix} 1 & {}^t\mathbf{0} \\ \hline 0 & R \end{pmatrix}.$$
Since both Q and R are orthogonal matrices, P is an orthogonal matrix by Exercises (Sect. 6.2) 6 (2). Now, we have

$$P^{-1}AP = \begin{pmatrix} 1 & {}^t\mathbf{0} \\ \hline 0 & R^{-1} \end{pmatrix} \begin{pmatrix} \lambda_1 & * \\ \hline 0 & B \end{pmatrix} \begin{pmatrix} 1 & {}^t\mathbf{0} \\ \hline 0 & R \end{pmatrix}$$

$$= \begin{pmatrix} \lambda_1 & * \cdots * \\ \hline 0 & R^{-1}BR \end{pmatrix} = \begin{pmatrix} \lambda_1 & * & \cdots & * \\ & \lambda_2 & & * \\ 0 & & \ddots & \\ & & 0 & \lambda_n \end{pmatrix}.$$

If $\left| P \right| = -1$, we replace the first column $\mathbf{p}_1$ of P with $-\mathbf{p}_1$. Thus we find an orthogonal matrix P of determinant 1 so that $P^{-1}AP$ is an upper triangular matrix. $\parallel$

Exercise 6.5 Show that every characteristic root of the following matrix A is a real number, and upper find the triangularization of the matrix A by an orthogonal matrix:

$$A = \begin{pmatrix} 1 & 0 & 0 \\ 2 & 3 & 4 \\ -2 & -2 & -3 \end{pmatrix}.$$

Answer. We are going to solve the exercise following the procedure given in the proof of Theorem 6.3.5. Since the characteristic polynomial of A is

$$g_A(t) = (t-1)^2(t+1),$$

the characteristic roots of A are $\lambda = 1, -1$ and they are all real numbers.

Let $\lambda = 1$. We solve the linear equation $A\mathbf{x} = \mathbf{x}$ and obtain the eigenspace of A

$$W(1; A) = \left\{ c_1 \begin{pmatrix} -1 \\ 1 \\ 0 \end{pmatrix} + c_2 \begin{pmatrix} -2 \\ 0 \\ 1 \end{pmatrix} \middle| c_1, c_2 \in \mathbf{R} \right\}$$

with the eigenvalue 1. Take a non-zero vector in this space and add two other vectors to make a basis of $\mathbf{R}^3$, for example, $\left\{ \begin{pmatrix} -1 \\ 1 \\ 0 \end{pmatrix}, \begin{pmatrix} 0 \\ 1 \\ 0 \end{pmatrix}, \begin{pmatrix} 0 \\ 0 \\ 1 \end{pmatrix} \right\}$. We orthonormalize this basis by the Schmidt orthonormalization (Theorem 6.2.1) in the Euclidean space $\mathbf{R}^3$ and obtain

$$\left\{ \begin{pmatrix} -1/\sqrt{2} \\ 1/\sqrt{2} \\ 0 \end{pmatrix}, \begin{pmatrix} 1/\sqrt{2} \\ 1/\sqrt{2} \\ 0 \end{pmatrix}, \begin{pmatrix} 0 \\ 0 \\ 1 \end{pmatrix} \right\}.$$

Then the matrix $Q = \begin{pmatrix} -1/\sqrt{2} & 1/\sqrt{2} & 0 \\ 1/\sqrt{2} & 1/\sqrt{2} & 0 \\ 0 & 0 & 1 \end{pmatrix}$ is an orthogonal matrix by Theorem 6.2.4 and we obtain

$$Q^{-1}AQ = {}^tQAQ = \begin{pmatrix} 1 & 2 & 2\sqrt{2} \\ 0 & 3 & 2\sqrt{2} \\ 0 & -2\sqrt{2} & -3 \end{pmatrix}.$$

Let $B = \begin{pmatrix} 3 & 2\sqrt{2} \\ -2\sqrt{2} & -3 \end{pmatrix}$. The characteristic polynomial of B is $g_B(t) = |tE_2 - B| = t^2 - 1$. Take the eigenvalue 1 of B and solve the linear equation $B\mathbf{x} = \mathbf{x}$. We obtain the eigenspace of B,

$$W(1\,;\,B) = \left\{ c \begin{pmatrix} -\sqrt{2} \\ 1 \end{pmatrix} \,\middle|\, c \in \mathbf{R} \right\}$$

with the eigenvalue 1. Take a basis of $\mathbf{R}^2$ containing the vector $\begin{pmatrix} -\sqrt{2} \\ 1 \end{pmatrix}$. For example,

we take a basis $\left\{ \begin{pmatrix} -\sqrt{2} \\ 1 \end{pmatrix}, \begin{pmatrix} 1 \\ 0 \end{pmatrix} \right\}$. We orthonormalize this basis by the Schmidt orthonormalization. As an orthonormal basis of the Euclidean space $\mathbf{R}^2$, we obtain

$$\left\{ \begin{pmatrix} -\sqrt{2}/\sqrt{3} \\ 1/\sqrt{3} \end{pmatrix}, \begin{pmatrix} 1/\sqrt{3} \\ \sqrt{2}/\sqrt{3} \end{pmatrix} \right\}.$$

The matrix $R = \begin{pmatrix} -\sqrt{2}/\sqrt{3} & 1/\sqrt{3} \\ 1/\sqrt{3} & \sqrt{2}/\sqrt{3} \end{pmatrix}$ is an orthogonal matrix of degree 2 by Theorem 6.2.4. Now, we let P by

$$P = Q \begin{pmatrix} 1 & 0 & 0 \\ \hline 0 & & \\ 0 & & R \end{pmatrix} = \begin{pmatrix} -1/\sqrt{2} & -1/\sqrt{3} & 1/\sqrt{6} \\ 1/\sqrt{2} & -1/\sqrt{3} & 1/\sqrt{6} \\ 0 & 1/\sqrt{3} & 2/\sqrt{6} \end{pmatrix}.$$

Since P is a product of two orthogonal matrices of degree 3, P is also an orthogonal matrix of degree 3. Thus we obtain an upper triangularization

$$P^{-1}AP = {}^tPAP = \begin{pmatrix} 1 & 0 & 2\sqrt{3} \\ 0 & 1 & -4\sqrt{2} \\ 0 & 0 & -1 \end{pmatrix}$$

of A by the orthogonal matrix P. ‖

A sketch of another solution. Take linearly independent vectors of $W(1\,;\,A)$ and $W(-1\,;\,A)$. The set of vectors $\left\{ \begin{pmatrix} -1 \\ 1 \\ 0 \end{pmatrix}, \begin{pmatrix} -2 \\ 0 \\ 1 \end{pmatrix}, \begin{pmatrix} 0 \\ -1 \\ 1 \end{pmatrix} \right\}$ is a basis of $\mathbf{R}^3$. We orthonormalize this basis and have an orthonormal basis

$$\left\{ \begin{pmatrix} -1/\sqrt{2} \\ 1/\sqrt{2} \\ 0 \end{pmatrix}, \begin{pmatrix} -1/\sqrt{3} \\ -1/\sqrt{3} \\ 1/\sqrt{3} \end{pmatrix}, \begin{pmatrix} 1/\sqrt{6} \\ 1/\sqrt{6} \\ 2/\sqrt{6} \end{pmatrix} \right\}$$

of the Euclidean space $\mathbf{R}^3$. Then the matrix $P = \begin{pmatrix} -1/\sqrt{2} & -1/\sqrt{3} & 1/\sqrt{6} \\ 1/\sqrt{2} & -1/\sqrt{3} & 1/\sqrt{6} \\ 0 & 1/\sqrt{3} & 2/\sqrt{6} \end{pmatrix}$ is an orthogonal matrix and we have

$$P^{-1}AP = \begin{pmatrix} 1 & 0 & 2\sqrt{3} \\ 0 & 1 & -4\sqrt{2} \\ 0 & 0 & -1 \end{pmatrix}. ‖$$

Theorem 6.3.6 *Let A be a real symmetric matrix of degree n and* λ, μ *be eigenvalues of A. Let* $\mathbf{u} \in \mathbf{R}^n$ *be an eigenvector of A with the eigenvalue* λ *and* $\mathbf{v} \in \mathbf{R}^n$ *an eigenvector of A with the eigenvalue* μ. *If* $\lambda \neq \mu$, *then the vectors* $\mathbf{u}$ *and* $\mathbf{v}$ *are orthogonal in the Euclidean space* $\mathbf{R}^n$.

Proof By definition, we have

$$\begin{aligned}\lambda(\mathbf{u}, \mathbf{v}) &= (\lambda\mathbf{u}, \mathbf{v}) = (A\mathbf{u}, \mathbf{v}) = {}^t(A\mathbf{u})\mathbf{v} = {}^t\mathbf{u}\,{}^tA\mathbf{v}\\ &= {}^t\mathbf{u}A\mathbf{v} = (\mathbf{u}, A\mathbf{v}) = (\mathbf{u}, \mu\mathbf{v}) = \mu(\mathbf{u}, \mathbf{v}).\end{aligned}$$

Therefore $(\lambda - \mu)(\mathbf{u}, \mathbf{v}) = 0$. Since $\lambda \neq \mu$, we have $(\mathbf{u}, \mathbf{v}) = 0$. ‖

Theorem 6.3.7 (Diagonalization of symmetric matrices) *Let A be a real symmetric matrix of degree n. Then there exists an orthogonal matrix P satisfying*

$$P^{-1}AP = \begin{pmatrix} \lambda_1 & & 0 \\ & \ddots & \\ 0 & & \lambda_n \end{pmatrix}.$$

Here $\lambda_1, \ldots, \lambda_n$ *are eigenvalues of A.*

Proof Since all characteristic values of real symmetric matrices are real numbers by Theorem 6.3.4, there exists an orthogonal matrix P by Theorem 6.3.5 so that

$$P^{-1}AP = \begin{pmatrix} \lambda_1 & & * \\ & \ddots & \\ 0 & & \lambda_n \end{pmatrix}.$$

Since ${}^tP = P^{-1}$ and ${}^tA = A$, we see that ${}^t(P^{-1}AP) = {}^tP\,{}^tA\,{}^t(P^{-1}) = P^{-1}AP$. Thus $P^{-1}AP$ is a symmetric matrix and we have

$$P^{-1}AP = \begin{pmatrix} \lambda_1 & & 0 \\ & \ddots & \\ 0 & & \lambda_n \end{pmatrix}. \quad \|$$

Exercise 6.6 Diagonalize the real symmetric matrix $A = \begin{pmatrix} 1 & 2 & -1 \\ 2 & -2 & 2 \\ -1 & 2 & 1 \end{pmatrix}$ by an orthogonal matrix.

Answer. Since the characteristic polynomial of A is

$$g_A(t) = \left| tE_3 - A \right| = (t - 2)^2(t + 4),$$

the eigenvalues of A are $2, -4$. For each eigenvalue, we calculate the eigenspace of A:

$$W(2\,;A) = \left\{ c_1 \begin{pmatrix} 2 \\ 1 \\ 0 \end{pmatrix} + c_2 \begin{pmatrix} -1 \\ 0 \\ 1 \end{pmatrix} \middle| c_1, c_2 \in \mathbf{R} \right\}$$

and

$$W(-4\,;A) = \left\{ c_3 \begin{pmatrix} 1 \\ -2 \\ 1 \end{pmatrix} \middle| c_3 \in \mathbf{R} \right\}.$$

We orthonormalize the basis $\left\{ \begin{pmatrix} 2 \\ 1 \\ 0 \end{pmatrix}, \begin{pmatrix} -1 \\ 0 \\ 1 \end{pmatrix} \right\}$ of $W(2\,;A)$ and obtain

$$\left\{ \begin{pmatrix} 2/\sqrt{5} \\ 1/\sqrt{5} \\ 0 \end{pmatrix}, \begin{pmatrix} -1/\sqrt{30} \\ 2/\sqrt{30} \\ 5/\sqrt{30} \end{pmatrix} \right\}.$$

We orthonormalize the basis $\left\{ \begin{pmatrix} 1 \\ -2 \\ 1 \end{pmatrix} \right\}$ of $W(-4\,;A)$ and obtain

$$\left\{ \begin{pmatrix} 1/\sqrt{6} \\ -2/\sqrt{6} \\ 1/\sqrt{6} \end{pmatrix} \right\}.$$

By Theorem 6.3.6, vectors in $W(2\,;A)$ and vectors in $W(-4\,;A)$ are orthogonal. Then

$$\left\{ \begin{pmatrix} 2/\sqrt{5} \\ 1/\sqrt{5} \\ 0 \end{pmatrix}, \begin{pmatrix} -1/\sqrt{30} \\ 2/\sqrt{30} \\ 5/\sqrt{30} \end{pmatrix}, \begin{pmatrix} 1/\sqrt{6} \\ -2/\sqrt{6} \\ 1/\sqrt{6} \end{pmatrix} \right\}$$

is an orthonormal basis of the Euclidean space $\mathbf{R}^3$. Therefore by the orthogonal matrix

$$P = \begin{pmatrix} 2/\sqrt{5} & -1/\sqrt{30} & 1/\sqrt{6} \\ 1/\sqrt{5} & 2/\sqrt{30} & -2/\sqrt{6} \\ 0 & 5/\sqrt{30} & 1/\sqrt{6} \end{pmatrix},$$

we have a diagonalization of A:

$$P^{-1}AP = \begin{pmatrix} 2 & 0 & 0 \\ 0 & 2 & 0 \\ 0 & 0 & -4 \end{pmatrix}. \quad \|$$

Exercises (Sect. 6.3)

1 Diagonalize the following symmetric matrices by orthogonal matrices:

(1) $\begin{pmatrix} 0 & 1 \\ 1 & 0 \end{pmatrix}$.

(2) $\begin{pmatrix} 1 & -2 \\ -2 & 1 \end{pmatrix}$.

(3) $\begin{pmatrix} 2 & 3 \\ 3 & 2 \end{pmatrix}$.

(4) $\begin{pmatrix} 0 & 0 & 1 \\ 0 & 1 & 0 \\ 1 & 0 & 0 \end{pmatrix}$.

(5) $\begin{pmatrix} 1 & 0 & 1 \\ 0 & 1 & 0 \\ 1 & 0 & 1 \end{pmatrix}$.

(6) $\begin{pmatrix} 1 & 1 & 0 \\ 1 & 4 & 3 \\ 0 & 3 & 1 \end{pmatrix}$.

2 Show that the characteristic roots of the following matrices are all real numbers, and find the upper triangularizations of these matrices by orthogonal matrices:

(1) $\begin{pmatrix} 1 & 2 & 0 \\ 0 & 2 & 0 \\ -2 & 4 & -1 \end{pmatrix}$.

(2) $\begin{pmatrix} 5 & -3 & 6 \\ 2 & 0 & 6 \\ -4 & 4 & -1 \end{pmatrix}$.

3 Show that a real square matrix A is diagonalizable by an orthogonal matrix if and only if A is a real symmetric matrix.

4 Let T be a transformation of $\mathbf{R}[x]_1$ defined by

$$T(a_0 + a_1 x) = a_1 + 3a_0 x.$$

Show that T is a self-adjoint transformation of the inner product space $\mathbf{R}[x]_1$ with the standard inner product.

5 Let $p_{\mathbf{u}}$ be the linear functional of V for $\mathbf{u} \in V$ defined just before Theorem 6.3.2. Show that if $\{\mathbf{u}_1, \ldots, \mathbf{u}_n\}$ is an orthonormal basis of V, then $\{p_{\mathbf{u}_1}, \ldots, p_{\mathbf{u}_n}\}$ is the dual basis of $\{\mathbf{u}_1, \ldots, \mathbf{u}_n\}$.

6 Let A be a square matrix of degree n and T_A the linear transformation associated with A. Then show that

$$(T_A)^* = T_{{}^t A}.$$

6.4 Quadratic Forms

Quadratic forms A homogeneous polynomial $q(x_1, \ldots, x_n)$ of n variables $x_1, \ldots, x_n$ of degree 2 with real coefficients a_{ij} $(i, j = 1, \ldots, n)$ given by

$$q(x_1, \ldots, x_n) = \sum_{i=1}^{n} \sum_{j=1}^{n} a_{ij} x_i x_j$$

is called a *quadratic form*.

Expressions of quadratic forms For a real square matrix $A = \left(a_{ij} \right)$ of degree n

and a variable vector $\mathbf{x} = \begin{pmatrix} x_1 \\ \vdots \\ x_n \end{pmatrix}$ of degree n, we define $A[\mathbf{x}]$ by

$$A[\mathbf{x}] = {}^t\mathbf{x}A\mathbf{x} = \left(x_1 \ \ldots \ x_n \right) A \begin{pmatrix} x_1 \\ \vdots \\ x_n \end{pmatrix}.$$

If we use this expression, the quadratic form $q(x_1, \ldots, x_n)$ is written as

$$q(x_1, \ldots, x_n) = \sum_{i=1}^{n} \sum_{j=1}^{n} a_{ij} x_i x_j = A[\mathbf{x}] = {}^t\mathbf{x}A\mathbf{x} \quad \left(\mathbf{x} = \begin{pmatrix} x_1 \\ \vdots \\ x_n \end{pmatrix}, \ A = \left(a_{ij} \right) \right).$$

We call the matrix A the *coefficient matrix* of the quadratic form $q(x_1, \ldots, x_n)$ and $q(x_1, \ldots, x_n) = A[\mathbf{x}]$ the *quadratic form associated with the matrix A*. We define real numbers b_{ij} $(1 \le i, j \le n)$ by

$$b_{ij} = b_{ji} = \frac{a_{ij} + a_{ji}}{2}.$$

Let $B = \left(b_{ij} \right)$. Then B is a real symmetric matrix of degree n and we easily see that

$$q(x_1, \ldots, x_n) = A[\mathbf{x}] = B[\mathbf{x}].$$

Therefore we may assume that *the coefficient matrix of a quadratic form is a real symmetric matrix* from the beginning.

Example 1 The quadratic form $q(x_1, x_2) = x_1^2 + 2x_1 x_2 + 3x_2^2$ is

$$q(x_1, x_2) = \left(x_1 \ x_2 \right) \begin{pmatrix} 1 & 1 \\ 1 & 3 \end{pmatrix} \begin{pmatrix} x_1 \\ x_2 \end{pmatrix} = A[\mathbf{x}]$$

$$\left(A = \begin{pmatrix} 1 & 1 \\ 1 & 3 \end{pmatrix}, \ \mathbf{x} = \begin{pmatrix} x_1 \\ x_2 \end{pmatrix} \right).$$

Orthogonal change of variables We let variable vectors $\mathbf{x}, \mathbf{y}, \mathbf{z}$ of degree n be

$$\mathbf{x} = \begin{pmatrix} x_1 \\ \vdots \\ x_n \end{pmatrix}, \quad \mathbf{y} = \begin{pmatrix} y_1 \\ \vdots \\ y_n \end{pmatrix}, \quad \mathbf{z} = \begin{pmatrix} z_1 \\ \vdots \\ z_n \end{pmatrix}.$$

We consider a *change of variables* $\mathbf{x}$ to $\mathbf{y}$ by $\mathbf{x} = P\mathbf{y}$ with a regular matrix P. The matrix P is called the *matrix of changing variables* from $\mathbf{x}$ to $\mathbf{y}$. When P is an orthogonal matrix, we call the change of variables $\mathbf{x} = P\mathbf{y}$ an *orthogonal change of variables*. By a change of variables $\mathbf{x} = P\mathbf{y}$, the quadratic form $A[\mathbf{x}]$ is transformed into the quadratic form $({}^t P A P)[\mathbf{y}]$ of variables $y_1, \ldots, y_n$, namely

$$A[\mathbf{x}] = ({}^t P A P)[\mathbf{y}].$$

Diagonal quadratic forms The quadratic form $q(x_1, \ldots, x_n)$ is called a *diagonal quadratic form* if $q(x_1, \ldots, x_n)$ is of the form

$$q(x_1, \ldots, x_n) = \lambda_1 x_1^2 + \cdots + \lambda_n x_n^2.$$

We shall transform quadratic forms into diagonal quadratic forms by an orthogonal change of variables. Such transformations are called the *diagonalizations of quadratic forms*. Let A be a real symmetric matrix and $q(x_1, \ldots, x_n) = A[\mathbf{x}]$ the quadratic form associated with A. Then by Theorem 6.3.7, there exists an orthogonal matrix P such that $D_1 = P^{-1}AP = {}^t PAP$ is a diagonal matrix. Denote the diagonal matrix D_1 by

$$D_1 = \begin{pmatrix} \mu_1 & & 0 \\ & \ddots & \\ 0 & & \mu_n \end{pmatrix}.$$

We shall exchange the order of diagonal elements of the matrix D_1 by an orthogonal matrix. For a permutation σ of n element, we define the square matrix Q_σ of degree n by the column partition $Q_\sigma = \begin{pmatrix} \mathbf{e}_{\sigma(1)} & \cdots & \mathbf{e}_{\sigma(n)} \end{pmatrix}$, where $\{\mathbf{e}_1, \ldots, \mathbf{e}_n\}$ is the standard basis of $\mathbf{R}^n$. Since $(\mathbf{e}_{\sigma(i)}, \mathbf{e}_{\sigma(j)}) = \delta_{i,j}$, Q_σ is an orthogonal matrix.

Example 2 Let $n = 3$ and $\sigma = \begin{pmatrix} 1 & 2 & 3 \end{pmatrix}$. Then the matrix Q_σ is

$$Q_\sigma = \begin{pmatrix} \mathbf{e}_{\sigma(1)} & \mathbf{e}_{\sigma(2)} & \mathbf{e}_{\sigma(3)} \end{pmatrix} = \begin{pmatrix} \mathbf{e}_2 & \mathbf{e}_3 & \mathbf{e}_1 \end{pmatrix} = \begin{pmatrix} 0 & 0 & 1 \\ 1 & 0 & 0 \\ 0 & 1 & 0 \end{pmatrix}.$$

Now, we assume that $\mu_{i_1}, \ldots, \mu_{i_t}$ are not zeros and $\mu_{i_{t+1}} = \cdots = \mu_{i_n} = 0$. Take a permutation

$$\sigma = \begin{pmatrix} 1 & \cdots & n \\ i_1 & \cdots & i_n \end{pmatrix} \in S_n,$$

and let $D = {}^t Q_\sigma D_1 Q_\sigma$. Since the multiplication of matrix Q_σ permutes the order of the diagonal components $\{\mu_1, \ldots, \mu_n\}$ into $\{\mu_{i_1}, \ldots, \mu_{i_n}\}$, we obtain the diagonal matrix D

$$D = {}^t Q_\sigma D_1 Q_\sigma = \begin{pmatrix} \mu_{i_1} & & 0 \\ & \ddots & \\ 0 & & \mu_{i_n} \end{pmatrix}.$$

We note that the diagonal components $\mu_{i_1}, \ldots \mu_{i_t}$ of D are not zeros and $\mu_{i_{t+1}} = \cdots = \mu_{i_n} = 0$. Writing $\lambda_j = \mu_{i_j}$ for $j = 1, \ldots, n$, we have the following theorem.

Theorem 6.4.1 (Diagonalization of quadratic forms) *Let* $q(x_1, \ldots, x_n) = A[\mathbf{x}]$ *be a quadratic form associated with a real symmetric matrix A of degree n. We take the eigenvalues $\lambda_1, \ldots, \lambda_n$ of A so that $\lambda_i \neq 0$ for $i = 1, \ldots, t$ and $\lambda_j = 0$ for $j = t+1, \ldots, n$. Then by a certain orthogonal change of variables $\mathbf{x} = P\mathbf{y}$, we have*

$$q(x_1, \ldots, x_n) = ({}^t PAP)[\mathbf{y}]$$
$$= \lambda_1 y_1^2 + \cdots + \lambda_t y_t^2 + \lambda_{t+1} y_{t+1}^2 + \cdots + \lambda_n y_n^2$$
$$= \lambda_1 y_1^2 + \cdots + \lambda_t y_t^2.$$

We note $t = \text{rank}(A)$.

Exercise 6.7 Diagonalize the following quadratic form by an orthogonal change of variables:

$$q(x_1, x_2, x_3) = x_1^2 + 4x_1 x_2 - 2x_1 x_3 - 2x_2^2 + 4x_2 x_3 + x_3^2.$$

Answer.   The quadratic form is expressed as

$$q(x_1, x_2, x_3) = A[\mathbf{x}] \qquad \left(A = \begin{pmatrix} 1 & 2 & -1 \\ 2 & -2 & 2 \\ -1 & 2 & 1 \end{pmatrix}, \quad \mathbf{x} = \begin{pmatrix} x_1 \\ x_2 \\ x_3 \end{pmatrix} \right).$$

By Exercise 6.6, we have ${}^t PAP = P^{-1}AP = \begin{pmatrix} 2 & 0 & 0 \\ 0 & 2 & 0 \\ 0 & 0 & -4 \end{pmatrix}$. Here the matrix

$$P = \begin{pmatrix} 2/\sqrt{5} & -1/\sqrt{30} & 1/\sqrt{6} \\ 1/\sqrt{5} & 2/\sqrt{30} & -2/\sqrt{6} \\ 0 & 5/\sqrt{30} & 1/\sqrt{6} \end{pmatrix}$$

is an orthogonal matrix. Let $D = {}^t PAP$. Then by the orthogonal change of variables $\mathbf{x} = P\mathbf{y}$, we obtain the diagonalization of the quadratic form

$$q(x_1, x_2, x_3) = ({}^t PAP)[\mathbf{y}] = D[\mathbf{y}] = 2y_1^2 + 2y_2^2 - 4y_3^2. \quad \|$$

Sylvester's law of inertia Let A and P be the same matrices considered in Theorem 6.4.1. We know that any quadratic form $q(x_1, \ldots, x_n) = A[\mathbf{x}]$ is transformed into a diagonal quadratic form $q(x_1, \ldots, x_n) = D[\mathbf{y}]$, $(D = {}^t PAP)$ by an orthogonal change of variables by Theorem 6.4.1. By this transformation, we can take eigenvalues of D so that

$$\lambda_i > 0 \quad \text{for} \quad i = 1, \ldots, r, \qquad \lambda_{r+j} < 0 \quad \text{for} \quad j = 1, \ldots, s$$

and

$$\lambda_{r+s+k} = 0 \quad \text{for} \quad k = 1, \ldots, n - r - s.$$

We let $t = r + s$. We are again going to transform the quadratic form $D[\mathbf{y}]$. Take a matrix (not necessarily orthogonal) of degree n

$$\Lambda = \begin{pmatrix} \sqrt{|\lambda_1|}^{-1} & & & & & 0 \\ & \ddots & & & & \\ & & \sqrt{|\lambda_{r+s}|}^{-1} & & & \\ & & & 1 & & \\ 0 & & & & \ddots & \\ & & & & & 1 \end{pmatrix}$$

and let $P_1 = P\Lambda$. Then we have

$${}^t P_1 A P_1 = {}^t(P\Lambda)A P \Lambda = \begin{pmatrix} 1 & & & & & & & \\ & \ddots & & & & & & \\ & & 1 & & & & & \\ & & & -1 & & & & \\ & & & & \ddots & & & \\ & & & & & -1 & & \\ & & & & & & 0 & \\ 0 & & & & & & & \ddots \\ & & & & & & & & 0 \end{pmatrix}$$

Therefore by the change of variables $\mathbf{y} = \Lambda\mathbf{z}$ or $\mathbf{x} = P\mathbf{y} = P\Lambda\mathbf{z} = P_1\mathbf{z}$, we have

$$q(x_1, \ldots, x_n) = A[\mathbf{x}] = D[\mathbf{y}] = ({}^t P_1 A P_1)[\mathbf{z}]$$
$$= z_1^2 + \cdots + z_r^2 - z_{r+1}^2 - \cdots - z_{r+s}^2.$$

Thus we see that any quadratic form $q(x_1, \ldots, x_n)$ is transformed into

$$q(x_1, \ldots, x_n) = z_1^2 + \cdots + z_r^2 - z_{r+1}^2 - \cdots - z_{r+s}^2$$

by a (not necessarily orthogonal) change of variables $\mathbf{x} = P\mathbf{z}$. Since the number $t = r + s$ is equal to $\text{rank}(A)$, it is uniquely determined by A. We are going to show that the numbers r and s are also uniquely determined by the quadratic form $q(x_1, \ldots, x_n)$ or the matrix A. Namely, they are independent of the choices of regular matrices P_1, which is *Sylvester's law of inertia* stated in the theorem below.

Theorem 6.4.2 (Sylvester's law of inertia) *Let $q(x_1, \ldots, x_n)$ be a quadratic form. When $q(x_1, \ldots, x_n)$ is transformed into*

$$q(x_1, \ldots, x_n) = y_1^2 + \cdots + y_r^2 - y_{r+1}^2 - y_{r+s}^2$$

by a change of variables $\mathbf{x} = P\mathbf{y}$, the pair of numbers (r, s) is uniquely determined and independent of the choice of P.

Proof We assume that the quadratic form $q(x_1, \ldots, x_n) = A[\mathbf{x}]$ is transformed into

$$\begin{aligned} q(x_1, \ldots, x_n) = ({}^t PAP)[\mathbf{y}] &= y_1^2 + \cdots + y_r^2 - y_{r+1}^2 - \cdots - y_{r+s}^2 \\ &= ({}^t QAQ)[\mathbf{z}] = z_1^2 + \cdots + z_u^2 - z_{u+1}^2 - \cdots - z_{u+v}^2 \end{aligned}$$

by changes of variables $\mathbf{x} = P\mathbf{y}$ and $\mathbf{x} = Q\mathbf{z}$ with regular matrices P and Q. We are going to show $r = u$ and $s = v$. We note that $r + s = u + v = \text{rank}(A)$. Thus we have only to show that $r = u$.

Now, we assume $r > u$ and we will see the contradiction. Note that $\mathbf{y} = P^{-1}\mathbf{x}$ and $\mathbf{z} = Q^{-1}\mathbf{x}$. Therefore y_i and z_j are linear combinations of $x_1, \ldots, x_n$. We shall solve the linear equation:

$$(*) \qquad\qquad \begin{cases} y_i = 0 & (r+1 \le i \le n), \\ z_j = 0 & (1 \le j \le u) \end{cases}$$

consisting of $n - r + u$ equations of variables $x_1, \ldots, x_n$. Since we see that

$$\text{the number of the equations in } (*)$$
$$= n - r + u < n = \text{the number of variables,}$$

the linear equation $(*)$ has a non-trivial solution $x_1 = a_1, \quad \ldots, \quad x_n = a_n$ by Theorem 2.3.3 (2). Putting this solution into the equalities $\mathbf{y} = P^{-1}\mathbf{x}$ and $\mathbf{z} = Q^{-1}\mathbf{x}$, we obtain

$$\begin{pmatrix} y_1 \\ \vdots \\ y_r \\ y_{r+1} \\ \vdots \\ y_n \end{pmatrix} = P^{-1} \begin{pmatrix} a_1 \\ \vdots \\ \vdots \\ \vdots \\ \vdots \\ a_n \end{pmatrix} = \begin{pmatrix} b_1 \\ \vdots \\ b_r \\ 0 \\ \vdots \\ 0 \end{pmatrix}, \quad \begin{pmatrix} z_1 \\ \vdots \\ z_u \\ z_{u+1} \\ \vdots \\ z_n \end{pmatrix} = Q^{-1} \begin{pmatrix} a_1 \\ \vdots \\ \vdots \\ \vdots \\ \vdots \\ a_n \end{pmatrix} = \begin{pmatrix} 0 \\ \vdots \\ 0 \\ c_{u+1} \\ \vdots \\ c_n \end{pmatrix}.$$

We note that $y_1, \ldots, y_r$; $z_{u+1}, \ldots, z_n$ do not appear in the linear equation $(*)$. Thus we have

$$\begin{aligned} y_1 &= b_1, \quad \ldots \quad, \quad y_r = b_r, & y_{r+1} &= \cdots = y_n = 0, \\ z_1 &= \cdots = z_u = 0, & z_{u+1} &= c_{u+1}, \ldots, z_n = c_n \end{aligned}$$

with some numbers $b_1, \ldots, b_r$; $c_{u+1}, \ldots, c_n$. Now, we substitute

$$x_1 = a_1, \ldots, x_n = a_n$$

for the quadratic forms $q(x_1, \ldots, x_n)$, then by the consideration above we have

$$q(a_1, \ldots, a_n) = b_1^2 + \cdots + b_r^2 + 0 + \cdots + 0$$
$$= 0 + \cdots + 0 - c_{u+1}^2 - \cdots - c_n^2.$$

Thus we obtain

$$b_1^2 + \cdots + b_r^2 = -c_{u+1}^2 - \cdots - c_n^2.$$

Since the left-hand side is non-negative and the right-hand side is non-positive, they must be equal to zero, namely, $b_1^2 + \cdots + b_r^2 = 0$. Therefore $b_1 = \cdots = b_r = 0$ and we obtain $P^{-1} \begin{pmatrix} a_1 \\ \vdots \\ a_n \end{pmatrix} = \begin{pmatrix} 0 \\ \vdots \\ 0 \end{pmatrix}$. Since P is a regular matrix, we see $a_1 = \cdots = a_n = 0$. This contradicts that the solution $x_1 = a_1, \ldots, x_n = a_n$ of the linear equation $(*)$ is a non-trivial solution. If we assume $r < u$, then we also obtain the contradiction by a similar argument as above. Therefore we obtain $r = u$. ‖

Canonical forms of quadratic forms When we transform the quadratic form $q(x_1, \ldots, x_n) = A[\mathbf{x}]$ to the form

$$q(x_1, \ldots, x_n) = y_1^2 + \cdots + y_r^2 - y_{r+1}^2 - \cdots - y_{r+s}^2,$$

we call the right-hand of the equality the *canonical form* of the quadratic form $q(x_1, \ldots, x_n)$. We note $r + s = \text{rank}\,(A)$.

Signatures of quadratic forms The pair of numbers (r, s) obtained above is called the *signature* of the quadratic form $q(x_1, \ldots, x_n)$. We write

$$\text{sgn}\,(q(x_1, \ldots, x_n)) = (r, s).$$

Exercise 6.8 Find the canonical form and the signature of the following quadratic form:

$$q(x_1, x_2, x_3) = x_1^2 - 2x_1x_2 + 2x_2^2 + 2x_2x_3 + x_3^2.$$

Answer.   The quadratic form is expressed as

$$q(x_1, x_2, x_3) = A[\mathbf{x}] \qquad \left(A = \begin{pmatrix} 1 & -1 & 0 \\ -1 & 2 & 1 \\ 0 & 1 & 1 \end{pmatrix}, \quad \mathbf{x} = \begin{pmatrix} x_1 \\ x_2 \\ x_3 \end{pmatrix} \right).$$

Since A is a real symmetric matrix, it is diagonalizable. The characteristic polynomial of A is

$$g_A(t) = |tE_3 - A| = t(t-1)(t-3),$$

and the eigenvalues of A are $1, 3, 0$. Therefore $\mathrm{sgn}(q(x_1, x_2, x_3)) = (2, 0)$. In this case, we easily obtain the canonical form of $q(x_1, x_2, x_3)$, which is

$$q(x_1, x_2, x_3) = (x_1 - x_2)^2 + (x_2 + x_3)^2 = y_1^2 + y_2^2$$
$$(y_1 = x_1 - x_2, \quad y_2 = x_2 + x_3).$$

By the relation $y_1 = x_1 - x_2$, $y_2 = x_2 + x_3$, we can find a matrix P which is the matrix of changing variables $\mathbf{x} = P\mathbf{y}$. We note that we can take y_3 to be an arbitrary variable as long as y_1, y_2, y_3 are independent variables. Then we have

$$\begin{pmatrix} y_1 \\ y_2 \\ y_3 \end{pmatrix} = P^{-1} \begin{pmatrix} x_1 \\ x_2 \\ x_3 \end{pmatrix} = \begin{pmatrix} 1 & -1 & 0 \\ 0 & 1 & 1 \\ * & * & * \end{pmatrix} \begin{pmatrix} x_1 \\ x_2 \\ x_3 \end{pmatrix}$$

as long as $P^{-1} = \begin{pmatrix} 1 & -1 & 0 \\ 0 & 1 & 1 \\ * & * & * \end{pmatrix}$ is a regular matrix. For example, take

$$P^{-1} = \begin{pmatrix} 1 & -1 & 0 \\ 0 & 1 & 1 \\ 0 & 0 & 1 \end{pmatrix}, \qquad P = \begin{pmatrix} 1 & 1 & -1 \\ 0 & 1 & -1 \\ 0 & 0 & 1 \end{pmatrix}.$$

Then

$$^tPAP = \begin{pmatrix} 1 & 0 & 0 \\ 0 & 1 & 0 \\ 0 & 0 & 0 \end{pmatrix}.$$

We note that the matrix P is not an orthogonal matrix. Since the third row of P^{-1} can be taken to be any row vector as long as P^{-1} is a regular matrix, P is not uniquely determined. ∥

Positive definite quadratic forms We say that a quadratic form $q(x_1, \ldots, x_n) = A[\mathbf{x}]$ is *positive definite* (or *positive*) if it satisfies

$$q(x_1, \ldots, x_n) > 0 \quad \text{for any} \quad (x_1, \ldots, x_n) \neq (0, \ldots, 0) \quad (x_i \in \mathbf{R}).$$

This property is equivalent to one of the following conditions:

(1) $\mathrm{sgn}(q(x_1, \ldots, x_n)) = (n, 0)$.
(2) Every eigenvalue of A is positive.

(3) If we define $(\mathbf{a}, \mathbf{b}) = {}^t\mathbf{a}A\mathbf{b}$ for vectors $\mathbf{a}, \mathbf{b} \in \mathbf{R}^n$, then $(\mathbf{a}, \mathbf{b})$ is an inner product of $\mathbf{R}^n$.

Primary submatrices For a square matrix

$$A = \begin{pmatrix} a_{11} & \cdots & a_{1k} & \cdots & a_{1n} \\ \vdots & & \vdots & & \vdots \\ a_{k1} & \cdots & a_{kk} & \cdots & a_{kn} \\ \vdots & & \vdots & & \vdots \\ a_{n1} & \cdots & a_{nk} & \cdots & a_{nn} \end{pmatrix},$$

we call $A_k = \begin{pmatrix} a_{11} & \cdots & a_{1k} \\ \vdots & & \vdots \\ a_{k1} & \cdots & a_{kk} \end{pmatrix}$ the *primary submatrix* of A of degree k for $k = 1, \ldots, n$.

Theorem 6.4.3 *The quadratic form* $q(x_1, \ldots, x_n) = A[\mathbf{x}]$ *is positive definite if and only if* $|A_k| > 0$ *for* $k = 1, \ldots, n$.

Proof ("Only if" part.) Let $q(x_1, \ldots, x_n) = A[\mathbf{x}]$ be a positive definite quadratic form. Since all eigenvalues of A are positive and $|A|$ is the product of all eigenvalues of A, we have $|A_n| = |A| > 0$. Note that A_k is also a real symmetric matrix. Define a quadratic form $q_k(x_1, \ldots, x_k)$ of k variables by

$$q_k(x_1, \ldots, x_k) = A_k[\mathbf{x}_k] \qquad \left(\mathbf{x}_k = \begin{pmatrix} x_1 \\ \vdots \\ x_k \end{pmatrix}\right)$$

for $k = 1, \ldots, n - 1$. For any $(x_1, \ldots, x_k) \neq (0, \ldots, 0)$, we see that

$$q_k(x_1, \ldots, x_k) = q(x_1, \ldots, x_k, 0, \ldots, 0) > 0,$$

by taking $x_{k+1} = \cdots = x_n = 0$. Therefore $q_k(x_1, \ldots, x_k)$ is also positive definite and thus $|A_k| > 0$.

("If" part.) We are going to show it by induction on n. If $n = 1$, then the assertion is clear. We assume that $n > 1$ and the assertion holds until $n - 1$. Let $A = \begin{pmatrix} A_{n-1} & \mathbf{a}'_{n-1} \\ {}^t\mathbf{a}'_{n-1} & a_{nn} \end{pmatrix}$ be a partition of the matrix A. Here A_{n-1} is the primary submatrix of A of degree $n - 1$ and $|A_{n-1}| > 0$ by assumption. Then A_{n-1} is a regular symmetric matrix. By Exercises (Sect. 3.4) 6, A_{n-1}^{-1} is also a symmetric matrix. Namely, ${}^tA_{n-1}^{-1} = A_{n-1}^{-1}$. Let

$$R = \begin{pmatrix} E_{n-1} & A_{n-1}^{-1}\mathbf{a}'_{n-1} \\ {}^t\mathbf{0}_{n-1} & 1 \end{pmatrix} \qquad \text{and} \qquad B = \begin{pmatrix} A_{n-1} & \mathbf{0}_{n-1} \\ {}^t\mathbf{0}_{n-1} & a_{nn} - A_{n-1}^{-1}[\mathbf{a}'_{n-1}] \end{pmatrix}.$$

Since $a_{nn} - A_{n-1}^{-1}[\mathbf{a}_{n-1}'] = a_{nn} - {}^t\mathbf{a}_{n-1}' A_{n-1}^{-1} \mathbf{a}_{n-1}'$, we have the equality

$$
{}^t R B R
$$

$$
= \left(\begin{array}{c|c} E_{n-1} & \mathbf{0}_{n-1} \\ \hline {}^t\mathbf{a}_{n-1}' A_{n-1}^{-1} & 1 \end{array} \right) \left(\begin{array}{c|c} A_{n-1} & \mathbf{0}_{n-1} \\ \hline {}^t\mathbf{0}_{n-1} & a_{nn} - A_{n-1}^{-1}[\mathbf{a}_{n-1}'] \end{array} \right) \left(\begin{array}{c|c} E_{n-1} & A_{n-1}^{-1}\mathbf{a}_{n-1}' \\ \hline {}^t\mathbf{0}_{n-1} & 1 \end{array} \right)
$$

$$
= \left(\begin{array}{c|c} A_{n-1} & \mathbf{0}_{n-1} \\ \hline {}^t\mathbf{a}_{n-1}' & a_{nn} - A_{n-1}^{-1}[\mathbf{a}_{n-1}'] \end{array} \right) \left(\begin{array}{c|c} E_{n-1} & A_{n-1}^{-1}\mathbf{a}_{n-1}' \\ \hline {}^t\mathbf{0}_{n-1} & 1 \end{array} \right)
$$

$$
= \left(\begin{array}{c|c} A_{n-1} & \mathbf{a}_{n-1}' \\ \hline {}^t\mathbf{a}_{n-1}' & a_{nn} \end{array} \right) = A.
$$

Thus we have $A[\mathbf{x}] = B[R\mathbf{x}]$. Since R is a real regular matrix and B is a real symmetric matrix, to see that $A[\mathbf{x}]$ is a positive definite quadratic form, we have only to show that $B[\mathbf{x}]$ is a positive definite quadratic form. For a variable vector $\mathbf{x} = \begin{pmatrix} x_1 \\ \vdots \\ x_n \end{pmatrix}$, we

let $\mathbf{x}_{n-1} = \begin{pmatrix} x_1 \\ \vdots \\ x_{n-1} \end{pmatrix}$. By the induction assumption, $A_{n-1}[\mathbf{x}_{n-1}]$ is a positive definite quadratic form of $n-1$ variables. We have

$$
B[\mathbf{x}] = A_{n-1}[\mathbf{x}_{n-1}] + (a_{nn} - A_{n-1}^{-1}[\mathbf{a}_{n-1}'])x_n^2
$$

by the definition of B. Therefore, if we show that $a_{nn} - A_{n-1}^{-1}[\mathbf{a}_{n-1}'] > 0$, we have $B[\mathbf{x}] > 0$ for any $\mathbf{x} \neq \mathbf{0}$. Now, we see that

$$
{}^t R^{-1} A R^{-1} = B = \left(\begin{array}{c|c} A_{n-1} & \mathbf{0}_{n-1} \\ \hline {}^t\mathbf{0}_{n-1} & a_{nn} - A_{n-1}^{-1}[\mathbf{a}_{n-1}'] \end{array} \right).
$$

Taking determinants of both sides, we obtain

$$
|A| = |A_{n-1}|(a_{nn} - A_{n-1}^{-1}[\mathbf{a}_{n-1}']).
$$

Since $|A| = |A_n| > 0$ and $|A_{n-1}| > 0$ by the assumption, we have

$$
a_{nn} - A_{n-1}^{-1}[\mathbf{a}_{n-1}'] > 0.
$$

Therefore we have $B[\mathbf{x}] > 0$ for any $\mathbf{x} \neq \mathbf{0}_n$. $\|$

Example 3 Take a quadratic form $q(x_1, x_2, x_3) = x_1^2 + 2x_1x_2 + 2x_2^2 + 2x_2x_3 + 2x_3^2$. We let $A = \begin{pmatrix} 1 & 1 & 0 \\ 1 & 2 & 1 \\ 0 & 1 & 2 \end{pmatrix}$. Then we have $q(x_1, x_2, x_3) = A[\mathbf{x}]$. Since

$$|A_1| = |(1)| = 1 > 0, \qquad |A_2| = \begin{vmatrix} 1 & 1 \\ 1 & 2 \end{vmatrix} = 1 > 0$$

and

$$|A_3| = |A| = 1 > 0,$$

it is a positive definite quadratic form by Theorem 6.4.3.

Positive semidefinite quadratic forms A quadratic form $q(x_1, \ldots, x_n) = A[\mathbf{x}]$ is called *positive semidefinite* (or *non-negative*) if it satisfies

$$q(x_1, \ldots, x_n) \geq 0 \quad \text{for any} \quad (x_1, \ldots, x_n) \ (x_i \in \mathbf{R}).$$

Negative definite and negative semidefinite quadratic forms A quadratic form $q(x_1, \ldots, x_n)$ is called *negative definite* (or *negative*) when $q(x_1, \ldots, x_n) < 0$ for any $(x_1, \ldots, x_n) \neq (0, \ldots, 0)$ $(x_i \in \mathbf{R})$. We also call a quadratic form $q(x_1, \ldots, x_n)$ *negative semidefinite* (or *non-positive*) if it satisfies $q(x_1, \ldots, x_n) \leq 0$ for any $(x_1, \ldots, x_n)(x_i \in \mathbf{R})$.

It is easy to see that a quadratic form $q(x_1, \ldots, x_n)$ is negative definite (resp. negative semidefinite) is equivalent to that $-q(x_1, \ldots, x_n)$ is positive definite (resp. positive semidefinite). The next theorem can be easily seen.

Theorem 6.4.4 *Let $q(x_1, \ldots, x_n)$ be a quadratic form of signature (r, s).*

(1) The quadratic form $q(x_1, \ldots, x_n)$ is positive definite if and only if $r = n$.
(2) The quadratic form $q(x_1, \ldots, x_n)$ is positive semidefinite if and only if $s = 0$.
(3) The quadratic form $q(x_1, \ldots, x_n)$ is negative definite if and only if $s = n$.
(4) The quadratic form $q(x_1, \ldots, x_n)$ is negative semidefinite if and only if $r = 0$.

Exercises (Sect. 6.4)
1 Diagonalize the following quadratic forms by orthogonal change of variables:

(1) $q(x_1, x_2) = x_1^2 + 4x_1x_2 + x_2^2$.
(2) $q(x_1, x_2, x_3) = 2x_1x_2 + 2x_2x_3$.
(3) $q(x_1, x_2, x_3) = x_1^2 + 2\sqrt{2}x_1x_2 + x_2^2 + 2\sqrt{2}x_2x_3 + x_3^2$.

2 Find the canonical forms of the following quadratic forms and their signatures:

(1) $q(x_1, x_2, x_3) = x_1^2 + 2x_1x_2 + x_2^2 - x_3^2$.
(2) $q(x_1, x_2, x_3, x_4) = x_1^2 + 2x_1x_2 + 2x_2^2 + 4x_3x_4$.
(3) $q(x_1, x_2, x_3, x_4) = x_1^2 + 4x_1x_3 + 3x_2^2 + 3x_3^2 - 6x_2x_4$.

3 Show that the following quadratic forms are positive definite:

(1) $q(x_1, x_2, x_3) = \begin{pmatrix} x_1 & x_2 & x_3 \end{pmatrix} \begin{pmatrix} 2 & 0 & -1 \\ 0 & 2 & 0 \\ -1 & 0 & 1 \end{pmatrix} \begin{pmatrix} x_1 \\ x_2 \\ x_3 \end{pmatrix}.$

(2) $q(x_1, x_2, x_3) = \begin{pmatrix} x_1 & x_2 & x_3 \end{pmatrix} \begin{pmatrix} 1 & 1 & 1 \\ 1 & 3 & -1 \\ 1 & -1 & 4 \end{pmatrix} \begin{pmatrix} x_1 \\ x_2 \\ x_3 \end{pmatrix}.$

4   Let A be a real symmetric matrix of degree n. Show that the quadratic form $q(x_1, \ldots, x_n) = A[\mathbf{x}]$ associated with the matrix A is negative definite if and only if

$$(-1)^k \, \big| A_k \big| > 0 \quad \text{for} \quad k = 1, \ldots, n.$$

Here A_k is the primary submatrix of A of degree k.

5   Show that the following quadratic forms are negative definite:

(1) $q(x_1, x_2) = \begin{pmatrix} x_1 & x_2 \end{pmatrix} \begin{pmatrix} -4 & 1 \\ 1 & -1 \end{pmatrix} \begin{pmatrix} x_1 \\ x_2 \end{pmatrix}.$

(2) $q(x_1, x_2, x_3) = \begin{pmatrix} x_1 & x_2 & x_3 \end{pmatrix} \begin{pmatrix} -1 & -1 & 1 \\ -1 & -3 & 5 \\ 1 & 5 & -12 \end{pmatrix} \begin{pmatrix} x_1 \\ x_2 \\ x_3 \end{pmatrix}.$

Chapter 7
Hermitian Inner Product Spaces

7.1 Hermitian Inner Products

In this chapter, we always consider vector spaces over the complex number field $\mathbf{C}$ and complex matrices. We define inner products called Hermitian inner products on vector spaces over $\mathbf{C}$. Hermitian inner products have properties similar to the inner products on vector spaces over the real number field $\mathbf{R}$ considered in Sect. 6.1.

Hermitian inner products Let V be a vector space over $\mathbf{C}$. A mapping of $V \times V$ into $\mathbf{C}$ given by $(\mathbf{u}, \mathbf{v})$ is called a *Hermitian inner product* on V if it satisfies the following four conditions:

(1) $(\mathbf{u}_1 + \mathbf{u}_2, \mathbf{v}) = (\mathbf{u}_1, \mathbf{v}) + (\mathbf{u}_2, \mathbf{v})$.

(2) $(c\mathbf{u}, \mathbf{v}) = c(\mathbf{u}, \mathbf{v})$.

(3) $(\mathbf{u}, \mathbf{v}) = \overline{(\mathbf{v}, \mathbf{u})}$.

(4)   If $\mathbf{u} \neq \mathbf{0}_V$, then $(\mathbf{u}, \mathbf{u}) > 0$.

Here $\mathbf{u}, \mathbf{u}_1, \mathbf{u}_2, \mathbf{v} \in V, c \in \mathbf{C}$ and $\overline{(\mathbf{v}, \mathbf{u})}$ is the complex conjugate of $(\mathbf{v}, \mathbf{u})$. The complex number $(\mathbf{u}, \mathbf{v})$ is called the Hermitian inner product of vectors $\mathbf{u}$ and $\mathbf{v}$ in V. By (1) and (3), we have

$$(\mathbf{u}, \mathbf{v}_1 + \mathbf{v}_2) = \overline{(\mathbf{v}_1 + \mathbf{v}_2, \mathbf{u})} = \overline{(\mathbf{v}_1, \mathbf{u})} + \overline{(\mathbf{v}_2, \mathbf{u})}$$
$$= (\mathbf{u}, \mathbf{v}_1) + (\mathbf{u}, \mathbf{v}_2).$$

By (2) and (3), we have

$$(\mathbf{u}, c\mathbf{v}) = \overline{(c\mathbf{v}, \mathbf{u})} = \overline{c(\mathbf{v}, \mathbf{u})} = \overline{c}\,\overline{(\mathbf{v}, \mathbf{u})} = \overline{c}\,(\mathbf{u}, \mathbf{v}).$$

Since $(\mathbf{0}_V, \mathbf{u}) = (0\,\mathbf{0}_V, \mathbf{u}) = 0\,(\mathbf{0}_V, \mathbf{u}) = 0$ and $(\mathbf{u}, \mathbf{0}_V) = \overline{(\mathbf{0}_V, \mathbf{u})} = 0$, we have

$$(\mathbf{u}, \mathbf{0}_V) = (\mathbf{0}_V, \mathbf{u}) = 0 \quad \text{for any} \quad \mathbf{u} \in V.$$

© The Author(s), under exclusive license to Springer Nature Singapore Pte Ltd. 2022
T. Miyake, *Linear Algebra*, https://doi.org/10.1007/978-981-16-6994-1_7

We note that $(\mathbf{u}, \mathbf{u})$ is a real number, since $(\mathbf{u}, \mathbf{u}) = \overline{(\mathbf{u}, \mathbf{u})}$ by (3).

Hermitian inner product spaces Vector spaces over $\mathbf{C}$ with Hermitian inner products are called *Hermitian inner product spaces*.

Example 1 For vectors $\mathbf{a} = \begin{pmatrix} a_1 \\ \vdots \\ a_n \end{pmatrix}$, $\mathbf{b} = \begin{pmatrix} b_1 \\ \vdots \\ b_n \end{pmatrix} \in \mathbf{C}^n$, we define

$$(\mathbf{a}, \mathbf{b}) = {}^t\mathbf{a}\overline{\mathbf{b}} = a_1\overline{b}_1 + \cdots + a_n\overline{b}_n.$$

Then $(\mathbf{a}, \mathbf{b})$ is a Hermitian inner product of $\mathbf{C}^n$.

Example 2 For $\mathbf{a}, \mathbf{b} \in \mathbf{C}^n$, we also define $(\mathbf{a}, \mathbf{b})$ by

$$(\mathbf{a}, \mathbf{b}) = a_1\overline{b}_1 + 2a_2\overline{b}_2 + \cdots + na_n\overline{b}_n.$$

Then $(\mathbf{a}, \mathbf{b})$ is also a Hermitian inner product of $\mathbf{C}^n$.

Hermitian space   The Hermitian inner product in Example 2 is called the *standard Hermitian inner product* of $\mathbf{C}^n$. We call the vector space $\mathbf{C}^n$ with the standard Hermitian inner product the *Hermitian space* of dimension n.

Norms of Hermitian inner product spaces   Let V be a Hermitian inner product space with a Hermitian inner product $(\ ,\ )$. For $\mathbf{u} \in V$, we let $\|\mathbf{u}\| = \sqrt{(\mathbf{u}, \mathbf{u})}$ and call it the *norm* or the *length* of a vector $\mathbf{u}$.

Theorem 7.1.1 *Let V be a Hermitian inner product space with a Hermitian inner product $(\ ,\ )$. The following statements hold for $\mathbf{u}, \mathbf{v} \in V$ and $c \in \mathbf{C}$:*

(1) $\|c\mathbf{u}\| = |c| \cdot \|\mathbf{u}\|$.

(2) $|(\mathbf{u}, \mathbf{v})| \leq \|\mathbf{u}\| \cdot \|\mathbf{v}\|$ *(the Schwarz inequality).*

(3) $\|\mathbf{u} + \mathbf{v}\| \leq \|\mathbf{u}\| + \|\mathbf{v}\|$ *(the triangular inequality).*

Proof   (1) and (3) can be proved similarly to Theorem 6.1.1(1) and (3). We shall show (2). If $\mathbf{u} = \mathbf{0}_V$, then both sides of (2) are zero. Therefore (2) holds. We assume $\mathbf{u} \neq \mathbf{0}_V$. By definition, we have

$$\|a\mathbf{u} + b\mathbf{v}\|^2 = |a|^2\|\mathbf{u}\|^2 + a\overline{b}(\mathbf{u}, \mathbf{v}) + \overline{a}b\overline{(\mathbf{u}, \mathbf{v})} + |b|^2\|\mathbf{v}\|^2 \quad (a, b \in \mathbf{C}).$$

Letting $a = -\overline{(\mathbf{u}, \mathbf{v})}$ and $b = \|\mathbf{u}\|^2$, we see that

$$\begin{aligned} \|a\mathbf{u} + b\mathbf{v}\|^2 &= |(\mathbf{u}, \mathbf{v})|^2\|\mathbf{u}\|^2 - 2\|\mathbf{u}\|^2|(\mathbf{u}, \mathbf{v})|^2 + \|\mathbf{u}\|^4\|\mathbf{v}\|^2 \\ &= \|\mathbf{u}\|^2(\|\mathbf{u}\|^2\|\mathbf{v}\|^2 - |(\mathbf{u}, \mathbf{v})|^2). \end{aligned}$$

Since $\|a\mathbf{u} + b\mathbf{v}\|^2 \geq 0$ for any $a, b \in \mathbf{C}$ and $\|\mathbf{u}\|^2 > 0$, we have

$$|(\mathbf{u}, \mathbf{v})|^2 \leq \|\mathbf{u}\|^2 \|\mathbf{v}\|^2.$$

Taking the non-negative square roots of both sides, we obtain (2). ‖

In this chapter, we consider only Hermitian inner product spaces V and the Hermitian space $\mathbf{C}^n$ unless otherwise stated.

Orthogonality of vectors We say that two vectors $\mathbf{u}$ and $\mathbf{v}$ in V are *orthogonal* if they satisfy $(\mathbf{u}, \mathbf{v}) = 0$. We denote by $\mathbf{u} \perp \mathbf{v}$ the orthogonality of vectors $\mathbf{u}$ and $\mathbf{v}$.

The following theorem can be proved similarly to Theorem 6.1.2.

Theorem 7.1.2 *If non-zero vectors $\mathbf{u}_1, \ldots, \mathbf{u}_n$ in V are orthogonal to each other, then they are linearly independent.*

Orthogonality of subspaces We say that subspaces W_1 and W_2 of V are *orthogonal* if $(\mathbf{w}_1, \mathbf{w}_2) = 0$ for any $\mathbf{w}_1 \in W_1$ and $\mathbf{w}_2 \in W_2$. The orthogonality of W_1 and W_2 is denoted by $W_1 \perp W_2$.

Orthogonal complements For a subspace W of V, we define

$$W^\perp = \{\mathbf{u} \in V \mid (\mathbf{u}, \mathbf{v}) = 0 \quad \text{for any} \quad \mathbf{v} \in W\}.$$

We call $W^\perp$ the *orthogonal complement* of W in V. Similarly to the case of the (real) inner product space, $W^\perp$ is a subspace of V. For the real case, see Exercises (Sect. 6.1) 6. Then by the definition, we have $W \perp W^\perp$ and $W^\perp$ is the maximum subspace of V which is orthogonal to W.

The next theorem can be shown similarly to Exercises (Sect. 6.2) 5.

Theorem 7.1.3 *For a subspace W, W_1, W_2 of V, we have the following statements:*

(1) $W \cap W^\perp = \{\mathbf{0}_V\}$. (2) $V = W \oplus W^\perp$.
(3) $(W^\perp)^\perp = W$. (4) $W_1 \subset W_2 \Leftrightarrow W_1^\perp \supset W_2^\perp$.
(5) $\dim(W) + \dim(W^\perp) = \dim(V)$.

Example 3 Let $W = \left\{ \begin{pmatrix} x_1 \\ x_2 \\ x_3 \end{pmatrix} \in \mathbf{C}^3 \ \middle|\ x_1 + ix_2 - 2x_3 = 0 \right\}$ be a subspace of $\mathbf{C}^3$.

Here $i = \sqrt{-1}$ is the imaginary unit. Solving the linear equation $x_1 + ix_2 - 2x_3 = 0$, we have

$$W = \left\{ c_1 \begin{pmatrix} -i \\ 1 \\ 0 \end{pmatrix} + c_2 \begin{pmatrix} 2 \\ 0 \\ 1 \end{pmatrix} \,\middle|\, c_1, c_2 \in \mathbf{C} \right\}.$$

Since $\mathbf{x} = \begin{pmatrix} x_1 \\ x_2 \\ x_3 \end{pmatrix} \in W^{\perp}$ satisfies $\left(\begin{pmatrix} x_1 \\ x_2 \\ x_3 \end{pmatrix}, \begin{pmatrix} -i \\ 1 \\ 0 \end{pmatrix} \right) = \left(\begin{pmatrix} x_1 \\ x_2 \\ x_3 \end{pmatrix}, \begin{pmatrix} 2 \\ 0 \\ 1 \end{pmatrix} \right) = 0,$ we obtain

$$W^{\perp} = \left\{ c_3 \begin{pmatrix} -1 \\ i \\ 2 \end{pmatrix} \,\middle|\, c_3 \in \mathbf{C} \right\}.$$

Then we see that $\dim (W) = 2$ and $\dim (W^{\perp}) = 1$. Therefore we have $\dim (W) + \dim (W^{\perp}) = 3 = \dim (\mathbf{C}^3)$.

Orthonormal bases A basis $\{\mathbf{u}_1, \ldots, \mathbf{u}_n\}$ of V is called an *orthonormal basis* of V if it satisfies

$$(\mathbf{u}_i, \mathbf{u}_j) = \delta_{ij} \qquad (i, j = 1, \ldots, n).$$

Here δ_{ij} is the Kronecker delta.

Standard bases of $\mathbf{C}^n$ The standard basis $\{\mathbf{e}_1, \ldots, \mathbf{e}_n\}$ of $\mathbf{C}^n$ is an orthonormal basis of $\mathbf{C}^n$.

The following theorem can be proved similarly to Theorem 6.2.1 (the Schmidt orthonormalization). Theorem 7.1.4 is also called the *Schmidt orthonormalization*.

Theorem 7.1.4 (Schmidt orthonormalization) *Let* $\{\mathbf{v}_1, \ldots, \mathbf{v}_n\}$ *be a basis of* V. *Then there exists an orthonormal basis* $\{\mathbf{u}_1, \ldots, \mathbf{u}_n\}$ *of* V *satisfying*

$$< \mathbf{u}_1, \ldots, \mathbf{u}_r > = < \mathbf{v}_1, \ldots, \mathbf{v}_r > \quad \text{for} \quad r = 1, \ldots, n.$$

In particular, any Hermitian inner product space has an orthonormal basis. Finding the orthonormal basis $\{\mathbf{u}_1, \ldots, \mathbf{u}_n\}$ *from the basis* $\{\mathbf{v}_1, \ldots, \mathbf{v}_n\}$ *is called the orthonormalization of the basis* $\{\mathbf{v}_1, \ldots, \mathbf{v}_n\}$.

Orthogonal projections Let W be a subspace of V. Since $V = W \oplus W^{\perp}$, any $\mathbf{v} \in V$ is expressed uniquely as

$$\mathbf{v} = \mathbf{w} + \mathbf{w}' \quad (\mathbf{w} \in W, \ \mathbf{w}' \in W^{\perp}).$$

We define the linear transformation $P_{V/W}$ of V by $P_{V/W}(\mathbf{v}) = \mathbf{w}$ and call it the *orthogonal projection* of V into W. Let $\{\mathbf{u}_1, \cdots, \mathbf{u}_r\}$ be an orthonormal basis of W. Then the vector $(\mathbf{v}, \mathbf{u}_1)\mathbf{u}_1 + \cdots + (\mathbf{v}, \mathbf{u}_r)\mathbf{u}_r$ belongs to W. Since $(\mathbf{u}_i, \mathbf{u}_j) = \delta_{ij}$, we have

$$(\mathbf{v} - \{(\mathbf{v}, \mathbf{u}_1)\mathbf{u}_1 + \cdots + (\mathbf{v}, \mathbf{u}_r)\mathbf{u}_r\}, \mathbf{u}_i) = (\mathbf{v}, \mathbf{u}_i) - (\mathbf{v}, \mathbf{u}_i) = 0$$

for $i = 1, \ldots, r$. Therefore if we put $\mathbf{w} = (\mathbf{v}, \mathbf{u}_1)\mathbf{u}_1 + \cdots + (\mathbf{v}, \mathbf{u}_r)\mathbf{u}_r$ and $\mathbf{w}' = \mathbf{v} - \mathbf{w}$, then $\mathbf{w} \in W$ and $\mathbf{w}' \in W^{\perp}$. As $V = W \oplus W^{\perp}$, we have

$$P_{V/W}(\mathbf{v}) = \mathbf{w} = \sum_{i=1}^{r} (\mathbf{v}, \mathbf{u}_i)\,\mathbf{u}_i \quad (\mathbf{v} \in V).$$

Exercise 7.1 Let $W = \left\{ c \begin{pmatrix} 1 \\ -i \end{pmatrix} \middle| c \in \mathbf{C} \right\}$ be a subspace of the Hermitian space $V = \mathbf{C}^2$. Find the orthogonal projection of V into W.

Answer. We orthonormalize a basis $\left\{ \begin{pmatrix} 1 \\ -i \end{pmatrix} \right\}$ of W. The orthonormalization of the vector $\begin{pmatrix} 1 \\ -i \end{pmatrix}$ is $\mathbf{a} = \dfrac{1}{\sqrt{2}} \begin{pmatrix} 1 \\ -i \end{pmatrix}$. Let $\mathbf{x} = \begin{pmatrix} x_1 \\ x_2 \end{pmatrix}$ be a vector in V. Then we have the orthogonal projection $P_{V/W}$ of V into W by

$$P_{V/W}(\mathbf{x}) = (\mathbf{x}, \mathbf{a})\,\mathbf{a} = \frac{x_1 + ix_2}{2} \begin{pmatrix} 1 \\ -i \end{pmatrix} \in W. \quad \|$$

Adjoint transformations Let T be a linear transformation of a Hermitian inner product space V. We call a linear transformation T^* of V satisfying

$$(T(\mathbf{u}), \mathbf{b}) = (\mathbf{u}, T^*(\mathbf{b})) \quad (\mathbf{u}, \mathbf{b} \in V)$$

the *adjoint transformation* of T. The existence and the uniqueness of adjoint transformations will be proved in Theorem 7.1.5 below.

Theorem 7.1.5 *Let T be a linear transformation of V. Then the following statements hold:*

(1)   The adjoint transformation T^* of T uniquely exists.

(2)   If T is an automorphism, then T^* is also an automorphism.

(3) $(T^*)^* = T.$

Proof   (1)   Let the dimension of V be n and $\{\mathbf{u}_1, \ldots, \mathbf{u}_n\}$ be an orthonormal basis of V. We denote by $A = \left(a_{ij} \right)$ the representation matrix of T with respect to the basis $\{\mathbf{u}_1, \ldots, \mathbf{u}_n\}$. We assume that the adjoint transformation T^* exists. Let $B = \left(b_{ij} \right)$ be the representation matrix of T^* with respect to the basis $\{\mathbf{u}_1, \ldots, \mathbf{u}_n\}$. Then we have the relation

$$(T(\mathbf{u}), \mathbf{v}) = (\mathbf{u}, T^*(\mathbf{v})).$$

Particularly, we have

$$(T(\mathbf{u}_i), \mathbf{u}_j) = (\mathbf{u}_i, T^*(\mathbf{u}_j))$$

for $i, j = 1, \ldots, n$. The left-hand side is

$$(T(\mathbf{u}_i), \mathbf{u}_j) = (a_{1i}\mathbf{u}_1 + \cdots + a_{ni}\mathbf{u}_n, \mathbf{u}_j) = a_{ji}.$$

The right-hand side is

$$(\mathbf{u}_i, T^*(\mathbf{u}_j)) = (\mathbf{u}_i, b_{1j}\mathbf{u}_1 + \cdots + b_{nj}\mathbf{u}_n) = \overline{b}_{ij}.$$

Therefore we have

$$a_{ji} = \overline{b}_{ij} \quad \text{for} \quad i, j = 1, \ldots, n.$$

Therefore if we define the linear transformation T^* by

$$T^*(\mathbf{u}_j) = \overline{a}_{j1}\mathbf{u}_1 + \cdots + \overline{a}_{jn}\mathbf{u}_n,$$

then T^* satisfies

$$(T(\mathbf{u}), \mathbf{v}) = (\mathbf{u}, T^*(\mathbf{v})) \quad \text{for} \quad \mathbf{u}, \mathbf{v} \in V.$$

Therefore T^* is an adjoint transformation of T. Furthermore, the representation matrix of T^* with respect to the basis $\{\mathbf{u}_1, \ldots, \mathbf{u}_n\}$ is $B = {}^t\overline{A}$ by the definition of T^*. The uniqueness of T^* is proved by the argument above.

(2) We assume that T is an automorphism. Then the representation matrix A is a regular matrix. Therefore the representation matrix ${}^t\overline{A}$ of T^* is also a regular matrix. Thus we see that the adjoint transformation T^* is also an automorphism.

(3) For any $\mathbf{u}, \mathbf{v} \in V$, we have the equality

$$(T(\mathbf{u}), \mathbf{v}) = (\mathbf{u}, T^*(\mathbf{v})) = \overline{(T^*(\mathbf{v}), \mathbf{u})}$$
$$= \overline{(\mathbf{v}, (T^*)^*(\mathbf{u}))} = ((T^*)^*(\mathbf{u}), \mathbf{v}).$$

Since the equality

$$(T(\mathbf{u}), \mathbf{v}) = ((T^*)^*(\mathbf{u}), \mathbf{v})$$

holds for any vector $\mathbf{v} \in V$, we see that $T(\mathbf{u}) = (T^*)^*(\mathbf{u})$ for any vector $\mathbf{u} \in V$ and therefore $(T^*)^* = T$. $\parallel$

Adjoint matrices For a square matrix A of degree n, we let $A^* = {}^t\overline{A}$ and call the matrix A^* the *adjoint matrix* of A. Then we have

$$(A\mathbf{x}, \mathbf{y}) = {}^t(A\mathbf{x})\overline{\mathbf{y}} = {}^t\mathbf{x}\,{}^tA\overline{\mathbf{y}} = {}^t\mathbf{x}\overline{A^*\mathbf{y}} = (\mathbf{x}, A^*\mathbf{y})$$

for $\mathbf{x}, \mathbf{y} \in \mathbf{C}^n$. We see that $(A^*)^* = A$, since $(A^*)^* = ({}^t\overline{A})^* = {}^t\overline{({}^t\overline{A})} = {}^t\overline{({}^t\overline{A})} = A$. As $A^* = {}^t\overline{A}$, we have

$$|A^*| = |\overline{A}| = \overline{|A|}.$$

Example 4 If $A = \begin{pmatrix} i & 1+2i \\ 3 & 5-i \end{pmatrix}$, then $A^* = \begin{pmatrix} -i & 3 \\ 1-2i & 5+i \end{pmatrix}$.

Theorem 7.1.6 *Let A be the representation matrix of a linear transformation T of V with respect to an orthonormal basis $\{\mathbf{u}_1, \ldots, \mathbf{u}_n\}$ of V.*

(1) The representation matrix of the adjoint transformation T^* of T with respect to the basis $\{\mathbf{u}_1, \ldots, \mathbf{u}_n\}$ is the adjoint matrix A^* of A.

(2) For a square matrix A of degree n, we have $(T_A)^* = T_{A^*}$. Here T_A and T_{A^*} are the linear transformations of $\mathbf{C}^n$ associated with A and A^*, respectively.

Proof (1) is already shown in the proof of Theorem 7.1.5 (1).

(2) Since the representation matrices of T_A and $(T_A)^*$ with respect to the standard basis of $\mathbf{C}^m$ are A and A^*, respectively, we have $(T_A)^* = T_{A^*}$. ∥

Exercise (Sect. 7.1)

1 Find the norms of the following vectors in $\mathbf{C}^3$:

(1) $\begin{pmatrix} 2+3i \\ -1+i \\ i \end{pmatrix}$.

(2) $\begin{pmatrix} 1+2i \\ -5-i \\ 2+i \end{pmatrix}$.

2 Find the Hermitian inner products of the following vectors in $\mathbf{C}^3$:

(1) $\begin{pmatrix} 1+i \\ 2-i \\ i \end{pmatrix}, \begin{pmatrix} 2-3i \\ 1+2i \\ 1-i \end{pmatrix}$.

(2) $\begin{pmatrix} i \\ 1-i \\ 3+2i \end{pmatrix}, \begin{pmatrix} 1+2i \\ 3-i \\ 3 \end{pmatrix}$.

(3) $\begin{pmatrix} 3i \\ 1+i \\ 1-i \end{pmatrix}, \begin{pmatrix} 2-i \\ -5 \\ 2+3i \end{pmatrix}$.

(4) $\begin{pmatrix} -2-2i \\ i \\ 1-i \end{pmatrix}, \begin{pmatrix} i \\ -2 \\ 1-i \end{pmatrix}$.

3 Find if the following vectors in $\mathbf{C}^3$ are orthogonal or not:

(1) $\begin{pmatrix} 1-i \\ 3 \\ 8-i \end{pmatrix}, \begin{pmatrix} 2 \\ -1+2i \\ -i \end{pmatrix}.$ (2) $\begin{pmatrix} -1+2i \\ -1 \\ 1-i \end{pmatrix}, \begin{pmatrix} 2i \\ 1+i \\ 2-i \end{pmatrix}.$

4 For the following subspaces W of $\mathbf{C}^3$, find the orthogonal complements $W^\perp$ of W:

(1) $W = \left\{ a \begin{pmatrix} 1 \\ i \\ 1 \end{pmatrix} \in \mathbf{C}^3 \,\middle|\, a \in \mathbf{C} \right\}.$

(2) $W = \left\{ a_1 \begin{pmatrix} -i \\ 2-i \\ 2 \end{pmatrix} + a_2 \begin{pmatrix} 1 \\ 1+i \\ 1-2i \end{pmatrix} \in \mathbf{C}^3 \,\middle|\, a_1, a_2 \in \mathbf{C} \right\}.$

5 We define $(\mathbf{a}, \mathbf{b})$ for vectors $\mathbf{a} = \begin{pmatrix} a_1 \\ a_2 \\ a_3 \end{pmatrix}$ and $\mathbf{b} = \begin{pmatrix} b_1 \\ b_2 \\ b_3 \end{pmatrix}$ in $\mathbf{C}^3$ in the following way. Find if $(\mathbf{a}, \mathbf{b})$ is a Hermitian inner product of $\mathbf{C}^3$ or not:

(1) $(\mathbf{a}, \mathbf{b}) = a_1 \bar{b}_3 + a_2 \bar{b}_2 + a_3 \bar{b}_1.$
(2) $(\mathbf{a}, \mathbf{b}) = 3a_1 \bar{b}_1 + 2a_2 \bar{b}_2 + a_3 \bar{b}_3.$

6 Let V be a Hermitian inner product space and $\{\mathbf{u}_1, \ldots, \mathbf{u}_n\}$ an orthonormal basis of V. Show that any vector $\mathbf{v} \in V$ is uniquely determined by the Hermitian inner products $(\mathbf{v}, \mathbf{u}_k)$ for $k = 1, \ldots, n$.

7 Let V be a Hermitian inner product space and W_1, W_2 subspaces of V. Show the following relations:

(1) $(W_1 + W_2)^\perp = W_1^\perp \cap W_2^\perp.$
(2) $(W_1 \cap W_2)^\perp = W_1^\perp + W_2^\perp.$

8 Let V be a Hermitian inner product space of dimension n and J a linear transformation of V. Show that J is the orthogonal projection of V into a subspace W of V if and only if J satisfies the following conditions:

(i) $J^2 = J.$ (ii) $J^* = J.$

9 Let V be a Hermitian inner product space and W_1, W_2 subspaces of V. We let $I_1 = P_{V/W_1}$ and $I_2 = P_{V/W_2}$ be the orthogonal projections of V into W_1 and W_2, respectively. Show that subspaces W_1 and W_2 are orthogonal if and only if $I_1 I_2 = O_V$.

10 Show the following:

(1)   For linear transformations T_1 and T_2 of V, $(T_1 T_2)^* = T_2^* T_1^*$ holds.
(2)   For an automorphism T of V, $(T^{-1})^* = (T^*)^{-1}$ holds.

11 Show the following:

(1) For square matrices A_1 and A_2 of degree n, $(A_1 A_2)^* = A_2^* A_1^*$ holds.
(2) For a regular matrix A, $(A^{-1})^* = (A^*)^{-1}$ holds.

7.2 Hermitian Transformations

Hermitian transformations A linear transformation T of V is called a *Hermitian transformation* or a *self-adjoint transformation* if $T^* = T$ is satisfied. In other words, T is a Hermitian transformation if and only if T satisfies

$$(T(\mathbf{u}), \mathbf{v}) = (\mathbf{u}, T(\mathbf{v})) \quad (\mathbf{u}, \mathbf{v} \in V).$$

For any linear transformation T of V, both TT^* and T^*T are Hermitian transformations. In fact,

$$(TT^*)^* = (T^*)^*T^* = TT^* \quad \text{and} \quad (T^*T)^* = T^*(T^*)^* = T^*T$$

by Theorem 7.1.5 (3) and Exercises (Sect. 7.1) 10 (1).

Hermitian matrices A complex square matrix A is called a *Hermitian matrix* if $A^* = A$. For any complex square matrix A, both AA^* and A^*A are Hermitian matrices.

Theorem 7.2.1 *(1) Let T be a linear transformation of V. Then T is a Hermitian transformation if and only if the representation matrix of T with respect to an orthonormal basis of V is a Hermitian matrix.*

(2) For a complex square matrix A of degree n, A is a Hermitian matrix if and only if T_A is a Hermitian transformation of $\mathbf{C}^n$. Here T_A is the linear transformation of $\mathbf{C}^n$ associated with A.

Proof The theorem is clear by Theorem 7.1.6. ∥

Theorem 7.2.2 *Eigenvalues of Hermitian transformations and Hermitian matrices are all real numbers.*

Proof The proof for Hermitian matrices is similar to the proof of Theorem 6.3.4. We give here the proof for Hermitian transformations. Let T be a Hermitian transformation. We let $\lambda\ (\in \mathbf{C})$ be an eigenvalue of T and $\mathbf{u}$ an eigenvector with the eigenvalue λ. Then $(T(\mathbf{u}), \mathbf{u}) = (\lambda\mathbf{u}, \mathbf{u}) = \lambda\|\mathbf{u}\|^2$. On the other hand, since T is a Hermitian transformation, we have

$$(T(\mathbf{u}), \mathbf{u}) = (\mathbf{u}, T(\mathbf{u})) = (\mathbf{u}, \lambda\mathbf{u}) = \overline{\lambda}\|\mathbf{u}\|^2.$$

Since $\|\mathbf{u}\| \neq 0$, we have $\overline{\lambda} = \lambda$. Therefore λ is a real number. ∥

Unitary transformations   A linear transformation T of V is called a *unitary transformation* if it does not change the Hermitian inner products:

$$(T(\mathbf{u}), T(\mathbf{v})) = (\mathbf{u}, \mathbf{v}) \quad (\mathbf{u}, \mathbf{v} \in V).$$

Unitary transformations are characterized by the following theorem.

Theorem 7.2.3 *T is a unitary transformation of V $\iff \|T(\mathbf{u})\| = \|\mathbf{u}\|$ ($\mathbf{u} \in V$). Therefore unitary transformations of V are automorphisms of V.*

Proof   ($\Rightarrow$)   If T is a unitary transformation, then $\|T(\mathbf{u})\|^2 = (T(\mathbf{u}), T(\mathbf{u})) = (\mathbf{u}, \mathbf{u}) = \|\mathbf{u}\|^2$ for any $\mathbf{u} \in V$. Therefore $\|T(\mathbf{u})\| = \|\mathbf{u}\|$.

($\Leftarrow$)   Assume that $\|T(\mathbf{u})\| = \|\mathbf{u}\|$ ($\mathbf{u} \in V$). For any vectors $\mathbf{u}, \mathbf{v} \in V$,

$$\|T(\mathbf{u}) + T(\mathbf{v})\|^2 = \|T(\mathbf{u})\|^2 + (T(\mathbf{u}), T(\mathbf{v})) + \overline{(T(\mathbf{u}), T(\mathbf{v}))} + \|T(\mathbf{v})\|^2$$

and

$$\|\mathbf{u} + \mathbf{v}\|^2 = \|\mathbf{u}\|^2 + (\mathbf{u}, \mathbf{v}) + \overline{(\mathbf{u}, \mathbf{v})} + \|\mathbf{v}\|^2$$

hold. Therefore we have $(T(\mathbf{u}), T(\mathbf{v})) + \overline{(T(\mathbf{u}), T(\mathbf{v}))} = (\mathbf{u}, \mathbf{v}) + \overline{(\mathbf{u}, \mathbf{v})}$. Hence $\mathrm{Re}\{(T(\mathbf{u}), T(\mathbf{v}))\} = \mathrm{Re}\{(\mathbf{u}, \mathbf{v})\}$. Take $i\mathbf{u}$ in place of $\mathbf{u}$. Since $\mathrm{Re}\{-i(\mathbf{u}, \mathbf{v})\} = \mathrm{Im}\{(\mathbf{u}, \mathbf{v})\}$ and $T(i\mathbf{u}) = iT(\mathbf{u})$, we have

$$\mathrm{Im}\{(T(\mathbf{u}), T(\mathbf{v}))\} = \mathrm{Re}\{-i(T(\mathbf{u}), T(\mathbf{v}))\}$$
$$= \mathrm{Re}\{-(T(i\mathbf{u}), T(\mathbf{v}))\} = \mathrm{Re}\{-(i\mathbf{u}, \mathbf{v})\} = \mathrm{Im}\{(\mathbf{u}, \mathbf{v})\}.$$

Then $(T(\mathbf{u}), T(\mathbf{v})) = (\mathbf{u}, \mathbf{v})$ for $\mathbf{u}, \mathbf{v} \in V$ and T is a unitary transformation of V. Since unitary transformations are one-to-one, they are automorphisms of V. ∥

Theorem 7.2.4 *For an automorphism T of V, the following statements are equivalent:*

(1)   The linear transformation T is a unitary transformation of V.
(2)   The relation $T^*T = I_V$ holds.
(3)   The relation $TT^* = I_V$ holds.
(4)   The inverse automorphism T^{-1} of T is equal to T^*.

Here I_V is the identity transformation of V.

Proof   (1) ⇔ (2)   Let T be a unitary transformation. Then $(\mathbf{u}, \mathbf{v}) = (T(\mathbf{u}), T(\mathbf{v})) = (\mathbf{u}, T^*T(\mathbf{v}))$ for any vectors $\mathbf{u}$ and $\mathbf{v}$ in V. Since this equality holds for any $\mathbf{u}, \mathbf{v} \in V$, we have $T^*T = I_V$. Conversely, if $T^*T = I_V$, then T satisfies $(T(\mathbf{u}), T(\mathbf{v})) = (\mathbf{u}, T^*T(\mathbf{v})) = (\mathbf{u}, \mathbf{v})$ $(\mathbf{u}, \mathbf{v} \in V)$. Therefore T is a unitary transformation.

(2) ⇔ (4)   We easily see that $T^*T = I_V$ is equivalent to $T^{-1} = T^*$.

(3) ⇔ (4)   We easily see that $TT^* = I_V$ is equivalent to $T^{-1} = T^*$.   ∥

Unitary matrices   A complex square matrix U of degree n is called a *unitary matrix* if it satisfies

$$U^*U = E_n.$$

Here U^* is the adjoint matrix of U, i.e. $U^* = {}^t\overline{U}$. By the relation $U^*U = E_n$, we see that a unitary matrix is a regular matrix.

Theorem 7.2.5 *For a square matrix U of degree n, the following statements are equivalent:*

(1)   The matrix U is a unitary matrix.
(2)   The matrix U satisfies $UU^* = E_n$.
(3)   The inverse matrix U^{-1} of U is equal to U^*.
(4)   The matrix U satisfies ${}^t U \overline{U} = E_n$.

Proof   (1) ⇔ (3) If U is a unitary matrix, then $U^*U = E_n$. Therefore U is a regular matrix and $U^{-1} = U^*$. Conversely, if $U^{-1} = U^*$, then we easily see that $U^*U = E_n$.

(2) ⇔ (3) If U satisfies $UU^* = E_n$, then U is a regular matrix and $U^{-1} = U^*$. Conversely, if $U^{-1} = U^*$, then we easily see that $UU^* = E_n$.

(1) ⇔ (4)   Take the complex conjugate of ${}^t\overline{U}U = U^*U = E_n$ and we have ${}^t U\overline{U} = E_n$. Conversely, if U satisfies ${}^t U\overline{U} = E_n$, then we have $U^*U = E_n$ by taking the complex conjugate of both sides.   ∥

Theorem 7.2.6 *For a square matrix U of degree n, we let $U = \begin{pmatrix} \mathbf{a}_1 & \cdots & \mathbf{a}_n \end{pmatrix}$ be the column partition of the matrix U. Then the following statements are equivalent:*

(1)   The matrix U is a unitary matrix.

(2)   The set $\{\mathbf{a}_1, \ldots, \mathbf{a}_n\}$ is an orthonormal basis of $\mathbf{C}^n$.

Proof   Since ${}^t U = \begin{pmatrix} {}^t\mathbf{a}_1 \\ \vdots \\ {}^t\mathbf{a}_n \end{pmatrix}$, we have

$$
{}^{t}U\overline{U} = \begin{pmatrix} {}^{t}\mathbf{a}_1 \\ \vdots \\ {}^{t}\mathbf{a}_n \end{pmatrix} \left(\overline{\mathbf{a}}_1 \cdots \overline{\mathbf{a}}_n \right) = \begin{pmatrix} {}^{t}\mathbf{a}_1 \overline{\mathbf{a}}_1 \cdots {}^{t}\mathbf{a}_1 \overline{\mathbf{a}}_n \\ \vdots \qquad \vdots \\ {}^{t}\mathbf{a}_n \overline{\mathbf{a}}_1 \cdots {}^{t}\mathbf{a}_n \overline{\mathbf{a}}_n \end{pmatrix}
$$

$$
= \begin{pmatrix} (\mathbf{a}_1, \mathbf{a}_1) \cdots (\mathbf{a}_1, \mathbf{a}_n) \\ \vdots \qquad \vdots \\ (\mathbf{a}_n, \mathbf{a}_1) \cdots (\mathbf{a}_n, \mathbf{a}_n) \end{pmatrix}.
$$

Here $(\mathbf{a}_i, \mathbf{a}_j)$ is the standard inner product of $\mathbf{a}_i$ and $\mathbf{a}_j$ for $i, j = 1, \ldots, n$. By Theorem 7.2.5, the matrix is a unitary matrix if and only if ${}^{t}U\overline{U} = E_n$. Now, if U is a unitary matrix, then we have

$$
E_n = {}^{t}U\overline{U} = \begin{pmatrix} (\mathbf{a}_1, \mathbf{a}_1) \cdots (\mathbf{a}_1, \mathbf{a}_n) \\ \vdots \qquad \vdots \\ (\mathbf{a}_n, \mathbf{a}_1) \cdots (\mathbf{a}_n, \mathbf{a}_n) \end{pmatrix}.
$$

Therefore we have

$$
(\mathbf{a}_i, \mathbf{a}_j) = \delta_{ij} \qquad (i, j = 1, \ldots, n)
$$

and $\{\mathbf{a}_1, \ldots, \mathbf{a}_n\}$ is an orthonormal basis of $\mathbf{C}^n$. Conversely, assume that $\{\mathbf{a}_1, \ldots, \mathbf{a}_n\}$ is an orthonormal basis of $\mathbf{C}^n$. Then $(\mathbf{a}_i, \mathbf{a}_j) = \delta_{ij}$ for $i, j = 1, \ldots, n$. Therefore

$$
{}^{t}U\overline{U} = \begin{pmatrix} (\mathbf{a}_1, \mathbf{a}_1) \cdots (\mathbf{a}_1, \mathbf{a}_n) \\ \vdots \qquad \vdots \\ (\mathbf{a}_n, \mathbf{a}_1) \cdots (\mathbf{a}_n, \mathbf{a}_n) \end{pmatrix} = E_n
$$

and U is a unitary matrix. ∥

Example 1 Let $U = \begin{pmatrix} \cos\theta & i\sin\theta \\ i\sin\theta & \cos\theta \end{pmatrix}$. Since

$$
U^{*}U = \begin{pmatrix} \cos\theta & -i\sin\theta \\ -i\sin\theta & \cos\theta \end{pmatrix} \begin{pmatrix} \cos\theta & i\sin\theta \\ i\sin\theta & \cos\theta \end{pmatrix} = E_2,
$$

U is a unitary matrix.

Theorem 7.2.7 *Let T be a linear transformation of V. Then the following are equivalent:*

(1) The linear transformation T is a unitary transformation of V.
(2) The representation matrix of T with respect to an orthonormal basis of V is a unitary matrix.

Proof   The assertion can be easily shown by Theorem 7.1.6, Theorem 7.2.4 and Theorem 7.2.5. In fact, since the representation matrix of $T^{*}T$ with respect to the basis is U^*U by Theorem 7.1.6, we see that $T^*T = I_V$ if and only if $U^*U = E$, which is nothing but the equivalence of (1) and (2).   ‖

Exercise 7.2 Show that the following matrix U is a unitary matrix.

$$U = \begin{pmatrix} 1/2 & i/\sqrt{2} & 1/2 \\ (1+i)/2 & 0 & -(1+i)/2 \\ i/2 & 1/\sqrt{2} & i/2 \end{pmatrix}.$$

Answer. We have only to see that $U^*U = E_3$. In fact,

U^*U

$$= \begin{pmatrix} 1/2 & (1-i)/2 & -i/2 \\ -i/\sqrt{2} & 0 & 1/\sqrt{2} \\ 1/2 & -(1-i)/2 & -i/2 \end{pmatrix} \begin{pmatrix} 1/2 & i/\sqrt{2} & 1/2 \\ (1+i)/2 & 0 & -(1+i)/2 \\ i/2 & 1/\sqrt{2} & i/2 \end{pmatrix}$$

$= E_3$. ‖

Let U be a square matrix of degree n. Since the representation matrix of the linear transformation T_U associated with U with respect to the standard basis of $\mathbf{C}^n$ is U, the following theorem follows from Theorem 7.2.7.

Theorem 7.2.8 *For a square matrix U, we have*

$$T_U \text{ is a unitary transformation} \iff U \text{ is a unitary matrix.}$$

Here T_U is the linear transformation associated with the matrix U.

Normal transformations   A linear transformation T of V is called a *normal transformation* if it satisfies

$$T^*T = TT^*.$$

Normal matrices   A complex square matrix A is called a *normal matrix* if it satisfies

$$A^*A = AA^*.$$

Example 2 Hermitian transformations are normal transformations. In fact, if T is a Hermitian transformation, then $T^* = T$. Therefore $T^*T = T^2 = TT^*$.

Example 3 Unitary transformations are also normal transformations. In fact, a unitary transformation T of V satisfies $T^*T = I_V$. Therefore $T^* = T^{-1}$ and we obtain

$$T^*T = I_V = TT^{-1} = TT^*.$$

We have the following theorem by Theorem 7.1.6.

Theorem 7.2.9 *(1) The following are equivalent:*

  (i) A linear transformation T of V is a normal transformation.
  *(ii) The representation matrix of T with respect to an orthonormal basis of V
  is a normal matrix.*

(2) The following are equivalent:

  (i) A complex square matrix A is a normal matrix.
  (ii) The linear transformation T_A is a normal transformation.

Example 4 Hermitian transformations are normal transformations by Example 2, and unitary transformations are also normal transformations by Example 3. However, there are normal transformations which are neither Hermitian transformations nor unitary transformations. For example, The matrix $A = \begin{pmatrix} 3 & i \\ 1 & 3 \end{pmatrix}$ is neither a Hermitian matrix nor a unitary matrix, but it is a normal matrix. Therefore the linear transformation T_A associated with A is neither a Hermitian transformation nor a unitary transformation, but it is a normal transformation of $\mathbf{C}^2$.

Example 5 Let U be a unitary matrix. Then the linear transformation T_U associated with U is a unitary transformation. The transformation $T_{2U} = 2T_U$ is no longer a unitary transformation, but it is still a normal transformation.

Theorem 7.2.10 *Let $\dim(V) = n$. If T_1 and T_2 are commutative linear transformations of V, then there exist n subspaces $W(1), \ldots, W(n)$ of V satisfying the following three conditions:*

*(1) Both linear transformations T_1 and T_2 map the subspace $W(k)$ into $W(k)$ itself
  for $k = 1, \ldots, n$.*
(2) $W(1) \subset \cdots \subset W(n-1) \subset W(n) = V$.
(3) $\dim(W(k)) = k$ for $k = 1, \ldots, n$.

Proof We are going to prove the theorem by induction on n.

If $n = 1$, then the theorem is trivial. We assume that the theorem holds if the dimension is less than or equal to $n - 1$. Since T_1 and T_2 are commutative, the adjoint transformations T_1^* and T_2^* are also commutative. In fact, we see that

$$T_1^* T_2^* = (T_2 T_1)^* = (T_1 T_2)^* = T_2^* T_1^*.$$

Then by Theorem 5.3.5 (1), there exists a common eigenvector $\mathbf{u}$ of T_1^* and T_2^*. We let

$$T_i^*(\mathbf{u}) = \lambda_i \mathbf{u} \quad \text{for} \quad i = 1, 2,$$

and put $W(n-1) = <\mathbf{u}>^{\perp}$. Then we have $\dim(W(n-1)) = n-1$ by Theorem 7.1.3(5). We see that the linear transformations T_1 and T_2 map the space $W(n-1)$ into itself. In fact, for a vector $\mathbf{v} \in W(n-1)$, we have

$$(\mathbf{u}, T_i(\mathbf{v})) = (T_i^*(\mathbf{u}), \mathbf{v}) = (\lambda_i \mathbf{u}, \mathbf{v})$$
$$= \lambda_i(\mathbf{u}, \mathbf{v}) = 0$$

for $i = 1, 2$. This implies that $T_i(\mathbf{v}) \in <\mathbf{u}>^{\perp} = W(n-1)$ for $i = 1, 2$. Therefore both T_1 and T_2 map the subspace $W(n-1)$ into $W(n-1)$.

Since $\dim(W(n-1)) = n-1$, we can use the induction assumption for the vector space $W(n-1)$ and the restrictions of commutative linear transformations T_1 and T_2 on $W(n-1)$. Hence there exist the subspaces $W(1), \ldots, W(n-2)$ of $W(n-1)$ that satisfy the following three conditions:

(1) The restrictions of the linear transformations T_1 and T_2 on $W(k)$ are linear transformations of $W(k)$ for $k = 1, \ldots, n-1$.
(2) $W(1) \subset \cdots \subset W(n-1)$.
(3) $\dim(W(k)) = k$ for $k = 1, \ldots, n-1$.

We let $W(n) = V$. Then the subspaces $W(1), \ldots, W(n-1), W(n)$ are the subspaces we ask for. ‖

Theorem 7.2.11 *Let T_1 and T_2 be commutative linear transformations of V. Then there exists an orthonormal basis of V such that the representation matrices of T_1 and T_2 with respect to the basis are both upper triangular matrices.*

Proof Let $W(k)$ be the subspaces of the vector space V for $k = 1, \ldots, n$ in Theorem 7.2.10. Let $W(0) = \{\mathbf{0}_V\}$. Since $\dim(W(k)) = k$, we can take linearly independent vectors $\mathbf{v}_1, \ldots, \mathbf{v}_n$ so that $\mathbf{v}_k \in W(k)$ but $\mathbf{v}_k \notin W(k-1)$. We note that T_1 and T_2 maps $W(k)$ into itself by Theorem 7.2.10(1). We orthonormalize the basis $\{\mathbf{v}_1, \ldots, \mathbf{v}_n\}$ by the Schmidt orthonormalization and obtain the orthonormal basis $\{\mathbf{u}_1, \ldots, \mathbf{u}_n\}$ of V. By the process of the Schmidt orthonormalization, we see that $\mathbf{u}_k \in W(k)$ for $k = 1, \ldots, n$. Let A_1 and A_2 be the representation matrices of T_1 and T_2, respectively. Since T_1 and T_2 map $W(k)$ into itself, both matrices A_1 and A_2 are upper triangular matrices. ‖

Theorem 7.2.12 *Let T be a normal transformation of V.*

(1) *For any eigenvalue λ of T, $\bar{\lambda}$ is an eigenvalue of T^* and $W(\bar{\lambda}; T^*) = W(\lambda; T)$.*
(2) *Eigenspaces of T with different eigenvalues are orthogonal to each other.*

Proof   (1) Since T and T^* are commutative, we see that T^* maps $W(\lambda; T)$ into $W(\lambda; T)$. Then for any vectors $\mathbf{u}, \mathbf{v} \in W(\lambda; T)$, we have

$$(\mathbf{u}, T^*(\mathbf{v})) = (T(\mathbf{u}), \mathbf{v}) = (\lambda\mathbf{u}, \mathbf{v}) = (\mathbf{u}, \overline{\lambda}\mathbf{v}),$$

Since $\mathbf{u}$ is any vector in $W(\lambda; T)$, we see that $T^*(\mathbf{v}) = \overline{\lambda}\mathbf{v}$. Therefore $\overline{\lambda}$ is an eigenvalue of T^* and $W(\lambda; T) \subset W(\overline{\lambda}; T^*)$. Since $(T^*)^* = T$ and $\overline{\overline{\lambda}} = \lambda$, we have $W(\overline{\lambda}; T^*) \subset W(\lambda; T)$. Thus we obtain $W(\overline{\lambda}; T^*) = W(\lambda; T)$.

(2) Let λ_1 and λ_2 be different eigenvalues of T, and $\mathbf{u}_1$ and $\mathbf{u}_2$ eigenvectors with the eigenvalues λ_1 and λ_2, respectively. Since $T^*(\mathbf{u}_2) = \overline{\lambda}_2\mathbf{u}_2$ by (1), we have

$$\begin{aligned}
\lambda_1(\mathbf{u}_1, \mathbf{u}_2) &= (T(\mathbf{u}_1), \mathbf{u}_2) = (\mathbf{u}_1, T^*(\mathbf{u}_2)) \\
&= (\mathbf{u}_1, \overline{\lambda}_2\mathbf{u}_2) = \lambda_2(\mathbf{u}_1, \mathbf{u}_2).
\end{aligned}$$

Since $(\lambda_1 - \lambda_2)(\mathbf{u}_1, \mathbf{u}_2) = 0$ and $\lambda_1 - \lambda_2 \neq 0$, we have $(\mathbf{u}_1, \mathbf{u}_2) = 0$. $\parallel$

Orthogonal decomposition of Hermitian inner product spaces   Let V be a Hermitian inner product space. If there are non-empty subspaces W_i of V that for $i = 1, \ldots, r$ satisfy the following conditions:

(i) $V = W_1 \oplus \cdots \oplus W_r$,
(ii)   subspaces W_i and W_j are orthogonal if $i \neq j$ for $i, j = 1, \ldots, r$,

then we call $V = W_1 \oplus \cdots \oplus W_r$ the *orthogonal decomposition* of V by the subspaces $W_1, \ldots, W_r$.

Now, we have the following theorem on the orthogonal decomposition of a Hermitian inner space by the eigenspaces of a normal transformation.

Theorem 7.2.13 *Let T be a transformation of V and $\lambda_1, \ldots, \lambda_r$ all the different eigenvalues of T. Then the following two statements are equivalent:*

(1)   The transformation T is a normal transformation.
(2)   We have the orthogonal decomposition

$$V = W(\lambda_1; T) \oplus \cdots \oplus W(\lambda_r; T).$$

Proof   (1) $\Rightarrow$ (2) Let T be a normal transformation of V. Since T and T^* are commutative, there exists an orthonormal basis $\{\mathbf{u}_1, \ldots, \mathbf{u}_n\}$ of V with respect to which the representation matrices A of T and A^* of T^* are both upper triangular matrices by Theorem 7.2.11. Since $A^* = {}^t\overline{A}$ is an upper triangular matrix, A is also a lower triangular matrix. Hence A is a diagonal matrix. Therefore the representation matrix A and A^* are diagonal matrices and $\mathbf{u}_1, \ldots, \mathbf{u}_n$ are eigenvectors of T. Changing the order of the vectors in the basis, we have

$$V = W(\lambda_1; T) \oplus \cdots \oplus W(\lambda_r; T).$$

Since $W(\lambda_i; T)$ and $W(\lambda_j; T)$ are orthogonal to each other if $i \neq j$ by Theorem 7.2.12 (2), this decomposition of V is the orthogonal decomposition.

(2) $\Rightarrow$ (1) Take an orthonormal basis of $W(\lambda_i; T)$ for $i = 1, \ldots, r$. Then the union of the orthonormal bases of $W(\lambda_i; T)$ for $i = 1, \ldots, r$ is an orthonormal basis of V and the representation matrix A of T with respect to the basis is a diagonal matrix. The representation matrix of T^* with respect to the same basis is the adjoint matrix A^* of A by Theorem 7.1.6 (1). Since A is a diagonal matrix, $A^* = {}^t\overline{A}$ is also a diagonal matrix and $AA^* = A^*A$ holds. Thus we have $TT^* = T^*T$. ‖

Example 6 Let $A = \begin{pmatrix} 0 & 1 \\ 1 & 0 \end{pmatrix}$, then A is a normal matrix. Therefore the linear transformation T_A associated with A is a normal transformation. Since the characteristic polynomial $g_A(t) = t^2 - 1 = (t-1)(t+1)$, the eigenvalues of A are 1 and -1. Solving the linear equations $Ax = x$ and $Ax = -x$, we have

$$W(1; T_A) = W(1; A) = \left\{ a\begin{pmatrix} 1 \\ 1 \end{pmatrix} \in \mathbf{C}^2 \,\middle|\, a \in \mathbf{C} \right\}$$

and

$$W(-1; T_A) = W(-1; A) = \left\{ b\begin{pmatrix} -1 \\ 1 \end{pmatrix} \in \mathbf{C}^2 \,\middle|\, b \in \mathbf{C} \right\}.$$

since they are orthogonal. We see that subspace s

$$\mathbf{C}^2 = W(1; A) \oplus W(-1; A)$$

is the orthogonal decomposition of $\mathbf{C}^2$.

Spectral resolution of normal transformations Let T be a normal transformation of V and $\lambda_1, \ldots, \lambda_r$ all the different eigenvalues of T. By Theorem 7.2.13, the subspaces $W(\lambda_i; T)$ are orthogonal to each other for $i = 1, \ldots, r$ and V has the orthogonal decomposition

$$V = W(\lambda_1; T) \oplus \cdots \oplus W(\lambda_r; T).$$

We denote by I_i the orthogonal projection of V into the subspace $W(\lambda_i; T)$ for $i = 1, \ldots, r$. Then we see that

$$(*) \qquad \begin{cases} I_V = I_1 + \cdots + I_r, \\ I_i I_j = \delta_{ij} I_i \quad \text{and} \quad I_i^* = I_i \quad \text{for} \quad i, j = 1, \dots, r, \end{cases}$$

by Exercises (Sect. 7.1) 8 and 9. Furthermore, we have

$$(**) \qquad\qquad T = \lambda_1 I_1 + \cdots + \lambda_r I_r.$$

In fact, for $\mathbf{u} = \mathbf{u}_1 + \cdots + \mathbf{u}_r \in V$ $(\mathbf{u}_i \in W(\lambda_i ; T))$, we have

$$T(\mathbf{u}) = T(\mathbf{u}_1) + \cdots + T(\mathbf{u}_r) = \lambda_1 \mathbf{u}_1 + \cdots + \lambda_r \mathbf{u}_r$$
$$= \lambda_1 I_1(\mathbf{u}) + \cdots + \lambda_r I_r(\mathbf{u}) = (\lambda_1 I_1 + \cdots + \lambda_r I_r)(\mathbf{u}).$$

We call $(**)$ together with orthogonal projections $I_1, \dots, I_r$ of V satisfying $(*)$ the *spectral resolution* of T. The spectral resolution of a normal transformation T is uniquely determined up to the order of W_i and $\lambda_i I_i$, since $\lambda_1, \dots, \lambda_r$ are all different eigenvalues and $I_i(V)$ is the eigenspace of T for $i = 1, \dots, r$.

The spectral resolution of the adjoint transformation T^* of T is

$$(***) \qquad\qquad T^* = \bar{\lambda}_1 I_1 + \cdots + \bar{\lambda}_r I_r$$

with orthogonal projections $I_1, \dots, I_r$ of V satisfying $(*)$. If there exists the spectral resolution of T, then it is easy to see that T is a normal transformation.

Thus we have the following theorem.

Theorem 7.2.14 *Let T be a linear transformation of V. Then the following are equivalent:*

(1) A linear transformation T is a normal transformation.
(2) A linear transformation T has the spectral resolution

$$T = \lambda_1 I_1 + \cdots + \lambda_r I_r.$$

Here $\lambda_1, \dots \lambda_r$ are all different complex numbers and $I_1, \dots, I_r$ are the orthogonal projections of V satisfying $(*)$.
When T is a normal transformation and $\lambda_1, \dots \lambda_r$ are all different eigenvalues of T, the spectral resolution of T^* is

$$T^* = \bar{\lambda}_1 I_1 + \cdots + \bar{\lambda}_r I_r.$$

Diagonalization of normal matrices by unitary matrices   Let A be a normal matrix of degree n, $\lambda_1, \dots, \lambda_r$ all the different eigenvalues of A. Then the linear transformation T_A of $\mathbf{C}^n$ associated with A is a normal transformation and the representation matrix of T_A with respect to the standard basis of $\mathbf{C}^n$ is A. Then by Theorem 7.2.13, $\mathbf{C}^n$ is orthogonally decomposed as

$$\mathbf{C}^n = W(\lambda_1\,;\,A) \oplus \cdots \oplus W(\lambda_r\,;\,A).$$

Let $\{\mathbf{w}_{i1}, \ldots, \mathbf{w}_{in_i}\}$ be an orthonormal basis of the subspace W_i for $i = 1, \ldots, r$. Then $n = n_1 + \cdots + n_r$ and

$$\{\mathbf{w}_{11}, \ldots, \mathbf{w}_{1n_1}, \ldots, \mathbf{w}_{r1}, \ldots, \mathbf{w}_{rn_r}\}$$

is an orthonormal basis of $\mathbf{C}^n$. Let

$$U = \left(\mathbf{w}_{11} \cdots \mathbf{w}_{1n_1} \cdots \mathbf{w}_{r1} \cdots \mathbf{w}_{rn_r} \right).$$

Since $\{\mathbf{w}_{11}, \ldots, \mathbf{w}_{1n_1}, \ldots, \mathbf{w}_{r1}, \ldots, \mathbf{w}_{rn_r}\}$ is an orthonormal basis, the matrix U is a unitary matrix. The matrix A is diagonalized as

$$U^{-1}AU = \lambda_1 E_{n_1} \oplus \cdots \oplus \lambda_r E_{n_r}$$

$$= \begin{pmatrix} \lambda_1 & & & & & \\ & \ddots & & & \mathbf{0} & \\ & & \lambda_1 & & & \\ & & & \ddots & & \\ & \mathbf{0} & & & \lambda_r & \\ & & & & & \ddots \\ & & & & & & \lambda_r \end{pmatrix}$$

by the unitary matrix U. Since $W(\overline{\lambda}_i\,;\,A^*) = W(\overline{\lambda}_i\,;\,T_{A^*}) = W(\lambda_i\,;\,T_A) = W(\lambda_i\,;\,A)$ for $i = 1, \ldots, r$, we have an orthogonal decomposition

$$\mathbf{C}^n = W(\overline{\lambda}_1\,;\,A^*) \oplus \cdots \oplus W(\overline{\lambda}_r\,;\,A^*),$$

and A^* is also diagonalized by the same unitary matrix U as

$$U^{-1}AU = \lambda_1 E_{n_1} \oplus \cdots \oplus \lambda_r E_{n_r}$$

$$= \begin{pmatrix} \overline{\lambda}_1 & \overset{n_1}{\ddots} & & & & \\ & \overline{\lambda}_1 & & & 0 & \\ & & \ddots & & & \\ & & & \overline{\lambda}_r & \overset{n_r}{\ddots} & \\ & 0 & & & \ddots & \\ & & & & & \overline{\lambda}_r \end{pmatrix}$$

Conversely, we assume that a square matrix A is diagonalizable by a unitary matrix U. Then there exists a diagonal matrix D such that

$$U^{-1}AU = D.$$

Thus we have

$$A = UDU^{-1}.$$

Since $U^* = U^{-1}$, we have $A^* = (UDU^{-1})^* = UD^*U^{-1}$ and

$$AA^* = (UDU^{-1})(UD^*U^{-1}) = UDD^*U^{-1}.$$

As D is a diagonal matrix, D^* is also a diagonal matrix and we have $DD^* = D^*D$. Then we have

$$AA^* = UDD^*U^{-1} = UD^*DU^{-1} = (UD^*U^{-1})(UDU^{-1}) = A^*A.$$

Therefore the matrix A is a normal matrix. Thus we have the following theorem.

Theorem 7.2.15 (Toeplitz) *The following two statements are equivalent for a square matrix A:*

(1)   The matrix A is a normal matrix.
(2)   The matrix A is diagonalizable by a unitary matrix.

Exercise 7.3 (1)   Show that the following matrix A is a normal matrix.

$$A = \begin{pmatrix} 2 & 0 & i \\ 0 & i & 0 \\ -i & 0 & 2 \end{pmatrix}.$$

(2)　Diagonalize A and A^* by a unitary matrix.

Answers.

(1)　Since $A^* = \begin{pmatrix} 2 & 0 & i \\ 0 & -i & 0 \\ -i & 0 & 2 \end{pmatrix}$, we have

$$AA^* = A^*A = \begin{pmatrix} 5 & 0 & 4i \\ 0 & 1 & 0 \\ -4i & 0 & 5 \end{pmatrix}.$$

Therefore A is a normal matrix.

(2)　We shall find a unitary matrix U. The characteristic polynomial of A is

$$g_A(t) = \begin{vmatrix} t-2 & 0 & -i \\ 0 & t-i & 0 \\ i & 0 & t-2 \end{vmatrix} = (t-3)(t-i)(t-1).$$

Therefore the characteristic roots λ of A are $3, i, 1$.

Let $\lambda = 3$. We solve the linear equation $A\mathbf{x} = 3\mathbf{x}$. Then the solutions are

$$\mathbf{x} = a_1 \begin{pmatrix} i \\ 0 \\ 1 \end{pmatrix} \quad (a_1 \in \mathbf{C}).$$

Let $\lambda = i$. We solve the linear equation $A\mathbf{x} = i\mathbf{x}$. Then the solutions are

$$\mathbf{x} = a_2 \begin{pmatrix} 0 \\ 1 \\ 0 \end{pmatrix} \quad (a_2 \in \mathbf{C}).$$

Let $\lambda = 1$. We solve the linear equation $A\mathbf{x} = \mathbf{x}$. Then the solutions are

$$\mathbf{x} = a_3 \begin{pmatrix} -i \\ 0 \\ 1 \end{pmatrix} \quad (a_3 \in \mathbf{C}).$$

Orthonormalizing the vectors $\left\{ \begin{pmatrix} i \\ 0 \\ 1 \end{pmatrix}, \begin{pmatrix} 0 \\ 1 \\ 0 \end{pmatrix}, \begin{pmatrix} -i \\ 0 \\ 1 \end{pmatrix} \right\}$, we have an orthonormal

basis $\left\{ \begin{pmatrix} i/\sqrt{2} \\ 0 \\ 1/\sqrt{2} \end{pmatrix}, \begin{pmatrix} 0 \\ 1 \\ 0 \end{pmatrix}, \begin{pmatrix} -i/\sqrt{2} \\ 0 \\ 1/\sqrt{2} \end{pmatrix} \right\}$ of $\mathbf{C}^3$. We let

$$U = \begin{pmatrix} i/\sqrt{2} & 0 & -i/\sqrt{2} \\ 0 & 1 & 0 \\ i/\sqrt{2} & 0 & i/\sqrt{2} \end{pmatrix},$$

then U is a unitary matrix. The matrices A and A^* are diagonalized as

$$U^{-1}AU = \begin{pmatrix} 3 & 0 & 0 \\ 0 & i & 0 \\ 0 & 0 & 1 \end{pmatrix} \quad \text{and} \quad U^{-1}A^*U = \begin{pmatrix} 3 & 0 & 0 \\ 0 & -i & 0 \\ 0 & 0 & 1 \end{pmatrix}. \quad \parallel$$

As a remark to Exercise 7.3, we note that A is neither a Hermitian matrix nor a unitary matrix. Furthermore, the unitary matrices we want are not uniquely determined.

Positive definite Hermitian transformations   We know that all eigenvalues of Hermitian transformations are real numbers (Theorem 7.2.2). When all eigenvalues of a Hermitian transformation T are positive, we call T a *positive definite Hermitian transformation*. We also call a Hermitian transformation T a *positive semidefinite* (or *non-negative*) *Hermitian transformation* when all the eigenvalues of T are non-negative.

Theorem 7.2.16 *For a Hermitian transformation T of V, the following assertions hold:*

(1) T is a positive definite Hermitian transformation

$$\Longleftrightarrow \quad (T(\mathbf{u}), \mathbf{u}) > 0 \text{ for any non-zero vector } \mathbf{u} \in V.$$

(2) If T is a positive definite Hermitian transformation, then T^{-1} is also a positive definite Hermitian transformation.

(3) T is a positive semidefinite Hermitian transformation
$$\Longleftrightarrow \quad (T(\mathbf{u}), \mathbf{u}) \geq 0 \text{ for any vector } \mathbf{u} \in V.$$

Proof   Since a Hermitian transformation T is a normal transformation, there exists an orthonormal basis $\{\mathbf{u}_1, \ldots, \mathbf{u}_n\}$ of V consisting of eigenvectors of T. We let $T(\mathbf{u}_i) = \lambda_i \mathbf{u}_i$ for $i = 1, \ldots, n$. Then if $\mathbf{u} = a_1 \mathbf{u}_1 + \cdots + a_n \mathbf{u}_n \in V$, then we have

$$(*) \qquad\qquad (T(\mathbf{u}), \mathbf{u}) = \lambda_1 |a_1|^2 + \cdots + \lambda_n |a_n|^2.$$

(1) By definition, T is a positive definite Hermitian transformation if and only if all the eigenvalues $\lambda_1, \ldots, \lambda_n$ of T are positive. Therefore we have

$$T \text{ is a positive definite Hermitian transformation}$$
$$\Longleftrightarrow \quad (T(\mathbf{u}), \mathbf{u}) > 0 \text{ if } \mathbf{u} \neq \mathbf{0}_V.$$

(2) Let T be a positive definite Hermitian transformation of V. Then T is an automorphism by (1). For vectors $\mathbf{u}$ and $\mathbf{v}$ in V, take vectors $\mathbf{u}', \mathbf{v}'$ satisfying $T(\mathbf{u}') = \mathbf{u}$, $T(\mathbf{v}') = \mathbf{v}$. Then we have

$$(T^{-1}(\mathbf{u}), \mathbf{v}) = (\mathbf{u}', T(\mathbf{v}')) = (T(\mathbf{u}'), \mathbf{v}') = (\mathbf{u}, T^{-1}(\mathbf{v})).$$

Therefore T^{-1} is a Hermitian transformation. If $\mathbf{u}$ is a non-zero vector in V, then $\mathbf{u}'$ is also a non-zero vector and we have

$$(T^{-1}(\mathbf{u}), \mathbf{u}) = (\mathbf{u}', T(\mathbf{u}')) = (T(\mathbf{u}'), \mathbf{u}') > 0.$$

Therefore T^{-1} is also a positive definite Hermitian transformation. (3) is clear by the equality $(*)$. $\parallel$

Positive definite Hermitian matrices   We also know that all eigenvalues of Hermitian matrices are real numbers (Theorem 7.2.2). When all eigenvalues of a Hermitian matrix A are positive, we call A a *positive definite Hermitian matrix*. A Hermitian matrix A is also called a *positive semidefinite* (or *non-negative*) *Hermitian matrix* when all the eigenvalues of A are non-negative.

The following theorem on positive definite Hermitian matrices and positive semidefinite Hermitian matrices can be easily shown by Theorem 7.2.16.

Theorem 7.2.17   *For a Hermitian matrix A of degree n and the linear transformation T_A associated with A, the following assertions hold:*

(1)   *A is a positive definite Hermitian matrix*
   $\Leftrightarrow$ *T_A is a positive definite Hermitian transformation*
   $\Leftrightarrow$ *$(A\mathbf{x}, \mathbf{x}) > 0$ for any non-zero vector $\mathbf{x} \in \mathbf{C}^n$.*
(2)   *A is a positive semidefinite Hermitian matrix*
   $\Leftrightarrow$ *T_A is a positive semidefinite Hermitian transformation*
   $\Leftrightarrow$ *$(A\mathbf{x}, \mathbf{x}) \geq 0$ for any vector $\mathbf{x} \in \mathbf{C}^n$.*

Example 7 The matrix $A = \begin{pmatrix} 5 & -i \\ i & 5 \end{pmatrix}$ is a positive definite Hermitian matrix. In fact, since

$$A^* = {}^t\overline{A} = \begin{pmatrix} 5 & -i \\ i & 5 \end{pmatrix} = A,$$

the matrix A is a Hermitian matrix. The characteristic polynomial is

$$g_A(t) = \begin{vmatrix} t-5 & i \\ -i & t-5 \end{vmatrix}$$
$$= t^2 - 10t + 24 = (t-4)(t-6).$$

Then the eigenvalues of A are 4 and 6. Since they are all positive numbers, A is a positive definite Hermitian matrix.

We have the following theorem for positive definite or positive semidefinite Hermitian transformations.

Theorem 7.2.18 *Let T be a positive definite (resp. positive semidefinite) Hermitian transformation of V and k (≥ 2) a positive integer. Then there uniquely exists a positive definite (resp. positive semidefinite) Hermitian transformation S of V satisfying*

$$T = S^k.$$

We call S the k-th root of the Hermitian transformation T and denote it by $\sqrt[k]{T}$. The square root $\sqrt{T}$ of the Hermitian transformation is the special case of the k-th root of the Hermitian transformation T for $k = 2$.

Proof Let $T = \lambda_1 I_1 + \cdots + \lambda_r I_r$ be the spectral resolution of T. Then we have $\lambda_i > 0$ (resp. $\lambda_i \geq 0$) for $i = 1, \ldots, r$. Now, we let

$$S = \sqrt[k]{\lambda_1} I_1 + \cdots + \sqrt[k]{\lambda_r} I_r.$$

Then S is also a positive definite (resp. positive semidefinite) Hermitian transformation and satisfies $T = S^k$.

We shall show the uniqueness of S. Let S' be another positive definite (resp. positive semidefinite) Hermitian transformation of V which satisfies $S'^k = T$. Let

$$S' = \mu_1 I_1' + \cdots + \mu_s I_s'$$

be the spectral resolution of S'. Then $\mu_1, \ldots, \mu_s$ are all different eigenvalues of S'. Since $I_i' I_j' = \delta_{ij} I_i'$ for $i, j = 1, \ldots, s$,

$$
\begin{aligned}
T = S'^k &= (\mu_1 I_1' + \cdots + \mu_s I_s')^k \\
&= \mu_1^k I_1' + \cdots + \mu_s^k I_s',
\end{aligned}
$$

which is the spectral resolution of T. Note that $T = \lambda_1 I_1 + \cdots + \lambda_r I_r$ is also the spectral resolution of T. By the uniqueness of the spectral resolution of T, we obtain that $r = s$, $I_1' = I_1, \ldots, I_r' = I_r$ and

$$\mu_1^k = \lambda_1, \quad \cdots, \quad \mu_r^k = \lambda_r$$

by changing the order of indices. Since $\mu_i \geq 0$, we have $\mu_i = \sqrt[k]{\lambda_i}$ for $i = 1, \ldots, r$. Therefore we have

$$S' = \sqrt[k]{\lambda_1} I_1 + \cdots + \sqrt[k]{\lambda_r} I_r = S. \quad \|$$

The next theorem is the restatement of Theorem 7.2.18 in terms of matrices.

Theorem 7.2.19 *Let A be a positive definite (resp. positive semidefinite) Hermitian matrix of degree n and k (≥ 2) a positive integer. Then there uniquely exists a positive definite (resp. positive semidefinite) Hermitian matrix B of degree n satisfying*

$$A = B^k.$$

We call B the k-th root of the Hermitian matrix A and denote it by $\sqrt[k]{A}$. The square root $\sqrt{A}$ of the Hermitian matrix is the special case of the k-th root of the Hermitian matrix A for $k = 2$.

Example 8 Let $A = \begin{pmatrix} 5 & -i \\ i & 5 \end{pmatrix}$ be the matrix considered in Example 7. Then A is a positive definite Hermitian matrix as we see in Example 7. We shall find the square root $\sqrt{A}$ of the matrix A. In Example 7, we see that the eigenvalues λ of A are 4 and 6. We solve the linear equation $Ax = \lambda x$.

Let $\lambda = 4$, then the eigenvectors are $x = c \begin{pmatrix} i \\ 1 \end{pmatrix}$ $(c \neq 0)$.

Let $\lambda = 6$, then the eigenvectors are $x = c \begin{pmatrix} -i \\ 1 \end{pmatrix}$ $(c \neq 0)$.

Orthonormalizing the basis $\left\{ \begin{pmatrix} i \\ 1 \end{pmatrix}, \begin{pmatrix} -i \\ 1 \end{pmatrix} \right\}$ of $\mathbf{C}^2$, we obtain an orthonormal basis $\left\{ \frac{1}{\sqrt{2}} \begin{pmatrix} i \\ 1 \end{pmatrix}, \frac{1}{\sqrt{2}} \begin{pmatrix} -i \\ 1 \end{pmatrix} \right\}$. Let $U = \frac{1}{\sqrt{2}} \begin{pmatrix} i & -i \\ 1 & 1 \end{pmatrix}$. Then U is a unitary matrix and $U^{-1} = \frac{1}{\sqrt{2}} \begin{pmatrix} -i & 1 \\ i & 1 \end{pmatrix}$. We have $U^{-1}AU = \begin{pmatrix} 4 & 0 \\ 0 & 6 \end{pmatrix}$. Therefore if we let

$$B = U \begin{pmatrix} 2 & 0 \\ 0 & \sqrt{6} \end{pmatrix} U^{-1} = \frac{1}{2} \begin{pmatrix} 2 + \sqrt{6} & (2 - \sqrt{6})i \\ -(2 - \sqrt{6})i & 2 + \sqrt{6} \end{pmatrix},$$

then $B = \sqrt{A}$. In fact,

$$B^2 = U \begin{pmatrix} 2 & 0 \\ 0 & \sqrt{6} \end{pmatrix} U^{-1} U \begin{pmatrix} 2 & 0 \\ 0 & \sqrt{6} \end{pmatrix} U^{-1}$$
$$= U \begin{pmatrix} 4 & 0 \\ 0 & 6 \end{pmatrix} U^{-1} = A.$$

We shall show that any automorphism can be uniquely expressed as a product of a positive definite Hermitian transformation and a unitary transformation and also any complex regular matrix can be uniquely expressed as a product of a positive definite Hermitian matrix and a unitary matrix.

Theorem 7.2.20 *(1) Let V be a Hermitian inner product. Any automorphism T of V is uniquely expressed as a product of a positive definite Hermitian transformation and a unitary transformation.*

(2) Any complex regular matrix A can be uniquely expressed as a product of a positive definite Hermitian matrix and a unitary matrix.

Proof We have only to prove (1) since (2) is only a restatement of (1) in terms of matrices. Since T is an automorphism, the adjoint transformation T^* of T is an automorphism by Theorem 7.1.5. Since TT^* is a Hermitian transformation and

$$(TT^*(\mathbf{u}), \mathbf{u}) = (T^*(\mathbf{u}), T^*(\mathbf{u})) > 0 \quad (\mathbf{u} (\neq \mathbf{0}_V) \in V),$$

TT^* is a positive definite Hermitian transformation of V by Theorem 7.2.16(1). By Theorem 7.2.18, there exists a positive definite Hermitian transformation $H = \sqrt{TT^*}$. Since H is a positive definite Hermitian transformation, H^{-1} is also a positive definite Hermitian transformation by Theorem 7.2.16(2). We let $U = H^{-1}T$. Then we see that

$$UU^* = (H^{-1}T)(H^{-1}T)^* = H^{-1}TT^*(H^{-1})^*$$
$$= H^{-1}H^2H^{-1} = I_V.$$

Here I_V is the identity transformation of V. By Theorem 7.2.4, U is a unitary transformation. Since $U = H^{-1}T$, we have

$$T = HU \quad \begin{pmatrix} H : \text{a positive definite Hermitian transformation,} \\ U : \text{a unitary transformation} \end{pmatrix}.$$

We shall show the uniqueness of H and U. Assume $T = H_1U_1 = H_2U_2$. Here H_1, H_2 are positive definite Hermitian transformations and U_1, U_2 are unitary transformations. Then we have

$$H_2 = H_1U_1U_2^{-1} \quad \text{and} \quad H_2 = H_2^* = (U_2^{-1})^*U_1^*H_1^* = U_2U_1^{-1}H_1.$$

Therefore we see that

$$H_2^2 = (H_1U_1U_2^{-1})(U_2U_1^{-1}H_1) = H_1^2.$$

Since both H_1 and H_2 are positive definite Hermitian transformations,

$$H_1 = \sqrt{H_1^2} = \sqrt{H_2^2} = H_2.$$

Thus we also obtain $U_1 = U_2$. $\quad \|$

Example 9 Let $A = \begin{pmatrix} 3-i & 3+i \\ -1+3i & -1-3i \end{pmatrix}$. We shall express the regular matrix A as a product of a positive definite Hermitian matrix and a unitary matrix. The matrix

$$AA^* = \begin{pmatrix} 3-i & 3+i \\ -1+3i & -1-3i \end{pmatrix}\begin{pmatrix} 3+i & -1-3i \\ 3-i & -1+3i \end{pmatrix} = \begin{pmatrix} 20 & -12 \\ -12 & 20 \end{pmatrix}$$

is a positive definite Hermitian matrix as we see in the proof of Theorem 7.2.20. Next, we shall find the square root of AA^*. Since the characteristic polynomial $g_{AA^*}(t) = t^2 - 40t + 256 = (t-8)(t-32)$, the eigenvalues λ of AA^* are 8 and 32. To find eigenvectors with the eigenvalue λ, we solve the linear equation $AA^*\mathbf{x} = \lambda\mathbf{x}$.

Let $\lambda = 8$. Then the eigenvectors are $\mathbf{x} = c\begin{pmatrix} 1 \\ 1 \end{pmatrix}$ $(c \neq 0)$.

Let $\lambda = 32$. Then the eigenvectors are $\mathbf{x} = c \begin{pmatrix} -1 \\ 1 \end{pmatrix}$ $(c \neq 0)$.

Orthonormalizing the basis $\left\{ \begin{pmatrix} 1 \\ 1 \end{pmatrix}, \begin{pmatrix} -1 \\ 1 \end{pmatrix} \right\}$ of $\mathbf{C}^2$, we obtain an orthonormal basis

$\left\{ \frac{1}{\sqrt{2}} \begin{pmatrix} 1 \\ 1 \end{pmatrix}, \frac{1}{\sqrt{2}} \begin{pmatrix} -1 \\ 1 \end{pmatrix} \right\}$. Then $U_0 = \begin{pmatrix} 1/\sqrt{2} & -1/\sqrt{2} \\ 1/\sqrt{2} & 1/\sqrt{2} \end{pmatrix}$ is a unitary matrix and

$$U_0^{-1} A A^* U_0 = \begin{pmatrix} 8 & 0 \\ 0 & 32 \end{pmatrix}.$$

holds. Since $U_0 \begin{pmatrix} \sqrt{8} & 0 \\ 0 & \sqrt{32} \end{pmatrix} U_0^{-1} U_0 \begin{pmatrix} \sqrt{8} & 0 \\ 0 & \sqrt{32} \end{pmatrix} U_0^{-1} = A A^*$, we obtain

$$\sqrt{A A^*} = U_0 \begin{pmatrix} \sqrt{8} & 0 \\ 0 & \sqrt{32} \end{pmatrix} U_0^{-1} = \begin{pmatrix} 3\sqrt{2} & -\sqrt{2} \\ -\sqrt{2} & 3\sqrt{2} \end{pmatrix}$$

Let $H = \sqrt{A A^*}$ and $U = H^{-1} A = \begin{pmatrix} 1/\sqrt{2} & 1/\sqrt{2} \\ i/\sqrt{2} & -i/\sqrt{2} \end{pmatrix}$. Then U is a unitary matrix.
Therefore $A = HU$, a product of a positive definite Hermitian matrix H and a unitary matrix U.

Exercises (Sect. 7.2)
1 For the matrices A below, answer the following questions:

(i)   Show that the matrices A are Hermitian matrices.

(ii)   Diagonalize A by unitary matrices.

(1) $\begin{pmatrix} 1 & i & 0 \\ -i & 1 & 0 \\ 0 & 0 & 2 \end{pmatrix}$.   (2) $\begin{pmatrix} 3 & 0 & i \\ 0 & 2 & 0 \\ -i & 0 & 1 \end{pmatrix}$.

2 For the matrices A below, answer the following questions:

(i)   Show that the matrices A are unitary matrices.

(ii)   Diagonalize A by unitary matrices.

(1) $\begin{pmatrix} \frac{1+i}{2} & \frac{-1+i}{2} & 0 \\ \frac{-1+i}{2} & \frac{1+i}{2} & 0 \\ 0 & 0 & -i \end{pmatrix}$.   (2) $\begin{pmatrix} \frac{1}{4}(1+3i) & 0 & \frac{\sqrt{3}}{4}(-1+i) \\ 0 & \frac{\sqrt{2}}{2}(1+i) & 0 \\ \frac{\sqrt{3}}{4}(-1+i) & 0 & \frac{1}{4}(3+i) \end{pmatrix}$.

3 For the matrices A below, answer the following questions:

(i)   Show that the matrices A are normal matrices.

(ii)   Diagonalize A by unitary matrices.

(1) $\begin{pmatrix} 0 & \sqrt{2}i \\ \sqrt{2} & 0 \end{pmatrix}.$

(2) $\begin{pmatrix} i & 0 & 2 \\ 0 & 2i & 0 \\ -2 & 0 & i \end{pmatrix}.$

4 For the matrices A below, answer the following questions:

(i)   Show that the linear transformations T_A of $\mathbf{C}^3$ associated with the matrices A are normal transformations.

(ii)   Find the spectral resolutions of T_A.

(1) $A = \begin{pmatrix} 3 & 0 & -\sqrt{2}i \\ 0 & 2i & 0 \\ \sqrt{2}i & 0 & 4 \end{pmatrix}.$

(2) $A = \begin{pmatrix} 6 & -\sqrt{2}i & -\sqrt{2}i \\ \sqrt{2}i & 5 & 1 \\ \sqrt{2}i & 1 & 5 \end{pmatrix}.$

5 Let T_1 and T_2 be commutative normal transformations of V. Show that there exists an orthonormal basis of V consisting of eigenvectors of both T_1 and T_2.

6 For a normal transformation T, show the following equivalences:

(1) T is a Hermitian transformation $\Leftrightarrow$ every eigenvalue of T is real.

(2) T is a unitary transformation $\Leftrightarrow$ the absolute value of every eigenvalue of T is equal to 1.

7 Express the following square matrices A as a product of a positive definite Hermitian matrix and a unitary matrix:

(1) $A = \begin{pmatrix} 0 & i \\ 2 & 0 \end{pmatrix}.$

(2) $A = \begin{pmatrix} 1-i & \sqrt{2} \\ \sqrt{2} & -i \end{pmatrix}.$

8 For a complex square matrix A of degree n, show the following assertions:

(1)   There uniquely exist Hermitian matrices H and K of degree n satisfying

$$A = H + iK.$$

(2)   The matrix A is a Hermitian matrix if and only if $K = O_n$.

(3)   The matrix A is a normal matrix if and only if $HK = KH$.

(4)   The matrix A is a unitary matrix if and only if $H^2 + K^2 = E_n$ and $HK = KH$.

Chapter 8
Jordan Normal Forms

8.1 Generalized Eigenspaces

Continued from the preceding chapter, we still only consider vector spaces over the complex number field $\mathbf{C}$ and complex matrices.

Generalized eigenspaces of linear transformations Let T be a linear transformation of a vector space V and λ an eigenvalue of T. We define the subset $\tilde{W}(\lambda; T)$ of V by

$$\tilde{W}(\lambda; T) = \{\, \mathbf{v} \in V \mid (T - \lambda I_V)^m(\mathbf{v}) = \mathbf{0}_V \text{ for some positive integer } m \,\}.$$

Here I_V is the identity transformation of V. The subset $\tilde{W}(\lambda; T)$ is a subspace of V (Theorem 8.1.1 below). We call the subspace $\tilde{W}(\lambda; T)$ the *generalized eigenspace* of T with the eigenvalue λ.

Theorem 8.1.1 *Let T be a linear transformation of a vector space V and λ an eigenvalue of T.*

(1) The subset $\tilde{W}(\lambda; T)$ is a subspace of V.
(2) The eigenspace $W(\lambda; T)$ is a subspace of $\tilde{W}(\lambda; T)$.

Proof (1) It is easy to see that the subset $\tilde{W}(\lambda; T)$ can be expressed as

$$\tilde{W}(\lambda; T) = \bigcup_{m=1}^{\infty} \mathrm{Ker}((T - \lambda I_V)^m).$$

Since $\mathrm{Ker}((T - \lambda I_V)^m)$ is a subspace of V for any positive integer m by Theorem 5.1.1 and

$$\mathrm{Ker}((T - \lambda I_V)^l) \subset \mathrm{Ker}((T - \lambda I_V)^{l+1}) \quad \text{for } l = 1, 2, \ldots,$$

the subset $\tilde{W}(\lambda; T)$ is a subspace of V.

© The Author(s), under exclusive license to Springer Nature Singapore Pte Ltd. 2022
T. Miyake, *Linear Algebra*, https://doi.org/10.1007/978-981-16-6994-1_8

(2) Since $W(\lambda\,;T) = \mathrm{Ker}(T - \lambda I_V)$, we have

$$W(\lambda\,;T) = \mathrm{Ker}(T - \lambda I_V) \subset \bigcup_{m=1}^{\infty} \mathrm{Ker}((T - \lambda I_V)^m) = \tilde{W}(\lambda\,;T). \quad \|$$

Generalized eigenspaces of square matrices For a square matrix A of degree n and an eigenvalue λ of A, we let

$$\tilde{W}(\lambda\,;A) = \{\, \mathbf{x} \in \mathbf{C}^n \mid (A - \lambda E_n)^m \mathbf{x} = \mathbf{0}_n \ \text{ for some positive integer } m \,\}.$$

Since $\tilde{W}(\lambda\,;A) = \tilde{W}(\lambda\,;T_A)$, $\tilde{W}(\lambda\,;A)$ is a subspace of $\mathbf{C}^n$ by Theorem 8.1.1. Here T_A is the linear transformation of $\mathbf{C}^n$ associated with A. We call $\tilde{W}(\lambda\,;A)$ the *generalized eigenspace* of A with the eigenvalue λ.

We have the following theorem.

Theorem 8.1.2 *Let A be a square matrix of degree n and λ an eigenvalue of A. Then the eigenspace $W(\lambda\,;A)$ is a subspace of $\tilde{W}(\lambda\,;A)$.*

Multiplicities of eigenvalues For a linear transformation T of a vector space V (resp. a square matrix A of degree n), we let $\{\lambda_1, \dots, \lambda_r\}$ be the set of all different eigenvalues of T (resp. of A). When the characteristic polynomial of T (resp. of A) is

$$g_T(t) = \prod_{i=1}^{r}(t - \lambda_i)^{n_i} \quad \left(\text{resp.} \quad g_A(t) = \prod_{i=1}^{r}(t - \lambda_i)^{n_i} \right),$$

we call the number n_i the *multiplicity of the eigenvalue* λ_i of T (resp. of A) for $i = 1, \dots, r$.

Theorem 8.1.3 *Let T be a linear transformation of a vector space V and $\{\lambda_1, \dots, \lambda_r\}$ the set of all different eigenvalues of T. Let $g_T(t) = \displaystyle\prod_{i=1}^{r}(t - \lambda_i)^{n_i}$ be the characteristic polynomial of T. Then we have the following statements:*

(1) The linear transformation T induces a linear transformation of $\tilde{W}(\lambda_i\,;T)$ and the linear transformation of $\tilde{W}(\lambda_i\,;T)$ induced by $T - \lambda_i I_V$ is a nilpotent transformation for $i = 1, \dots, r$.
(2) $V = \tilde{W}(\lambda_1\,;T) \oplus \cdots \oplus \tilde{W}(\lambda_r\,;T)$.
(3) $\tilde{W}(\lambda_i\,;T) = \mathrm{Ker}((T - \lambda_i I_V)^{n_i})$ for $i = 1, \dots, r$.
(4) $\dim(\tilde{W}(\lambda_i\,;T)) = n_i$ for $i = 1, \dots, r$.

Proof (1) Since $\tilde{W}(\lambda_i\,;T) = \displaystyle\bigcup_{m=1}^{\infty} \mathrm{Ker}((T - \lambda_i I_V)^m)$ and $\tilde{W}(\lambda_i\,;T)$ is finite-dimensional, there exists a positive number m satisfying

$$\tilde{W}(\lambda_i : T) = \mathrm{Ker}((T - \lambda_i I_V)^m).$$

We note that the transformations T and $(T - \lambda_i I_V)^m$ are commutative. Then for any vector $\mathbf{v} \in \tilde{W}(\lambda_i ; T)$, we have

$$(T - \lambda_i I_V)^m (T(\mathbf{v})) = T((T - \lambda_i I_V)^m (\mathbf{v}))$$
$$= T(\mathbf{0}_V) = \mathbf{0}_V$$

for $i = 1, \ldots, r$. Therefore we see that T induces a linear transformation of $\tilde{W}(\lambda_i ; T)$ and the transformation of $\tilde{W}(\lambda_i ; T)$ induced by $T - \lambda_i I_V$ is a nilpotent transformation for $i = 1, \ldots, r$.

(2), (3) We let $f_i(t) = \prod_{j \neq i}(t - \lambda_j)^{n_j}$ for $i = 1, \ldots, r$. Since the polynomials $f_1(t), \ldots, f_r(t)$ have no common divisor, there exist polynomials $g_1(t), \ldots, g_r(t)$ satisfying

$$(*) \qquad 1 = g_1(t) f_1(t) + \cdots + g_r(t) f_r(t)$$

by Theorem 5.4.5. Substituting T for t, we obtain

$$I_V = g_1(T) f_1(T) + \cdots + g_r(T) f_r(T).$$

Let

$$I_i = g_i(T) f_i(T) \quad \text{for} \quad i = 1, \ldots, r.$$

Then we have

$$I_V = I_1 + \cdots + I_r.$$

We shall show that the linear transformations $I_1, \ldots, I_r$ satisfy conditions (i), (ii) and (iii) of Theorem 5.4.2. Condition (i) is already shown above. If $i \neq j$, then $f_i(t) f_j(t)$ is divisible by the characteristic polynomial $g_T(t)$ of T. Since $g_T(T) = O_V$, we have $f_i(T) f_j(T) = O_V$ and the equality

$$I_i I_j = g_i(T) f_i(T) g_j(T) f_j(T) = g_i(T) g_j(T) f_i(T) f_j(T) = O_V \quad \text{for} \quad i, j = 1, \ldots, r.$$

This is condition (ii). Since $I_V = I_1 + \cdots + I_r$, we see that

$$I_i = I_i I_V = I_i(I_1 + \cdots + I_r) = I_i I_i,$$

and condition (iii) is satisfied. Then by Theorem 5.4.2, we have

$$V = I_1(V) \oplus \cdots \oplus I_r(V).$$

Therefore to prove (2) and (3), we have only to show

$$I_i(V) = \text{Ker}\,((T - \lambda_i I_V)^{n_i}) = \tilde{W}(\lambda_i ; T) \quad \text{for} \quad i = 1, \ldots, r.$$

Let $\mathbf{w}_i$ be a vector in $I_i(V)$. Then there exists a vector $\mathbf{v}_i \in V$ satisfying $\mathbf{w}_i = I_i(\mathbf{v}_i) = g_i(T)f_i(T)(\mathbf{v}_i)$. Since $(t - \lambda_i)^{n_i} f_i(t) = g_T(t)$ and $g_T(T) = O_V$, we have

$$(**) \qquad (T - \lambda_i I_V)^{n_i}(\mathbf{w}_i) = (T - \lambda_i I_V)^{n_i} g_i(T)f_i(T)(\mathbf{v}_i)$$
$$= g_i(T)g_T(T)(\mathbf{v}_i) = \mathbf{0}_V.$$

Therefore $\mathbf{w}_i \in \mathrm{Ker}\,((T - \lambda_i I_V)^{n_i})$ and we have

$$I_i(V) \subset \mathrm{Ker}\,((T - \lambda_i I_V)^{n_i}) \subset \tilde{W}(\lambda_i\,;T) \qquad \text{for}\quad i = 1, \ldots, r.$$

Conversely, we shall show that $\tilde{W}(\lambda_i\,;T) \subset I_i(V)$ for $i = 1, \ldots, r$. Let $\mathbf{w}_i \in \tilde{W}(\lambda_i\,;T)$. We note that for any positive integer m, $(t - \lambda_i)^m$ and $g_i(t)f_i(t)$ have no common divisor. In fact, if $g_i(t)f_i(t)$ is divisible by $t - \lambda_i$, then 1 must be divisible by $t - \lambda_i$ by the equality $(*)$. By Theorem 5.4.5, there exist polynomials $p(t)$ and $q(t)$ satisfying

$$p(t)(t - \lambda_i)^m + q(t)g_i(t)f_i(t) = 1.$$

Substituting T for t, we have

$$p(T)(T - \lambda_i I_V)^m + q(T)g_i(T)f_i(T) = I_V.$$

Since any positive integer can be taken for m, we take a large integer m so that $\mathbf{w}_i \in \mathrm{Ker}\,((T - \lambda_i I_V)^m)$. Then we have

$$(T - \lambda_i I_V)^m(\mathbf{w}_i) = \mathbf{0}_V.$$

Since $I_i = g_i(T)f_i(T)$ commutes with $q(T)$, we have

$$\begin{aligned}
\mathbf{w}_i &= I_V(\mathbf{w}_i) \\
&= (p(T)(T - \lambda_i I_V)^m + q(T)g_i(T)f_i(T))(\mathbf{w}_i) \\
&= q(T)g_i(T)f_i(T)(\mathbf{w}_i) \\
&= g_i(T)f_i(T)q(T)(\mathbf{w}_i) \\
&= I_i\,(q(T)(\mathbf{w}_i)) \in I_i(V).
\end{aligned}$$

Therefore we have

$$I_i(V) = \mathrm{Ker}\,((T - \lambda_i I_V)^{n_i}) = \tilde{W}(\lambda_i\,;T)) \qquad \text{for}\quad i = 1, \ldots, r.$$

(4) Let $l_i = \dim(\tilde{W}(\lambda_i\,;T))$ and T_i the restriction of T to the subspace $\tilde{W}(\lambda_i\,;T)$ for $i = 1, \ldots, r$. Since T_i has only one eigenvalue λ_i by (3), the characteristic polynomial of T_i is $g_{T_i}(t) = (t - \lambda_i)^{l_i}$. Therefore by (2) and Theorem 5.4.3 (2), we have

$$g_T(t) = \prod_{i=1}^{r} g_{T_i}(t) = \prod_{i=1}^{r} (t - \lambda_i)^{l_i}.$$

On the other hand, since $g_T(t) = \displaystyle\prod_{i=1}^{r}(t - \lambda_i)^{n_i}$, we see that

$$l_i = n_i \quad \text{for} \quad i = 1, \ldots, r$$

by comparing both sides of the equality. ‖

We restate Theorem 8.1.3 in terms of matrices.

Theorem 8.1.4 *Let A be a square matrix of degree n and $\{\lambda_1, \ldots, \lambda_r\}$ the set of all different eigenvalues of A. Let the characteristic polynomial of A be $g_A(t) = \displaystyle\prod_{i=1}^{r}(t - \lambda_i)^{n_i}$. Then we have the following statements:*

(1) The linear transformation T_A associated with A induces a linear transformation of $\tilde{W}(\lambda_i\,;\,A)$ and the restriction of $T_A - \lambda_i T_i$ to $\tilde{W}(\lambda_i\,;\,A)$ is a mail potent transformation for $i = 1, \ldots, r$.
(2) $\mathbf{C}^n = \tilde{W}(\lambda_1\,;\,A) \oplus \cdots \oplus \tilde{W}(\lambda_r\,;\,A)$.
(3) $\tilde{W}(\lambda_i\,;\,A) = \mathrm{Ker}((A - \lambda_i E_n)^{n_i})$ for $i = 1, \ldots, r$.
(4) $\dim(\tilde{W}(\lambda_i\,;\,A)) = n_i$ for $i = 1, \ldots, r$.

Exercise 8.1 Let $A = \begin{pmatrix} 1 & -2 & 0 \\ 0 & 1 & 0 \\ -2 & -4 & 3 \end{pmatrix}$.

(1) Verify that the characteristic polynomial of A is

$$g_A(t) = (t - 1)^2(t - 3)$$

and the eigenvalues of A are $\lambda_1 = 1$ and $\lambda_2 = 3$.
(2) Find the multiplicities of the eigenvalue λ_1 and the eigenvalue λ_2.
(3) We let $f_i(t) = \displaystyle\prod_{j \neq i}(t - \lambda_j)^{n_j}$ for $i = 1, 2$. Here n_i is the multiplicity of the eigenvalue λ_i for $i = 1, 2$. Then show that

$$f_1(t) = t - 3 \quad \text{and} \quad f_2(t) = (t - 1)^2,$$

and find the polynomial $g_1(t)$ and $g_2(t)$ satisfying

$$g_1(t)f_1(t) + g_2(t)f_2(t) = 1.$$

(4) Find the subspaces $\tilde{W}(1\,;\,A)$ and $\tilde{W}(3\,;\,A)$ and verify

$$\mathbf{C}^3 = \tilde{W}(1\,;\,A) \oplus \tilde{W}(3\,;\,A).$$

Answers. (1) The characteristic polynomial $g_A(t)$ is

$$g_A(t) = |tE_3 - A| = (t-3)(t-1)^2.$$

Therefore the eigenvalues of A are $\lambda_1 = 1$ and $\lambda_2 = 3$.

(2) The multiplicity n_1 of the eigenvalue $\lambda_1 = 1$ is 2 and the multiplicity n_2 of the eigenvalue $\lambda_2 = 3$ is 1.

(3) By definition,

$$f_1(t) = \prod_{j \neq 1}(t-\lambda_j)^{n_j} = t-3 \quad \text{for} \quad f_2(t) = \prod_{j \neq 2}(t-\lambda_j)^{n_j} = (t-1)^2.$$

To obtain the polynomials $g_1(t)$ and $g_2(t)$, we use the Euclidean algorithm for polynomials. Since $\deg(f_1(t)) = 1$ and $\deg(f_2(t)) = 2$, we divide $f_2(t)$ by $f_1(t)$ and obtain $f_2(t) = (t+1)f_1(t) + 4$. Therefore

$$4 = -(t+1)f_1(t) + f_2(t).$$

We divide both sides of this equality by the constant 4 and put

$$g_1(t) = -\frac{1}{4}(t+1) \quad \text{and} \quad g_2(t) = \frac{1}{4}.$$

Then we obtain the equality $g_1(t)f_1(t) + g_2(t)f_2(t) = 1$. (If the remainder of $f_1(t)$ divided by $f_2(t)$ is not a constant, then we must continue the same processes until the remainder is a constant. See Exercises (Sect. 8.1) 2 (3)).

(4) We shall find $\tilde{W}(\lambda_i ; A)$ for $i = 1, 2$. Let I_i be the linear transformation of $\mathbf{C}^3$ defined by $I_i = g_i(A)f_i(A)$ for $i = 1, 2$ as in the proof of Theorem 8.1.3 (2). Since we know

$$\tilde{W}(\lambda_i ; A) = I_i(\mathbf{C}^3) = \text{Im}\,(g_i(A)f_i(A)) \quad \text{for} \quad i = 1, 2$$

in the proof of Theorem 8.1.3 (2), we have only to find $\text{Im}\,(g_i(A)f_i(A))$ for $i = 1, 2$.

Let $i = 1$. Then $\lambda_1 = 1$. Since $g_1(t)f_1(t) = -\frac{1}{4}(t+1)(t-3)$, we have

$$g_1(A)f_1(A) = -\frac{1}{4}(A+E_3)(A-3E_3)$$

$$= \begin{pmatrix} 1 & 0 & 0 \\ 0 & 1 & 0 \\ 1 & 1 & 0 \end{pmatrix}.$$

Therefore we have

$$\tilde{W}(1\,;A) = \mathrm{Im}\,(g_1(A)f_1(A))$$
$$= \{\,g_1(A)f_1(A)\mathbf{x}\mid \mathbf{x}\in\mathbf{C}^3\,\}$$
$$= \left\{ \begin{pmatrix} 1 & 0 & 0 \\ 0 & 1 & 0 \\ 1 & 1 & 0 \end{pmatrix} \begin{pmatrix} x_1 \\ x_2 \\ x_3 \end{pmatrix} \;\middle|\; x_1, x_2, x_3 \in \mathbf{C} \right\}$$
$$= \left\{ \begin{pmatrix} x_1 \\ x_2 \\ x_1 + x_2 \end{pmatrix} \;\middle|\; x_1, x_2 \in \mathbf{C} \right\}$$
$$= \left\{ x_1 \begin{pmatrix} 1 \\ 0 \\ 1 \end{pmatrix} + x_2 \begin{pmatrix} 0 \\ 1 \\ 1 \end{pmatrix} \;\middle|\; x_1, x_2 \in \mathbf{C} \right\}.$$

Let $i = 2$. Then $\lambda_2 = 3$. Since $g_2(t)f_2(t) = \dfrac{1}{4}(t-1)^2$, we have

$$g_2(A)f_2(A) = \frac{1}{4}\begin{pmatrix} 0 & -2 & 0 \\ 0 & 0 & 0 \\ -2 & -4 & 2 \end{pmatrix}^2 = \begin{pmatrix} 0 & 0 & 0 \\ 0 & 0 & 0 \\ -1 & -1 & 1 \end{pmatrix}.$$

Therefore we have

$$\tilde{W}(3\,;A) = \mathrm{Im}\,(g_2(A)f_2(A))$$
$$= \{\,g_2(A)f_2(A)\mathbf{x}\mid \mathbf{x}\in\mathbf{C}^3\,\}$$
$$= \left\{ \begin{pmatrix} 0 & 0 & 0 \\ 0 & 0 & 0 \\ -1 & -1 & 1 \end{pmatrix} \begin{pmatrix} x_1 \\ x_2 \\ x_3 \end{pmatrix} \;\middle|\; x_1, x_2, x_3 \in \mathbf{C} \right\}$$
$$= \left\{ \begin{pmatrix} 0 \\ 0 \\ -x_1 - x_2 + x_3 \end{pmatrix} \;\middle|\; x_1, x_2, x_3 \in \mathbf{C} \right\}$$
$$= \left\{ x \begin{pmatrix} 0 \\ 0 \\ 1 \end{pmatrix} \;\middle|\; x \in \mathbf{C} \right\}.$$

Now, to verify $\mathbf{C}^3 = \tilde{W}(3\,;A) \oplus \tilde{W}(1\,;A)$, we shall show directly that any vector $\begin{pmatrix} x_1 \\ x_2 \\ x_3 \end{pmatrix} \in \mathbf{C}^3$ is uniquely expressed as a linear combination of vectors in $\tilde{W}(1\,;A)$ and $\tilde{W}(3\,;A)$. In fact, we have

$$\begin{pmatrix} x_1 \\ x_2 \\ x_3 \end{pmatrix} = x_1 \begin{pmatrix} 1 \\ 0 \\ 1 \end{pmatrix} + x_2 \begin{pmatrix} 0 \\ 1 \\ 1 \end{pmatrix} + (-x_1 - x_2 + x_3) \begin{pmatrix} 0 \\ 0 \\ 1 \end{pmatrix}.$$

Therefore $\mathbf{C}^3$ is a direct sum of subspaces $\tilde{W}(1\,;A)$ and $\tilde{W}(3\,;A)$:

$$\mathbf{C}^3 = \tilde{W}(1\,;A) \oplus \tilde{W}(3\,;A). \quad \|$$

Another answer to (4).   To obtain only $\tilde{W}(\lambda\,;A)$ for eigenvalues λ of A, this way will be a little easier than the above way using $\mathrm{Im}(I_i)$. We have

$$\tilde{W}(\lambda_i\,;A) = \mathrm{Ker}\,((A - \lambda_i E_3)^{n_i})$$

by Theorem 8.1.4 (3).

The multiplicity of the eigenvalue $\lambda_1 = 1$ is 2. Then

$$\tilde{W}(1\,;A) = \mathrm{Ker}\,((A - E_3)^2).$$

Solving the linear equation $(A - E_3)^2\mathbf{x} = \mathbf{0}_3$, we have

$$\tilde{W}(1\,;A) = \left\{ x_1 \begin{pmatrix} 1 \\ 0 \\ 1 \end{pmatrix} + x_2 \begin{pmatrix} -1 \\ 1 \\ 0 \end{pmatrix} \middle| x_1, x_2 \in \mathbf{C} \right\}.$$

Next, the multiplicity of the eigenvalue $\lambda_1 = 3$ is 1. Then

$$\tilde{W}(3\,;A) = \mathrm{Ker}(A - 3E_3).$$

Solving the linear equation $(A - 3E_3)\mathbf{x} = \mathbf{0}_3$, we have

$$\tilde{W}(3\,;A) = \left\{ x \begin{pmatrix} 0 \\ 0 \\ 1 \end{pmatrix} \middle| x \in \mathbf{C} \right\}. \quad \|$$

Exercises (Sect. 8.1)

1 For the following matrices A, find the characteristic polynomial $g_A(t)$ of the matrix A, the eigenvalues of A and the generalized eigenspace $\tilde{W}(\lambda\,;A)$ for each eigenvalue λ of A:

(1) $A = \begin{pmatrix} 1 & 0 & -2 \\ 2 & 1 & -1 \\ 0 & 0 & 1 \end{pmatrix}$.

(2) $A = \begin{pmatrix} 2 & 0 & 0 \\ 2 & 5 & 3 \\ 0 & -6 & -4 \end{pmatrix}$.

(3) $A = \begin{pmatrix} 4 & 0 & 6 \\ 0 & 1 & 0 \\ -3 & 0 & -5 \end{pmatrix}$.

(4) $A = \begin{pmatrix} i & 0 & 1 \\ -1+i & 1 & 2-i \\ 0 & 0 & i \end{pmatrix}$.

(5) $A = \begin{pmatrix} i & -2+i & 0 \\ 0 & 2 & 0 \\ -2+i & -3+i & 2 \end{pmatrix}$.

(6) $A = \begin{pmatrix} 2 & 2 & 1 & 0 \\ 0 & 5 & 3 & 0 \\ 0 & -6 & -4 & 0 \\ 0 & 0 & 0 & -1 \end{pmatrix}$.

(7) $A = \begin{pmatrix} i & -1 & 0 & 1+i \\ 0 & i & 0 & 0 \\ 0 & 1 & i & 0 \\ 0 & 0 & 0 & 1 \end{pmatrix}$.

(8) $A = \begin{pmatrix} i & 0 & 0 & 0 \\ 1-i & 1-i & -1 & -i \\ 0 & 1-i & 1 & 1-i \\ -1+i & -1+2i & 1 & 2i \end{pmatrix}$.

2 For the following polynomials $f_1(t)$ and $f_2(t)$, find a pair of the polynomials $g_1(t)$ and $g_2(t)$ satisfying $g_1(t)f_1(t) + g_2(t)f_2(t) = 1$:

$$\begin{aligned} &(1) && f_1(t) = (t+1)^2, && f_2(t) = (t-2)^2. \\ &(2) && f_1(t) = t+3, && f_2(t) = (t+1)^2. \\ &(3) && f_1(t) = t^3, && f_2(t) = t^2 + 2t - 3. \end{aligned}$$

3 Let $A_1, \ldots, A_r$ be non-zero square matrices and $A = A_1 \oplus \cdots \oplus A_r$. Let $f_1(t)$, $\ldots, f_r(t)$ be polynomials and $B_i = f_i(A_i)$ for $i = 1, \ldots, r$. We also let $B = B_1 \oplus \cdots \oplus B_r$. Show that if any pair of the characteristic polynomials $g_{A_i}(t)$ of A_i has no common divisor, then there exists a polynomial $f(t)$ satisfying

$$B = f(A).$$

8.2 Jordan Normal Forms

If all square matrices were diagonalizable, we would be very happy and linear algebra would be much simpler. But unfortunately that is not true. Nevertheless, all complex square matrices are similar to quite simple matrices called matrices of Jordan normal form. The purpose of this section is to explain matrices of Jordan normal form.

Jordan cells Let λ be a complex number. The following square matrix of degree n is called the *Jordan cell* of degree n with the eigenvalue λ:

$$J(\lambda; n) = \left.\begin{pmatrix} \lambda & 1 & & 0 \\ & \lambda & 1 & \\ & & \ddots & \ddots \\ & & & \lambda & 1 \\ 0 & & & & \lambda \end{pmatrix}\right\} n \quad (n \geq 2), \qquad J(\lambda; 1) = \begin{pmatrix} \lambda \end{pmatrix}.$$

Example 1 The matrix $J(i; 2) = \begin{pmatrix} i & 1 \\ 0 & i \end{pmatrix}$ is the Jordan cell of degree 2 with the eigenvalue i.

Matrices of Jordan normal form Direct sums of Jordan cells are called *matrices of Jordan normal form.*

Example 2 The following matrix is a matrix of Jordan normal form:

$$A = \begin{pmatrix} i & 1 & 0 & 0 & 0 & 0 \\ 0 & i & 0 & 0 & 0 & 0 \\ 0 & 0 & 2i & 1 & 0 & 0 \\ 0 & 0 & 0 & 2i & 0 & 0 \\ 0 & 0 & 0 & 0 & 3 & 0 \\ 0 & 0 & 0 & 0 & 0 & 3 \end{pmatrix}.$$

In fact, the matrix A is a direct sum of Jordan cells:

$$A = J(i \, ; 2) \oplus J(2i \, ; 2) \oplus J(3 \, ; 1) \oplus J(3 \, ; 1).$$

Jordan normal forms of linear transformations Let T be a linear transformation of a vector space V. If a representation matrix A of T is a matrix of Jordan normal form, then we call A the *Jordan normal form* of T. The Jordan normal form of T uniquely exists up to the order of Jordan cells (Theorem 8.2.3 below).

Example 3 Let T be a transformation of a vector space V of dimension 2. If $T \neq O_V$ and $T^2 = O_V$, then there exists a vector $\mathbf{v} \in V$ such that $T(\mathbf{v}) \neq \mathbf{0}_V$. We see that the set of vectors $\{T(\mathbf{v}), \mathbf{v}\}$ is a basis of V by Exercises (Sect. 5.2) 4 (1). The representation matrix of T with respect to the basis $\{T(\mathbf{v}), \mathbf{v}\}$ of V is

$$\begin{pmatrix} 0 & 1 \\ 0 & 0 \end{pmatrix} = J(0; 2),$$

which is the Jordan normal form of T.

Jordan normal forms of square matrices Let A be a square matrix of degree n. If A is similar to a matrix B of Jordan normal form, we call the matrix B the *Jordan normal form* of A. Let T_A be the linear transformation of $\mathbf{C}^n$ associated with A. The Jordan normal form of A is nothing but the Jordan normal form of T_A since the representation matrix of T_A with respect to the standard basis of $\mathbf{C}^n$ is A and all representation matrices of T_A are similar to the matrix A.

Theorem 8.2.1 *Let T be a linear transformation of a vector space V of dimension n over $\mathbf{C}$ and A a representation matrix of T. Then the following three conditions are equivalent:*

(1) The linear transformation T is a nilpotent transformation.
(2) Any representation matrix A of T is a nilpotent matrix.
(3) Every eigenvalue of T is 0.
(4) Every eigenvalue of A is 0.

Proof Let $g_T(t)$ and $g_A(t)$ be the characteristic polynomials of T and A, respectively.

(1) $\Leftrightarrow$ (2) holds since $T^m = O_V$ and $A^m = O_n$ are equivalent.

(1) $\Rightarrow$ (3) Let λ be an eigenvalue of T and $\mathbf{u} \in V$ an eigenvector of T with the eigenvalue λ. If $T^m = O_V$, then

$$\mathbf{0}_V = T^m(\mathbf{u}) = \lambda^m \mathbf{u}.$$

Therefore we have $\lambda = 0$.

(3) $\Rightarrow$ (4) Since every eigenvalue of T is 0, we see that $g_T(t) = t^n$. As $g_A(t) = g_T(t) = t^n$, we see that every eigenvalue of A is 0.

(4) $\Rightarrow$ (2) Since every eigenvalue of A is 0, we have $g_A(t) = t^n$. Therefore $A^n = O_n$ by Theorem 5.3.2 (Cayley–Hamilton). This implies that A is a nilpotent matrix. $\quad\|$

For nilpotent transformations, we have the following theorem.

Theorem 8.2.2 *Let T be a nilpotent transformation of a vector space V of dimension n. We assume that $T^m = O_V$ and $T^{m-1} \neq O_V$. Then T has a representation matrix A which is a direct sum of Jordan cells $J(0; k_i)$:*

$$A = J(0; k_1) \oplus \cdots \oplus J(0; k_r), \quad J(0; k_i) = \begin{pmatrix} 0 & 1 & & 0 \\ & \ddots & \ddots & \\ & & \ddots & 1 \\ 0 & & & 0 \end{pmatrix} \Bigg\} k_i.$$

The matrix A uniquely exists up to the order of Jordan cells. The matrix A is the Jordan normal form of the nilpotent transformation T.

Proof Since the proof is rather long, we assume $m = 3$ to understand the idea of the proof quickly. The proof for the general case will be given at the end of the section.

If $T = O_V$, then the representation matrix is the zero matrix, which is the direct sum of n Jordan cells $J(0; 1)$.

Let T be a non-zero nilpotent transformation of V. We assume that $T^3 = O_V$ and $T^2 \neq O_V$. We let $V(0) = \{\mathbf{0}_V\}$ and

$$V(k) = \mathrm{Ker}\,(T^k) = \{\mathbf{u} \in V \mid T^k(\mathbf{u}) = \mathbf{0}_V\} \quad \text{for} \quad k = 1, 2, 3.$$

Since $T^3 = O_V$, we see that $V(3) = V$. We easily see that

$$\{\mathbf{0}_V\} = V(0) \subset V(1) \subset V(2) \subset V(3) = V.$$

As $T^2 \neq O_V$, we see $V(2) \neq V(3) = V$. Let $t_3 = \dim(V(3)) - \dim(V(2)) > 0$. Take linearly independent vectors $\mathbf{u}_{3,1}, \ldots, \mathbf{u}_{3,t_3}$ in $V = V(3)$ so that

$$V = V(3) = <\mathbf{u}_{3,1}, \ldots, \mathbf{u}_{3,t_3}> \oplus V(2).$$

Then we see that the vectors $T(\mathbf{u}_{3,1}), \ldots, T(\mathbf{u}_{3,t_3})$ in $V(2)$ are linearly independent. To see the linear independence, we let

$$\mathbf{v} = a_1 T(\mathbf{u}_{3,1}) + \cdots + a_{t_3} T(\mathbf{u}_{3,t_3}) \quad \text{and} \quad \mathbf{u} = a_1 \mathbf{u}_{3,1} + \cdots + a_{t_3} \mathbf{u}_{3,t_3}.$$

Here $a_i \in \mathbf{C}$ for $i = 1, \ldots, t_3$. We note $T(\mathbf{u}) = \mathbf{v}$. Assume that $\mathbf{v} = \mathbf{0}_V$. As $T(\mathbf{u}) = \mathbf{v} = \mathbf{0}_V$, we have $\mathbf{u} \in V(1)$. Since $V(1) \subset V(2)$, we see that

$$\mathbf{u} \in < \mathbf{u}_{3,1}, \ldots, \mathbf{u}_{3,t_3} > \cap V(2) = \{\mathbf{0}_V\}.$$

Then we have $\mathbf{u} = \mathbf{0}_V$. As $\mathbf{u}_{3,1}, \ldots, \mathbf{u}_{3,t_3}$ are linearly independent, we have $a_1 = \cdots = a_{t_3} = 0$. Therefore $T(\mathbf{u}_{3,1}), \ldots, T(\mathbf{u}_{3,t_3})$ are linearly independent. Since $T^3 = O_V$, $T(\mathbf{u}_{3,1}), \ldots, T(\mathbf{u}_{3,t_3}) \in V(2)$. Furthermore, we see that

$$(*) \qquad\qquad < T(\mathbf{u}_{3,1}), \ldots, T(\mathbf{u}_{3,t_3}) > \cap V(1) = \{\mathbf{0}_V\}.$$

In fact, if $\mathbf{v} = T(\mathbf{u}) \in V(1)$, then $\mathbf{u} \in V(2)$. Since $< \mathbf{u}_{3,1}, \ldots, \mathbf{u}_{3,t_3} > \cap V(2) = \{\mathbf{0}_V\}$, we have $\mathbf{u} = \mathbf{0}_V$. Therefore $\mathbf{v} = T(\mathbf{u}) = \mathbf{0}_V$ and this implies $(*)$. Now, we take linearly independent vectors $\mathbf{u}_{2,1}, \ldots, \mathbf{u}_{2,t_2}$ in $V(2)$ so that

$$V(2) = < T(\mathbf{u}_{3,1}), \ldots, T(\mathbf{u}_{3,t_3}) > \oplus < \mathbf{u}_{2,1}, \ldots, \mathbf{u}_{2,t_2} > \oplus V(1).$$

Since the vectors $T(\mathbf{u}_{3,1}), \ldots, T(\mathbf{u}_{3,t_3})$ are linearly independent, the vectors $T(\mathbf{u}_{3,1}), \ldots, T(\mathbf{u}_{3,t_3}), \mathbf{u}_{2,1}, \ldots, \mathbf{u}_{2,t_2}$ are linearly independent. We shall show that the vectors $T^2(\mathbf{u}_{3,1}), \ldots, T^2(\mathbf{u}_{3,t_3}), T(\mathbf{u}_{2,1}), \ldots, T(\mathbf{u}_{2,t_2})$ in $V(1)$ are linearly independent. Let $a_i, b_j \in \mathbf{C}$ for $i = 1, \ldots, t_3$ and $j = 1, \ldots, t_2$ and

$$a_1 T^2(\mathbf{u}_{3,1}) + \cdots + a_{t_3} T^2(\mathbf{u}_{3,t_3}) + b_1 T(\mathbf{u}_{2,1}) + \cdots + b_{t_2} T(\mathbf{u}_{2,t_2}) = \mathbf{0}_V$$

be a linear relation of these vectors. Let

$$\mathbf{w} = a_1 T(\mathbf{u}_{3,1}) + \cdots + a_{t_3} T(\mathbf{u}_{3,t_3}) + b_1 \mathbf{u}_{2,1} + \cdots + b_{t_2} \mathbf{u}_{2,t_2}.$$

Then $T(\mathbf{w}) = \mathbf{0}_V$. Therefore we have

$$\mathbf{w} \in (< T(\mathbf{u}_{3,1}), \ldots, T(\mathbf{u}_{3,t_3}) > \oplus < \mathbf{u}_{2,1}, \ldots, (\mathbf{u}_{2,t_2} >) \cap V(1) = \{\mathbf{0}_V\}.$$

Thus we see $\mathbf{w} = \mathbf{0}_V$, namely,

$$a_1 T(\mathbf{u}_{3,1}) + \cdots + a_{t_3} T(\mathbf{u}_{3,t_3}) + b_1 \mathbf{u}_{2,1} + \cdots + b_{t_2} \mathbf{u}_{2,t_2} = \mathbf{0}_V.$$

Since the vectors $T(\mathbf{u}_{3,1}), \ldots, T(\mathbf{u}_{3,t_3}), \mathbf{u}_{2,1}, \ldots, \mathbf{u}_{2,t_2}$ are linearly independent, we have $a_1 = \cdots = a_{t_3} = b_1 = \cdots = b_{t_2} = 0$. Therefore the vectors

$$T^2(\mathbf{u}_{3,1}), \ldots, T^2(\mathbf{u}_{3,t_3}), T(\mathbf{u}_{2,1}), \ldots, T(\mathbf{u}_{2,t_2})$$

are linearly independent.

Next, take linearly independent vectors $\mathbf{u}_{1,1}, \ldots, \mathbf{u}_{1,t_1}$ in $V(1)$ so that

$$V(1) = < T^2(\mathbf{u}_{3,1}), \ldots, T^2(\mathbf{u}_{3,t_3}) >$$
$$\oplus < T(\mathbf{u}_{2,1}), \ldots, T(\mathbf{u}_{2,t_2}) > \oplus < \mathbf{u}_{1,1}, \ldots, \mathbf{u}_{1,t_1} > .$$

Then we see that the vectors

$$\begin{cases} \mathbf{u}_{3,1}, & \ldots, & \mathbf{u}_{3,t_3}, \\ T(\mathbf{u}_{3,1}), & \ldots, & T(\mathbf{u}_{3,t_3}), & \mathbf{u}_{2,1}, & \ldots, & \mathbf{u}_{2,t_2}, \\ T^2(\mathbf{u}_{3,1}), & \ldots, & T^2(\mathbf{u}_{3,t_3}), & T(\mathbf{u}_{2,1}), & \ldots, & T(\mathbf{u}_{2,t_2}), & \mathbf{u}_{1,1}, \ldots, \mathbf{u}_{1,t_1} \end{cases}$$

are linearly independent and these vectors generate the vector space V. Therefore the set of these vectors is a basis of V. We change the order of vectors of the basis. Let

$$\begin{aligned} U(3\,;\,j) &= < T^2(\mathbf{u}_{3,j}), T(\mathbf{u}_{3,j}), \mathbf{u}_{3,j} > & \text{for} \quad j = 1, \ldots, t_3, \\ U(2\,;\,j) &= < T(\mathbf{u}_{2,j}), \mathbf{u}_{2,j} > & \text{for} \quad j = 1, \ldots, t_2, \\ U(1\,;\,j) &= < \mathbf{u}_{1,j} > & \text{for} \quad j = 1, \ldots, t_1. \end{aligned}$$

Then we see that

$$V = U(3\,;\,1) \oplus \cdots \oplus U(3\,;\,t_3) \oplus U(2\,;\,1) \oplus \cdots \oplus U(2\,;\,t_2)$$
$$\oplus U(1\,;\,1) \oplus \cdots \oplus U(1\,;\,t_1).$$

We note that T maps the subspaces $U(i\,;\,j)$ into itself for $i = 1, 2, 3$ and $j = 1, \ldots, t_i$. The representation matrix of the restriction of T to the subspace $U(3\,;\,j)$ with respect to the basis $\{T^2(\mathbf{u}_{3,j}), T(\mathbf{u}_{3,j}), \mathbf{u}_{3,j}\}$, the representation matrix of the restriction of T to the subspace $U(2\,;\,j)$ with respect to the basis $\{T(\mathbf{u}_{2,j}), \mathbf{u}_{2,j}\}$ and the representation matrix of the restriction of T to the subspace $U(1\,;\,j)$ with respect to the basis $\{\mathbf{u}_{1,j}\}$ are

$$J(0\,;\,3) = \begin{pmatrix} 0 & 1 & 0 \\ 0 & 0 & 1 \\ 0 & 0 & 0 \end{pmatrix}, \quad J(0\,;\,2) = \begin{pmatrix} 0 & 1 \\ 0 & 0 \end{pmatrix} \quad \text{and} \quad J(0\,;\,1) = \begin{pmatrix} 0 \end{pmatrix},$$

respectively. Then the representation matrix of T with respect to the basis

$$\{T^2(\mathbf{u}_{3,1}), T(\mathbf{u}_{3,1}), \mathbf{u}_{3,1}, \ldots, T^2(\mathbf{u}_{3,t_3}), T(\mathbf{u}_{3,t_3}), \mathbf{u}_{3,t_3};$$
$$T(\mathbf{u}_{2,1}), \mathbf{u}_{2,1}, \ldots, T(\mathbf{u}_{2,t_2}), \mathbf{u}_{2,t_2}; \mathbf{u}_{1,1}, \ldots, \mathbf{u}_{1,t_1}\}$$

is

$$A = \underbrace{J(0\,;\,3) \oplus \cdots \oplus J(0\,;\,3)}_{t_3} \oplus \underbrace{J(0\,;\,2) \oplus \cdots \oplus J(0\,;\,2)}_{t_2} \oplus \underbrace{J(0\,;\,1) \oplus \cdots \oplus J(0\,;\,1)}_{t_1},$$

which is a matrix of Jordan normal form.

We shall show the uniqueness of the Jordan normal form of T. Since $T^3 = O_V$, we may assume that

$$B = \underbrace{J(0;3) \oplus \cdots \oplus J(0;3)}_{s_3} \oplus \underbrace{J(0;2) \oplus \cdots \oplus J(0;2)}_{s_2} \oplus \underbrace{J(0;1) \oplus \cdots \oplus J(0;1)}_{s_1}$$

is another Jordan normal form of T. Since $B^l = J(0;3)^l \oplus \cdots \oplus J(0;1)^l$ and

$$\dim (\mathrm{Ker}\,(J(0;k)^l)) = \begin{cases} l & \text{for } l \le k, \\ k & \text{for } l > k, \end{cases}$$

we see that

$$\begin{cases} \dim (V(3)) = \dim (\mathrm{Ker}\,(B^3)) = 3s_3 + 2s_2 + s_1, \\ \dim (V(2)) = \dim (\mathrm{Ker}\,(B^2)) = 2s_3 + 2s_2 + s_1, \\ \dim (V(1)) = \dim (\mathrm{Ker}\,(B)) \ \ = s_3 + s_2 + s_1. \end{cases}$$

Then we have the linear equation

$$\begin{cases} s_3 = \dim (V(3)) - \dim (V(2)), \\ s_3 + s_2 = \dim (V(2)) - \dim (V(1)), \\ s_3 + s_2 + s_1 = \dim (V(1)). \end{cases}$$

Since $\begin{vmatrix} 1 & 0 & 0 \\ 1 & 1 & 0 \\ 1 & 1 & 1 \end{vmatrix} = 1 \ne 0$, this linear equation has a unique solution. On the other hand, t_1, t_2, t_3 also satisfy the relation

$$\begin{cases} t_3 = \dim (V(3)) - \dim (V(2)), \\ t_3 + t_2 = \dim (V(2)) - \dim (V(1)), \\ t_3 + t_2 + t_1 = \dim (V(1)). \end{cases}$$

Therefore we have $s_1 = t_1$, $s_2 = t_2$ and $s_3 = t_3$ and the Jordan normal form of T is uniquely determined up to the order of Jordan cells. ‖

Exercise 8.2 Let $A = \begin{pmatrix} 0 & 0 & 0 & 0 & 0 \\ 0 & 0 & 1 & 1 & 0 \\ 0 & 0 & 0 & 1 & 0 \\ 0 & 0 & 0 & 0 & 0 \\ 1 & 0 & 0 & 0 & 0 \end{pmatrix}$. Verify that A is a nilpotent matrix and find the Jordan normal form of the nilpotent transformation T_A associated with the matrix A.

Answer.   Let $V = \mathbf{C}^5$. Since the characteristic polynomial of A is $g_A(t) = t^5$, A is a nilpotent matrix. Then T_A is a nilpotent transformation by Theorem 8.2.1. Since we easily see that $A^3 = O_5$ and $A^2 \ne O_5$, we have

$$T_A^3 = O_V \quad \text{and} \quad T_A^2 \ne O_V.$$

We let

$$V(0) = \{\mathbf{0}_5\}$$

and

$$V(k) = \text{Ker}(T_A^k) = \text{Ker}(A^k) = \{\mathbf{x} \in \mathbf{C}^5 \mid A^k\mathbf{x} = \mathbf{0}_5\}$$

for $k = 1, 2, 3$. Since $A^3 = O_5$, we have $V(3) = \mathbf{C}^5$. We also let

$$t_3 = \dim(V(3)) - \dim(V(2)).$$

We shall find the subspace $V(2)$. Calculate the reduced matrix of A^2:

$$A^2 = \begin{pmatrix} 0\,0\,0\,0\,0 \\ 0\,0\,0\,1\,0 \\ 0\,0\,0\,0\,0 \\ 0\,0\,0\,0\,0 \\ 0\,0\,0\,0\,0 \end{pmatrix} \rightarrow \begin{pmatrix} 0\,0\,0\,1\,0 \\ 0\,0\,0\,0\,0 \\ 0\,0\,0\,0\,0 \\ 0\,0\,0\,0\,0 \\ 0\,0\,0\,0\,0 \end{pmatrix}.$$

Solving the linear equation $A^2\mathbf{x} = \mathbf{0}_5$, we have $\dim(V(2)) = 4$ and

$$V(2) = \left\{ \begin{pmatrix} a_1 \\ a_2 \\ a_3 \\ 0 \\ a_4 \end{pmatrix} \in \mathbf{C}^5 \;\middle|\; a_1, a_2, a_3, a_4 \in \mathbf{C} \right\}.$$

Therefore we have $t_3 = 5 - 4 = 1$. Take a vector $\mathbf{u}_{3,1} \in V(3)$ which is not in $V(2)$. For example, take the vector $\mathbf{u}_{3,1} = \begin{pmatrix} 0 \\ 0 \\ 0 \\ 1 \\ 0 \end{pmatrix} \in V(3)$. Then we see that

$$V(3) = <\mathbf{u}_{3,1}> \oplus V(2).$$

Next, we shall find the subspace $V(1)$. We calculate the reduced matrix of A:

$$A = \begin{pmatrix} 0\,0\,0\,0\,0 \\ 0\,0\,1\,1\,0 \\ 0\,0\,0\,1\,0 \\ 0\,0\,0\,0\,0 \\ 1\,0\,0\,0\,0 \end{pmatrix} \longrightarrow \begin{pmatrix} 1\,0\,0\,0\,0 \\ 0\,0\,1\,0\,0 \\ 0\,0\,0\,1\,0 \\ 0\,0\,0\,0\,0 \\ 0\,0\,0\,0\,0 \end{pmatrix}.$$

Solving the linear equation $A\mathbf{x} = \mathbf{0}_5$, we have $\dim(V(1)) = 2$ and

$$V(1) = \left\{ b_1 \begin{pmatrix} 0 \\ 1 \\ 0 \\ 0 \\ 0 \end{pmatrix} + b_2 \begin{pmatrix} 0 \\ 0 \\ 0 \\ 0 \\ 1 \end{pmatrix} \middle| b_1, b_2 \in \mathbf{C} \right\}.$$

Therefore $\dim(V(2)) - \dim(V(1)) = 4 - 2 = 2$. Take a vector $\mathbf{u}_{2,1} \in V(2)$ so that

$$V(2) = < T_A(\mathbf{u}_{3,1}) > \oplus < \mathbf{u}_{2,1} > \oplus V(1).$$

Since $T_A(\mathbf{u}_{3,1}) = \begin{pmatrix} 0 \\ 1 \\ 1 \\ 0 \\ 0 \end{pmatrix}$, we can take $\mathbf{u}_{2,1} = \begin{pmatrix} 1 \\ 0 \\ 0 \\ 0 \\ 0 \end{pmatrix} \in V(2)$. Then the vectors

$T_A^2(\mathbf{u}_{3,1}) = \begin{pmatrix} 0 \\ 1 \\ 0 \\ 0 \\ 0 \end{pmatrix}$ and $T_A(\mathbf{u}_{2,1}) = \begin{pmatrix} 0 \\ 0 \\ 0 \\ 0 \\ 1 \end{pmatrix}$ in $V(1)$ are linearly independent. Since

$\dim(V(1)) = 2$, the set $\{T_A^2(\mathbf{u}_{3,1}), T_A(\mathbf{u}_{2,1})\}$ is a basis of $V(1)$. Therefore

$$\{T_A^2(\mathbf{u}_{3,1}), T_A(\mathbf{u}_{3,1}), \mathbf{u}_{3,1}, T_A(\mathbf{u}_{2,1}), \mathbf{u}_{2,1}\}$$

is a basis of $V = \mathbf{C}^5$. We take a representation matrix of T_A with respect to this basis and obtain the Jordan normal form of T_A:

$$B = J(0\,;3) \oplus J(0\,;2) = \left(\begin{array}{ccc|cc} 0 & 1 & 0 & 0 & 0 \\ 0 & 0 & 1 & 0 & 0 \\ 0 & 0 & 0 & 0 & 0 \\ \hline 0 & 0 & 0 & 0 & 1 \\ 0 & 0 & 0 & 0 & 0 \end{array} \right). \quad \|$$

Remark. The matrix $B = \left(\begin{array}{ccc|cc} 0 & 1 & 0 & 0 & 0 \\ 0 & 0 & 1 & 0 & 0 \\ 0 & 0 & 0 & 0 & 0 \\ \hline 0 & 0 & 0 & 0 & 1 \\ 0 & 0 & 0 & 0 & 0 \end{array} \right)$ is of course the Jordan normal form of

A. In fact, take

$$P = \begin{pmatrix} T_A^2(\mathbf{u}_{3,1}) & T_A(\mathbf{u}_{3,1}) & \mathbf{u}_{3,1} & T_A(\mathbf{u}_{2,1}) & \mathbf{u}_{2,1} \end{pmatrix}$$

$$= \begin{pmatrix} 0 & 0 & 0 & 0 & 1 \\ 1 & 1 & 0 & 0 & 0 \\ 0 & 1 & 0 & 0 & 0 \\ 0 & 0 & 1 & 0 & 0 \\ 0 & 0 & 0 & 1 & 0 \end{pmatrix}.$$

Then we have $B = P^{-1}AP$.

Now, we shall give the Jordan normal forms of general linear transformations.

Theorem 8.2.3 *Let T be a linear transformation of a vector space V and $\{\lambda_1, \ldots, \lambda_r\}$ the set of all different eigenvalues of T.*

(1) The Jordan normal form of T is of the form

$$J(\lambda_1 ; k_{1,1}) \oplus \cdots \oplus J(\lambda_1 ; k_{1,m_1}) \oplus \cdots$$
$$\oplus J(\lambda_r ; k_{r,1}) \oplus \cdots \oplus J(\lambda_r ; k_{r,m_r}).$$

(2) The Jordan normal form of a linear transformation T is uniquely determined up to the order of Jordan normal cells.

(3) The minimal polynomial $p_T(t)$ of T is

$$p_T(t) = \prod_{i=1}^{r} (t - \lambda_i)^{l_i} \qquad (l_i = \max\{k_{i,1}, \ldots k_{i,m_i}\}).$$

Proof Let $\tilde{W}(\lambda_i ; T)$ be the generalized eigenspace of T with the eigenvalue λ_i for $i = 1, \ldots, r$. Then by Theorem 8.1.3 (2), we have

$$V = \tilde{W}(\lambda_1 ; T) \oplus \cdots \oplus \tilde{W}(\lambda_r ; T).$$

We denote by I_i and T_i the identity transformation of $\tilde{W}(\lambda_i ; T)$ and the restriction of T to $\tilde{W}(\lambda_i ; T)$, respectively. Since $T_i - \lambda_i I_i$ is a nilpotent transformation of $\tilde{W}(\lambda_i ; T)$ for $i = 1, \ldots, r$ by Theorem 8.1.3 (3), the linear transformation

$$T - (\lambda_1 I_1 \oplus \cdots \oplus \lambda_r I_r) = (T_1 - \lambda_1 I_1) \oplus \cdots \oplus (T_r - \lambda_r I_r)$$

is also a nilpotent transformation. The representation matrix of the nilpotent transformation $T_i - \lambda_i I_i$ with respect to a suitable basis of $\tilde{W}(\lambda_i ; T)$ is $J(0 ; k_{i,1}) \oplus \cdots \oplus J(0 ; k_{i,m_i})$ by Theorem 8.2.2. Since $J(\lambda_i ; k_{i,j}) = \lambda_i E_{k_{i,j}} + J(0 ; k_{i,j})$, the representation matrix of T_i with respect to the same basis is the direct sum of the Jordan cells

$$J(\lambda_i ; k_{i,1}) \oplus \cdots \oplus J(\lambda_i ; k_{i,m_i})$$

for $i = 1, \ldots, r$. Therefore the representation matrix of T with respect to a suitable basis of V is a direct sum of Jordan cells

$$J(\lambda_1 ; k_{1,1}) \oplus \cdots \oplus J(\lambda_r ; k_{r,m_r}).$$

Thus we obtain (1).

(2) The uniqueness of the Jordan normal form of T up to the order of Jordan cells comes from the uniqueness of the Jordan normal form up to the order of Jordan cells of nilpotent transformations.

(3) Since the Jordan normal form of T_i is the direct sum of $J(\lambda_i ; k_{i,j})$, the minimal polynomial of T_i is $p_{T_i}(t) = (t - \lambda_i)^{l_i}$ $(l_i = \max\{k_{i,1}, \ldots k_{i,m_i}\})$. By Theorem 5.4.8, we see that the minimal polynomial of $T = T_1 \oplus \cdots \oplus T_r$ is

$$p_T(t) = \prod_{i=1}^{r}(t - \lambda_i)^{l_i} \qquad (l_i = \max\{k_{i,1}, \ldots k_{i,m_i}\}). \quad \|$$

As for the Jordan normal form of square matrices, we have the following theorem.

Theorem 8.2.4 *Let A be a square matrix of degree n. Then there exists a Jordan normal form of A. The Jordan normal form of A is uniquely determined up to the order of Jordan cells. The Jordan normal form of A is equal to the Jordan normal form of the linear transformation T_A associated with A.*

Proof Let A be a square matrix and B the Jordan normal form of the linear transformation T_A. Then B is a representation matrix of T_A with respect to a certain basis of $\mathbf{C}^n$. Let P be the matrix of changing bases from the standard basis to the basis. Then we have $B = P^{-1}AP$. $\quad \|$

Exercise 8.3 Find the Jordan normal form of the following matrix:

$$A = \begin{pmatrix} 2 & 0 & 0 \\ 0 & 1 & 0 \\ 1 & 0 & 2 \end{pmatrix}.$$

Answer. We follow the procedure in the proof of Theorems 8.2.2 and 8.2.3. Since the characteristic polynomial of A is $g_A(t) = (t - 1)(t - 2)^2$, the eigenvalues of A are $\lambda = 1, 2$.

Let $\lambda = 1$. The reduced matrix of $A - E_3 = \begin{pmatrix} 1 & 0 & 0 \\ 0 & 0 & 0 \\ 1 & 0 & 1 \end{pmatrix}$ is $\begin{pmatrix} 1 & 0 & 0 \\ 0 & 0 & 1 \\ 0 & 0 & 0 \end{pmatrix}$ and

$$\tilde{W}(1 ; A) = W(1 ; A) = \left\{ a \begin{pmatrix} 0 \\ 1 \\ 0 \end{pmatrix} \in \mathbf{C}^3 \, \middle| \, a \in \mathbf{C} \right\}$$

by Theorem 8.1.4 (3). We take $\mathbf{u}_1 = \begin{pmatrix} 0 \\ 1 \\ 0 \end{pmatrix}$.

Let $\lambda = 2$. The reduced matrix of $A - 2E_3 = \begin{pmatrix} 0 & 0 & 0 \\ 0 & -1 & 0 \\ 1 & 0 & 0 \end{pmatrix}$ is $\begin{pmatrix} 1 & 0 & 0 \\ 0 & 1 & 0 \\ 0 & 0 & 0 \end{pmatrix}$ and

$$W(2\,;A) = \left\{ b \begin{pmatrix} 0 \\ 0 \\ 1 \end{pmatrix} \in \mathbb{C}^3 \,\middle|\, b \in \mathbb{C} \right\}.$$

We shall solve the linear equation $(A - 2E_3)^2 \mathbf{x} = \mathbf{0}_3$. The reduced matrix of $(A - 2E_3)^2 = \begin{pmatrix} 0 & 0 & 0 \\ 0 & 1 & 0 \\ 0 & 0 & 0 \end{pmatrix}$ is $\begin{pmatrix} 0 & 1 & 0 \\ 0 & 0 & 0 \\ 0 & 0 & 0 \end{pmatrix}$. Solving the linear equation $(A - 2E_3)^2 \mathbf{x} = \mathbf{0}_3$, we have

$$\tilde{W}(2\,;A) = \mathrm{Ker}\,((A - 2E_3)^2) = \left\{ b \begin{pmatrix} 0 \\ 0 \\ 1 \end{pmatrix} + c \begin{pmatrix} 1 \\ 0 \\ 0 \end{pmatrix} \in \mathbb{C}^3 \,\middle|\, b, c \in \mathbb{C} \right\}$$

by Theorem 8.1.4 (3). Let $\mathbf{u}_2 = \begin{pmatrix} 0 \\ 0 \\ 1 \end{pmatrix}$ and $\mathbf{u}_3 = \begin{pmatrix} 1 \\ 0 \\ 0 \end{pmatrix}$. Then $A\mathbf{u}_2 = 2\mathbf{u}_2$ and $A\mathbf{u}_3 = \mathbf{u}_2 + 2\mathbf{u}_3$. Thus we have the equality

$$\left(A\mathbf{u}_1 \ A\mathbf{u}_2 \ A\mathbf{u}_3 \right) = \left(\mathbf{u}_1 \ \mathbf{u}_2 \ \mathbf{u}_3 \right) \begin{pmatrix} 1 & 0 & 0 \\ 0 & 2 & 1 \\ 0 & 0 & 2 \end{pmatrix}.$$

Therefore, if we take a regular matrix

$$P = \left(\mathbf{u}_1 \ \mathbf{u}_2 \ \mathbf{u}_3 \right) = \begin{pmatrix} 0 & 0 & 1 \\ 1 & 0 & 0 \\ 0 & 1 & 0 \end{pmatrix},$$

then we obtain the Jordan normal form of A:

$$B = P^{-1}AP = \begin{pmatrix} 1 & 0 & 0 \\ 0 & 2 & 1 \\ 0 & 0 & 2 \end{pmatrix} = J(1\,;1) \oplus J(2\,;2). \quad \|$$

Another answer. This answer is simpler but not applicable to all cases. Since the characteristic polynomial of A in Exercise 8.3 is $g_A(t) = (t - 1)(t - 2)^2$, the eigenvalues of A are $\lambda = 1, 2$ and the Jordan normal form of A is either

$$J(1\,;1) \oplus J(2\,;1) \oplus J(2\,;1) \quad \text{or} \quad J(1\,;1) \oplus J(2\,;2).$$

Thus the minimal polynomial $p_A(t)$ is either $g_A(t)$ or $(t-1)(t-2)$. When we substitute A for the polynomial $(t-1)(t-2)$, we obtain $(A - E_3)(A - 2E_3) =$

$\begin{pmatrix} 0 & 0 & 0 \\ 0 & 0 & 0 \\ 1 & 0 & 0 \end{pmatrix} \neq O_3$. Then $p_A(t) = g_A(t) = (t-1)(t-2)^2$. Therefore the Jordan nor-

mal form of A is

$$J(1\,;1) \oplus J(2\,;2). \quad \|$$

Theorem 8.2.5 *Let A be a square matrix of degree n and s the number of Jordan cells with eigenvalue 0 of the Jordan normal form of A. Then* $\mathrm{rank}\,(A) = n - s$.

Proof Let B be the Jordan normal form of A. Since B is similar to A, we have $\mathrm{rank}\,(A) = \mathrm{rank}\,(B)$. Let

$$B = \left(\bigoplus_{\lambda_i \neq 0} J(\lambda_i\,;m_i) \right) \oplus \left(\bigoplus_{j=1}^{s} J(0\,;n_j) \right).$$

Then $n = \sum\limits_{\lambda_i \neq 0} m_i + \sum\limits_{j=1}^{s} n_j$. If $\lambda \neq 0$, then $J(\lambda\,;m)$ is a regular matrix and rank $(J(\lambda\,;m)) = m$. On the other hand, we see $\mathrm{rank}\,(J(0\,;m)) = m - 1$. Therefore

$$\mathrm{rank}\,(A) = \mathrm{rank}\,(B) = \mathrm{rank}\,\left(\bigoplus_{\lambda_i \neq 0} J(\lambda_i\,;m_i) \right) + \mathrm{rank}\,\left(\bigoplus_{j=1}^{s} J(0\,;n_j) \right)$$

$$= \sum_{\lambda_i \neq 0} \mathrm{rank}\,(J(\lambda_i\,;m_i)) + \sum_{j=1}^{s} \mathrm{rank}\,(J(0\,;n_j))$$

$$= \sum_{\lambda_i \neq 0} m_i + \sum_{j=1}^{s} (n_j - 1)$$

$$= n - s. \quad \|$$

Proof of Theorem 8.2.2 in the general case.

If $T = O_V$, then the representation matrix is the zero matrix, which is a direct sum of n Jordan cells $J(0\,;1)$. Let T be a non-zero nilpotent transformation of V. We assume that $T^m = O_V$ and $T^{m-1} \neq O_V$. If $m = 1$, then $T = O_V$. So, we have $m \geq 2$. Let

$$V(0) = \{0_V\} \quad \text{and} \quad V(k) = \mathrm{Ker}\,(T^k) = \{\,\mathbf{u} \in V \mid T^k(\mathbf{u}) = \mathbf{0}_V\,\}$$

for $k = 1, \ldots, m$. Since $T^m = O_V$, we see that $V(m) = V$. We easily see that

$$\{0\} = V(0) \subset V(1) \subset \cdots \subset V(m-1) \subset V(m) = V.$$

As $T^{m-1} \neq O_V$, we have $V(m-1) \neq V(m) = V$. Then $t_m = \dim(V(m)) - \dim(V(m-1)) > 0$. Take linearly independent vectors $\mathbf{u}_{m,1}, \ldots, \mathbf{u}_{m,t_m}$ in $V(m)$ so that

$$V = V(m) = <\mathbf{u}_{m,1}, \ldots, \mathbf{u}_{m,t_m}> \oplus V(m-1).$$

Then the vectors $T(\mathbf{u}_{m,1}), \ldots, T(\mathbf{u}_{m,t_m})$ in $V(m-1)$ are linearly independent. To see the linear independence, we let

$$\mathbf{v} = a_1 T(\mathbf{u}_{m,1}) + \cdots + a_{t_m} T(\mathbf{u}_{m,t_m})$$

and

$$\mathbf{u} = a_1 \mathbf{u}_{m,1} + \cdots + a_{t_m} \mathbf{u}_{m,t_m}.$$

Here $a_i \in \mathbf{C}$ for $i = 1, \ldots, t_m$. Assume that $\mathbf{v} = a_1 T(\mathbf{u}_{m,1}) + \cdots + a_{t_m} T(\mathbf{u}_{m,t_m}) = \mathbf{0}_V$. Then $T(\mathbf{u}) = \mathbf{v} = \mathbf{0}_V$. Since $V(1) \subset V(m-1)$, we see that

$$\mathbf{u} \in <\mathbf{u}_{m,1}, \ldots, \mathbf{u}_{m,t_m}> \cap V(m-1) = \{\mathbf{0}_V\}.$$

Thus we have $\mathbf{u} = \mathbf{0}_V$. As $\mathbf{u}_{m,1}, \ldots, \mathbf{u}_{m,t_m}$ are linearly independent, we have $a_1 = \cdots = a_{t_m} = 0$. Therefore $T(\mathbf{u}_{m,1}), \ldots, T(\mathbf{u}_{m,t_m})$ are linearly independent. Furthermore, we have

$$(*) \qquad <T(\mathbf{u}_{m,1}), \ldots, T(\mathbf{u}_{m,t_m})> \cap V(m-2) = \{\mathbf{0}_V\}.$$

In fact, if $\mathbf{v} = T(\mathbf{u}) = a_1 T(\mathbf{u}_{m,1}) + \cdots + a_{t_m} T(\mathbf{u}_{m,t_m}) \in V(m-2)$, then we see that $T^{m-1}(\mathbf{u}) = T^{m-2}(\mathbf{v}) = \mathbf{0}_V$. Therefore $\mathbf{u} \in V(m-1)$. Since $<\mathbf{u}_{m,1}, \ldots, \mathbf{u}_{m,t_m}> \cap V(m-1) = \{\mathbf{0}_V\}$, we see that $\mathbf{u} = \mathbf{0}_V$. Then $\mathbf{v} = T(\mathbf{u}) = \mathbf{0}_V$ and this implies the equality $(*)$.

Next, we take linearly independent vectors $\mathbf{u}_{m-1,1}, \ldots, \mathbf{u}_{m-1,t_{m-1}}$ in $V(m-1)$ so that they satisfy

$$V(m-1) = <T(\mathbf{u}_{m,1}), \ldots, T(\mathbf{u}_{m,t_m})>$$
$$\oplus <\mathbf{u}_{m-1,1}, \ldots, \mathbf{u}_{m-1,t_{m-1}}> \oplus V(m-2).$$

Then vectors

$$T(\mathbf{u}_{m,1}), \ldots, T(\mathbf{u}_{m,t_m}), \mathbf{u}_{m-1,1}, \ldots, \mathbf{u}_{m-1,t_{m-1}}$$

in $V(m-1)$ are linearly independent. To see that the vectors

$$T^2(\mathbf{u}_{m,1}), \ldots, T^2(\mathbf{u}_{m,t_m}), T(\mathbf{u}_{m-1,1}), \ldots, T(\mathbf{u}_{m-1,t_{m-1}})$$

in $V(m-2)$ are linearly independent, we let

$$a_1 T^2(\mathbf{u}_{m,1}) + \cdots + a_{t_m} T^2(\mathbf{u}_{m,t_m})$$
$$+ b_1 T(\mathbf{u}_{m-1,1}) + \cdots + b_{t_{m-1}} T(\mathbf{u}_{m-1,t_{m-1}}) = \mathbf{0}_V$$

be a linear relation of these vectors. Here $a_i, b_j \in \mathbf{C}$ for $i = 1, \ldots, t_m$ and $j = 1, \ldots, t_{m-1}$. Let

$$\mathbf{w} = a_1 T(\mathbf{u}_{m,1}) + \cdots + a_{t_m} T(\mathbf{u}_{m,t_m}) + b_1 \mathbf{u}_{m-1,1} + \cdots + b_{t_{m-1}} \mathbf{u}_{m-1,t_{m-1}}.$$

Then $T(\mathbf{w}) = \mathbf{0}_V$ and we have

$$\mathbf{w} \in (\ < T(\mathbf{u}_{m,1}), \ldots, T(\mathbf{u}_{m,t_m}) > \oplus < \mathbf{u}_{m-1,1}, \ldots, \mathbf{u}_{m-1,t_{m-1}} >)$$
$$\cap V(m-2) = \{\mathbf{0}_V\}.$$

Therefore $\mathbf{w} = \mathbf{0}_V$. We note that $V(1) \subset V(m-2)$, since $m > 2$. As the vectors

$$T(\mathbf{u}_{m,1}), \ldots, T(\mathbf{u}_{m,t_m}), \mathbf{u}_{m-1,1}, \ldots, \mathbf{u}_{m-1,t_{m-1}}$$

are linearly independent, we have

$$a_1 = \cdots = a_{t_m} = b_1 = \cdots = b_{t_{m-1}} = 0.$$

This implies that the vectors

$$T^2(\mathbf{u}_{m,1}), \ \ldots, \ T^2(\mathbf{u}_{m,t_m}), \ T(\mathbf{u}_{m-1,1}), \ \ldots, \ T(\mathbf{u}_{m-1,t_{m-1}})$$

are linearly independent. Take linearly independent vectors $\mathbf{u}_{m-2,1}, \ldots, \mathbf{u}_{m-2,t_{m-2}}$ in $V(m-2)$ so that

$$V(m-2) = < T^2(\mathbf{u}_{m,1}), \ldots, T^2(\mathbf{u}_{m,t_m}) >$$
$$\oplus < T(\mathbf{u}_{m-1,1}), \ldots, T(\mathbf{u}_{m-1,t_{m-1}}) >$$
$$\oplus < \mathbf{u}_{m-2,1}, \ldots, \mathbf{u}_{m-2,t_{m-2}} > \oplus V(m-3).$$

Repeating this procedure, we take vectors $\mathbf{u}_{k,1}, \ldots \mathbf{u}_{k,t_k}$ in $V(k)$ similarly for $k = m-3, \ldots, 1$. Then the set of vectors

$$
\left\{
\begin{array}{l}
\mathbf{u}_{m,1}, \quad\quad \ldots, \ \mathbf{u}_{m,t_m}, \\
T(\mathbf{u}_{m,1}), \quad \ldots, \ T(\mathbf{u}_{m,t_m}), \quad\quad \mathbf{u}_{m-1,1}, \quad\quad \ldots\ldots, \ \mathbf{u}_{m-1,t_{m-1}}, \\
T^2(\mathbf{u}_{m,1}), \quad \ldots, \ T^2(\mathbf{u}_{m,t_m}), \quad T(\mathbf{u}_{m-1,1}), \quad \ldots\ldots T(\mathbf{u}_{m-1,t_{m-1}}), \\
\quad\quad\quad\quad\quad\quad\quad\quad\quad\quad\quad\quad\quad\quad\quad\quad\quad \mathbf{u}_{m-2,1}, \ldots, \mathbf{u}_{m-2,t_{m-2}}, \\
\ldots\ldots\ldots \\
T^{m-1}(\mathbf{u}_{m,1}), \ldots, T^{m-1}(\mathbf{u}_{m,t_m}), \ T^{m-2}(\mathbf{u}_{m-1,1}), \ldots\ldots, T^{m-2}(\mathbf{u}_{m-1,t_{m-1}}), \\
\quad\quad\quad\quad\quad\quad\quad\quad\quad\quad\quad\quad\quad\quad\quad\quad\quad \ldots, \mathbf{u}_{1,1}, \ldots, \mathbf{u}_{1,t_1}
\end{array}
\right.
$$

is a basis of V as we see in the case of $m = 3$. Let $k = 1, \ldots, m$. For $i = 1, \ldots, t_k$, we consider subspaces

$$U(k\,;i) = < T^{k-1}(\mathbf{u}_{k,i}), \ldots, T(\mathbf{u}_{k,i}), \mathbf{u}_{k,i} >$$

of V. Then we see that

$$V = U(m\,;1) \oplus \cdots \oplus U(m\,;t_m)$$
$$\oplus U(m-1\,;1) \oplus \cdots \oplus U(m-1\,;t_{m-1})$$
$$\oplus \cdots\cdots \oplus U(1\,;1) \oplus \cdots \oplus U(1\,;t_1).$$

Since the vectors $T^{k-1}(\mathbf{u}_{k,i}), \ldots, T(\mathbf{u}_{k,i}), \mathbf{u}_{k,i}$ are linearly independent, the set of vectors $\{T^{k-1}(\mathbf{u}_{k,i}), \ldots, T(\mathbf{u}_{k,i}), \mathbf{u}_{k,i}\}$ is a basis of $U(k\,;i)$ for $k = 1, \ldots, m$ and $i = 1, \ldots, t_k$. The linear transformation T induces a linear transformation $T_{k,i}$ of the subspace $U(k\,;i)$. The representation matrix of $T_{k,i}$ with respect to the above basis of $U(k\,;i)$ is

$$J(0\,;k) = \left. \begin{pmatrix} 0 & 1 & & 0 \\ & 0 & \ddots & \\ & & \ddots & 1 \\ 0 & & & 0 \end{pmatrix} \right\} k$$

Therefore the representation matrix of T with respect to the basis

$$\{T^{m-1}(\mathbf{u}_{m,1}), T^{m-2}(\mathbf{u}_{m,1}), \ldots, \mathbf{u}_{m,1}, \ldots$$
$$\ldots, T^{m-1}(\mathbf{u}_{m,t_m}), T^{m-2}(\mathbf{u}_{m,t_m}), \ldots, \mathbf{u}_{m,t_m};$$
$$T^{m-2}(\mathbf{u}_{m-1,1}), \ldots, \mathbf{u}_{m-1,1}, \ldots\ldots, T^{m-2}(\mathbf{u}_{m-1,t_{m-1}}), \ldots, \mathbf{u}_{m-1,t_{m-1}};$$
$$\ldots\ldots\ldots;$$
$$T(\mathbf{u}_{2,1}), \mathbf{u}_{2,1}, \ldots\ldots, T(\mathbf{u}_{2,t_2}), \mathbf{u}_{2,t_2},$$
$$\mathbf{u}_{1,1}, \ldots\ldots, \mathbf{u}_{1,t_1}\}$$

is a matrix of Jordan normal form

$$\underbrace{J(0\,;m) \oplus \cdots \oplus J(0\,;m)}_{t_m} \oplus \underbrace{J(0\,;m-1) \oplus \cdots \oplus J(0\,;m-1)}_{t_{m-1}} \oplus$$
$$\cdots\cdots \oplus \underbrace{J(0\,;2) \oplus \cdots \oplus J(0\,;2)}_{t_2} \oplus \underbrace{J(0\,;1) \oplus \cdots \oplus J(0\,;1)}_{t_1}.$$

Thus we obtain the Jordan normal form of T.

We shall show the uniqueness of the Jordan normal form of T. Since $J(0\,;k)^m \neq O_k$ for $k > m$, we may assume that

$$B = \underbrace{J(0\,;m) \oplus \cdots \oplus J(0\,;m)}_{s_m} \oplus \underbrace{J(0\,;m-1) \oplus \cdots \oplus J(0\,;m-1)}_{s_{m-1}} \oplus$$
$$\cdots\cdots \oplus \underbrace{J(0\,;2) \oplus \cdots \oplus J(0\,;2)}_{s_2} \oplus \underbrace{J(0\,;1) \oplus \cdots \oplus J(0\,;1)}_{s_1}$$

is another Jordan normal form of T. By a similar argument in the case of $m = 3$, we see that

$$\begin{cases} \dim(V(m)) & = ms_m + (m-1)s_{m-1} + \cdots + 2s_2 + s_1, \\ \dim(V(m-1)) = (m-1)s_m + (m-1)s_{m-1} + \cdots + 2s_2 + s_1, \\ \qquad \cdots\cdots\cdots, \\ \dim(V(1)) & = s_m + s_{m-1} + \cdots + s_2 + s_1, \end{cases}$$

thus $s_1, \ldots, s_m$ satisfy

$$\begin{cases} s_m & = \dim(V(m)) - \dim(V(m-1)), \\ s_m + s_{m-1} & = \dim(V(m-1)) - \dim(V(m-2)), \\ \qquad \cdots\cdots\cdots, \\ s_m + s_{m-1} + \cdots + s_1 = \dim(V(1)). \end{cases}$$

On the other hand, we see that $t_1, \ldots, t_m$ also satisfy

$$\begin{cases} t_m & = \dim(V(m)) - \dim(V(m-1)), \\ t_m + t_{m-1} & = \dim(V(m-1)) - \dim(V(m-2)), \\ \qquad \cdots\cdots\cdots, \\ t_m + t_{m-1} + \cdots + t_1 = \dim(V(1)). \end{cases}$$

Then we have

$$\begin{pmatrix} 1 & & & 0 \\ 1 & 1 & & \\ \vdots & \vdots & \ddots & \\ 1 & 1 & \cdots & 1 \end{pmatrix} \begin{pmatrix} s_m - t_m \\ s_{m-1} - t_{m-1} \\ \vdots \\ s_1 - t_1 \end{pmatrix} = \begin{pmatrix} 0 \\ 0 \\ \vdots \\ 0 \end{pmatrix}.$$

Since the square matrix $\begin{pmatrix} 1 & & & 0 \\ 1 & 1 & & \\ \vdots & \vdots & \ddots & \\ 1 & 1 & \cdots & 1 \end{pmatrix}$ of degree m is a regular matrix, we have

$s_1 = t_1, s_2 = t_2, \ldots, s_m = t_m$. Therefore the Jordan normal form of T is uniquely determined up to the order of Jordan cells.   ‖

Exercises (Sect. 8.2)

1 Find the minimal polynomials $p_A(t)$ and the Jordan normal forms of the following matrices A:

(1) $A = \begin{pmatrix} 1 & 1 & 0 & 2 \\ 1 & 0 & 0 & 1 \\ 0 & 0 & 0 & 0 \\ -1 & 0 & 0 & -1 \end{pmatrix}.$

(2) $A = \begin{pmatrix} 2 & -1 & 0 & 2 \\ 0 & 0 & 0 & 0 \\ 1 & -1 & 0 & 1 \\ -2 & 1 & 0 & -2 \end{pmatrix}.$

(3) $A =$
$\begin{pmatrix} 0\,0\,0 & 1\,0\,0 \\ 0\,0\,0 & 2\,1\,0 \\ 1\,0\,0 & 0\,0\,1 \\ 0\,0\,0 & 0\,0\,0 \\ 0\,0\,1 & 0\,0\,0 \\ 0\,0\,0 & -1\,0\,0 \end{pmatrix}$.

(4) $A =$
$\begin{pmatrix} 0\,0 & 0\,0\,0\,1 \\ 0\,0 & -1\,0\,0\,0 \\ 1\,0 & 0\,0\,0\,0 \\ -2\,0 & 0\,0\,0\,0 \\ 0\,0 & 0\,0\,0\,1 \\ 0\,0 & 0\,0\,0\,0 \end{pmatrix}$.

2 Find the minimal polynomials $p_A(t)$ and the Jordan normal forms of the following matrices A:

(1) $A = \begin{pmatrix} 2\,1\,0 \\ 0\,3\,0 \\ 0\,0\,3 \end{pmatrix}$.   (2) $A = \begin{pmatrix} 3\,0\,0 \\ 1\,3\,0 \\ 1\,1\,3 \end{pmatrix}$.   (3) $A = \begin{pmatrix} 0\,0\,-1 \\ 0\,i\,\,0 \\ 1\,0\,\,0 \end{pmatrix}$.

(4) $A = \begin{pmatrix} 2\,0\,-3\,\,1 \\ 0\,2\,\,\,0\,\,0 \\ 0\,0\,-1\,0 \\ 0\,0\,\,\,0\,2 \end{pmatrix}$.

(5) $A = \begin{pmatrix} 1 & 0\,0\,0\,1 \\ 1 & 2\,0\,0\,0 \\ -1 & -1\,1\,0\,0 \\ 0 & 1\,1\,1\,0 \\ 0 & 0\,0\,0\,2 \end{pmatrix}$.

(6) $A = \begin{pmatrix} 2 & 0\,0 & 0 & 0 \\ -1 & -1\,0 & -3 & 4 \\ 0 & 0\,1 & 0 & 0 \\ 1 & 0\,0 & 2 & -3 \\ 0 & 0\,0 & 0 & -1 \end{pmatrix}$.

(7) $A = \begin{pmatrix} 3 & 0\,0 & 0 & 4 \\ 0 & 1\,0 & -2 & 1 \\ -3+i & 0\,i & 0 & -4 \\ 0 & 0\,0 & 3 & -2 \\ 0 & 0\,0 & 0 & 1 \end{pmatrix}$.

3 Find the ranks and the minimal polynomials $p_A(t)$ of the following matrices A:

(1) $A = J(0\,;1) \oplus J(0\,;2) \oplus J(0\,;3) \oplus J(2\,;2) \oplus J(2\,;3)$.
(2) $A = J(3\,;2) \oplus J(3\,;2) \oplus J(3\,;1) \oplus J(i\,;2)$.
(3) $A = J(0\,;2) \oplus J(0\,;2) \oplus J(1\,;3) \oplus J(-2\,;1) \oplus J(-2\,;3)$.
(4) $A = J(0\,;1) \oplus J(2\,;3) \oplus J(i\,;2) \oplus J(-i\,;1)$.

4 Show that the following two statements are equivalent for a square matrix A:

(1) The minimal polynomial $p_A(t)$ of A is equal to the characteristic polynomial $g_A(t)$ of A.

(2) For each eigenvalue λ of A, the number of the Jordan cell with the eigenvalue λ appearing in the Jordan normal form of A is one.

5 Show that if a square matrix A of degree n is semisimple and nilpotent, then $A = O_n$.

6 Let A be a square matrix of degree n. Then show the following statements:

(1) There uniquely exists a pair of a semisimple matrix H and a nilpotent matrix N satisfying $A = H + N$ and $HN = NH$. Furthermore, both H and N are polynomials of A.

(2) If A is a real matrix, then the matrices H and N are also real matrices.

Answers to Exercises

1.1

1　(1) (2, 3) or 2×3.　(2) -2.　(3) $\left(-1\ 3\ -2\right)$.　(4) $\begin{pmatrix} -2 \\ 5 \end{pmatrix}$.

(5) $\begin{pmatrix} -1\ 1 \\ 3\ 2 \\ -2\ 5 \end{pmatrix}$.

2　(1) (3, 4) or 3×4.　(2) 0.　(3) $\left(3\ -1\ 2\ -5\right)$.　(4) $\begin{pmatrix} -3 \\ 2 \\ 2 \end{pmatrix}$.

(5) $\begin{pmatrix} 2 & 3 & 18 \\ 4 & -1 & 0 \\ -3 & 2 & 2 \\ 8 & -5 & 12 \end{pmatrix}$.

3　(1) (4, 3) or 4×3.　(2) 5.　(3) $\left(9\ 7\ -1\right)$.　(4) $\begin{pmatrix} -1 \\ 2 \\ 7 \\ 0 \end{pmatrix}$.

(5) $\begin{pmatrix} 4 & -1 & 9 & 5 \\ -1 & 2 & 7 & 0 \\ 2 & 0 & -1 & 3 \end{pmatrix}$.

4　(1) $a = 4, b = 4, c = 5, d = 1$.　(2) $a = 2, b = 3, c = 1, d = 1$.

5　(1) $a = 3, b = 3, c = 1$ and $A = \begin{pmatrix} 1 & 3 & 3 \\ 3 & -2 & 1 \\ 3 & 1 & 0 \end{pmatrix}$.

(2) $a = 1, b = 3, c = 2$ and $A = \begin{pmatrix} 2 & 1 & 1 \\ 1 & 3 & 2 \\ 1 & 2 & 5 \end{pmatrix}$.

(3) $a = 3$, $b = 1$, $c = -1$ and $A = \begin{pmatrix} 3 & 4 & 2 \\ 4 & 5 & -1 \\ 2 & -1 & -1 \end{pmatrix}$.

6 (1) $a = 5$, $b = 2$, $c = -3$, $d = 5$ and $A = \begin{pmatrix} 0 & -5 & 3 \\ 5 & 0 & -3 \\ -3 & 3 & 0 \end{pmatrix}$.

(2) $a = -4$, $b = 3$, $c = 0$, $d = 1$ and $A = \begin{pmatrix} 0 & -3 & -1 \\ 3 & 0 & 1 \\ 1 & -1 & 0 \end{pmatrix}$.

(3) $a = 0$, $b = 2$, $c = -4$, $d = 6$ and $A = \begin{pmatrix} 0 & 4 & -3 \\ -4 & 0 & -4 \\ 3 & 4 & 0 \end{pmatrix}$.

7 (1) $A = \begin{pmatrix} 1 & 2 & 0 \\ 0 & 1 & 2 \\ 0 & 0 & 1 \end{pmatrix}$. (2) $A = \begin{pmatrix} 1 & 0 & 1 \\ 0 & 2 & 0 \\ 1 & 0 & 1 \end{pmatrix}$. (3) $A = \begin{pmatrix} 3 & 0 & 0 \\ 0 & 2 & 0 \\ 0 & 1 & 2 \end{pmatrix}$.

8   Let $A = \left(a_{ij} \right)_{n \times n}$ be an alternating matrix of degree n. Then $a_{ii} = -a_{ii}$ for $i = 1, \ldots, n$. Therefore we have $a_{ii} = 0$ for $i = 1, \ldots, n$.

1.2

1 (1) $\begin{pmatrix} 4 & 7 \\ 22 & 7 \end{pmatrix}$. (2) $\begin{pmatrix} 7 & -4 \\ -25 & 22 \end{pmatrix}$. (3) $\begin{pmatrix} 6 & 10 & -5 \\ 0 & -9 & 27 \end{pmatrix}$.

(4) $\begin{pmatrix} 6 & 2 & -4 \\ -3 & -1 & 2 \\ 12 & 4 & -8 \end{pmatrix}$. (5) $(-3) = -3$. (6) O_3.

(7) $\begin{pmatrix} 5 & -1 & 6 \\ -2 & 1 & 3 \\ 3 & -1 & 2 \end{pmatrix}$. (8) $\begin{pmatrix} 3 & 1 & 0 \\ -1 & 1 & -1 \\ 3 & 0 & 2 \end{pmatrix}$. (9) $\begin{pmatrix} 0 & -3 & 0 \\ 2 & -1 & 0 \\ -1 & 3 & 0 \end{pmatrix}$.

(10) $\begin{pmatrix} 0 & 2 & 1 & -1 \\ -1 & -1 & 1 & 2 \\ 1 & 0 & 0 & 1 \\ -1 & -1 & 0 & 1 \end{pmatrix}$. (11) $\begin{pmatrix} 1 & 4 & 0 \\ -2 & -2 & 1 \\ 2 & 3 & 0 \end{pmatrix}$.

(12) $\begin{pmatrix} -2 & -1 & 1 & -1 \\ -3 & 0 & 1 & -2 \\ -4 & 0 & -2 & -1 \\ -2 & -1 & 1 & -1 \end{pmatrix}$. (13) $\begin{pmatrix} -12 & -12 & 17 \\ 13 & -14 & 22 \\ -16 & -13 & -13 \end{pmatrix}$.

2 $AC = \begin{pmatrix} 4 & 0 & 2 \\ 2 & 0 & 1 \\ -2 & 0 & -1 \end{pmatrix}$, $BD = \begin{pmatrix} 4 & 17 \\ 7 & 16 \\ -1 & 4 \end{pmatrix}$, $CA = (3) = 3$, $CB = (6 \ 5)$,

$D^2 = \begin{pmatrix} 1 & 18 \\ -6 & 13 \end{pmatrix}$.

3 $AC = \begin{pmatrix} 1 & -2 \\ 1 & 2 \\ -2 & 0 \end{pmatrix}$, $BD = \begin{pmatrix} 0 & 0 & 0 \\ -1 & 0 & 1 \\ 0 & 0 & 0 \end{pmatrix}$, $DA = (-4 \ 1)$,

$C^2 = \begin{pmatrix} 1 & 0 \\ 3 & 4 \end{pmatrix}$, $DB = (0) = 0$.

4 (1) $A^2 = \begin{pmatrix} 0 & 0 & 1 \\ 0 & 0 & 0 \\ 0 & 0 & 0 \end{pmatrix}$, $A^n = O_3 \ (n \geq 3)$.

(2) $A^{3k-2} = \begin{pmatrix} 0 & 0 & 1 \\ 1 & 0 & 0 \\ 0 & 1 & 0 \end{pmatrix}$, $A^{3k-1} = \begin{pmatrix} 0 & 1 & 0 \\ 0 & 0 & 1 \\ 1 & 0 & 0 \end{pmatrix}$, $A^{3k} = E_3 \ (k = 1, 2, \ldots)$.

(3) $A^n = \begin{pmatrix} a^n & 0 & 0 \\ 0 & b^n & 0 \\ 0 & 0 & c^n \end{pmatrix}$. (4) $A^n = \begin{pmatrix} a^n & (a^{n-1} + a^{n-2} + \cdots + a + 1)\,b \\ 0 & 1 \end{pmatrix}$.

5 (1) Not commutative. (2) If $a = c$, then they are commutative. If $a \neq c$, then they are not commutative.

6 $a = 1$, $b = c = -1$, $d = 6$.

7 The left-hand side $= (E_n + A + \cdots + A^{m-1}) - (A + A^2 + \cdots + A^m) = E_n - A^m = E_n - O_n = E_n$.

8 We assume that $A^l = O_n$ and $B^m = O_n$. Let $t = \max\{l, m\}$. Then $(AB)^t = A^t B^t = O_n O_n = O_n$. $(A + B)^{2t} = \sum_{k=0}^{2t} \binom{2t}{k} A^{2t-k} B^k$. We know that either $k \geq t$ or $2t - k \geq t$ holds for $k = 0, \ldots, 2t$. Then either $A^{2t-k} = O_n$ or $B^k = O_n$ for $k = 0, \ldots, 2t$. Therefore, we see that $A^{2t-k} B^k = O_n$ for $k = 0, \ldots, 2t$. Thus we have $(A + B)^{2t} = O_n$.

9 For a square matrix A, we put $S = \frac{1}{2}(A + {}^tA)$ and $Q = \frac{1}{2}(A - {}^tA)$. Then we easily see that S is a symmetric matrix and Q is an alternating matrix. Since $A = S + Q$, we see that A is a sum of a symmetric matrix and an alternating matrix. Let $A = S + Q$ with a symmetric matrix S and an alternating matrix Q. Then ${}^tA = {}^tS + {}^tQ = S - Q$. Thus $S = \frac{1}{2}(A + {}^tA)$ and $Q = \frac{1}{2}(A - {}^tA)$. Therefore S and Q are uniquely determined.

1.3

1 (1) $\left(\begin{array}{cc|cc} 5 & 4 & 4 & 2 \\ 13 & 10 & 5 & 3 \\ \hline 0 & 0 & 4 & 1 \\ 0 & 0 & 1 & 0 \end{array} \right)$. (2) $\left(\begin{array}{cc|cc} 0 & 1 & 0 & 0 \\ 10 & 7 & 0 & 0 \\ \hline 2 & 3 & 4 & 1 \\ 2 & 3 & -1 & -1 \end{array} \right)$.

(3) $\left(\begin{array}{cc|cc} -1 & 0 & 0 & 0 \\ \hline 0 & -1 & -2 & 0 \\ 0 & 3 & -2 & 0 \\ \hline 0 & 0 & 0 & 2 \end{array}\right).$
(4) $\left(\begin{array}{cc|c|c} 2 & 0 & 0 & 1 \\ \hline 0 & 1 & 8 & 0 \\ 0 & -3 & -4 & 0 \\ \hline 0 & 0 & 0 & -3 \end{array}\right).$

2 (1) $2\mathbf{a}_1 + \mathbf{a}_2 + 3\mathbf{a}_3$.   (2) $-\mathbf{a}_1 + 4\mathbf{a}_2 - 2\mathbf{a}_3$.

3 (1) $3\,{}^t\mathbf{b}_1 - 2\,{}^t\mathbf{b}_2 + 4\,{}^t\mathbf{b}_3$.   (2) ${}^t\mathbf{b}_1 + 5\,{}^t\mathbf{b}_2 - {}^t\mathbf{b}_3$.

4 (1) $\left(3\mathbf{a}_1 + \mathbf{a}_3 \;\; \mathbf{a}_1 - 4\mathbf{a}_2 + \mathbf{a}_3 \right)$.   (2) $\left(-2\mathbf{a}_1 + \mathbf{a}_2 \;\; -\mathbf{a}_2 + \mathbf{a}_3 \;\; 3\mathbf{a}_1 + 2\mathbf{a}_2 - 2\mathbf{a}_3 \right)$.

5 (1) $\left(\begin{array}{c} {}^t\mathbf{b}_1 - {}^t\mathbf{b}_2 + {}^t\mathbf{b}_3 \\ {}^t\mathbf{b}_1 - {}^t\mathbf{b}_3 \end{array} \right)$.   (2) $\left(\begin{array}{c} 2\,{}^t\mathbf{b}_1 + {}^t\mathbf{b}_2 - {}^t\mathbf{b}_3 \\ 3\,{}^t\mathbf{b}_2 + {}^t\mathbf{b}_3 \\ {}^t\mathbf{b}_1 + {}^t\mathbf{b}_2 \end{array} \right)$.

6 By Exercise 1.6, $AB = \left(\begin{array}{cc} A_1 B_1 & O_{m,n} \\ O_{n,m} & A_2 B_2 \end{array} \right)$ and $BA = \left(\begin{array}{cc} B_1 A_1 & O_{m,n} \\ O_{n,m} & B_2 A_2 \end{array} \right)$. Since $A_1 B_1 = B_1 A_1$ and $A_2 B_2 = B_2 A_2$, we have $AB = BA$.

7 $\left(\begin{array}{cc} E_m & kA \\ O_{n,m} & E_n \end{array} \right)$.

1.4

1 (1) $\begin{pmatrix} 2 & 3 \\ 1 & -1 \end{pmatrix} \begin{pmatrix} x_1 \\ x_2 \end{pmatrix} = \begin{pmatrix} -1 \\ 2 \end{pmatrix}$. The coefficient matrix is $\begin{pmatrix} 2 & 3 \\ 1 & -1 \end{pmatrix}$.

The augmented matrix is $\left(\begin{array}{cc|c} 2 & 3 & -1 \\ 1 & -1 & 2 \end{array} \right)$.

(2) $\begin{pmatrix} 1 & 2 & -1 \\ -1 & 0 & 3 \\ 0 & 1 & -2 \end{pmatrix} \begin{pmatrix} x_1 \\ x_2 \\ x_3 \end{pmatrix} = \begin{pmatrix} 2 \\ 8 \\ -4 \end{pmatrix}$. The coefficient matrix is $\begin{pmatrix} 1 & 2 & -1 \\ -1 & 0 & 3 \\ 0 & 1 & -2 \end{pmatrix}$.

The augmented matrix is $\left(\begin{array}{ccc|c} 1 & 2 & -1 & 2 \\ -1 & 0 & 3 & 8 \\ 0 & 1 & -2 & -4 \end{array} \right)$.

(3) $\begin{pmatrix} 2 & 0 & 3 \\ 1 & -1 & -2 \\ 2 & 1 & 1 \end{pmatrix} \begin{pmatrix} x_1 \\ x_2 \\ x_3 \end{pmatrix} = \begin{pmatrix} 8 \\ -1 \\ 3 \end{pmatrix}$. The coefficient matrix is $\begin{pmatrix} 2 & 0 & 3 \\ 1 & -1 & -2 \\ 2 & 1 & 1 \end{pmatrix}$.

The augmented matrix is $\left(\begin{array}{ccc|c} 2 & 0 & 3 & 8 \\ 1 & -1 & -2 & -1 \\ 2 & 1 & 1 & 3 \end{array} \right)$.

2 (1) $\begin{cases} 2x_1 + x_2 + 3x_3 = 1, \\ \quad\;\; - x_2 + 2x_3 = 2, \\ x_1 \qquad\;\; - x_3 = -2. \end{cases}$ (2) $\begin{cases} 3x_1 \qquad\;\; + x_3 = -1, \\ x_1 - x_2 + 2x_3 = 0. \end{cases}$

(3) $\begin{cases} x_1 - x_2 + 2x_3 + x_4 = 0, \\ x_1 + x_2 + 3x_3 - 2x_4 = -2, \\ \quad\;\; 2x_2 - x_3 + x_4 = 2. \end{cases}$

3 (1) $\mathbf{a} = -(3/4)\mathbf{b}_1 + (1/4)\mathbf{b}_2$. (2) The vector $\mathbf{a}$ is not a linear combination of $\mathbf{b}_1$ and $\mathbf{b}_2$. (3) $\mathbf{a} = 2\mathbf{b}_1 - \mathbf{b}_2$.

4 (1) $a = -1$. (2) $a = 3b$. (3) $a - 3b + 2 = 0$.

5 (1) $\mathbf{c} = \begin{pmatrix} 1 & -3 \end{pmatrix} \begin{pmatrix} \mathbf{b}_1 \\ \mathbf{b}_2 \end{pmatrix} = \begin{pmatrix} 1 & -3 \end{pmatrix} \begin{pmatrix} 2 & 3 \\ -1 & 4 \end{pmatrix} \begin{pmatrix} \mathbf{a}_1 \\ \mathbf{a}_2 \end{pmatrix} = \begin{pmatrix} 5 & -9 \end{pmatrix} \begin{pmatrix} \mathbf{a}_1 \\ \mathbf{a}_2 \end{pmatrix}$.
Then $\mathbf{c} = 5\mathbf{a}_1 - 9\mathbf{a}_2$.

(2) $\mathbf{c} = \begin{pmatrix} 2 & 1 \end{pmatrix} \begin{pmatrix} \mathbf{b}_1 \\ \mathbf{b}_2 \end{pmatrix} = \begin{pmatrix} 2 & 1 \end{pmatrix} \begin{pmatrix} -1 & -4 \\ 3 & 1 \end{pmatrix} \begin{pmatrix} \mathbf{a}_1 \\ \mathbf{a}_2 \end{pmatrix} = \begin{pmatrix} 1 & -7 \end{pmatrix} \begin{pmatrix} \mathbf{a}_1 \\ \mathbf{a}_2 \end{pmatrix}$.
Then $\mathbf{c} = \mathbf{a}_1 - 7\mathbf{a}_2$.

(3) $\begin{pmatrix} \mathbf{c}_1 \\ \mathbf{c}_2 \end{pmatrix} = \begin{pmatrix} 1 & 2 \\ -1 & 1 \end{pmatrix} \begin{pmatrix} \mathbf{b}_1 \\ \mathbf{b}_2 \end{pmatrix} = \begin{pmatrix} 1 & 2 \\ -1 & 1 \end{pmatrix} \begin{pmatrix} 3 & -2 \\ -1 & 2 \end{pmatrix} \begin{pmatrix} \mathbf{a}_1 \\ \mathbf{a}_2 \end{pmatrix} = \begin{pmatrix} 1 & 2 \\ -4 & 4 \end{pmatrix} \begin{pmatrix} \mathbf{a}_1 \\ \mathbf{a}_2 \end{pmatrix}$. Then
$\mathbf{c}_1 = \mathbf{a}_1 + 2\mathbf{a}_2$, $\mathbf{c}_2 = -4\mathbf{a}_1 + 4\mathbf{a}_2$.

2.1

1 (1) $\begin{cases} x_1 = 1, \\ x_2 = -1. \end{cases}$ (2) $\begin{cases} x_1 = 2, \\ x_2 = -3. \end{cases}$ (3) $\begin{cases} x_1 = 3, \\ x_2 = 1. \end{cases}$ (4) $\begin{cases} x_1 = -1, \\ x_2 = 2. \end{cases}$

(5) $\begin{cases} x_1 = 1, \\ x_2 = 2, \\ x_3 = 3. \end{cases}$ (6) $\begin{cases} x_1 = 1, \\ x_2 = -1, \\ x_3 = 1. \end{cases}$ (7) $\begin{cases} x_1 = 1, \\ x_2 = 1, \\ x_3 = 2. \end{cases}$ (8) $\begin{cases} x_1 = -1, \\ x_2 = 0, \\ x_3 = 1. \end{cases}$

(9) $\begin{cases} x_1 = 1, \\ x_2 = 2, \\ x_3 = -1. \end{cases}$ (10) $\begin{cases} x_1 = 2, \\ x_2 = 1, \\ x_3 = -1, \\ x_4 = 1. \end{cases}$

2 (1) $\begin{cases} x_1 = 1, \\ x_2 = 2. \end{cases}$ (2) $\begin{cases} x_1 = 3, \\ x_2 = -1. \end{cases}$ (3) $\begin{cases} x_1 = -1, \\ x_2 = 1. \end{cases}$ (4) $\begin{cases} x_1 = 2, \\ x_2 = 1. \end{cases}$

(5) $\begin{cases} x_1 = -1, \\ x_2 = 0, \\ x_3 = 1. \end{cases}$ (6) $\begin{cases} x_1 = 1, \\ x_2 = -1, \\ x_3 = 2. \end{cases}$ (7) $\begin{cases} x_1 = -1, \\ x_2 = 2, \\ x_3 = 1. \end{cases}$ (8) $\begin{cases} x_1 = 2, \\ x_2 = 1, \\ x_3 = 0. \end{cases}$

(9) $\begin{cases} x_1 = 1, \\ x_2 = -1, \\ x_3 = 1, \\ x_4 = -1. \end{cases}$

3 (V)$\Rightarrow$(IV): ② $\times (-1)$.   (IV)$\Rightarrow$(III): Exchange ① and ②.
(III)$\Rightarrow$(II): ② + ① $\times (-2)$.   (II)$\Rightarrow$(I): ① + ② $\times 2$.

2.2

1 (1) $\begin{pmatrix} 0 & 0 & 1 \\ 0 & 0 & 0 \end{pmatrix}$. (2) $\begin{pmatrix} 1 & 0 & -5 \\ 0 & 1 & 1 \\ 0 & 0 & 0 \end{pmatrix}$. (3) $\begin{pmatrix} 0 & 1 & 0 \\ 0 & 0 & 1 \\ 0 & 0 & 0 \end{pmatrix}$.

(4) $\begin{pmatrix} 1\,0\,0 & 1 \\ 0\,1\,0 & -1/2 \\ 0\,0\,1 & 1 \end{pmatrix}$.   (5) A reduced matrix.   (6) $\begin{pmatrix} 1\,0\,0\,0 \\ 0\,1\,0\,0 \\ 0\,0\,1\,0 \end{pmatrix}$.

(7) $\begin{pmatrix} 1\,0\,0\,0 \\ 0\,0\,0\,1 \\ 0\,0\,0\,0 \end{pmatrix}$.   (8) $\begin{pmatrix} 0\,1\,0\,0 \\ 0\,0\,1\,0 \\ 0\,0\,0\,1 \end{pmatrix}$.   (9) $\begin{pmatrix} 1\,0 & 0\,3 \\ 0\,1 & -1\,2 \\ 0\,0 & 0\,0 \end{pmatrix}$.

(10) $\begin{pmatrix} 1\,1\,0\,0 \\ 0\,0\,1\,0 \\ 0\,0\,0\,1 \end{pmatrix}$.   (11) $\begin{pmatrix} 1\,0\,0\,0 \\ 0\,1\,1\,0 \\ 0\,0\,0\,1 \end{pmatrix}$.   (12) $\begin{pmatrix} 0\,1\,0\,0 & 2 \\ 0\,0\,1\,0 & -4 \\ 0\,0\,0\,1 & -1 \\ 0\,0\,0\,0 & 0 \end{pmatrix}$.

2   Let A be a reduced square matrix of degree 2. First $a_{11} = 0$ or 1. Assume that $a_{11} = 0$. Then $a_{12} = 0$ or 1. If $a_{12} = 0$, then $A = \begin{pmatrix} 0\,0 \\ 0\,0 \end{pmatrix}$. If $a_{12} = 1$, then $A = \begin{pmatrix} 0\,1 \\ 0\,0 \end{pmatrix}$. Assume that $a_{11} = 1$. Then $a_{21} = 0$. Furthermore, $a_{22} = 0$ or 1. If $a_{22} = 0$, then we can take any number for a_{12} and $A = \begin{pmatrix} 1\,* \\ 0\,0 \end{pmatrix}$. If $a_{22} = 1$, then we have $A = \begin{pmatrix} 1\,0 \\ 0\,1 \end{pmatrix}$.

3   Using an argument similar to that for question 2, we obtain all the reduced 2×4 matrices:

$\begin{pmatrix} 0\,0\,0\,0 \\ 0\,0\,0\,0 \end{pmatrix}$, $\begin{pmatrix} 0\,0\,0\,1 \\ 0\,0\,0\,0 \end{pmatrix}$, $\begin{pmatrix} 0\,0\,1\,0 \\ 0\,0\,0\,1 \end{pmatrix}$, $\begin{pmatrix} 0\,0\,1\,* \\ 0\,0\,0\,0 \end{pmatrix}$,

$\begin{pmatrix} 0\,1\,0\,* \\ 0\,0\,1\,* \end{pmatrix}$, $\begin{pmatrix} 0\,1\,*\,0 \\ 0\,0\,0\,1 \end{pmatrix}$, $\begin{pmatrix} 0\,1\,*\,* \\ 0\,0\,0\,0 \end{pmatrix}$, $\begin{pmatrix} 1\,0\,*\,* \\ 0\,1\,*\,* \end{pmatrix}$,

$\begin{pmatrix} 1\,*\,0\,* \\ 0\,0\,1\,* \end{pmatrix}$, $\begin{pmatrix} 1\,*\,*\,0 \\ 0\,0\,0\,1 \end{pmatrix}$, $\begin{pmatrix} 1\,*\,*\,* \\ 0\,0\,0\,0 \end{pmatrix}$.

4   Using an argument similar to that for question 2, we obtain all the reduced square matrices of degree 3:

$\begin{pmatrix} 0\,0\,0 \\ 0\,0\,0 \\ 0\,0\,0 \end{pmatrix}$, $\begin{pmatrix} 0\,0\,1 \\ 0\,0\,0 \\ 0\,0\,0 \end{pmatrix}$, $\begin{pmatrix} 0\,1\,0 \\ 0\,0\,1 \\ 0\,0\,0 \end{pmatrix}$, $\begin{pmatrix} 0\,1\,* \\ 0\,0\,0 \\ 0\,0\,0 \end{pmatrix}$,

$\begin{pmatrix} 1\,0\,0 \\ 0\,1\,0 \\ 0\,0\,1 \end{pmatrix}$, $\begin{pmatrix} 1\,0\,* \\ 0\,1\,* \\ 0\,0\,0 \end{pmatrix}$, $\begin{pmatrix} 1\,*\,0 \\ 0\,0\,1 \\ 0\,0\,0 \end{pmatrix}$, $\begin{pmatrix} 1\,*\,* \\ 0\,0\,0 \\ 0\,0\,0 \end{pmatrix}$.

5   (1) $\begin{pmatrix} 1\,0 \\ 0\,1 \end{pmatrix}$, rank $= 2$.   (2) $\begin{pmatrix} 1\,0 & 5 \\ 0\,1 & -4 \end{pmatrix}$, rank $= 2$.

(3) $\begin{pmatrix} 1\,0 & -1 \\ 0\,1 & 0 \end{pmatrix}$, rank $= 2$.   (4) $\begin{pmatrix} 1\,0\,0 & 0 \\ 0\,1\,0 & -1/2 \\ 0\,0\,1 & 1/2 \end{pmatrix}$, rank $= 3$.

(5) $\begin{pmatrix} 1&0&0&3 \\ 0&1&0&1 \\ 0&0&1&0 \end{pmatrix}$, rank = 3. (6) $\begin{pmatrix} 1&0&1&1 \\ 0&1&3&1 \\ 0&0&0&0 \end{pmatrix}$, rank = 2.

(7) $\begin{pmatrix} 1&2&0&1/2 \\ 0&0&1&1/2 \\ 0&0&0&0 \end{pmatrix}$, rank = 2. (8) $\begin{pmatrix} 1&0&0&0&1 \\ 0&1&0&-1&3 \\ 0&0&1&0&-1 \\ 0&0&0&0&0 \end{pmatrix}$, rank = 3.

(9) $\begin{pmatrix} 1&0&1&2&0 \\ 0&1&-1&-1&0 \\ 0&0&0&0&1 \\ 0&0&0&0&0 \end{pmatrix}$, rank = 3. (10) $\begin{pmatrix} 1&2&0&-1&1 \\ 0&0&1&1&2 \\ 0&0&0&0&0 \\ 0&0&0&0&0 \end{pmatrix}$, rank = 2.

(11) $\begin{pmatrix} 1&0&1&0&0 \\ 0&1&2&0&0 \\ 0&0&0&1&0 \\ 0&0&0&0&1 \end{pmatrix}$, rank = 4. (12) $\begin{pmatrix} 1&2&3&0&1 \\ 0&0&0&1&1 \\ 0&0&0&0&0 \\ 0&0&0&0&0 \end{pmatrix}$, rank = 2.

2.3

1 (1) $\begin{pmatrix} x_1 \\ x_2 \\ x_3 \end{pmatrix} = \begin{pmatrix} -2 \\ 0 \\ 1 \end{pmatrix} + c\begin{pmatrix} 1 \\ 1 \\ 0 \end{pmatrix}$ $(c \in \mathbf{R})$. (2) $\begin{pmatrix} x_1 \\ x_2 \\ x_3 \end{pmatrix} = \begin{pmatrix} 1 \\ 3 \\ 0 \end{pmatrix} + c\begin{pmatrix} -3 \\ -1 \\ 1 \end{pmatrix}$

$(c \in \mathbf{R})$. (3) No solution. (4) No solution.

(5) $\begin{pmatrix} x_1 \\ x_2 \\ x_3 \\ x_4 \end{pmatrix} = \begin{pmatrix} 1 \\ 1 \\ 0 \\ -1 \end{pmatrix} + c\begin{pmatrix} -3 \\ -2 \\ 1 \\ 0 \end{pmatrix}$ $(c \in \mathbf{R})$. (6) No solution.

(7) $\begin{pmatrix} x_1 \\ x_2 \\ x_3 \\ x_4 \\ x_5 \end{pmatrix} = \begin{pmatrix} 2 \\ 1 \\ 0 \\ 0 \\ 0 \end{pmatrix} + c_1\begin{pmatrix} -3 \\ 2 \\ 1 \\ 0 \\ 0 \end{pmatrix} + c_2\begin{pmatrix} -1 \\ 0 \\ 0 \\ 1 \\ 0 \end{pmatrix} + c_3\begin{pmatrix} 2 \\ -2 \\ 0 \\ 0 \\ 1 \end{pmatrix}$ $(c_1, c_2, c_3 \in \mathbf{R})$.

(8) $\begin{pmatrix} x_1 \\ x_2 \\ x_3 \\ x_4 \\ x_5 \end{pmatrix} = \begin{pmatrix} -1 \\ 2 \\ 0 \\ 1 \\ 0 \end{pmatrix} + c_1\begin{pmatrix} 2 \\ -1 \\ 1 \\ 0 \\ 0 \end{pmatrix} + c_2\begin{pmatrix} -1 \\ 2 \\ 0 \\ 0 \\ 1 \end{pmatrix}$ $(c_1, c_2 \in \mathbf{R})$.

2 (1) $\begin{pmatrix} x_1 \\ x_2 \\ x_3 \end{pmatrix} = c \begin{pmatrix} -1 \\ -2 \\ 1 \end{pmatrix}$ $(c \in \mathbf{R})$. (2) $\begin{pmatrix} x_1 \\ x_2 \\ x_3 \end{pmatrix} = c \begin{pmatrix} -1 \\ 1 \\ 1 \end{pmatrix}$ $(c \in \mathbf{R})$.

(3) $\begin{pmatrix} x_1 \\ x_2 \\ x_3 \\ x_4 \end{pmatrix} = c_1 \begin{pmatrix} -2 \\ 1 \\ 1 \\ 0 \end{pmatrix} + c_2 \begin{pmatrix} 1 \\ -1 \\ 0 \\ 1 \end{pmatrix}$ $(c_1, c_2 \in \mathbf{R})$.

(4) $\begin{pmatrix} x_1 \\ x_2 \\ x_3 \\ x_4 \\ x_5 \end{pmatrix} = c_1 \begin{pmatrix} 2 \\ 1 \\ 0 \\ 0 \\ 0 \end{pmatrix} + c_2 \begin{pmatrix} -1 \\ 0 \\ -1 \\ 1 \\ 0 \end{pmatrix} + c_3 \begin{pmatrix} -2 \\ 0 \\ -1 \\ 0 \\ 1 \end{pmatrix}$ $(c_1, c_2, c_3 \in \mathbf{R})$.

3 (1) $a + 2b = 1$. (2) $a \neq 2$.

4 Since $A(\mathbf{x}_0 + \mathbf{x}_1) = A\mathbf{x}_0 + A\mathbf{x}_1 = \mathbf{b} + \mathbf{0} = \mathbf{b}$, $\mathbf{x}_0 + \mathbf{x}_1$ is a solution of $(*)$. Conversely, if $\mathbf{x}$ is a solution of $(*)$, then $A(\mathbf{x} - \mathbf{x}_0) = A\mathbf{x} - A\mathbf{x}_0 = \mathbf{b} - \mathbf{b} = \mathbf{0}$. Therefore, if we put $\mathbf{x}_1 = \mathbf{x} - \mathbf{x}_0$, then $\mathbf{x}_1$ is a solution of $(**)$ and $\mathbf{x} = \mathbf{x}_0 + \mathbf{x}_1$.

2.4

1 (1) $\begin{pmatrix} 1 & -1 \\ 1 & -2 \end{pmatrix}$. (2) $\begin{pmatrix} -5 & -6 \\ -1 & -1 \end{pmatrix}$. (3) $\begin{pmatrix} -3/2 & -5/2 \\ 1/2 & 1/2 \end{pmatrix}$. (4) $\begin{pmatrix} -1 & 0 & -2 \\ 0 & 1 & 0 \\ -1 & 0 & -1 \end{pmatrix}$.

(5) $\begin{pmatrix} 0 & 0 & 1 \\ 0 & 1 & -1 \\ 1 & -1 & 0 \end{pmatrix}$. (6) $\begin{pmatrix} 1 & -1 & 1 \\ 1 & -2 & 2 \\ 1 & -1 & 0 \end{pmatrix}$. (7) $\begin{pmatrix} -1 & 0 & -2 \\ -1 & -1 & 0 \\ -4 & -3 & -3 \end{pmatrix}$.

(8) $\begin{pmatrix} 3/2 & 1/2 & 1 \\ -1 & 1/2 & -1/2 \\ 1/2 & 0 & 1/2 \end{pmatrix}$. (9) $\begin{pmatrix} -1/3 & 1/3 & 1/3 \\ 1/3 & 2/3 & -1/3 \\ -2/3 & 5/3 & -1/3 \end{pmatrix}$.

(10) $\begin{pmatrix} 1 & -1 & 0 & 0 \\ 0 & 1 & -1 & 0 \\ 0 & 0 & 1 & -1 \\ 0 & 0 & 0 & 1 \end{pmatrix}$. (11) $\begin{pmatrix} 1 & 0 & -1 & 0 \\ -1 & -3 & 2 & -2 \\ -1 & 0 & 2 & 0 \\ 0 & 1 & 0 & 1 \end{pmatrix}$.

2 (1) Since $A^{-1} = \begin{pmatrix} 1 & -2 \\ 3 & -7 \end{pmatrix}$, we have $\mathbf{x} = \begin{pmatrix} 1 & -2 \\ 3 & -7 \end{pmatrix}\begin{pmatrix} 3 \\ 4 \end{pmatrix} = \begin{pmatrix} -5 \\ -19 \end{pmatrix}$.

(2) Since $A^{-1} = \begin{pmatrix} 1 & 0 & 2 \\ 1 & 1 & 4 \\ -1 & 1 & -1 \end{pmatrix}$, we have $\mathbf{x} = \begin{pmatrix} 5 \\ 11 \\ -2 \end{pmatrix}$.

(3) Since $A^{-1} = \begin{pmatrix} -1 & -2 & 1 \\ 0 & -1 & -1 \\ -1 & -2 & 0 \end{pmatrix}$, we have $\mathbf{x} = \begin{pmatrix} -8 \\ 1 \\ -5 \end{pmatrix}$.

(4) Since $A^{-1} = \begin{pmatrix} 5/2 & -2 & -1/2 \\ -1 & 1 & 0 \\ -9/2 & 4 & 3/2 \end{pmatrix}$, we have $\mathbf{x} = \begin{pmatrix} 7 \\ -3 \\ -11 \end{pmatrix}$.

(5) Since $A^{-1} = \begin{pmatrix} 1 & -1 & 2 \\ -1 & 1 & -1 \\ 2 & -3 & 7 \end{pmatrix}$, we have $\mathbf{x} = \begin{pmatrix} a-b+2c \\ -a+b-c \\ 2a-3b+7c \end{pmatrix}$.

3 (1) Since $A^{-1}A = E_n$, A^{-1} is a regular matrix and A is the inverse matrix of A^{-1} by Theorem 2.4.1. Therefore $(A^{-1})^{-1} = A$.

(2) Taking the transpositions of both sides of $A^{-1}A = E_n$, we obtain ${}^tA\,{}^t(A^{-1}) = {}^tE_n = E_n$. Therefore ${}^tA$ is a regular matrix by Theorem 2.4.1 and $({}^tA)^{-1} = {}^t(A^{-1})$.

(3) Since $AB(B^{-1}A^{-1}) = ABB^{-1}A^{-1} = AA^{-1} = E_n$, AB is a regular matrix and $(AB)^{-1} = B^{-1}A^{-1}$ by Theorem 2.4.1.

4 (1) Multiplying $AB = BA$ by A^{-1} from the left and from the right, we obtain $A^{-1}ABA^{-1} = A^{-1}BAA^{-1}$. Therefore $BA^{-1} = A^{-1}B$.

(2) Taking the inverse matrices of both sides of $AB = BA$, we obtain $(AB)^{-1} = (BA)^{-1}$. Since $(AB)^{-1} = B^{-1}A^{-1}$ and $(BA)^{-1} = A^{-1}B^{-1}$, we have $B^{-1}A^{-1} = A^{-1}B^{-1}$.

(3) Taking the transpositions of both sides of $AB = BA$, we obtain ${}^t(AB) = {}^t(BA)$. Since ${}^t(AB) = {}^tB\,{}^tA$ and ${}^t(BA) = {}^tA\,{}^tB$, we have ${}^tB\,{}^tA = {}^tA\,{}^tB$.

5 Assume that A is a regular matrix. Multiplying $AB = O$ by A^{-1} from the left, we obtain $B = A^{-1}AB = A^{-1}O = O$. This contradicts the assumption $B \neq O$. Therefore A is not a regular matrix.

6 We assume that $A^m = O_n$. Then $(E_n - A)(E_n + A + \cdots + A^{m-1}) = E_n$ by Exercises (§1.2)7. Therefore $E_n - A$ is a regular matrix and $(E_n - A)^{-1} = E_n + A + \cdots + A^{m-1}$ by Theorem 2.4.1. Taking $-A$ in place of A, we see that $E_n + A$ is also a regular matrix and $(E_n + A)^{-1} = E_n - A + \cdots + (-1)^{m-1}A^{m-1}$.

7 We shall show that X is a regular matrix. Let P be a square matrix of degree m and S a square matrix of degree n. We also let Q be an $m \times n$ matrix and R an $n \times m$ matrix. Assume that $X\begin{pmatrix} P & Q \\ R & S \end{pmatrix} = \begin{pmatrix} E_m & O_{m,n} \\ O_{n,m} & E_n \end{pmatrix}$. Since

$X\begin{pmatrix} P & Q \\ R & S \end{pmatrix} = \begin{pmatrix} A & B \\ O_{n,m} & D \end{pmatrix}\begin{pmatrix} P & Q \\ R & S \end{pmatrix} = \begin{pmatrix} AP + BR & AQ + BS \\ DR & DS \end{pmatrix}$, we see that

$AP + BR = E_m$, $AQ + BS = O_{m,n}$, $DR = O_{n,m}$, $DS = E_n$, namely, $P = A^{-1}$, $Q = -A^{-1}BD^{-1}$, $R = O_{n,m}$, $S = D^{-1}$ and $X\begin{pmatrix} A^{-1} & -A^{-1}BD^{-1} \\ O_{n,m} & D^{-1} \end{pmatrix} = \begin{pmatrix} E_m & O_{m,n} \\ O_{n,m} & E_n \end{pmatrix} = E_{m+n}$. Thus we see that X is a regular matrix and $X^{-1} = \begin{pmatrix} A^{-1} & -A^{-1}BD^{-1} \\ O_{n,m} & D^{-1} \end{pmatrix}$ by Theorem 2.4.1. By a similar argument, the matrices Y

and Z are also regular matrices, and we have $Y^{-1} = \begin{pmatrix} A^{-1} & O_{m,n} \\ -D^{-1}CA^{-1} & D^{-1} \end{pmatrix}$ and

$Z^{-1} = \begin{pmatrix} O_{n,m} & D^{-1} \\ A^{-1} & -A^{-1}BD^{-1} \end{pmatrix}$.

3.1 For 1 through 4, the answers are not unique.

1 (1) ε.   (2) $\begin{pmatrix} 1\,2\,3\,4 \\ 3\,1\,2\,4 \end{pmatrix} (= (1\,3\,2))$.   (3) $\begin{pmatrix} 1\,2\,3\,4 \\ 4\,3\,1\,2 \end{pmatrix} (= (1\,4\,2\,3))$.

   (4) $\begin{pmatrix} 1\,2\,3\,4\,5 \\ 1\,2\,3\,5\,4 \end{pmatrix} (= (4\,5))$.   (5) $\begin{pmatrix} 1\,2\,3\,4 \\ 3\,4\,2\,1 \end{pmatrix} (= (1\,3\,2\,4))$.

   (6) $\begin{pmatrix} 1\,2\,3\,4 \\ 4\,3\,1\,2 \end{pmatrix} (= (1\,4\,2\,3))$.   (7) $\begin{pmatrix} 1\,2\,3\,4 \\ 3\,4\,2\,1 \end{pmatrix} (= (1\,3\,2\,4))$.

   (8) $\begin{pmatrix} 1\,2\,3\,4\,5 \\ 4\,2\,5\,1\,3 \end{pmatrix} \left(= \begin{pmatrix} 1\,3\,4\,5 \\ 4\,5\,1\,3 \end{pmatrix}\right)$.

2 (1) $(2\,7\,3\,6)(1\,4\,5)$.   (2) $(4\,8\,7\,6)(1\,3\,5\,2)$.
   (3) $(5\,6)(2\,8\,4)(1\,7\,3)$.   (4) $(4\,7\,8\,6)(3\,5)(1\,9\,2)$.
   (5) $(4\,9)(2\,5\,3\,7)(1\,8\,6)$.   (6) $(3\,9\,7\,8)(2\,6)(1\,4\,5\,10)$.

3 (1) $(1\,4)(1\,2)(1\,3)(1\,6)$.
   (2) $(2\,7)(2\,4)(2\,6)(2\,3)(2\,5)$.
   (3) $(1\,4)(1\,6)(1\,7)(1\,3)(1\,5)(1\,2)$.
   (4) $(1\,8)(1\,7)(1\,6)(1\,5)(1\,4)(1\,3)(1\,2)$.

4   Let σ be a cyclic permutation of length r. We denote it by $\sigma = (a_1\,a_2\,\cdots\,a_r)$. Since σ is expressed as a product of $r-1$ transpositions $\sigma = (a_1\,a_r)(a_1\,a_{r-1})\cdots(a_1\,a_2)$, we have $\mathrm{sgn}(\sigma) = (-1)^{r-1}$.

5   We denote the given permutations by σ.
   (1) $\sigma = (2\,4)(1\,3)$, $\mathrm{sgn}(\sigma) = (-1)^2 = 1$.
   (2) $\sigma = (1\,3)(1\,4)(1\,2)$, $\mathrm{sgn}(\sigma) = (-1)^3 = -1$.
   (3) $\sigma = (2\,3)(2\,4)(1\,5)$, $\mathrm{sgn}(\sigma) = (-1)^3 = -1$.
   (4) $\sigma = (3\,6)(2\,7)(2\,4)$, $\mathrm{sgn}(\sigma) = (-1)^3 = -1$.
   (5) $\sigma = (3\,5)(2\,4)(1\,6)$, $\mathrm{sgn}(\sigma) = (-1)^3 = -1$.
   (6) $\sigma = (3\,4)(1\,5)(1\,2)(1\,6)$, $\mathrm{sgn}(\sigma) = (-1)^4 = 1$.
   (7) $\sigma = (1\,4)(1\,2)(1\,5)(1\,3)(1\,6)$, $\mathrm{sgn}(\sigma) = (-1)^5 = -1$.
   (8) $\sigma = (2\,5)(2\,3)(2\,6)(1\,4)$, $\mathrm{sgn}(\sigma) = (-1)^4 = 1$.
   (9) $\sigma = (2\,6)(2\,7)(1\,4)(1\,5)(1\,3)$, $\mathrm{sgn}(\sigma) = (-1)^5 = -1$.
   (10) $\sigma = (1\,5)(1\,4)(1\,6)(1\,3)(1\,7)(1\,2)$, $\mathrm{sgn}(\sigma) = (-1)^6 = 1$.
   (11) $\sigma = (3\,9)(3\,5)(2\,6)(2\,8)(1\,7)(1\,4)$, $\mathrm{sgn}(\sigma) = (-1)^6 = 1$.

6 $\#(S_4) = 4! = 24$. $S_4 = \{\varepsilon, (1\ 2\ 3), (1\ 2\ 4), (1\ 3\ 2), (1\ 3\ 4),$
$(1\ 4\ 2), (1\ 4\ 3), (2\ 3\ 4), (2\ 4\ 3), (1\ 2)(3\ 4),$
$(1\ 3)(2\ 4), (1\ 4)(2\ 3), (1\ 2), (1\ 3), (1\ 4), (2\ 3), (2\ 4),$
$(3\ 4), (1\ 2\ 3\ 4), (1\ 2\ 4\ 3), (1\ 3\ 2\ 4), (1\ 3\ 4\ 2),$
$(1\ 4\ 2\ 3), (1\ 4\ 3\ 2)\}$.

7 $\#(A_4) = 4!/2 = 12$. $A_4 = \{\varepsilon, (1\ 2\ 3), (1\ 2\ 4), (1\ 3\ 2), (1\ 3\ 4),$
$(1\ 4\ 2), (1\ 4\ 3), (2\ 3\ 4), (2\ 4\ 3), (1\ 2)(3\ 4),$
$(1\ 3)(2\ 4), (1\ 4)(2\ 3)\}$.

8 (1) $(\sigma f)(x_1, x_2, x_3) = x_1 x_2 + 2x_1 + 3x_2 x_3$.

(2) $(\sigma f)(x_1, x_2, x_3) = 2x_1 x_2 + x_3^2 + x_2 x_3$.

(3) $(\sigma f)(x_1, x_2, x_3) = 2x_1 x_2 + x_3^2 + x_1 x_3$.

(4) $(\sigma f)(x_1, x_2, x_3) = (x_3 - x_1)(x_3 - x_2)(x_1 - x_2) = f(x_1, x_2, x_3)$.

9 (1) $(\sigma f)(x_1, x_2, x_3, x_4) = x_4^3 x_3 - x_3 x_2 + 2x_4 x_3 x_1 + 3x_3 x_2 x_1^2$
$= 3x_1^2 x_2 x_3 + 2x_1 x_3 x_4 - x_2 x_3 + x_3 x_4^3$.

(2) $(\sigma f)(x_1, x_2, x_3, x_4) = x_3 x_2 x_1 - x_2 x_4 - 2x_3^2 x_4 + 3x_3 x_2 x_4 + 4x_2 x_1^2$
$= 4x_1^2 x_2 + x_1 x_2 x_3 + 3x_2 x_3 x_4 - x_2 x_4 - 2x_3^2 x_4$.

3.2

1 (1) -2.   (2) 1.   (3) $ad - bc$.   (4) -2.   (5) -1.

2 (1) -2.   (2) -774.   (3) -12720.   (4) 81.   (5) 76.   (6) 0.   (7) $-11/864$.
(8) 600.   (9) 88.   (10) -8910.   (11) 192.   (12) 16.   (13) -19.
(14) $(-1)^m$ $(n = 2m)$, $(-1)^m$ $(n = 2m + 1)$.

3.3

1 (1) -1470.   (2) -1344.   (3) -57.   (4) -16.   (5) 72.   (6) 1310.
(7) -630.
(8) -1620.   (9) -2988.   (10) 252.   (11) -528.   (12) 18.   (13) 1680.
(14) -105.

2 Taking the determinants of both sides of $AA^{-1} = E$, we have $|A| \cdot |A^{-1}| = 1$.
Therefore $|A| \neq 0$ and $|A^{-1}| = |A|^{-1}$.

3 $\begin{vmatrix} a & b \\ b & a \end{vmatrix} \cdot \begin{vmatrix} c & d \\ d & c \end{vmatrix} = \left| \begin{pmatrix} a & b \\ b & a \end{pmatrix} \begin{pmatrix} c & d \\ d & c \end{pmatrix} \right| = \begin{vmatrix} ac + bd & ad + bc \\ bc + ad & bd + ac \end{vmatrix} = (ac + bd)^2 -$
$(ad + bc)^2$. On the other hand, $\begin{vmatrix} a & b \\ b & a \end{vmatrix} \cdot \begin{vmatrix} c & d \\ d & c \end{vmatrix} = (a^2 - b^2)(c^2 - d^2)$.

4 By exchanging the 1st column with the $(n + 1)$-th column,..., and the
n-th column with the $(2n)$-th column, we have $\begin{vmatrix} A & B \\ C & O_n \end{vmatrix} = (-1)^n \begin{vmatrix} B & A \\ O_n & C \end{vmatrix} =$
$(-1)^n |B| \cdot |C|$ by Theorem 3.3.4.

5 By exchanging the adjacent columns, we move the $(s+1)$-th column to the position of the 1st column. Then the determinant changes to $(-1)^s$ times the previous determinant. By exchanging the adjacent columns, we also move the $(s+2)$-th column to the position of the second column. Then the determinant changes to $(-1)^s$ times the previous determinant. Continuing this procedure until moving the $(s+r)$-th column to the r-th column, we have $\begin{vmatrix} O_{r,s} & B \\ C & D \end{vmatrix} =$

$(-1)^{rs} \begin{vmatrix} B & O_{r,s} \\ D & C \end{vmatrix} = (-1)^{rs} |B| \cdot |C|$ by Theorem 3.3.4.

6 By adding $-\{\text{the } (n+1)\text{-th row}\}$ to the 1st row, ..., and $-\{\text{the } (2n)\text{-th row}\}$ to the n-th row, we have $\begin{vmatrix} A & B \\ B & A \end{vmatrix} = \begin{vmatrix} A-B & B-A \\ B & A \end{vmatrix}$. Next add the 1st column to the $(n+1)$-th column, ..., and the n-th column to the $(2n)$-th column. We see that $\begin{vmatrix} A & B \\ B & A \end{vmatrix} = \begin{vmatrix} A-B & O_n \\ B & A+B \end{vmatrix} = |A-B| \cdot |A+B|$ by Theorem 3.3.4.

7 Taking the determinants of both sides of $A^m = E_n$, we have $|A|^m = 1$ by Theorem 3.3.5. Therefore $|A| = \pm 1$. Since m is odd, we have $|A| = 1$.

3.4

1 (1) $\tilde{A} = \begin{pmatrix} 2 & 40 \\ -14 & -19 \\ -6 & -39 \end{pmatrix}$, $A^{-1} = \dfrac{1}{18}\tilde{A}$. (2) $\tilde{A} = \begin{pmatrix} -3 & 9 & 6 \\ 1 & -2 & -1 \\ 5 & -10 & -8 \end{pmatrix}$,

$A^{-1} = \dfrac{1}{3}\tilde{A}$.

(3) $\tilde{A} = \begin{pmatrix} 6 & 9 & -3 \\ 14 & 16 & -4 \\ 2 & 1 & -1 \end{pmatrix}$, $A^{-1} = -\dfrac{1}{6}\tilde{A}$.

(4) $\tilde{A} = \begin{pmatrix} bc & 0 & 0 \\ -cd & ac & 0 \\ df-be & -af & ab \end{pmatrix}$, $A^{-1} = \dfrac{1}{abc}\tilde{A}$ if $abc \neq 0$.

(5) $\tilde{A} = \begin{pmatrix} x & 0 & -x \\ -2(x-1) & x & -(x-1)(x-2) \\ 2(2x-1) & -x & (2x-1)(x-2) \end{pmatrix}$, $A^{-1} = \dfrac{1}{x^2}\tilde{A}$ if $x \neq 0$.

2 (1) $\mathbf{x} = \dfrac{1}{9}\begin{pmatrix} 7 \\ 5 \\ 3 \end{pmatrix} = \begin{pmatrix} 7/9 \\ 5/9 \\ 1/3 \end{pmatrix}$. (2) $\mathbf{x} = \dfrac{1}{-2}\begin{pmatrix} -4 \\ -3 \\ 1 \end{pmatrix} = \begin{pmatrix} 2 \\ 3/2 \\ -1/2 \end{pmatrix}$.

(3) $\mathbf{x} = \dfrac{1}{14}\begin{pmatrix} -2 \\ 23 \\ 6 \end{pmatrix} = \begin{pmatrix} -1/7 \\ 23/14 \\ 3/7 \end{pmatrix}$. (4) $\mathbf{x} = \dfrac{1}{-8}\begin{pmatrix} -8 \\ -16 \\ -24 \end{pmatrix} = \begin{pmatrix} 1 \\ 2 \\ 3 \end{pmatrix}$.

3 We denote the given matrices by A.

(1) $|A| = -6 \begin{vmatrix} 2 & 3 \\ 1 & -2 \end{vmatrix} + 5 \begin{vmatrix} 2 & 1 \\ 1 & 3 \end{vmatrix} (= (-6)(-4-3) + 5(6-1) = 67)$.

(2) $|A| = 3 \begin{vmatrix} 0 & 2 \\ -2 & 4 \end{vmatrix} - \begin{vmatrix} 1 & 2 \\ -2 & 4 \end{vmatrix} + (-1) \begin{vmatrix} 1 & 2 \\ 0 & 2 \end{vmatrix} (= 3 \cdot 4 - 8 - 2 = 2)$.

(3) $|A| = a \begin{vmatrix} 1 & 0 \\ 4 & -1 \end{vmatrix} - b \begin{vmatrix} -2 & 0 \\ 3 & -1 \end{vmatrix} + c \begin{vmatrix} -2 & 1 \\ 3 & 4 \end{vmatrix} (= -a - 2b - 11c)$.

(4) $|A| = -x \begin{vmatrix} 3 & 2 \\ 2 & 1 \end{vmatrix} + y \begin{vmatrix} 1 & -1 \\ 2 & 1 \end{vmatrix} - z \begin{vmatrix} 1 & -1 \\ 3 & 2 \end{vmatrix} (= x + 3y - 5z)$.

4 We denote the given matrices by A. We note that calculations are not unique and we can obtain the determinant by other cofactor expansions.

(1) By the cofactor expansion with respect to the first column, we have $|A| = -5 \begin{vmatrix} -3 & 2 \\ 0 & 7 \end{vmatrix} = (-5)(-3) \cdot 7 = 105$.

(2) By the cofactor expansion with respect to the second column, we have $|A| = -2 \begin{vmatrix} -7 & 1 \\ 0 & -6 \end{vmatrix} = (-2)(-7)(-6) = -84$.

(3) By the cofactor expansion with respect to the second column, we have $|A| = -(-4) \begin{vmatrix} 0 & 2 & -1 \\ 1 & 0 & 3 \\ 1 & 0 & 2 \end{vmatrix}$. Furthermore, by the cofactor expansion with respect to the second column, we have $|A| = 4(-2) \begin{vmatrix} 1 & 3 \\ 1 & 2 \end{vmatrix} = 4(-2)(-1) = 8$.

(4) By the cofactor expansion with respect to the third column, we have $|A| = -6 \begin{vmatrix} 3 & 0 & -5 \\ 4 & -7 & 1 \\ -1 & 0 & 2 \end{vmatrix} = (-6)(-7) \begin{vmatrix} 3 & -5 \\ -1 & 2 \end{vmatrix} = (-6)(-7)(6-5) = 42$.

5 We show that $|\tilde{A}| = |A|^{n-1}$. If $A = O_n$, then $\tilde{A} = O_n$ and $|\tilde{A}| = 0 = |A|^{n-1}$. If $A (\neq O_n)$ is not a regular matrix, then $A\tilde{A} = |A| E_n = O_n$. Since $A \neq O_n$, $\tilde{A}$ is not a regular matrix. Therefore $|\tilde{A}| = 0 = |A|^{n-1}$. Next, we assume that A is a regular matrix. Taking the determinants of both sides of $A\tilde{A} = |A| E_n$, we have $|A| \cdot |\tilde{A}| = |A|^n$. Since $|A| \neq 0$, we have $|\tilde{A}| = |A|^{n-1}$.

6 If we exchange columns and rows of a matrix A by transposition, we see that $({}^tA)_{ij} = {}^t(A_{ji})$. Therefore, if A is a symmetric matrix, then we have $A_{ij} = ({}^tA)_{ij} = {}^t(A_{ji})$. Then $|A_{ij}| = |{}^t(A_{ji})| = |A_{ji}|$ and $\tilde{a}_{ji} = (-1)^{j+i} |A_{ij}| = (-1)^{i+j} |A_{ji}| = \tilde{a}_{ij}$. Thus we see that $\tilde{A}$ is a symmetric matrix. Furthermore, assume that A is a regular symmetric matrix. Since $A^{-1} = (1/d)\tilde{A}$ $(d = |A|)$ and $\tilde{A}$ is a symmetric matrix, A^{-1} is also a symmetric matrix.

7   Let A be an alternating matrix of degree n. Then $A_{ij} = (-{}^t A)_{ij} = -{}^t (A_{ji})$. Thus we have $\tilde{a}_{ji} = (-1)^{j+i} \left| A_{ij} \right| = (-1)^{i+j} \left| -{}^t (A_{ji}) \right| = (-1)^{n-1} \tilde{a}_{ij}$. Therefore if the degree of A is even, then $\tilde{A}$ is an alternating matrix, but if the degree of A is odd, then $\tilde{A}$ is a symmetric matrix.

8   Let A be an alternating matrix of degree n. Since ${}^t A = -A$, we have $\left| A \right| = \left| {}^t A \right| = \left| -A \right| = (-1)^n \left| A \right|$. Therefore if n is odd, then $\left| A \right| = 0$.

3.5

1 $R(f(x), g(x)) = \begin{vmatrix} a_2 & a_1 & a_0 & 0 \\ 0 & a_2 & a_1 & a_0 \\ b_2 & b_1 & b_0 & 0 \\ 0 & b_2 & b_1 & b_0 \end{vmatrix} = \begin{vmatrix} a_2 & a_1 & a_0 & 0 \\ 0 & a_2 & a_1 & a_0 \\ 0 & b_1 - a_1 b_2/a_2 & b_0 - a_0 b_2/a_2 & 0 \\ 0 & b_2 & b_1 & b_0 \end{vmatrix}$

$= a_2 \begin{vmatrix} a_2 & a_1 & a_0 \\ b_1 - a_1 b_2/a_2 & b_0 - a_0 b_2/a_2 & 0 \\ b_2 & b_1 & b_0 \end{vmatrix} = \begin{vmatrix} a_2 & a_1 & a_0 \\ a_2 b_1 - a_1 b_2 & a_2 b_0 - a_0 b_2 & 0 \\ b_2 & b_1 & b_0 \end{vmatrix}$

$= a_0 \{(a_2 b_1 - a_1 b_2) b_1 - (a_2 b_0 - a_0 b_2) b_2\} + \{a_2 (a_2 b_0 - a_0 b_2) - a_1 (a_2 b_1 - a_1 b_2)\} b_0$

$= (a_2 b_1 - a_1 b_2)(a_0 b_1 - a_1 b_0) + (a_2 b_0 - a_0 b_2)^2.$

2   If $f(x) = 0$ has a multiple root α, then $f(x) = (x - \alpha)^2 g(x)$. Since $f'(x) = 2(x - \alpha)g(x) + (x - \alpha)^2 g'(x)$, we have $f'(\alpha) = 0$. Conversely, we assume that $f(x) = 0$ and $f'(x) = 0$ have a common root α. Then $f(x) = (x - \alpha)g(x)$. Since $f'(x) = g(x) + (x - \alpha)g'(x)$, we have $0 = f'(\alpha) = g(\alpha)$. Therefore $g(x) = (x - \alpha)h(x)$. Then $f(x) = (x - \alpha)g(x) = (x - \alpha)^2 h(x)$ and α is a multiple root of $f(x) = 0$.

3   Since $f(x) = x^3 + px + q$, we have $f'(x) = 3x^2 + p$. Therefore

$R(f(x), f'(x)) = \begin{vmatrix} 1 & 0 & p & q & 0 \\ 0 & 1 & 0 & p & q \\ 3 & 0 & p & 0 & 0 \\ 0 & 3 & 0 & p & 0 \\ 0 & 0 & 3 & 0 & p \end{vmatrix} = \begin{vmatrix} 1 & 0 & p & q & 0 \\ 0 & 1 & 0 & p & q \\ 0 & 0 & -2p & -3q & 0 \\ 0 & 3 & 0 & p & 0 \\ 0 & 0 & 3 & 0 & p \end{vmatrix}$

$= \begin{vmatrix} 1 & 0 & p & q \\ 0 & -2p & -3q & 0 \\ 3 & 0 & p & 0 \\ 0 & 3 & 0 & p \end{vmatrix} = \begin{vmatrix} 1 & 0 & p & q \\ 0 & -2p & -3q & 0 \\ 0 & 0 & -2p & -3q \\ 0 & 3 & 0 & p \end{vmatrix} = \begin{vmatrix} -2p & -3q & 0 \\ 0 & -2p & -3q \\ 3 & 0 & p \end{vmatrix}$

$= -2p \begin{vmatrix} -2p & -3q \\ 0 & p \end{vmatrix} + 3 \begin{vmatrix} -3q & 0 \\ -2p & -3q \end{vmatrix} = 4p^3 + 27q^2.$

Then $D(f(x)) = -R(f(x), f'(x)) = -4p^3 - 27q^2$ by Theorem 3.5.2 (1).

4   They are essentially the Vandermonde determinants.
(1) -240.   (2) -2880.   (3) 12000.   (4) -3456.

5   (1) $\begin{vmatrix} 1 & 1 & 1 \\ a & b & c \\ bc & ca & ab \end{vmatrix} = \begin{vmatrix} 1 & 0 & 0 \\ a & b-a & c-a \\ bc & (a-b)c & (a-c)b \end{vmatrix} = \begin{vmatrix} b-a & c-a \\ (a-b)c & (a-c)b \end{vmatrix}$

$$= (a - b)(c - a) \begin{vmatrix} -1 & 1 \\ c & -b \end{vmatrix} = (a - b)(b - c)(c - a).$$

(2) By the rule of Sarrus, we have $-(a^3 + b^3 + c^3 - 3abc)$. We also have

$$\begin{vmatrix} a & b & c \\ b & c & a \\ c & a & b \end{vmatrix} = \begin{vmatrix} a+b+c & a+b+c & a+b+c \\ b & c & a \\ c & a & b \end{vmatrix} = (a+b+c) \begin{vmatrix} 1 & 1 & 1 \\ b & c & a \\ c & a & b \end{vmatrix} = (a+b+c) \begin{vmatrix} 1 & 0 & 0 \\ b & c-b & a-b \\ c & a-c & b-c \end{vmatrix}$$

$$= (a+b+c) \begin{vmatrix} c-b & a-b \\ a-c & b-c \end{vmatrix} = (a+b+c)(-(b-c)^2 - (a-b)(a-c)) = -(a+b+c)(a^2 + b^2 +$$

$c^2 - ab - bc - ca)$. Therefore, we have the factorization: $a^3 + b^3 + c^3 - 3abc = (a + b + c)(a^2 + b^2 + c^2 - ab - bc - ca)$.

(3) $\begin{vmatrix} a & a+b & a-b \\ a-b & a & a+b \\ a+b & a-b & a \end{vmatrix} = \begin{vmatrix} 3a & 3a & 3a \\ a-b & a & a+b \\ a+b & a-b & a \end{vmatrix} = 3a \begin{vmatrix} 1 & 1 & 1 \\ a-b & a & a+b \\ a+b & a-b & a \end{vmatrix}$

$$= 3a \begin{vmatrix} 1 & 0 & 0 \\ a-b & b & 2b \\ a+b & -2b & -b \end{vmatrix} = 3a \begin{vmatrix} b & 2b \\ -2b & -b \end{vmatrix} = 3ab^2 \begin{vmatrix} 1 & 2 \\ -2 & -1 \end{vmatrix} = 9ab^2.$$

(4) $\begin{vmatrix} 1 & x & x & x \\ 1 & 2 & y & y \\ 1 & 2 & 3 & z \\ 1 & 2 & 3 & 4 \end{vmatrix} = \begin{vmatrix} 1 & x & x & x \\ 1 & 2 & y & y \\ 1 & 2 & 3 & z \\ 0 & 0 & 0 & 4-z \end{vmatrix} = \begin{vmatrix} 1 & x & x & x \\ 1 & 2 & y & y \\ 0 & 0 & 3-y & z-y \\ 0 & 0 & 0 & 4-z \end{vmatrix}$

$$= \begin{vmatrix} 1 & x & x & x \\ 0 & 2-x & y-x & y-x \\ 0 & 0 & 3-y & z-y \\ 0 & 0 & 0 & 4-z \end{vmatrix} = (2-x)(3-y)(4-z).$$

(5) $\begin{vmatrix} x & 0 & 0 & 1 \\ 0 & x & 0 & 2 \\ 0 & 0 & x & 3 \\ 1 & 2 & 3 & x \end{vmatrix} = \begin{vmatrix} 0 & 0 & 0 & 1 \\ -2x & x & 0 & 2 \\ -3x & 0 & x & 3 \\ 1-x^2 & 2 & 3 & x \end{vmatrix} = - \begin{vmatrix} -2x & x & 0 \\ -3x & 0 & x \\ 1-x^2 & 2 & 3 \end{vmatrix}$

$$= - \begin{vmatrix} 0 & x & 0 \\ -3x & 0 & x \\ 5-x^2 & 2 & 3 \end{vmatrix} = x \begin{vmatrix} -3x & x \\ 5-x^2 & 3 \end{vmatrix} = x \begin{vmatrix} 0 & x \\ 14-x^2 & 3 \end{vmatrix} = x^4 - 14x^2.$$

(6) $\begin{vmatrix} 2 & -1 & 0 & 0 \\ -3 & x & -1 & 0 \\ 4 & 0 & x & -1 \\ 5 & 0 & 0 & x \end{vmatrix} = 2 \begin{vmatrix} x & -1 & 0 \\ 0 & x & -1 \\ 0 & 0 & x \end{vmatrix} + \begin{vmatrix} -3 & -1 & 0 \\ 4 & x & -1 \\ 5 & 0 & x \end{vmatrix}$

$$= 2x^3 - 3 \begin{vmatrix} x & -1 \\ 0 & x \end{vmatrix} + \begin{vmatrix} 4 & -1 \\ 5 & x \end{vmatrix} = 2x^3 - 3x^2 + 4x + 5.$$

6 (1)
$$\begin{vmatrix} 1 & 1 & 1 & 1 \\ x & a & a & a \\ x & y & b & b \\ x & y & z & c \end{vmatrix} = \begin{vmatrix} 1 & 1 & 1 & 0 \\ x & a & a & 0 \\ x & y & b & 0 \\ x & y & z & c-z \end{vmatrix} = \begin{vmatrix} 1 & 1 & 0 & 0 \\ x & a & 0 & 0 \\ x & y & b-y & 0 \\ x & y & z-y & c-z \end{vmatrix}$$

$$= \begin{vmatrix} 1 & 0 & 0 & 0 \\ x & a-x & 0 & 0 \\ x & y-x & b-y & 0 \\ x & y-x & z-y & c-z \end{vmatrix} = -(x-a)(y-b)(z-c).$$

(2)
$$\begin{vmatrix} a & b & b & b \\ a & b & a & a \\ a & a & b & a \\ b & b & b & a \end{vmatrix} = \begin{vmatrix} a & b & b & b \\ 0 & 0 & a-b & a-b \\ 0 & a-b & 0 & a-b \\ b-a & 0 & 0 & a-b \end{vmatrix} = (a-b)^3 \begin{vmatrix} a & b & b & b \\ 0 & 0 & 1 & 1 \\ 0 & 1 & 0 & 1 \\ -1 & 0 & 0 & 1 \end{vmatrix}$$

$$= (a-b)^3 \begin{vmatrix} a & b & b & 0 \\ 0 & 0 & 1 & 1 \\ 0 & 1 & 0 & 0 \\ -1 & 0 & 0 & 1 \end{vmatrix} = -(a-b)^3 \begin{vmatrix} a & b & 0 \\ 0 & 1 & 1 \\ -1 & 0 & 1 \end{vmatrix} = -(a-b)^4.$$

(3) Case $a = 0$.
$$\begin{vmatrix} 0 & a & b & c \\ -a & 0 & d & e \\ -b & -d & 0 & f \\ -c & -e & -f & 0 \end{vmatrix} = \begin{vmatrix} 0 & 0 & b & c \\ 0 & 0 & d & e \\ -b & -d & 0 & f \\ -c & -e & -f & 0 \end{vmatrix}$$

$$= (-1)^2 \begin{vmatrix} b & c & 0 & 0 \\ d & e & 0 & 0 \\ 0 & f & -b & -d \\ -f & 0 & -c & -e \end{vmatrix} = \begin{vmatrix} b & c \\ d & e \end{vmatrix} \cdot \begin{vmatrix} b & d \\ c & e \end{vmatrix} = (be-cd)^2$$

$$= (af - be + cd)^2.$$

Case $a \neq 0$.
$$\begin{vmatrix} 0 & a & b & c \\ -a & 0 & d & e \\ -b & -d & 0 & f \\ -c & -e & -f & 0 \end{vmatrix} = \begin{vmatrix} 0 & a & 0 & 0 \\ -a & 0 & d & e \\ -b & -d & bd/a & f+cd/a \\ -c & -e & -f+be/a & ce/a \end{vmatrix}$$

$$= -a \begin{vmatrix} -a & d & e \\ -b & bd/a & f+cd/a \\ -c & -f+be/a & ce/a \end{vmatrix}$$

$$= -a \begin{vmatrix} -a & 0 & 0 \\ -b & 0 & f+cd/a-be/a \\ -c & -f+be/a-cd/a & 0 \end{vmatrix}$$

$$= a^2 \begin{vmatrix} 0 & f+cd/a-be/a \\ -f+be/a-cd/a & 0 \end{vmatrix} = a^2(f+cd/a-be/a)^2$$

$$= (af - be + cd)^2.$$

(4) We denote the determinant by $f_n(x)$. We show the equality by induction on n. Assume $n = 2$. Then $f_2(x) = \begin{vmatrix} 1+x^2 & x \\ x & 1+x^2 \end{vmatrix} = x^4 + x^2 + 1$. We assume

that the equalities hold for the degree less than n. Then

$$f_n(x) = (1+x^2)f_{n-1}(x) - x \begin{vmatrix} x & 0 & 0 & \cdots & \cdots & 0 \\ x & 1+x^2 & x & \ddots & & \vdots \\ 0 & \ddots & \ddots & \ddots & \ddots & \vdots \\ \vdots & \ddots & \ddots & \ddots & \ddots & 0 \\ \vdots & \ddots & \ddots & \ddots & 1+x^2 & x \\ 0 & \cdots & \cdots & 0 & x & 1+x^2 \end{vmatrix} \text{(degree } n-1)$$

$$= (1+x^2)f_{n-1}(x) - x^2 f_{n-2}(x)$$
$$= \{(x^{2(n-1)} + x^{2(n-2)} + \cdots + 1) + (x^{2n} + x^{2(n-1)} + \cdots + x^2)\}$$
$$\qquad\qquad\qquad\qquad -(x^{2(n-1)} + x^{2(n-2)} + \cdots + x^2)$$
$$= x^{2n} + x^{2(n-1)} + \cdots + 1.$$

(5) $\begin{vmatrix} x & a & b & c \\ a & x & b & c \\ a & b & x & c \\ a & b & c & x \end{vmatrix} = \begin{vmatrix} x & a & b & c \\ a & x & b & c \\ a & b & x & c \\ 0 & 0 & c-x & x-c \end{vmatrix} = \begin{vmatrix} x & a & b & c \\ a & x & b & c \\ 0 & b-x & x-b & 0 \\ 0 & 0 & c-x & x-c \end{vmatrix}$

$$= \begin{vmatrix} x & a & b & c \\ a-x & x-a & 0 & 0 \\ 0 & b-x & x-b & 0 \\ 0 & 0 & c-x & x-c \end{vmatrix} = (x-a)(x-b)(x-c) \begin{vmatrix} x & a & b & c \\ -1 & 1 & 0 & 0 \\ 0 & -1 & 1 & 0 \\ 0 & 0 & -1 & 1 \end{vmatrix}$$

$$= (x-a)(x-b)(x-c) \begin{vmatrix} x+a+b+c & a & b & c \\ 0 & 1 & 0 & 0 \\ 0 & -1 & 1 & 0 \\ 0 & 0 & -1 & 1 \end{vmatrix}$$

$$= (x-a)(x-b)(x-c)(x+a+b+c) \begin{vmatrix} 1 & 0 & 0 \\ -1 & 1 & 0 \\ 0 & -1 & 1 \end{vmatrix}$$

$$= (x-a)(x-b)(x-c)(x+a+b+c).$$

(6) $\begin{vmatrix} 0 & a^2 & b^2 & 1 \\ a^2 & 0 & c^2 & 1 \\ b^2 & c^2 & 0 & 1 \\ 1 & 1 & 1 & 0 \end{vmatrix} = \begin{vmatrix} 0 & a^2 & b^2 & 1 \\ 0 & -a^2 & c^2-a^2 & 1 \\ 0 & c^2-b^2 & -b^2 & 1 \\ 1 & 1 & 1 & 0 \end{vmatrix} = -\begin{vmatrix} a^2 & b^2 & 1 \\ -a^2 & c^2-a^2 & 1 \\ c^2-b^2 & -b^2 & 1 \end{vmatrix}$

$$= -\begin{vmatrix} a^2 & b^2 & 1 \\ -2a^2 & c^2-a^2-b^2 & 0 \\ c^2-b^2-a^2 & -2b^2 & 0 \end{vmatrix} = -\begin{vmatrix} -2a^2 & c^2-a^2-b^2 \\ c^2-b^2-a^2 & -2b^2 \end{vmatrix}$$

$$= -4a^2b^2 + (a^2+b^2-c^2)^2 = (a^2+b^2-c^2+2ab)(a^2+b^2-c^2-2ab)$$
$$= ((a+b)^2 - c^2)((a-b)^2 - c^2) = (a+b+c)(a+b-c)(a-b+c)(a-b-c)$$
$$= -(a+b+c)(-a+b+c)(a-b+c)(a+b-c).$$

$$(7) \begin{vmatrix} A & -B \\ B & A \end{vmatrix} = \begin{vmatrix} A+iB & -B+iA \\ B & A \end{vmatrix} = \begin{vmatrix} A+iB & -B+iA-i(A+iB) \\ B & A-iB \end{vmatrix}$$

$$= \begin{vmatrix} A+iB & O_n \\ B & A-iB \end{vmatrix} = |A+iB| \cdot |A-iB| = \text{abs}\left(|A+iB|^2\right).$$

4.1

1 (1) A subspace. (2) Not a subspace. (3) A subspace. (4) Not a subspace.

(5) A subspace. (6) A subspace.

2 (1) A subspace. (2) Not a subspace. (3) A subspace. (4) Not a subspace.

(5) A subspace. (6) A subspace.

3 We check the three conditions of subspaces.

(i) Since $\mathbf{0} \in W_1$ and $\mathbf{0} \in W_2$, we have $\mathbf{0} \in W_1 \cap W_2$.

(ii) Assume that $\mathbf{u}, \mathbf{v} \in W_1 \cap W_2$. Since $\mathbf{u}, \mathbf{v} \in W_1$, we have $\mathbf{u} + \mathbf{v} \in W_1$. Similarly, $\mathbf{u} + \mathbf{v} \in W_2$. Therefore $\mathbf{u} + \mathbf{v} \in W_1 \cap W_2$.

(iii) Assume that $\mathbf{u} \in W_1 \cap W_2$ and $c \in \mathbf{R}$. Since $\mathbf{u} \in W_1$, we have $c\mathbf{u} \in W_1$. Similarly, $c\mathbf{u} \in W_2$. Therefore we have $c\mathbf{u} \in W_1 \cap W_2$.

4 We assume that $W_2 \not\subset W_1$ and $W_1 \not\subset W_2$. Then there exists a vector $\mathbf{u} \in W_1$ but $\mathbf{u} \notin W_2$, and also there exists a vector $\mathbf{v} \in W_2$ but $\mathbf{v} \notin W_1$. Since $W_1 \cup W_2$ is a subspace of V, $\mathbf{u} + \mathbf{v} \in W_1 \cup W_2$. If $\mathbf{u} + \mathbf{v} \in W_1$, then $\mathbf{v} = (\mathbf{u} + \mathbf{v}) - \mathbf{u} \in W_1$. This contradicts $\mathbf{v} \notin W_1$. If $\mathbf{u} + \mathbf{v} \in W_2$, then $\mathbf{u} = (\mathbf{u} + \mathbf{v}) - \mathbf{v} \in W_2$. This contradicts $\mathbf{u} \notin W_2$. Therefore $W_1 \cup W_2$ is not a subspace of V, which contradicts the assumption.

4.2

1 (1) Linearly independent. Let $A = \begin{pmatrix} 1 & 0 & 0 \\ 1 & 1 & 0 \\ 1 & 1 & 1 \end{pmatrix}$, $\mathbf{x} = \begin{pmatrix} x_1 \\ x_2 \\ x_3 \end{pmatrix}$. Since $|A| = 1 \neq 0$, A is a regular matrix. We see that $x_1 \begin{pmatrix} 1 \\ 1 \\ 1 \end{pmatrix} + x_2 \begin{pmatrix} 0 \\ 1 \\ 1 \end{pmatrix} + x_3 \begin{pmatrix} 0 \\ 0 \\ 1 \end{pmatrix} = A\mathbf{x}$. As A is a regular matrix, $A\mathbf{x} = \mathbf{0}_3$ has only the trivial solution by Theorem 2.4.2.

(2) Linearly dependent, since $-3 \begin{pmatrix} 3 \\ 2 \\ 1 \end{pmatrix} + 2 \begin{pmatrix} 2 \\ 1 \\ 3 \end{pmatrix} + \begin{pmatrix} 5 \\ 4 \\ -3 \end{pmatrix} = \begin{pmatrix} 0 \\ 0 \\ 0 \end{pmatrix}$.

(3) Linearly independent. Since $\begin{vmatrix} 2 & 1 & 1 \\ 1 & 1 & 2 \\ 1 & 2 & 1 \end{vmatrix} = -4 \neq 0$, $\begin{pmatrix} 2 & 1 & 1 \\ 1 & 1 & 2 \\ 1 & 2 & 1 \end{pmatrix}$ is a regular matrix. Then these vectors are linearly independent by the same reasoning as in (1).

(4) Linearly dependent, since the matrix $A = \begin{pmatrix} 2 & 3 & 5 & 2 \\ 4 & 1 & 1 & 0 \\ 1 & 2 & 1 & 3 \end{pmatrix}$ is 3×4 matrix, the

linear equation $A\mathbf{x} = \mathbf{0}_3$ has non-trivial solutions by Theorem 2.3.3 (2).

(5) Linearly dependent, since $2 \begin{pmatrix} 2 \\ 1 \\ 1 \\ 4 \end{pmatrix} - 3 \begin{pmatrix} 3 \\ 2 \\ 1 \\ 1 \end{pmatrix} + \begin{pmatrix} 5 \\ 4 \\ 1 \\ -5 \end{pmatrix} = \begin{pmatrix} 0 \\ 0 \\ 0 \\ 0 \end{pmatrix}$.

(6) Linearly independent. Since $\begin{vmatrix} 1 & 1 & 2 & -2 \\ 0 & 1 & 1 & 0 \\ 2 & 0 & 3 & -1 \\ 4 & 3 & 0 & 1 \end{vmatrix} = 30 \neq 0$, $\begin{pmatrix} 1 & 1 & 2 & -2 \\ 0 & 1 & 1 & 0 \\ 2 & 0 & 3 & -1 \\ 4 & 3 & 0 & 1 \end{pmatrix}$ is a regu-

lar matrix. Then these vectors are linearly independent by the same reasoning as
in (1).

(7) Linearly independent. Since $\big(f_1(x) \ f_2(x) \ f_3(x) \big) = \big(1 \ x \ x^2 \big) \times$

$\begin{pmatrix} 1 & 2 & -1 \\ 1 & -1 & 2 \\ 1 & 2 & 1 \end{pmatrix}$ and the vectors $1, \ x, \ x^2$ are linearly independent, we have only to

show the linear independence of $\begin{pmatrix} 1 \\ 1 \\ 1 \end{pmatrix}, \begin{pmatrix} 2 \\ -1 \\ 2 \end{pmatrix}, \begin{pmatrix} -1 \\ 2 \\ 1 \end{pmatrix}$ by Theorem 4.2.6 (2).

Since $\begin{vmatrix} 1 & 2 & -1 \\ 1 & -1 & 2 \\ 1 & 2 & 1 \end{vmatrix} = -6 \neq 0$, $\begin{pmatrix} 1 & 2 & -1 \\ 1 & -1 & 2 \\ 1 & 2 & 1 \end{pmatrix}$ is a regular matrix. Then these vec-

tors are linearly independent by the same reasoning as in (1).

(8) Linearly independent. We see that $\big(f_1(x) \ f_2(x) \ f_3(x) \big) = \big(1 \ x \ x^2 \big) \times$

$\begin{pmatrix} 1 & 5 & 2 \\ 1 & 5 & -1 \\ -1 & -1 & 2 \end{pmatrix}$. Since $\begin{vmatrix} 1 & 5 & 2 \\ 1 & 5 & -1 \\ -1 & -1 & 2 \end{vmatrix} = 12 \neq 0$, these vectors are linearly inde-

pendent by the same reasoning as in (7).

2 (1) $\begin{pmatrix} 1 \\ -2 \end{pmatrix} - 3 \begin{pmatrix} 1 \\ 1 \end{pmatrix} + \begin{pmatrix} 2 \\ 5 \end{pmatrix} = \begin{pmatrix} 0 \\ 0 \end{pmatrix}$.

(2) $-\begin{pmatrix} 2 \\ 1 \\ 1 \end{pmatrix} - 3 \begin{pmatrix} -1 \\ 1 \\ 2 \end{pmatrix} + \begin{pmatrix} -1 \\ 4 \\ 7 \end{pmatrix} = \begin{pmatrix} 0 \\ 0 \\ 0 \end{pmatrix}$.

(3) $-2 \begin{pmatrix} 1 \\ -2 \\ 1 \end{pmatrix} + \begin{pmatrix} 0 \\ -2 \\ 2 \end{pmatrix} - 3 \begin{pmatrix} -2 \\ 2 \\ -1 \end{pmatrix} + \begin{pmatrix} -4 \\ 4 \\ -3 \end{pmatrix} = \begin{pmatrix} 0 \\ 0 \\ 0 \end{pmatrix}$.

(4) $-2 f_1(x) + f_2(x) + f_3(x) = f_{00}(x)$.

(5) $f_1(x) - f_2(x) - 2 f_3(x) + f_4(x) = f_{00}(x)$.

3 (1) $\mathbf{b} = 6\mathbf{a}_1 + \mathbf{a}_2 - 4\mathbf{a}_3$. (2) $\mathbf{b} = -\mathbf{a}_1 + 2\mathbf{a}_2 + \mathbf{a}_3$. (3) $\mathbf{b} = 2\mathbf{a}_1 - \mathbf{a}_2 + 3\mathbf{a}_3$.

4 (1) $g(x) = 2f_1(x) - f_2(x) + 2f_3(x)$.   (2) $g(x) = -f_1(x) + f_2(x) - 2f_3(x)$.
(3) $g(x) = 2f_1(x) - f_2(x) + f_3(x)$.

5 (1) $\left(\mathbf{v}_1\ \mathbf{v}_2\ \mathbf{v}_3 \right) = \left(\mathbf{u}_1\ \mathbf{u}_2\ \mathbf{u}_3 \right) \begin{pmatrix} 2 & 1 & 1 \\ 1 & -1 & 2 \\ -3 & 1 & 4 \end{pmatrix}$.

(2) $\left(\mathbf{v}_1\ \mathbf{v}_2\ \mathbf{v}_3 \right) = \left(\mathbf{u}_1\ \mathbf{u}_2\ \mathbf{u}_3 \right) \begin{pmatrix} 1 & 2 & -1 \\ -1 & 2 & -7 \\ 2 & 0 & 6 \end{pmatrix}$.

(3) $\left(\mathbf{v}_1\ \mathbf{v}_2\ \mathbf{v}_3 \right) = \left(\mathbf{u}_1\ \mathbf{u}_2\ \mathbf{u}_3 \right) \begin{pmatrix} -1 & 1 & 3 \\ -2 & 1 & 2 \\ -1 & -2 & -3 \end{pmatrix}$.

6 (1) Since $\left(\mathbf{v}_1\ \mathbf{v}_2\ \mathbf{v}_3 \right) = \left(\mathbf{u}_1\ \mathbf{u}_2\ \mathbf{u}_3 \right) \begin{pmatrix} 2 & 1 & 1 \\ 1 & -1 & 2 \\ -3 & 1 & 4 \end{pmatrix}$, we have only to find the

linear independence of $\begin{pmatrix} 2 \\ 1 \\ -3 \end{pmatrix}, \begin{pmatrix} 1 \\ -1 \\ 1 \end{pmatrix}, \begin{pmatrix} 1 \\ 2 \\ 4 \end{pmatrix}$. Linearly independent.

(2) Linearly dependent.   (3) Linearly independent.

7 (1) We note $\left(\mathbf{v}_1\ \mathbf{v}_2\ \mathbf{v}_3 \right) = \left(\mathbf{u}_1\ \mathbf{u}_2\ \mathbf{u}_3 \right) P$, $P = \begin{pmatrix} 1 & 1 & -1 \\ -1 & 0 & 1 \\ 1 & -2 & 0 \end{pmatrix}$. Then

$\left(\mathbf{u}_1\ \mathbf{u}_2\ \mathbf{u}_3 \right) = \left(\mathbf{v}_1\ \mathbf{v}_2\ \mathbf{v}_3 \right) P^{-1}$. Since $P^{-1} = \begin{pmatrix} 2 & 2 & 1 \\ 1 & 1 & 0 \\ 2 & 3 & 1 \end{pmatrix}$, we have

$\mathbf{w} = \left(\mathbf{u}_1\ \mathbf{u}_2\ \mathbf{u}_3 \right) \begin{pmatrix} -2 \\ 1 \\ 4 \end{pmatrix} = \left(\mathbf{v}_1\ \mathbf{v}_2\ \mathbf{v}_3 \right) \begin{pmatrix} 2 & 2 & 1 \\ 1 & 1 & 0 \\ 2 & 3 & 1 \end{pmatrix} \begin{pmatrix} -2 \\ 1 \\ 4 \end{pmatrix} = 2\mathbf{v}_1 - \mathbf{v}_2 + 3\mathbf{v}_3$.

(2) $\mathbf{w} = -\mathbf{v}_1 + 2\mathbf{v}_2 - 3\mathbf{v}_3$.

8 (1) False. For example, take the vectors $\mathbf{u}_1 = \begin{pmatrix} 1 \\ 0 \end{pmatrix}$, $\mathbf{u}_2 = \begin{pmatrix} 0 \\ 1 \end{pmatrix}$, $\mathbf{u}_3 = \begin{pmatrix} 1 \\ 1 \end{pmatrix}$.
Then $\mathbf{u}_1$ and $\mathbf{u}_2$, $\mathbf{u}_2$ and $\mathbf{u}_3$ and $\mathbf{u}_1$ and $\mathbf{u}_3$ are linearly independent. But the vectors
$\mathbf{u}_1, \mathbf{u}_2, \mathbf{u}_3$ are linearly dependent.
(2) True. Let $\mathbf{v}_1 = \mathbf{u}_1$, $\mathbf{v}_2 = \mathbf{u}_1 + \mathbf{u}_2$, $\mathbf{v}_3 = \mathbf{u}_1 + \mathbf{u}_2 + \mathbf{u}_3$. Then $\left(\mathbf{v}_1\ \mathbf{v}_2\ \mathbf{v}_3 \right) =$
$\left(\mathbf{u}_1\ \mathbf{u}_2\ \mathbf{u}_3 \right) \begin{pmatrix} 1 & 1 & 1 \\ 0 & 1 & 1 \\ 0 & 0 & 1 \end{pmatrix}$. Therefore we see that $\left(\mathbf{u}_1\ \mathbf{u}_2\ \mathbf{u}_3 \right) =$

$\left(\mathbf{v}_1\ \mathbf{v}_2\ \mathbf{v}_3 \right) \begin{pmatrix} 1 & -1 & 0 \\ 0 & 1 & -1 \\ 0 & 0 & 1 \end{pmatrix}$. Since the vectors $\begin{pmatrix} 1 \\ 0 \\ 0 \end{pmatrix}, \begin{pmatrix} -1 \\ 1 \\ 0 \end{pmatrix}, \begin{pmatrix} 0 \\ -1 \\ 1 \end{pmatrix}$ are lin-

early independent, the vectors $\mathbf{u}_1, \mathbf{u}_2, \mathbf{u}_3$ are linearly independent by Theorem
4.2.6 (2).
(3) True. Let $1 \leq r \leq n - 1$, and we assume that $\mathbf{u}_1, \ldots, \mathbf{u}_r$ are linearly depen-

dent. Then they satisfy a non-trivial linear relation $c_1\mathbf{u}_1 + \cdots + c_r\mathbf{u}_r = \mathbf{0}_r$ $(c_i \in \mathbf{R})$. Therefore, $c_1\mathbf{u}_1 + \cdots + c_r\mathbf{u}_r + 0\mathbf{u}_{r+1} + \cdots + 0\mathbf{u}_n = \mathbf{0}_r$ is a non-trivial linear relation of $\mathbf{u}_1, \ldots, \mathbf{u}_n$, which contradicts the assumption. Therefore $\mathbf{u}_1, \ldots, \mathbf{u}_r$ are linearly independent.

4.3

1 (1) Note that $\left(\mathbf{a}_1 \ \mathbf{a}_2 \ \mathbf{a}_3 \right) = \left(\mathbf{e}_1 \ \mathbf{e}_2 \ \mathbf{e}_3 \right) \begin{pmatrix} 2 & -1 & 1 \\ 3 & 1 & 2 \\ 1 & 1 & 1 \end{pmatrix}$. Since $\begin{vmatrix} 2 & -1 & 1 \\ 3 & 1 & 2 \\ 1 & 1 & 1 \end{vmatrix} = 1 \neq 0$,

the vectors $\mathbf{a}_1$, $\mathbf{a}_2$, $\mathbf{a}_3$ are linearly independent.

(2) Since $\begin{vmatrix} 2 & 1 & 1 \\ -1 & -3 & 1 \\ 1 & 2 & -1 \end{vmatrix} = 3 \neq 0$, the vectors $\mathbf{a}_1$, $\mathbf{a}_2$, $\mathbf{a}_3$ are linearly independent.

(3) Note that $\left(f_1(x) \ f_2(x) \ f_3(x) \right) = \left(1 \ x \ x^2 \right) \begin{pmatrix} 1 & -1 & 1 \\ -2 & 1 & -1 \\ 3 & 2 & 1 \end{pmatrix}$.

Since $\begin{vmatrix} 1 & -1 & 1 \\ -2 & 1 & -1 \\ 3 & 2 & 1 \end{vmatrix} = -3 \neq 0$, the vectors $f_1(x)$, $f_2(x)$, $f_3(x)$ are linearly independent.

(4) Since $\begin{vmatrix} 3 & 1 & 0 \\ 1 & 1 & -2 \\ -1 & -2 & 3 \end{vmatrix} = -4 \neq 0$, the vectors $f_1(x)$, $f_2(x)$, $f_3(x)$ are linearly independent.

2 (1) $\left(\mathbf{a}_1 \ \mathbf{a}_2 \ \mathbf{a}_3 \ \mathbf{a}_4 \right) = \left(\mathbf{e}_1 \ \mathbf{e}_2 \ \mathbf{e}_3 \ \mathbf{e}_4 \right) \begin{pmatrix} 1 & -1 & 1 & -1 \\ 0 & 1 & 1 & 2 \\ 2 & 0 & 4 & 2 \end{pmatrix}$. The reduced matrix of

$\begin{pmatrix} 1 & -1 & 1 & -1 \\ 0 & 1 & 1 & 2 \\ 2 & 0 & 4 & 2 \end{pmatrix}$ is $\begin{pmatrix} 1 & 0 & 2 & 1 \\ 0 & 1 & 1 & 2 \\ 0 & 0 & 0 & 0 \end{pmatrix}$. (i) $r = 2$. (ii) Linearly independent vectors:

$\mathbf{a}_1, \mathbf{a}_2$. (iii) $\mathbf{a}_3 = 2\mathbf{a}_1 + \mathbf{a}_2$, $\mathbf{a}_4 = \mathbf{a}_1 + 2\mathbf{a}_2$.

(2) (i) $r = 3$. (ii) Linearly independent vectors: $\mathbf{a}_1, \mathbf{a}_2, \mathbf{a}_3$. (iii) $\mathbf{a}_4 = \mathbf{a}_1 - \mathbf{a}_2 + \mathbf{a}_3$.

(3) (i) $r = 3$. (ii) Linearly independent vectors: $\mathbf{a}_1, \mathbf{a}_2, \mathbf{a}_5$. (iii) $\mathbf{a}_3 = \mathbf{a}_1 + 2\mathbf{a}_2$, $\mathbf{a}_4 = \mathbf{a}_1 + \mathbf{a}_2$.

(4) $\left(f_1(x) \ f_2(x) \ f_3(x) \ f_4(x) \ f_5(x) \right) = \left(1 \ x \ x^2 \ x^3 \right) \begin{pmatrix} 1 & 0 & 2 & 0 & 0 \\ 0 & 1 & 0 & 1 & -1 \\ 1 & 0 & 2 & 1 & 1 \\ 0 & 1 & 0 & 1 & -1 \end{pmatrix}$.

The reduced matrix of $\begin{pmatrix} 1 & 0 & 2 & 0 & 0 \\ 0 & 1 & 0 & 1 & -1 \\ 1 & 0 & 2 & 1 & 1 \\ 0 & 1 & 0 & 1 & -1 \end{pmatrix}$ is $\begin{pmatrix} 1 & 0 & 2 & 0 & 0 \\ 0 & 1 & 0 & 0 & -2 \\ 0 & 0 & 0 & 1 & 1 \\ 0 & 0 & 0 & 0 & 0 \end{pmatrix}$. (i) $r = 3$. (ii) Lin-

early independent vectors: $f_1(x)$, $f_2(x)$, $f_4(x)$. (iii) $f_3(x) = 2 f_1(x)$, $f_5(x) = -2 f_2(x) + f_4(x)$.

(5) (i) $r = 3$. (ii) Linearly independent vectors: $f_1(x)$, $f_2(x)$, $f_5(x)$. (iii) $f_3(x) = f_1(x) + f_2(x)$, $f_4(x) = 2 f_1(x) + f_2(x)$.

3 (1) Let $a_1, \ldots, a_m$ be the columns of A. Since $AB = \begin{pmatrix} a_1 & \cdots & a_m \end{pmatrix} B$, the column vectors of AB are linear combinations of $a_1, \ldots, a_m$. Thus we have rank$(AB) = $ rank({the column vectors of AB}) $\le$ rank({$a_1, \ldots, a_m$}) $=$ rank(A) by Theorem 4.3.1.

(2) We consider row vectors of matrices instead of column vectors, or we take transposed matrices of AB and B, and apply (1) and Theorem 4.3.4.

4 (1) By Exercises ($\S 4.3$) 3 (2) above, we have rank$(PA) \le$ rank(A). We also have rank$(A) = $ rank$(P^{-1}PA) \le$ rank(PA) by Exercises ($\S 4.3$) 3 (2). Then rank$(PA) = $ rank(A).

(2) By Exercises ($\S 4.3$) 3 (1), we have rank$(AQ) \le$ rank(A). We also have rank$(A) = $ rank$(AQQ^{-1}) \le$ rank(AQ) by Exercises ($\S 4.3$) 3 (1). Then rank$(AQ) = $ rank(A).

5 Take r linearly independent column vectors of A, say $a_{i_1}, \ldots, a_{i_r}$. We put $B = \begin{pmatrix} a_{i_1} & \cdots & a_{i_r} \end{pmatrix}$. Since rank$(B) = r$, there exist r linearly independent row vectors of B, say $b_{j_1}, \ldots, b_{j_r}$. If we let $C = \begin{pmatrix} b_{j_1} \\ \vdots \\ b_{j_r} \end{pmatrix}$, then C is the regular submatrix of A of degree r.

6 (1) The reduced matrix of A is $\begin{pmatrix} 1 & 0 & 2 & 1 & 0 \\ 0 & 1 & 1 & 1 & 0 \\ 0 & 0 & 0 & 0 & 1 \\ 0 & 0 & 0 & 0 & 0 \end{pmatrix}$. Therefore $r = 3$ and the

column vectors $\begin{pmatrix} 1 \\ -1 \\ -2 \\ 1 \end{pmatrix}$, $\begin{pmatrix} 2 \\ 1 \\ -1 \\ -1 \end{pmatrix}$, $\begin{pmatrix} 1 \\ 0 \\ -1 \\ 2 \end{pmatrix}$ are linearly independent. Put $B = \begin{pmatrix} 1 & 2 & 1 \\ -1 & 1 & 0 \\ -2 & -1 & -1 \\ 1 & -1 & 2 \end{pmatrix}$. We shall find linearly independent row vectors of B. Since the

reduced matrix of $^t B$ is $\begin{pmatrix} 1 & 0 & -1 & 0 \\ 0 & 1 & 1 & 0 \\ 0 & 0 & 0 & 1 \end{pmatrix}$, the linearly independent row vectors of

B are the first row, the second row and the fourth row. Taking those rows of B, we have a submatrix $C = \begin{pmatrix} 1 & 2 & 1 \\ -1 & 1 & 0 \\ 1 & -1 & 2 \end{pmatrix}$ of A, which is a regular matrix of degree 3.

(2) $r = 4$, $C = \begin{pmatrix} -1 & 2 & 1 & 1 \\ 1 & 1 & 1 & 0 \\ 2 & 0 & -1 & 0 \\ -1 & 1 & 0 & -1 \end{pmatrix}$. (3) $r = 3$, $C = \begin{pmatrix} 1 & 1 & 0 \\ -2 & 1 & 1 \\ 1 & 1 & -1 \end{pmatrix}$.

4.4

1 (1) Since $\dim(\mathbf{R}^3) = 3$, we have only to show that $\mathbf{a}_1, \mathbf{a}_2, \mathbf{a}_3$ are linearly independent. To see the linear independence, we show that the matrix $A = \begin{pmatrix} -3 & 1 & 1 \\ 1 & -2 & -1 \\ -2 & 1 & 2 \end{pmatrix}$ is a regular matrix. To see that A is a regular matrix, we have only to see that $|A| \neq 0$. Since $|A| = 6 \neq 0$, the set $\{\mathbf{a}_1, \mathbf{a}_2, \mathbf{a}_3\}$ is a basis of $\mathbf{R}^3$.

(2) Since $\dim(\mathbf{R}^3) = 3$ and $\begin{vmatrix} 2 & -1 & 1 \\ -1 & 0 & 3 \\ 1 & -1 & 1 \end{vmatrix} = 3 \neq 0$, the set $\{\mathbf{a}_1, \mathbf{a}_2, \mathbf{a}_3\}$ is a basis of $\mathbf{R}^3$.

(3) Since $\dim(\mathbf{R}^4) = 4$ and $\begin{vmatrix} 1 & 0 & 2 & 1 \\ -1 & 1 & -1 & -1 \\ 2 & -1 & 1 & 2 \\ 0 & 2 & 1 & -1 \end{vmatrix} = 2 \neq 0$, the set $\{\mathbf{a}_1, \mathbf{a}_2, \mathbf{a}_3, \mathbf{a}_4\}$ is a basis of $\mathbf{R}^4$.

2 (1) We see that $\dim(\mathbf{R}[x]_2) = 3$, and $\left(f_1(x)\ f_2(x)\ f_3(x) \right) = \left(1\ x\ x^2 \right) A$ $\left(A = \begin{pmatrix} 1 & -1 & 1 \\ -1 & 2 & -2 \\ 1 & 2 & -1 \end{pmatrix} \right)$. Therefore we have only to show that the matrix A is a regular matrix. Since $|A| = 1 \neq 0$, A is a regular matrix and the set $\{f_1(x),\ f_2(x),\ f_3(x)\}$ is a basis of $\mathbf{R}[x]_2$ by Theorem 4.4.6.

(2) Since $\left(f_1(x)\ f_2(x)\ f_3(x) \right) = \left(1\ x\ x^2 \right) A$ $\left(A = \begin{pmatrix} 1 & 0 & -2 \\ 1 & 1 & 0 \\ 1 & 2 & -1 \end{pmatrix} \right)$ and $|A| = -3 \neq 0$, the set $\{f_1(x),\ f_2(x),\ f_3(x)\}$ is a basis of $\mathbf{R}[x]_2$.

(3) Since $\left(f_1(x)\ f_2(x)\ f_3(x) \right) = \left(1\ x\ x^2 \right) A$ $\left(A = \begin{pmatrix} 3 & -1 & 1 \\ 1 & -1 & 0 \\ -1 & 2 & 3 \end{pmatrix} \right)$ and $|A| = -5 \neq 0$, the set $\{f_1(x),\ f_2(x),\ f_3(x)\}$ is a basis of $\mathbf{R}[x]_2$.

3   (1) $\dim(W) = 2$. Basis $\left\{ \begin{pmatrix} -1 \\ 0 \\ 1 \\ 0 \\ 0 \end{pmatrix}, \begin{pmatrix} -2 \\ 0 \\ 0 \\ 1 \\ 1 \end{pmatrix} \right\}$.

(2) $\dim(W) = 2$. Basis $\left\{ \begin{pmatrix} 0 \\ -5 \\ 3 \\ 1 \\ 0 \end{pmatrix}, \begin{pmatrix} -3 \\ 7 \\ -2 \\ 0 \\ 1 \end{pmatrix} \right\}$.

(3) $\dim(W) = 1$. Basis $\left\{ \begin{pmatrix} -1/9 \\ 5/9 \\ 1 \end{pmatrix} \right\}$.

(4) $\dim(W) = 2$. Basis $\left\{ \begin{pmatrix} -3/2 \\ 5/2 \\ 1 \\ 0 \end{pmatrix}, \begin{pmatrix} 1 \\ -2 \\ 0 \\ 1 \end{pmatrix} \right\}$.

(5) $\dim(W) = 2$. Basis $\{-1 + x^2, -x + x^3\}$. Let $f(x) = a_0 + a_1 x + a_2 x^2 + a_3 x^3$. Then we have $f(1) = a_0 + a_1 + a_2 + a_3 = 0$ and $f(-1) = a_0 - a_1 + a_2 - a_3 = 0$. We have only to solve the linear equation $\begin{cases} a_0 + a_1 + a_2 + a_3 = 0, \\ a_0 - a_1 + a_2 - a_3 = 0. \end{cases}$

(6) $\dim(W) = 2$. Basis $\{1 - 2x + x^2, 2 - 3x + x^3\}$.

4   (1) Since $\begin{pmatrix} v_1 & v_2 & v_3 \end{pmatrix} = \begin{pmatrix} u_1 & u_2 & u_3 \end{pmatrix} A$ $\left(A = \begin{pmatrix} 2 & 1 & 1 \\ 1 & 2 & 1 \\ -1 & 1 & 1 \end{pmatrix} \right)$ and $|A| = 3 \neq$

0, $\{v_1, v_2, v_3\}$ is a basis of V by Theorem 4.4.6.

(2) Not a basis.   (3) A basis.

5   Let $\{u_1, \ldots, u_n\}$ be a basis of V. Take n linearly independent vectors among $v_1, \ldots, v_r, u_1, \ldots, u_n$ from the beginning. Then the set of the n linearly independent vectors is a basis of V containing the vectors $v_1, \ldots, v_r$.

6   (1) Since $\left\{ \begin{pmatrix} 1 \\ 0 \\ 0 \end{pmatrix}, \begin{pmatrix} 0 \\ 1 \\ 0 \end{pmatrix}, \begin{pmatrix} 0 \\ 0 \\ 1 \end{pmatrix} \right\}$ is a basis of $\mathbf{R}^3$, we have only to find linearly

independent vectors among the vectors $\begin{pmatrix} 1 \\ 2 \\ 1 \end{pmatrix}, \begin{pmatrix} 0 \\ 2 \\ 1 \end{pmatrix}, \begin{pmatrix} 1 \\ 0 \\ 0 \end{pmatrix}, \begin{pmatrix} 0 \\ 1 \\ 0 \end{pmatrix}, \begin{pmatrix} 0 \\ 0 \\ 1 \end{pmatrix}$

from the beginning. Then we obtain a basis $\left\{ \begin{pmatrix} 1 \\ 2 \\ 1 \end{pmatrix}, \begin{pmatrix} 0 \\ 2 \\ 1 \end{pmatrix}, \begin{pmatrix} 0 \\ 1 \\ 0 \end{pmatrix} \right\}$ of $\mathbf{R}^3$.

(2) $\left\{ \begin{pmatrix} 1 \\ 1 \\ 2 \\ -1 \end{pmatrix}, \begin{pmatrix} 0 \\ 2 \\ 4 \\ -2 \end{pmatrix}, \begin{pmatrix} 0 \\ 1 \\ 0 \\ 0 \end{pmatrix}, \begin{pmatrix} 0 \\ 0 \\ 1 \\ 0 \end{pmatrix} \right\}.$

7 Let $\dim(W) = \dim(V) = n$ and $\{\mathbf{u}_1, \ldots, \mathbf{u}_n\}$ be a basis of W. Since $\dim(V) = n$, $\{\mathbf{u}_1, \ldots, \mathbf{u}_n\}$ is also a basis of V. Therefore $V = W$.

5.1

1 We have only to verify the two conditions of linear mappings.

(1) A linear mapping. (2) Not a linear mapping. (3) A linear mapping.

(4) A linear mapping. (5) If $a = 0$, then T is a linear mapping. If $a \neq 0$, then T is not a linear mapping.

2 (1) (i) $\text{null}(A) = 2$. A basis of $\text{Ker}(A)$: $\left\{ \begin{pmatrix} -2 \\ 1 \\ 0 \\ 0 \end{pmatrix}, \begin{pmatrix} 1 \\ 0 \\ -1 \\ 1 \end{pmatrix} \right\}.$

(ii) $\text{rank}(A) = 2$. A basis of $\text{Im}(A)$: $\left\{ \begin{pmatrix} 2 \\ 0 \\ 1 \end{pmatrix}, \begin{pmatrix} 3 \\ 1 \\ 1 \end{pmatrix} \right\}.$

(2) (i) $\text{null}(A) = 2$. A basis of $\text{Ker}(A)$: $\left\{ \begin{pmatrix} -2 \\ 2 \\ 1 \\ 0 \end{pmatrix}, \begin{pmatrix} -4 \\ 3 \\ 0 \\ 1 \end{pmatrix} \right\}.$

(ii) $\text{rank}(A) = 2$. A basis of $\text{Im}(A)$: $\left\{ \begin{pmatrix} 1 \\ 0 \\ 2 \end{pmatrix}, \begin{pmatrix} 1 \\ 2 \\ 4 \end{pmatrix} \right\}.$

(3) (i) $\text{null}(A) = 1$. A basis of $\text{Ker}(A)$: $\left\{ \begin{pmatrix} 2 \\ -1 \\ 1 \\ 0 \end{pmatrix} \right\}.$

(ii) $\text{rank}(A) = 3$. A basis of $\text{Im}(A)$: $\left\{ \begin{pmatrix} 1 \\ 0 \\ -1 \end{pmatrix}, \begin{pmatrix} 3 \\ 1 \\ -2 \end{pmatrix}, \begin{pmatrix} 0 \\ 0 \\ 1 \end{pmatrix} \right\}.$

(4) (i) $\text{null}(A) = 2$. A basis of $\text{Ker}(A)$: $\left\{ \begin{pmatrix} -3 \\ -1 \\ 1 \\ 0 \end{pmatrix}, \begin{pmatrix} -2 \\ -1 \\ 0 \\ 1 \end{pmatrix} \right\}.$

(ii) rank$(A) = 2$. A basis of Im(A): $\left\{ \begin{pmatrix} 1 \\ 1 \\ -2 \\ 1 \end{pmatrix}, \begin{pmatrix} -2 \\ -2 \\ 4 \\ -1 \end{pmatrix} \right\}$.

(5) (i) null$(A) = 3$. A basis of Ker(A): $\left\{ \begin{pmatrix} -3 \\ -1 \\ 1 \\ 0 \\ 0 \end{pmatrix}, \begin{pmatrix} 1 \\ -1 \\ 0 \\ 1 \\ 0 \end{pmatrix}, \begin{pmatrix} 2 \\ -3 \\ 0 \\ 0 \\ 1 \end{pmatrix} \right\}$.

(ii) rank$(A) = 2$. A basis of Im(A): $\left\{ \begin{pmatrix} 0 \\ -1 \\ 1 \\ 1 \end{pmatrix}, \begin{pmatrix} 1 \\ -2 \\ 1 \\ -1 \end{pmatrix} \right\}$.

3 Let $\{e_1, \ldots, e_n\}$ and $\{e'_1, \ldots, e'_m\}$ be the standard bases of K^n and K^m, respectively. Put $T(e_j) = a_{1j}e'_1 + \cdots + a_{mj}e'_m$ for $j = 1, \ldots, n$, then T is expressed as $T(\mathbf{x}) = A\mathbf{x}$ for $\mathbf{x} \in K^n$ with the $m \times n$ matrix $A = (a_{ij})$. Conversely, if $T(\mathbf{x}) = A\mathbf{x}$ with an $m \times n$ matrix $A = (a_{ij})$, then we see that $T(e_j) = a_{1j}e'_1 + \cdots + a_{mj}e'_m$. Therefore the matrix A is uniquely determined.

4 (1) Since Ker$(T) = \left\{ a \begin{pmatrix} 2 \\ 1 \end{pmatrix} \middle| a \in \mathbf{R} \right\} \neq \{\mathbf{0}_2\}$, T is not a one-to-one mapping.

Since $T\left(\begin{pmatrix} a \\ 0 \end{pmatrix} \right) = a$ for any $a \in \mathbf{R}$, T is a surjective mapping.

(2) Since the solution of the linear equation $\begin{pmatrix} 2 & 1 \\ 1 & 1 \\ 1 & -3 \end{pmatrix} \begin{pmatrix} x_1 \\ x_2 \end{pmatrix} = \begin{pmatrix} 0 \\ 0 \\ 0 \end{pmatrix}$ is only

$\begin{pmatrix} 0 \\ 0 \end{pmatrix}$, we have Ker$(T) = \{\mathbf{0}_2\}$. Therefore T is a one-to-one mapping. Since rank$(T) = 2 < 3 = \dim(\mathbf{R}^3)$, T is not a surjective mapping.

5 Since $\dim(U) = \dim(V)$, we have only to see that T is either one-to-one or surjective by Theorem 5.1.4 (3).

(1) Let $A = \begin{pmatrix} -1 & 1 & -2 \\ 2 & 1 & -1 \\ 1 & 0 & 1 \end{pmatrix}$. Since $|A| = -2 \neq 0$, the linear equation $A\mathbf{x} = \mathbf{0}_3$ has only the trivial solution. Then T is a one-to-one mapping. Therefore T is an isomorphism.

(2) We see that $\left(T(1)\ T(x)\ T(x^2) \right) = \left(\begin{pmatrix} 2 \\ 0 \\ 0 \end{pmatrix} \begin{pmatrix} 1 \\ 2 \\ 0 \end{pmatrix} \begin{pmatrix} 1 \\ 0 \\ 2 \end{pmatrix} \right)$. Since $\begin{pmatrix} 2 \\ 0 \\ 0 \end{pmatrix}$,

$\begin{pmatrix} 1 \\ 2 \\ 0 \end{pmatrix}$, $\begin{pmatrix} 1 \\ 0 \\ 2 \end{pmatrix}$ are linearly independent, the set of these vectors is a basis of $\mathbf{R}^3$.

Then T is a surjective mapping. Therefore T is an isomorphism.

6   Let $\mathbf{x} \in K^n$. Then $T_A T_B(\mathbf{x}) = T_A(T_B(\mathbf{x})) = T_A(B\mathbf{x}) = A(B\mathbf{x}) = (AB)\mathbf{x} = T_{AB}(\mathbf{x})$.

7   (1) Since $T_A T_B = T_{AB}$ by Exercise 6 above, the product $ST : \mathbf{R}^2 \to \mathbf{R}^3$ is the linear mapping T_{AB} associated with the matrix $AB = \begin{pmatrix} -3 & -1 \\ 5 & -1 \\ -1 & 7 \end{pmatrix}$.

(2) $ST(f(x)) = S(T(f(x))) = S\left(\begin{pmatrix} f(1) \\ f(-1) \end{pmatrix} \right) = \begin{pmatrix} 2 & 1 \\ 3 & 1 \end{pmatrix} \begin{pmatrix} f(1) \\ f(-1) \end{pmatrix}$
$= \begin{pmatrix} 2f(1) + f(-1) \\ 3f(1) + f(-1) \end{pmatrix}$.

(3) $ST(f(x)) = S(T(f(x))) = S(f(x) - f'(x)) = \begin{pmatrix} f(0) - f'(0) \\ f(1) - f'(1) \end{pmatrix}$.

5.2

1   (1) The representation matrix with respect to the standard bases is $\begin{pmatrix} 1 & -2 \\ 1 & 1 \end{pmatrix}$.

Therefore the representation matrix we ask for is $\begin{pmatrix} 2 & 3 \\ 1 & 1 \end{pmatrix}^{-1} \begin{pmatrix} 1 & -2 \\ 1 & 1 \end{pmatrix} \begin{pmatrix} 1 & 2 \\ 0 & 1 \end{pmatrix} = \begin{pmatrix} 2 & 9 \\ -1 & -6 \end{pmatrix}$ by Theorem 5.2.5.

(2) $\begin{pmatrix} 1 & 2 \\ 2 & 3 \end{pmatrix}^{-1} \begin{pmatrix} 2 & 4 & 1 \\ 1 & 5 & 3 \end{pmatrix} \begin{pmatrix} 1 & 1 & 0 \\ 0 & 2 & 1 \\ 1 & 2 & 1 \end{pmatrix} = \begin{pmatrix} -1 & -2 & 1 \\ 2 & 7 & 2 \end{pmatrix}$.

(3) The representation matrix with respect to the standard bases $\{1, x, x^2\}$ of $\mathbf{R}[x]_2$ and $\{1, x\}$ of $\mathbf{R}[x]_1$ is $A = \begin{pmatrix} 0 & 2 & 0 \\ 1 & 0 & 4 \end{pmatrix}$. The matrices of changing bases are $P = \begin{pmatrix} 1 & 1 & 0 \\ 0 & 1 & 1 \\ 0 & 0 & 1 \end{pmatrix}$ and $Q = \begin{pmatrix} 1 & 0 \\ -1 & 1 \end{pmatrix}$, respectively. Therefore the representation matrix we ask for is $Q^{-1}AP = \begin{pmatrix} 0 & 2 & 2 \\ 1 & 3 & 6 \end{pmatrix}$ by Theorem 5.2.5.

(4) $\begin{pmatrix} 1 & 1 & 0 \\ 1 & 0 & 1 \\ 0 & 1 & 0 \end{pmatrix}^{-1} \begin{pmatrix} 2 & -1 & 1 \\ 1 & 1 & 1 \\ 0 & 1 & -1 \end{pmatrix} \begin{pmatrix} 0 & 0 & 1 \\ 0 & 1 & 0 \\ 1 & 0 & 0 \end{pmatrix} = \begin{pmatrix} 2 & -2 & 2 \\ -1 & 1 & 0 \\ -1 & 3 & -1 \end{pmatrix}$.

2   (1) $\begin{pmatrix} 1 & 1 & 1 \\ 1 & 0 & 1 \\ 1 & 1 & 0 \end{pmatrix}^{-1} \begin{pmatrix} 0 & 3 & -1 \\ -1 & 2 & 1 \\ 2 & 1 & 1 \end{pmatrix} \begin{pmatrix} 1 & 1 & 1 \\ 1 & 0 & 1 \\ 1 & 1 & 0 \end{pmatrix} = \begin{pmatrix} 4 & 4 & 1 \\ 0 & -1 & 2 \\ -2 & -4 & 0 \end{pmatrix}$.

(2) $\begin{pmatrix} 1 & 0 & 1 \\ 0 & 2 & 1 \\ -1 & 1 & -1 \end{pmatrix}^{-1} \begin{pmatrix} 3 & 1 & 2 \\ -2 & 0 & -1 \\ 1 & 2 & -2 \end{pmatrix} \begin{pmatrix} 1 & 0 & 1 \\ 0 & 2 & 1 \\ -1 & 1 & -1 \end{pmatrix} = \begin{pmatrix} 10 & 17 & 17 \\ 4 & 6 & 7 \\ -9 & -13 & -15 \end{pmatrix}.$

(3) $\begin{pmatrix} 1 & 2 & 0 \\ 1 & 1 & 1 \\ 0 & 0 & -1 \end{pmatrix}^{-1} \begin{pmatrix} 1 & 0 & 0 \\ 1 & 2 & 1 \\ 1 & 0 & 1 \end{pmatrix} \begin{pmatrix} 1 & 2 & 0 \\ 1 & 1 & 1 \\ 0 & 0 & -1 \end{pmatrix} = \begin{pmatrix} 7 & 10 & 0 \\ -3 & -4 & 0 \\ -1 & -2 & 1 \end{pmatrix}.$

(4) $\begin{pmatrix} 1 & 0 & 1 \\ 0 & 1 & 0 \\ 1 & 1 & 2 \end{pmatrix}^{-1} \begin{pmatrix} 1 & -1 & 1 \\ 0 & 1 & 0 \\ 1 & -1 & 3 \end{pmatrix} \begin{pmatrix} 1 & 0 & 1 \\ 0 & 1 & 0 \\ 1 & 1 & 2 \end{pmatrix} = \begin{pmatrix} 0 & -1 & -1 \\ 0 & 1 & 0 \\ 2 & 1 & 4 \end{pmatrix}.$

3 (1) Since $\left(T(1)\ T(x)\ T(x^2) \right) = \left(1\ x\ x^2 \right) \begin{pmatrix} 2 & 0 & 0 \\ 0 & 1 & 0 \\ 0 & 0 & 0 \end{pmatrix}$, we have (i) $\text{null}(T) = 1$

and a basis of $\text{Ker}(T)$ is $\{x^2\}$, (ii) $\text{rank}(T) = 2$ and a basis of $\text{Im}(T)$ is $\{1, x\}$.

(2) We note that $\left(T(1)\ T(x)\ T(x^2) \right) = \left(1\ x\ x^2 \right) \begin{pmatrix} 1 & 0 & 0 \\ 1 & 0 & 0 \\ 0 & 0 & -1 \end{pmatrix}$. Since the reduced

matrix of $\begin{pmatrix} 1 & 0 & 0 \\ 1 & 0 & 0 \\ 0 & 0 & -1 \end{pmatrix}$ is $\begin{pmatrix} 1 & 0 & 0 \\ 0 & 0 & 1 \\ 0 & 0 & 0 \end{pmatrix}$, we have (i) $\text{null}(T) = 1$ and a basis of $\text{Ker}(T)$

is $\{x\}$, (ii) $\text{rank}(T) = 2$ and a basis of $\text{Im}(T)$ is $\{1 + x, x^2\}$.

(3) We note that $\left(T(1)\ T(x)\ T(x^2) \right) = \left(e_1\ e_2\ e_3 \right) \begin{pmatrix} 1 & -2 & 1 \\ -1 & 0 & -1 \\ 0 & 2 & 0 \end{pmatrix}$. Since the

reduced matrix of $\begin{pmatrix} 1 & -2 & 1 \\ -1 & 0 & -1 \\ 0 & 2 & 0 \end{pmatrix}$ is $\begin{pmatrix} 1 & 0 & 1 \\ 0 & 1 & 0 \\ 0 & 0 & 0 \end{pmatrix}$, we have (i) $\text{null}(T) = 1$ and

a basis of $\text{Ker}(T)$ is $\{1 - x^2\}$, (ii) $\text{rank}(T) = 2$ and a basis of $\text{Im}(T)$ is

$\left\{ \begin{pmatrix} 1 \\ -1 \\ 0 \end{pmatrix}, \begin{pmatrix} -1 \\ 0 \\ 1 \end{pmatrix} \right\}.$

(4) We note that $\left(T(e_1)\ T(e_2)\ T(e_3) \right) = \left(1\ x\ x^2 \right) \begin{pmatrix} 1 & -1 & 0 \\ 0 & 1 & -1 \\ -1 & 0 & 1 \end{pmatrix}$. Since the

reduced matrix of $\begin{pmatrix} 1 & -1 & 0 \\ 0 & 1 & -1 \\ -1 & 0 & 1 \end{pmatrix}$ is $\begin{pmatrix} 1 & 0 & -1 \\ 0 & 1 & -1 \\ 0 & 0 & 0 \end{pmatrix}$, we have (i) $\text{null}(T) = 1$

and a basis of $\text{Ker}(T)$ is $\left\{ \begin{pmatrix} 1 \\ 1 \\ 1 \end{pmatrix} \right\}$, (ii) $\text{rank}(T) = 2$ and a basis of $\text{Im}(T)$ is

$\{1 - x^2, 1 - x\}.$

4 (1) Since $\dim(V) = n$, we have only to show that these vectors are linearly independent. Let $c_{n-1}T^{n-1}(\mathbf{u}) + \cdots + c_1 T(\mathbf{u}) + c_0\mathbf{u} = \mathbf{0}_V$ be a linear relation of $T^{n-1}(\mathbf{u}), \ldots, T(\mathbf{u}), \mathbf{u}$. Applying T^{n-1} to both sides, we obtain $c_0 T^{n-1}(\mathbf{u}) = \mathbf{0}_V$. Since $T^{n-1}(\mathbf{u}) \neq \mathbf{0}_V$, we see that $c_0 = 0$. Then we have $c_{n-1}T^{n-1}(\mathbf{u}) + \cdots + c_1 T(\mathbf{u}) = \mathbf{0}_V$. Applying T^{n-2} to both sides, we see that $c_1 = 0$. Continuing this procedure, we obtain $c_0 = c_1 = \cdots = c_{n-1} = 0$. Therefore $T^{n-1}(\mathbf{u}), \ldots, T(\mathbf{u}), \mathbf{u}$ are linearly independent.

(2)
$$\begin{pmatrix} 0 & 1 & 0 & & 0 \\ & \ddots & \ddots & \ddots & \\ & & \ddots & \ddots & 0 \\ & & & \ddots & 1 \\ 0 & & & & 0 \end{pmatrix} \quad \text{(degree } n\text{)}.$$

5 (1) Since $\big(T(1)\ T(x)\ T(x^2) \big) = \big(1\ x\ x^2 \big) \begin{pmatrix} 4 & 1 & 1 \\ 0 & 2 & 0 \\ 0 & 0 & 1 \end{pmatrix}$, the representation matrix

of T with respect to the standard basis $\{1, x, x^2\}$ of $\mathbf{R}[x]_2$ is $A = \begin{pmatrix} 4 & 1 & 1 \\ 0 & 2 & 0 \\ 0 & 0 & 1 \end{pmatrix}$. As

$|A| = 8 \neq 0$, A is a regular matrix. Then by Theorem 5.2.2, T is an automorphism of $\mathbf{R}[x]_2$.

(2) The representation matrix of T with respect to the standard basis of $\mathbf{R}^3$ is A by Theorem 5.2.1. Since $|A| = -6 \neq 0$, A is a regular matrix and T is an automorphism of $\mathbf{R}^3$ by Theorem 5.2.2.

(3) Since $\big(T(1)\ T(x)\ T(x^2) \big) = \big(\mathbf{e}_1\ \mathbf{e}_2\ \mathbf{e}_3 \big) \begin{pmatrix} 1 & 1 & 1 \\ 1 & 0 & 0 \\ 1 & -1 & 1 \end{pmatrix}$, the representation

matrix of T with respect to the standard basis $\{1, x, x^2\}$ of $\mathbf{R}[x]_2$ and the standard

basis $\{\mathbf{e}_1, \mathbf{e}_2, \mathbf{e}_3\}$ of $\mathbf{R}^3$ is $A = \begin{pmatrix} 1 & 1 & 1 \\ 1 & 0 & 0 \\ 1 & -1 & 1 \end{pmatrix}$. As $|A| = -2 \neq 0$, A is a regular

matrix. Then by Theorem 5.2.2, T is an isomorphism of $\mathbf{R}[x]_2$ into $\mathbf{R}^3$.

5.3

1 (1) $\begin{pmatrix} 17 & -1 \\ -4 & 22 \end{pmatrix}$. (2) O_2.

2 (1) $g_A(t) = (t - 2)(t - 3)$, $\lambda = 2, 3$.

$W(2\,;A) = \left\{ c\begin{pmatrix} 1 \\ 1 \end{pmatrix} \middle| c \in \mathbf{R} \right\}$, $W(3\,;A) = \left\{ c\begin{pmatrix} 3 \\ 2 \end{pmatrix} \middle| c \in \mathbf{R} \right\}$.

(2) $g_A(t) = (t + 1)(t - 4)$, $\lambda = -1, 4$.

$W(-1\,;A) = \left\{ c\begin{pmatrix} -2 \\ 3 \end{pmatrix} \middle| c \in \mathbf{R} \right\}$, $W(4\,;A) = \left\{ c\begin{pmatrix} 1 \\ 1 \end{pmatrix} \middle| c \in \mathbf{R} \right\}$.

(3) $g_A(t) = (t-1)(t-2)(t-3)$, $\lambda = 1, 2, 3.$

$$W(1\,;A) = \left\{ c\begin{pmatrix} -2 \\ -1 \\ 1 \end{pmatrix} \middle| c \in \mathbf{R} \right\}, \quad W(2\,;A) = \left\{ c\begin{pmatrix} -3 \\ -1 \\ 1 \end{pmatrix} \middle| c \in \mathbf{R} \right\},$$

$$W(3\,;A) = \left\{ c\begin{pmatrix} 1 \\ 1 \\ 0 \end{pmatrix} \middle| c \in \mathbf{R} \right\}.$$

(4) $g_A(t) = (t-1)^2(t-2)$, $\lambda = 1, 2.$

$$W(1\,;A) = \left\{ c\begin{pmatrix} -1 \\ 1 \\ 1 \end{pmatrix} \middle| c \in \mathbf{R} \right\}, \quad W(2\,;A) = \left\{ c\begin{pmatrix} -2 \\ 1 \\ 3 \end{pmatrix} \middle| c \in \mathbf{R} \right\}.$$

(5) $g_A(t) = (t+1)^2(t-2)^2$, $\lambda = -1, 2.$

$$W(-1\,;A) = \left\{ c_1\begin{pmatrix} 0 \\ -1 \\ 1 \\ 0 \end{pmatrix} + c_2\begin{pmatrix} 0 \\ 0 \\ 0 \\ 1 \end{pmatrix} \middle| c_1, c_2 \in \mathbf{R} \right\},$$

$$W(2\,;A) = \left\{ c_1\begin{pmatrix} 0 \\ 1 \\ 0 \\ 1 \end{pmatrix} + c_2\begin{pmatrix} -1 \\ 0 \\ 1 \\ 0 \end{pmatrix} \middle| c_1, c_2 \in \mathbf{R} \right\}.$$

(6) $g_A(t) = (t+1)(t-1)(t-2)^2$, $\lambda = -1, 1, 2.$

$$W(-1\,;A) = \left\{ c\begin{pmatrix} -1 \\ 0 \\ 1 \\ 1 \end{pmatrix} \middle| c \in \mathbf{R} \right\}, \quad W(1\,;A) = \left\{ c\begin{pmatrix} -1 \\ 1 \\ 0 \\ 0 \end{pmatrix} \middle| c \in \mathbf{R} \right\},$$

$$W(2\,;A) = \left\{ c_1\begin{pmatrix} 1 \\ 0 \\ 1 \\ 0 \end{pmatrix} + c_2\begin{pmatrix} -1 \\ 0 \\ 0 \\ 1 \end{pmatrix} \middle| c_1, c_2 \in \mathbf{R} \right\}.$$

3   (1) We note that $\big(T(1)\ T(x)\ T(x^2) \big) = \big(1\ x\ x^2 \big)\begin{pmatrix} 1 & 1 & 1 \\ 0 & -1 & -2 \\ 0 & 0 & 1 \end{pmatrix}.$

$g_T(t) = (t+1)(t-1)^2$, $\lambda = -1, 1.$ $W(-1\,;T) = \{ c(-1+2x) \mid c \in \mathbf{R} \}$,
$W(1\,;T) = \{ c_1 + c_2(-x+x^2) \mid c_1, c_2 \in \mathbf{R} \}.$

(2) We note that $\big(T(1)\ T(x)\ T(x^2) \big) = \big(1\ x\ x^2 \big)\begin{pmatrix} 1 & 1 & 0 \\ 0 & 2 & 2 \\ 0 & 0 & 4 \end{pmatrix}.$

$g_T(t) = (t-1)(t-2)(t-4)$, $\lambda = 1, 2, 4.$
$W(1\,;T) = \{ c \mid c \in \mathbf{R} \}$, $W(2\,;T) = \{ c(1+x) \mid c \in \mathbf{R} \}$,
$W(4\,;T) = \{ c(1+3x+3x^2) \mid c \in \mathbf{R} \}.$

4 Since $g_A(t) = t^2 + t + 1$, we see that $A^2 + A + E_2 = O_2$, $A^3 = E_2$.

(1) $A^{20} = (A^3)^6 A^2 = A^2 = -A - E_2 = \begin{pmatrix} -3 & -1 \\ 7 & 2 \end{pmatrix}$.

(2) $A^{11} + A^7 - 2E_2 = (A^3)^3 A^2 + (A^3)^2 A - 2E_2 = A^2 + A - 2E_2 = -A - E_2 + A - 2E_2 = -3E_2$.

5.4

1 (1) Let $\mathbf{a} = \begin{pmatrix} a_1 \\ a_2 \end{pmatrix}$, $\mathbf{a}_1 = \begin{pmatrix} 1 \\ 1 \end{pmatrix}$ and $\mathbf{a}_2 = \begin{pmatrix} 1 \\ -1 \end{pmatrix}$. We have only to show that

$x_1 \mathbf{a}_1 + x_2 \mathbf{a}_2 = \mathbf{a}$ has a unique solution. Since the matrix $\begin{pmatrix} 1 & 1 \\ 1 & -1 \end{pmatrix}$ is a regular matrix, the equation $x_1 \mathbf{a}_1 + x_2 \mathbf{a}_2 = \mathbf{a}$ has a unique solution. Therefore $\mathbf{R}^2 = W_1 \oplus W_2$.

(2) Let $\mathbf{a} = \begin{pmatrix} a_1 \\ a_2 \\ a_3 \end{pmatrix}$, $\mathbf{a}_1 = \begin{pmatrix} 1 \\ 0 \\ -1 \end{pmatrix}$, $\mathbf{a}_2 = \begin{pmatrix} 0 \\ 1 \\ 1 \end{pmatrix}$ and $\mathbf{a}_3 = \begin{pmatrix} 1 \\ 2 \\ 0 \end{pmatrix}$. We have only

to show that $x_1 \mathbf{a}_1 + x_2 \mathbf{a}_2 + x_3 \mathbf{a}_3 = \mathbf{a}$ has a unique solution. Since the matrix $\begin{pmatrix} 1 & 0 & 1 \\ 0 & 1 & 2 \\ -1 & 1 & 0 \end{pmatrix}$ is a regular matrix, the equation $x_1 \mathbf{a}_1 + x_2 \mathbf{a}_2 + x_3 \mathbf{a}_3 = \mathbf{a}$ has a

unique solution. Therefore $\mathbf{R}^3 = W_1 \oplus W_2$.

(3) We see that $W_1 = \left\{ c_1 \begin{pmatrix} -1 \\ 0 \\ 1 \\ 0 \end{pmatrix} + c_2 \begin{pmatrix} -1 \\ 0 \\ 0 \\ 1 \end{pmatrix} \middle| c_1, c_2 \in \mathbf{R} \right\}$ and

$W_2 = \left\{ c_1 \begin{pmatrix} -1 \\ 1 \\ 0 \\ 0 \end{pmatrix} + c_2 \begin{pmatrix} 1 \\ 0 \\ 1 \\ 0 \end{pmatrix} \middle| c_1, c_2 \in \mathbf{R} \right\}$. Since $\begin{pmatrix} -1 & -1 & -1 & 1 \\ 0 & 0 & 1 & 0 \\ 1 & 0 & 0 & 1 \\ 0 & 1 & 0 & 0 \end{pmatrix}$ is a regular matrix, we have $\mathbf{R}^4 = W_1 \oplus W_2$ similarly to (1) and (2).

2 (1) Let $f(x) = c_0 + c_1 x + c_2 x^2$. Then $f(1) = c_0 + c_1 + c_2$, $f(0) = c_0$, $f(-1) = c_0 - c_1 + c_2$. Therefore $W_1 = \{c_0 + c_1 x + c_2 x^2 \in \mathbf{R}[x]_2 \mid c_0 + c_1 + c_2 = 0\} = \{a_1(-1 + x) + a_2(-1 + x^2) \mid a_1, a_2 \in \mathbf{R}\}$ and $W_2 = \{c_0 + c_1 x + c_2 x^2 \in \mathbf{R}[x]_2 \mid c_0 = 0, c_0 - c_1 + c_2 = 0\} = \{a(x + x^2) \mid a \in \mathbf{R}\}$. To see the uniqueness of the linear combination of $f(x) \in \mathbf{R}[x]_2$ by $-1 + x$, $-1 + x^2$, $x + x^2$, we have only to see that any vector in $\mathbf{R}^3$ is uniquely expressed as a linear combination of the vectors $\begin{pmatrix} -1 \\ 1 \\ 0 \end{pmatrix}$, $\begin{pmatrix} -1 \\ 0 \\ 1 \end{pmatrix}$, $\begin{pmatrix} 0 \\ 1 \\ 1 \end{pmatrix}$, which can be shown similarly to Exercise 1 above.

(2) Let $f(x) = c_0 + c_1 x + c_2 x^2 + c_3 x^3$. Then $f(-1) = c_0 - c_1 + c_2 - c_3$, $f(0) = c_0$, $f(1) = c_0 + c_1 + c_2 + c_3$, $f(2) = c_0 + 2c_1 + 4c_2 + 8c_3$. Therefore $W_1 = \{a_1(x + x^2) + a_2(-x + x^3) \mid a_1, a_2 \in \mathbf{R}\}$ and $W_2 = \{a_1(2 - 3x + x^2) + $

$a_2(6 - 7x + x^3) \mid a_1, a_2 \in \mathbf{R}\}$. The direct sum $V = W_1 \oplus W_2$ can be shown similarly to (1) above.

3 (1) Since $A = \begin{pmatrix} 2 & 1 & 0 \\ 0 & 2 & 1 \\ 0 & 0 & 2 \end{pmatrix} \oplus (2) \oplus (5)$, we have $p_A(t) = (t - 2)^3(t - 5)$ by Theorem 5.4.9.

(2) Since $A = \begin{pmatrix} 3 & 1 & 0 \\ 0 & 3 & 1 \\ 0 & 0 & 3 \end{pmatrix} \oplus (3) \oplus (3)$, we have $p_A(t) = (t - 3)^3$ by Theorem 5.4.9.

4 It is easy to see that I_i is a linear transformation of V. Let $\mathbf{v} = \mathbf{w}_1 + \cdots + \mathbf{w}_r$ ($\mathbf{w}_i \in W_i$). Then $\mathbf{v} = I_1(\mathbf{v}) + \cdots + I_r(\mathbf{v})$. Therefore we have $I = I_1 + \cdots + I_r$. If $i \neq j$, then we have $I_i I_j(\mathbf{v}) = I_i(\mathbf{w}_j) = \mathbf{0}_V$. Therefore $I_i I_j = O_V$. Furthermore we have $I_i I_i(\mathbf{v}) = I_i(\mathbf{w}_i) = \mathbf{w}_i = I_i(\mathbf{v})$. Therefore $I_i I_i = I_i$. Thus $I_1, \ldots, I_r$ satisfy conditions (i), (ii) and (iii) of Theorem 5.4.2.

5.5

1 (1) $g_{T_A}(t) = g_A(t) = (t + 2)(t - 1)(t - 3)$. The eigenvalues of T_A are -2, 1 and 3. $W(-2; T_A) = \left\{ c \begin{pmatrix} 0 \\ 1 \\ 0 \end{pmatrix} \middle| c \in \mathbf{R} \right\}$, $W(1; T_A) = \left\{ c \begin{pmatrix} -1 \\ 1 \\ 0 \end{pmatrix} \middle| c \in \mathbf{R} \right\}$ and $W(3; T_A) = \left\{ c \begin{pmatrix} 1 \\ -1 \\ 1 \end{pmatrix} \middle| c \in \mathbf{R} \right\}$. Since $\dim(W(-2; T_A)) + \dim(W(1; T_A)) + \dim(W(3; T_A)) = 3 = \dim(\mathbf{R}^3)$, we see that $\mathbf{R}^3 = W(-2; T_A) \oplus W(1; T_A) \oplus W(3; T_A)$ and T_A is diagonalizable by Theorem 5.5.3. The basis of $\mathbf{R}^3$ consisting of eigenvectors of T_A is $\left\{ \begin{pmatrix} 0 \\ 1 \\ 0 \end{pmatrix}, \begin{pmatrix} -1 \\ 1 \\ 0 \end{pmatrix}, \begin{pmatrix} 1 \\ -1 \\ 1 \end{pmatrix} \right\}$.

(2) $g_{T_A}(t) = g_A(t) = (t - 1)^2(t - 2)$. The eigenvalues of T_A are 1 and 2. $W(1; T_A) = \left\{ c_1 \begin{pmatrix} -1 \\ 0 \\ 1 \end{pmatrix} + c_2 \begin{pmatrix} 0 \\ 1 \\ 0 \end{pmatrix} \middle| c_1, c_2 \in \mathbf{R} \right\}$ and $W(2; T_A) = \left\{ c \begin{pmatrix} 1 \\ 1 \\ 0 \end{pmatrix} \middle| c \in \mathbf{R} \right\}$. Since $\dim(W(1; T_A)) + \dim(W(2; T_A)) = 3 = \dim(\mathbf{R}^3)$, we see that $\mathbf{R}^3 = W(1; T_A) \oplus W(2; T_A)$ and T_A is diagonalizable by Theorem 5.5.3. The basis of $\mathbf{R}^3$ consisting of eigenvectors of T_A is $\left\{ \begin{pmatrix} -1 \\ 0 \\ 1 \end{pmatrix}, \begin{pmatrix} 0 \\ 1 \\ 0 \end{pmatrix}, \begin{pmatrix} 1 \\ 1 \\ 0 \end{pmatrix} \right\}$.

2 We denote the given matrices by A.

(1) $P = \begin{pmatrix} 1 & 2 \\ 1 & 1 \end{pmatrix}$, $P^{-1}AP = \begin{pmatrix} 1 & 0 \\ 0 & 4 \end{pmatrix}$.

(2) $P = \begin{pmatrix} 2 & 3 \\ 1 & 1 \end{pmatrix}$, $P^{-1}AP = \begin{pmatrix} -2 & 0 \\ 0 & 3 \end{pmatrix}$.

(3) $P = \begin{pmatrix} \sqrt{3} & -\sqrt{3} \\ 1 & 1 \end{pmatrix}$, $P^{-1}AP = \begin{pmatrix} 2-\sqrt{3} & 0 \\ 0 & 2+\sqrt{3} \end{pmatrix}$.

(4) $P = \begin{pmatrix} -1/2 & -1/2 & -1/2 \\ 1 & 0 & 1/2 \\ 0 & 1 & 1 \end{pmatrix}$, $P^{-1}AP = \begin{pmatrix} 1 & 0 & 0 \\ 0 & 1 & 0 \\ 0 & 0 & 3 \end{pmatrix}$.

(5) The matrix A is not diagonalizable.

(6) $P = \begin{pmatrix} 2 & 1 & 0 \\ 1 & 0 & -1 \\ 0 & 0 & 1 \end{pmatrix}$, $P^{-1}AP = \begin{pmatrix} 1 & 0 & 0 \\ 0 & 2 & 0 \\ 0 & 0 & 2 \end{pmatrix}$.

(7) $P = \begin{pmatrix} -2 & 1 & 3 \\ -2 & 1 & 4 \\ 1 & 0 & 1 \end{pmatrix}$, $P^{-1}AP = \begin{pmatrix} -1 & 0 & 0 \\ 0 & 1 & 0 \\ 0 & 0 & 2 \end{pmatrix}$.

(8) $P = \begin{pmatrix} 0 & -2 & -1 & 0 \\ -1 & 0 & -1 & 0 \\ 0 & 1 & 0 & 0 \\ 1 & 0 & 1 & 1 \end{pmatrix}$, $P^{-1}AP = \begin{pmatrix} 1 & 0 & 0 & 0 \\ 0 & 1 & 0 & 0 \\ 0 & 0 & 2 & 0 \\ 0 & 0 & 0 & -2 \end{pmatrix}$.

(9) $P = \begin{pmatrix} 0 & 1 & 2 & 0 \\ 0 & 0 & -1 & 0 \\ 1 & 0 & 1 & 0 \\ 0 & 1 & 2 & 1 \end{pmatrix}$, $P^{-1}AP = \begin{pmatrix} -1 & 0 & 0 & 0 \\ 0 & -1 & 0 & 0 \\ 0 & 0 & 2 & 0 \\ 0 & 0 & 0 & 3 \end{pmatrix}$.

3 (1) Since $\left(T(1)\ T(x)\ T(x^2)\right) = \left(1\ x\ x^2\right)\begin{pmatrix} 1 & 0 & 0 \\ 1 & 2 & 1 \\ 1 & 1 & 2 \end{pmatrix}$, we obtain that the eigenvalues of T are 1 and 3. Solving linear equations, we have $W(1; T) = \{c_1(-1+x) + c_2(-1+x^2) \mid c_1, c_2 \in \mathbf{R}\}$ and $W(3; T) = \{c(x+x^2) \mid c \in \mathbf{R}\}$. Since $\dim(W(1; T)) + \dim(W(3; T)) = 3 = \dim(\mathbf{R}[x]_2)$, the linear transformation T is diagonalizable and $\{-1+x, -1+x^2, x+x^2\}$ is a basis of $\mathbf{R}[x]_2$ consisting of the eigenvectors of T.

(2) Since $\left(T(1)\ T(x)\ T(x^2)\right) = \left(1\ x\ x^2\right)\begin{pmatrix} 1 & 0 & 0 \\ 1 & 1 & 0 \\ 1 & 1 & 2 \end{pmatrix}$, we obtain that the eigenvalues of T are 1 and 2. Solving linear equations, we have $W(1; T) = \{c(-x+x^2) \mid c \in \mathbf{R}\}$ and $W(2; T) = \{cx^2 \mid c \in \mathbf{R}\}$. Since $\dim(W(1; T)) + \dim(W(2; T)) = 2 \neq 3 = \dim(\mathbf{R}[x]_2)$, the linear transformation T is not diagonalizable.

(3) Since $\left(T(1) \ T(x) \ T(x^2) \right) = \left(1 \ x \ x^2 \right) \begin{pmatrix} 0 & 0 & 0 \\ 1 & 1 & 0 \\ 1 & 1 & 3 \end{pmatrix}$, we obtain that the eigen-

values of T are 0, 1 and 3. Solving linear equations, we have $W(0; T) = \{c(-1 + x) | c \in \mathbf{R}\}$, $W(1; T) = \{c(-2x + x^2) \mid c \in \mathbf{R}\}$ and $W(3; T) = \{cx^2 | c \in \mathbf{R}\}$. Since $\dim(W(0; T)) + \dim(W(1; T)) + \dim(W(3; T)) = 3 = \dim(\mathbf{R}[x]_2)$, the linear transformation T is diagonalizable and $\{-1 + x, -2x + x^2, x^2\}$ is a basis of $\mathbf{R}[x]_2$ consisting of the eigenvectors of T.

4   (1) Since $P^{-1}AP = \begin{pmatrix} 1 & 0 \\ 0 & 4 \end{pmatrix}$ with $P = \begin{pmatrix} 1 & 2 \\ 1 & 1 \end{pmatrix}$, we have $A^n = P \begin{pmatrix} 1 & 0 \\ 0 & 4 \end{pmatrix}^n P^{-1}$

$= P \begin{pmatrix} 1 & 0 \\ 0 & 4^n \end{pmatrix} P^{-1} = P \begin{pmatrix} 1 & 0 \\ 0 & 2^{2n} \end{pmatrix} P^{-1} = \begin{pmatrix} 2^{2n+1} - 1 & 2 - 2^{2n+1} \\ 2^{2n} - 1 & 2 - 2^{2n} \end{pmatrix}.$

(2) Since $P^{-1}AP = \begin{pmatrix} -2 & 0 \\ 0 & 3 \end{pmatrix}$ with $P = \begin{pmatrix} 2 & 3 \\ 1 & 1 \end{pmatrix}$, we have $A^n = P \begin{pmatrix} -2 & 0 \\ 0 & 3 \end{pmatrix}^n$

$P^{-1} = P \begin{pmatrix} (-2)^n & 0 \\ 0 & 3^n \end{pmatrix} P^{-1} = \begin{pmatrix} 3^{n+1} + (-2)^{n+1} & 6((-2)^n - 3^n) \\ 3^n - (-2)^n & -6((-2)^{n-1} + 3^{n-1}) \end{pmatrix}.$

5   This exercise is a restatement of Theorem 5.5.7 in terms of matrices. Let A be a non-zero nilpotent matrix of degree n satisfying $A^k = O_n$ $(k \geq 1)$. If A is diagonalized by a regular matrix P and $P^{-1}AP = \begin{pmatrix} \lambda_1 & & 0 \\ & \ddots & \\ 0 & & \lambda_n \end{pmatrix}$, then

$\begin{pmatrix} \lambda_1^k & & 0 \\ & \ddots & \\ 0 & & \lambda_n^k \end{pmatrix} = (P^{-1}AP)^k = P^{-1}A^k P = O_n$. Therefore $\lambda_1 = \cdots = \lambda_n = 0$.

This implies that $P^{-1}AP = O_n$. Therefore $A = PO_nP^{-1} = O_n$, which contradicts the assumption that $A \neq O_n$.

6   (1) (i) $AB = BA = \begin{pmatrix} 3 & 3 & 4 \\ 4 & 2 & 4 \\ 3 & -3 & 2 \end{pmatrix}.$

(ii) Since $g_A(t) = (t + 1)(t - 1)(t - 2)$ and $g_A(t)$ has no multiple root, we have $p_A(t) = g_A(t)$ by Theorem 5.4.6. We also see that $g_B(t) = (t - 1)(t - 2)(t - 3)$ and $g_B(t)$ has no multiple root and $p_B(t) = g_B(t)$ by Theorem 5.4.6. Since both $p_A(t)$ and $p_B(t)$ have no multiple roots, both A and B are diagonalizable by Theorem 5.5.4.

(iii)   We   see   that $W(-1; A) = W(1; B) = \left\{ c_1 \begin{pmatrix} -1 \\ 0 \\ 1 \end{pmatrix} \middle| c_1 \in \mathbf{R} \right\}$,

$W(1; A) = W(2; B) = \left\{ c_2 \begin{pmatrix} -1 \\ -1 \\ 1 \end{pmatrix} \middle| c_2 \in \mathbf{R} \right\}$ and $W(2; A) = W(3; B) = \left\{ c_1 \begin{pmatrix} 1 \\ 1 \\ 0 \end{pmatrix} \middle| c_3 \in \mathbf{R} \right\}.$

Let $P = \begin{pmatrix} -1 & -1 & 1 \\ 0 & -1 & 1 \\ 1 & 1 & 0 \end{pmatrix}$. Then $P^{-1}AP = \begin{pmatrix} -1 & 0 & 0 \\ 0 & 1 & 0 \\ 0 & 0 & 2 \end{pmatrix}$ and $P^{-1}BP = \begin{pmatrix} 1 & 0 & 0 \\ 0 & 2 & 0 \\ 0 & 0 & 3 \end{pmatrix}$. We note that the triplet of matrices P, $P^{-1}AP$, $P^{-1}BP$ is not unique.

(2) (i) $AB = BA = \begin{pmatrix} 10 & -4 & 4 \\ 11 & -5 & 4 \\ 3 & -3 & 2 \end{pmatrix}$.

(ii) The characteristic polynomial of A is $g_A(t) = (t+1)(t-2)^2$. Solving linear equations, we have dim $(W(-1; A)) = 1$ and dim $(W(2; A)) = 2$. Since dim $(\mathbf{R}^3) = $ dim $(W(-1; A)) + $ dim $(W(2; A))$, A is diagonalizable by Theorem 5.5.4. The characteristic polynomial of B is $g_B(t) = (t-1)^2(t-3)$. Solving linear equations, we have dim $(W(1; B)) = 2$ and dim $(W(3; B)) = 1$. Since dim $(\mathbf{R}^3) = $ dim $(W(1; B)) + $ dim $(W(3; B))$, B is also diagonalizable by Theorem 5.5.4.

(iii) We shall find a matrix P such that both $P^{-1}AP$ and $P^{-1}BP$ are diagonal matrices. Solving linear equations, we have $W(-1; A) = \left\{ c_1 \begin{pmatrix} 0 \\ 1 \\ 1 \end{pmatrix} \middle| c_1 \in \mathbf{R} \right\}$

and $B \begin{pmatrix} 0 \\ 1 \\ 1 \end{pmatrix} = \begin{pmatrix} 0 \\ 1 \\ 1 \end{pmatrix}$. We also have $W(3; B) = \left\{ c_2 \begin{pmatrix} 1 \\ 1 \\ 0 \end{pmatrix} \middle| c_2 \in \mathbf{R} \right\}$ and

$A \begin{pmatrix} 1 \\ 1 \\ 0 \end{pmatrix} = 2 \begin{pmatrix} 1 \\ 1 \\ 0 \end{pmatrix}$. We shall find another eigenvector of A with the eigenvalue 2 which is also an eigenvector of B with the eigenvalue 1. We note that

$W(1; B) = \left\{ c_3 \begin{pmatrix} 1 \\ 2 \\ 0 \end{pmatrix} + c_4 \begin{pmatrix} -1 \\ 0 \\ 2 \end{pmatrix} \middle| c_3, c_4 \in \mathbf{R} \right\}$. Since the matrices A and B are commutative, A induces a transformation of $W(1; B)$. Applying A to the subspace $W(1; B)$, we obtain an eigenvector $\begin{pmatrix} 1 \\ 1 \\ -1 \end{pmatrix}$ of A with the eigenvalue

2. Therefore, if we let $P = \begin{pmatrix} 0 & 1 & 1 \\ 1 & 1 & 1 \\ 1 & -1 & 0 \end{pmatrix}$, then we have $P^{-1}AP = \begin{pmatrix} -1 & 0 & 0 \\ 0 & 2 & 0 \\ 0 & 0 & 2 \end{pmatrix}$

and $P^{-1}BP = \begin{pmatrix} 1 & 0 & 0 \\ 0 & 1 & 0 \\ 0 & 0 & 3 \end{pmatrix}$. We note that the triplet of matrices P, $P^{-1}AP$, $P^{-1}BP$ is not unique.

7   Since $g_A(t) = |tE_n - A| = (t - a_{11}) \cdots (t - a_{nn}) +$ (a polynomial of t whose degree is equal to $n - 2$ or less) by the definition of the determinant, we have $a_{n-1} = -(a_{11} + \cdots + a_{nn})$ by comparing the terms of degree $n - 1$. Substituting $t = 0$ for $g_A(t)$, we have $a_0 = g_A(0) = |0E_n - A| = |-A| = (-1)^n |A|$.

8   Let $A = (a_{ij})$ and $B = (b_{ij})$. Since $\operatorname{tr}(AB) = \sum_{i=1}^{m} \left(\sum_{k=1}^{n} a_{ik} b_{ki} \right)$ and

$\operatorname{tr}(BA) = \sum_{k=1}^{n} \left(\sum_{i=1}^{m} b_{ki} a_{ik} \right)$, we see that $\operatorname{tr}(AB) = \operatorname{tr}(BA)$.

9   Let $B = P^{-1}AP$. Then $\operatorname{tr}(B) = \operatorname{tr}(P^{-1}AP) = \operatorname{tr}(APP^{-1}) = \operatorname{tr}(A)$ by Exercise 8 above.

10   We shall show the contraposition. The matrix A is not a regular matrix if and only if the linear equation $A\mathbf{x} = \mathbf{0}$ has a non-trivial solution by Theorem 2.4.2. The linear equation $A\mathbf{x} = \mathbf{0}$ has a non-trivial solution if and only if A has an eigenvalue 0. So, A has the characteristic root 0. Thus the assertion is proved.

11 Since $p_A(t) = p_T(t)$ by Theorem 5.4.7(3), we see the assertion by Theorem 5.5.3 and Theorem 5.5.4.

5.6

1   (1) We denote by $\{f_1, f_2\}$ the dual basis of the standard basis $\{\mathbf{e}_1, \mathbf{e}_2\}$ of $\mathbf{R}^2$. Let $\{g_1, g_2\}$ be the dual basis of $\{\mathbf{a}_1, \mathbf{a}_2\}$. Since $\begin{pmatrix} \mathbf{a}_1 & \mathbf{a}_2 \end{pmatrix} = \begin{pmatrix} \mathbf{e}_1 & \mathbf{e}_2 \end{pmatrix} \begin{pmatrix} 1 & 1 \\ -1 & 0 \end{pmatrix}$, we have $\begin{pmatrix} g_1 & g_2 \end{pmatrix} = \begin{pmatrix} f_1 & f_2 \end{pmatrix} {}^t\begin{pmatrix} 1 & 1 \\ -1 & 0 \end{pmatrix}^{-1} = \begin{pmatrix} f_1 & f_2 \end{pmatrix} \begin{pmatrix} 0 & 1 \\ -1 & 1 \end{pmatrix}$ by Theorem 5.6.4. In other words, $g_1 = -f_2$ and $g_2 = f_1 + f_2$.

(2) We denote by $\{f_1, f_2, f_3\}$ the dual basis of the standard basis $\{\mathbf{e}_1, \mathbf{e}_2, \mathbf{e}_3\}$ of $\mathbf{R}^3$. Let $\{g_1, g_2, g_3\}$ be the dual basis of $\{\mathbf{a}_1, \mathbf{a}_2, \mathbf{a}_3\}$. Since $\begin{pmatrix} \mathbf{a}_1 & \mathbf{a}_2 & \mathbf{a}_3 \end{pmatrix} = \begin{pmatrix} \mathbf{e}_1 & \mathbf{e}_2 & \mathbf{a}_3 \end{pmatrix} \times \begin{pmatrix} 2 & 0 & 1 \\ 1 & 1 & 0 \\ 0 & -1 & 1 \end{pmatrix}$, we have $\begin{pmatrix} g_1 & g_2 & g_3 \end{pmatrix} = \begin{pmatrix} f_1 & f_2 & f_3 \end{pmatrix} {}^t\begin{pmatrix} 2 & 0 & 1 \\ 1 & 1 & 0 \\ 0 & -1 & 1 \end{pmatrix}^{-1} = \begin{pmatrix} f_1 & f_2 & f_3 \end{pmatrix} \begin{pmatrix} 1 & -1 & 1 \\ -1 & 2 & 2 \\ -1 & 1 & 2 \end{pmatrix}$ by Theorem 5.6.4. In other words, $g_1 = f_1 - f_2 - f_3$, $g_2 = -f_1 + 2f_2 + f_3$ and $g_3 = f_1 + 2f_2 + 2f_3$.

2   (1) Let $\{\mathbf{e}_1, \mathbf{e}_2\}$ be the standard basis of $\mathbf{R}^2$. Then the representation matrix of T with respect to the standard basis is $\begin{pmatrix} 3 & -1 \\ 2 & 1 \end{pmatrix}$. Since $\begin{pmatrix} \mathbf{a}_1 & \mathbf{a}_2 \end{pmatrix} = \begin{pmatrix} \mathbf{e}_1 & \mathbf{e}_2 \end{pmatrix} \begin{pmatrix} 2 & -1 \\ -1 & 1 \end{pmatrix}$, the representation matrix of the linear transformation T with respect to the basis $\{\mathbf{a}_1, \mathbf{a}_2\}$ is $\begin{pmatrix} 2 & -1 \\ -1 & 1 \end{pmatrix}^{-1} \begin{pmatrix} 3 & -1 \\ 2 & 1 \end{pmatrix} \begin{pmatrix} 2 & -1 \\ -1 & 1 \end{pmatrix} = \begin{pmatrix} 10 & -5 \\ 13 & -6 \end{pmatrix}$

by Theorem 5.2.6.

(2) Let $\{g_1, g_2\}$ be the dual basis of $\{\mathbf{a}_1, \mathbf{a}_2\}$. Then $\left(T^*(g_1) \ T^*(g_2) \right) =$

$\left(g_1 \ g_2 \right) {}^t\!\begin{pmatrix} 10 & -5 \\ 13 & -6 \end{pmatrix}$ by Theorem 5.6.5. Therefore the representation matrix of

the dual transformation T^* with respect to the basis $\{g_1, g_2\}$ of V^* is

${}^t\!\begin{pmatrix} 10 & -5 \\ 13 & -6 \end{pmatrix} = \begin{pmatrix} 10 & 13 \\ -5 & -6 \end{pmatrix}$.

3 (1) Since $\left(T(1) \ T(x) \ T(x^2) \right) = \left(1 \ x \ x^2 \right) \begin{pmatrix} 2 & 0 & 0 \\ 1 & 4 & 1 \\ 1 & 1 & 5 \end{pmatrix}$, we have $A = \begin{pmatrix} 2 & 0 & 0 \\ 1 & 4 & 1 \\ 1 & 1 & 5 \end{pmatrix}$.

(2) Since $\left(f_1(x) \ f_2(x) \ f_3(x) \right) = \left(1 \ x \ x^2 \right) \begin{pmatrix} 1 & 0 & -1 \\ 2 & 1 & 0 \\ 0 & 1 & 1 \end{pmatrix}$ and $\begin{vmatrix} 1 & 0 & -1 \\ 2 & 1 & 0 \\ 0 & 1 & 1 \end{vmatrix}$

$= -1 \neq 0$, we have $\begin{pmatrix} 1 & 0 & -1 \\ 2 & 1 & 0 \\ 0 & 1 & 1 \end{pmatrix}$ is a regular matrix and $\{ f_1(x), \ f_2(x), \ f_3(x) \}$

is a basis of $\mathbf{R}[x]_2$.

(3) Let $P = \begin{pmatrix} 1 & 0 & -1 \\ 2 & 1 & 0 \\ 0 & 1 & 1 \end{pmatrix}$. Then $B = P^{-1}AP = \begin{pmatrix} 4 & -1 & -2 \\ 1 & 7 & 4 \\ 2 & -1 & 0 \end{pmatrix}$ by Theorem

5.2.6.

(4) The representation matrix of T^* is ${}^t\!B = \begin{pmatrix} 4 & 1 & 2 \\ -1 & 7 & -1 \\ -2 & 4 & 0 \end{pmatrix}$ by Theorem 5.6.5.

4 Since W satisfies the three conditions of Theorem 4.1.1, W is a subspace of $M_{n \times n}(\mathbf{R})$. $\dim(W) = 1 + 2 + \cdots + n = \frac{n(n+1)}{2}$.

5 Since W satisfies the three conditions of Theorem 4.1.1, W is a subspace of $M_{n \times n}(\mathbf{R})$. Let $X \in W$. Then each column of X is a solution of the linear equation $A\mathbf{x} = \mathbf{0}_m$. Since the dimension of the solution space of $A\mathbf{x} = \mathbf{0}_m$ is $n - \text{rank}(A) = n - r$ by Theorem 4.4.3, we have $\dim(W) = n(n-r)$.

6 Since W satisfies the three conditions of Theorem 4.1.1, W is a subspace of $M_{n \times n}(\mathbf{R})$. Since a matrix $A = \left(a_{ij} \right)$ is a symmetric matrix if and only if $a_{ij} = a_{ji}$ for $i, j = 1, \ldots, n$, we have $\dim(V) = 1 + 2 + \cdots + n = \frac{n(n+1)}{2}$.

7 Since W satisfies the three conditions of Theorem 4.1.1, W is a subspace of $M_{n \times n}(\mathbf{R})$. A matrix $A = \left(a_{ij} \right)$ is an alternating matrix if and only if $a_{ji} = -a_{ij}$ for $i = 1, \ldots, n \, ; \, j = 1, \ldots, n$. Then we have $a_{ii} = 0$ for $i = 1, \ldots, n$ and A is determined by a_{ij} for $1 \leq j < i \leq n$. Therefore $\dim(V) = 1 + 2 + \cdots + (n - 1) = \frac{n(n-1)}{2}$.

8   Since the set $\mathbf{C}^n$ have the addition and the scalar multiplication of vector space by taking $\mathbf{R}$ as the basic field in the definition of vector spaces, $\mathbf{C}^n$ is a vector space over $\mathbf{R}$. Let $\mathbf{e}_1, \ldots, \mathbf{e}_n$ be the unit vectors in $\mathbf{R}^n \subset \mathbf{C}^n$. Then it is easy to see that $\{\mathbf{e}_1, i\mathbf{e}_1, \ldots, \mathbf{e}_n, i\mathbf{e}_n\}$ is a basis of the vector space $\mathbf{C}^n$ over $\mathbf{R}$. Here $i = \sqrt{-1}$, the imaginary unit. Therefore $\dim(\mathbf{C}^n) = 2n$.

9   We shall show that (i) $V/U = W_1/U + W_2/U$ and (ii) $(W_1/U) \cap (W_2/U) = \{\mathbf{0}_{V/U}\}$. (i) is clear. To see (ii), we assume that $\mathrm{cl}(\mathbf{v}) \in W_1/U \cap W_2/U$. Then $\mathbf{v} \in W_1 \cap W_2 = U$. Therefore $\mathrm{cl}(\mathbf{v}) = \mathbf{0}_{V/U}$.

6.1

1   (1) 6.   (2) 3.   (3) 18/5.

2   (1) $3\sqrt{2}$.   (2) $\sqrt{14}$.   (3) $4\sqrt{10}/5$.

3   (1) $a = 1$.   (2) $a = -1/3$.

4   (1) $\pm \begin{pmatrix} 1/\sqrt{26} \\ 3/\sqrt{26} \\ 4/\sqrt{26} \end{pmatrix}$.   (2) $(\sqrt{10}/4)(1 - 3x^2)$.

5   (1) $\|\mathbf{u} + \mathbf{v}\|^2 + \|\mathbf{u} - \mathbf{v}\|^2 = \{(\mathbf{u}, \mathbf{u}) + 2(\mathbf{u}, \mathbf{v}) + (\mathbf{v}, \mathbf{v})\} + \{(\mathbf{u}, \mathbf{u}) - 2(\mathbf{u}, \mathbf{v}) + (\mathbf{v}, \mathbf{v})\} = 2\{(\mathbf{u}, \mathbf{u}) + (\mathbf{v}, \mathbf{v})\} = 2(\|\mathbf{u}\|^2 + \|\mathbf{v}\|^2)$.

(2) Since $\|\mathbf{u} + \mathbf{v}\|^2 = (\mathbf{u}, \mathbf{u}) + 2(\mathbf{u}, \mathbf{v}) + (\mathbf{v}, \mathbf{v}) = \|\mathbf{u}\|^2 + \|\mathbf{v}\|^2 + 2(\mathbf{u}, \mathbf{v})$, we have $\|\mathbf{u} + \mathbf{v}\|^2 = \|\mathbf{u}\|^2 + \|\mathbf{v}\|^2 \Leftrightarrow (\mathbf{u}, \mathbf{v}) = 0$.

(3) Since $(\mathbf{u} + \mathbf{v}, \mathbf{u} - \mathbf{v}) = \|\mathbf{u}\|^2 - \|\mathbf{v}\|^2$, we have $(\mathbf{u} + \mathbf{v}, \mathbf{u} - \mathbf{v}) = 0 \Leftrightarrow \|\mathbf{u}\|^2 = \|\mathbf{v}\|^2 \Leftrightarrow \|\mathbf{u}\| = \|\mathbf{v}\|$.

(4) We have $\|\mathbf{u} + \mathbf{v}\|^2 - \|\mathbf{u}\|^2 - \|\mathbf{v}\|^2 = 2(\mathbf{u}, \mathbf{v})$. Dividing both sides of the equality by 2, we have the equality we want to show.

6   We have only to verify the three conditions of Theorem 4.1.1 for the subset $W^\perp$ of V.

7   (1) We see that $W = \left\{ c_1 \begin{pmatrix} -1 \\ 1 \\ 0 \end{pmatrix} + c_2 \begin{pmatrix} -1 \\ 0 \\ 1 \end{pmatrix} \middle| c_1, c_2 \in \mathbf{R} \right\}$. Solving the

linear equation $\begin{cases} -x_1 + x_2 = 0, \\ -x_2 + x_3 = 0 \end{cases}$, we have $W^\perp = \left\{ c \begin{pmatrix} 1 \\ 1 \\ 1 \end{pmatrix} \middle| c \in \mathbf{R} \right\}$.

Then $\dim(W^\perp) = 1$ and we have a basis $\left\{ \begin{pmatrix} 1 \\ 1 \\ 1 \end{pmatrix} \right\}$ of $W^\perp$.

(2) Since $W = \left\{ c \begin{pmatrix} 5 \\ -3 \\ 1 \end{pmatrix} \middle| c \in \mathbf{R} \right\}$, we have $W^\perp = \{\mathbf{x} \in \mathbf{R}^3 \mid 5x_1 - 3x_2 + x_3 = 0\}$. Solving the linear equation $5x_1 - 3x_2 + x_3 = 0$, we have

$$W^\perp = \left\{ c_1 \begin{pmatrix} 3 \\ 5 \\ 0 \end{pmatrix} + c_2 \begin{pmatrix} -1 \\ 0 \\ 5 \end{pmatrix} \,\middle|\, c_1, c_2 \in \mathbf{R} \right\}. \text{ Then } \dim(W^\perp) = 2 \text{ and we have}$$

a basis $\left\{ \begin{pmatrix} 3 \\ 5 \\ 0 \end{pmatrix}, \begin{pmatrix} -1 \\ 0 \\ 5 \end{pmatrix} \right\}$ of $W^\perp$.

(3) Since $W = \left\{ c \begin{pmatrix} 3 \\ 2 \\ 1 \end{pmatrix} \,\middle|\, c \in \mathbf{R} \right\}$, we have $W^\perp = \{\mathbf{x} \in \mathbf{R}^3 \mid 3x_1 +$

$2x_2 + x_3 = 0\}$. Solving the linear equation $3x_1 + 2x_2 + x_3 = 0$, we have

$$W^\perp = \left\{ c_1 \begin{pmatrix} -2 \\ 3 \\ 0 \end{pmatrix} + c_2 \begin{pmatrix} -1 \\ 0 \\ 3 \end{pmatrix} \,\middle|\, c_1, c_2 \in \mathbf{R} \right\}. \text{ Then } \dim(W^\perp) = 2 \text{ and we}$$

have a basis $\left\{ \begin{pmatrix} -2 \\ 3 \\ 0 \end{pmatrix}, \begin{pmatrix} -1 \\ 0 \\ 3 \end{pmatrix} \right\}$ of $W^\perp$.

8 (1) $W_1 = \left\{ c_1 \begin{pmatrix} -1 \\ 1 \\ 0 \\ 0 \end{pmatrix} + c_2 \begin{pmatrix} -1 \\ 0 \\ 1 \\ 0 \end{pmatrix} + c_3 \begin{pmatrix} -1 \\ 0 \\ 0 \\ 1 \end{pmatrix} \,\middle|\, c_1, c_2, c_3 \in \mathbf{R} \right\}$,

$W_2 = \left\{ c \begin{pmatrix} 1 \\ 1 \\ 1 \\ 1 \end{pmatrix} \,\middle|\, c \in \mathbf{R} \right\}$. Since the vector $\begin{pmatrix} 1 \\ 1 \\ 1 \\ 1 \end{pmatrix}$ is orthogonal to the vectors

$\begin{pmatrix} -1 \\ 1 \\ 0 \\ 0 \end{pmatrix}, \begin{pmatrix} -1 \\ 0 \\ 1 \\ 0 \end{pmatrix}, \begin{pmatrix} -1 \\ 0 \\ 0 \\ 1 \end{pmatrix}$, the subspaces W_1 and W_2 are orthogonal.

(2) $W_1 = \left\{ c_1 \begin{pmatrix} -1 \\ 0 \\ 1 \\ 0 \end{pmatrix} + c_2 \begin{pmatrix} -1 \\ 2 \\ 0 \\ 1 \end{pmatrix} \,\middle|\, c_1, c_2 \in \mathbf{R} \right\}$,

$W_2 = \left\{ c_1 \begin{pmatrix} 2 \\ 1 \\ 2 \\ 0 \end{pmatrix} + c_2 \begin{pmatrix} 0 \\ -1 \\ 0 \\ 2 \end{pmatrix} \,\middle|\, c_1, c_2 \in \mathbf{R} \right\}$. Since the vectors $\begin{pmatrix} -1 \\ 0 \\ 1 \\ 0 \end{pmatrix}, \begin{pmatrix} -1 \\ 2 \\ 0 \\ 1 \end{pmatrix}$

are orthogonal to the vectors $\begin{pmatrix} 2 \\ 1 \\ 2 \\ 0 \end{pmatrix}, \begin{pmatrix} 0 \\ -1 \\ 0 \\ 2 \end{pmatrix}$, the subspaces W_1 and W_2 are

orthogonal.

(3) $W_1 = \left\{ c_1 \begin{pmatrix} 2 \\ -1 \\ 1 \\ 0 \end{pmatrix} + c_2 \begin{pmatrix} 1 \\ -1 \\ 0 \\ 1 \end{pmatrix} \middle| c_1, c_2 \in \mathbf{R} \right\}$,

$W_2 = \left\{ c_1 \begin{pmatrix} -1 \\ -1 \\ 1 \\ 0 \end{pmatrix} + c_2 \begin{pmatrix} 1 \\ 2 \\ 0 \\ 1 \end{pmatrix} \middle| c_1, c_2 \in \mathbf{R} \right\}$. Since the vectors $\begin{pmatrix} 2 \\ -1 \\ 1 \\ 0 \end{pmatrix}, \begin{pmatrix} 1 \\ -1 \\ 0 \\ 1 \end{pmatrix}$

are orthogonal to the vectors $\begin{pmatrix} -1 \\ -1 \\ 1 \\ 0 \end{pmatrix}, \begin{pmatrix} 1 \\ 2 \\ 0 \\ 1 \end{pmatrix}$, the subspaces W_1 and W_2 are

orthogonal.

6.2

1 (1) $\left\{ \begin{pmatrix} 1/\sqrt{2} \\ 1/\sqrt{2} \\ 0 \end{pmatrix}, \begin{pmatrix} 0 \\ 0 \\ 1 \end{pmatrix}, \begin{pmatrix} 1/\sqrt{2} \\ -1/\sqrt{2} \\ 0 \end{pmatrix} \right\}$.

(2) $\left\{ \begin{pmatrix} 2/\sqrt{6} \\ 1/\sqrt{6} \\ 1/\sqrt{6} \end{pmatrix}, \begin{pmatrix} 0 \\ -1/\sqrt{2} \\ 1/\sqrt{2} \end{pmatrix}, \begin{pmatrix} -1/\sqrt{3} \\ 1/\sqrt{3} \\ 1/\sqrt{3} \end{pmatrix} \right\}$.

(3) $\left\{ \begin{pmatrix} 1/\sqrt{3} \\ 1/\sqrt{3} \\ 1/\sqrt{3} \end{pmatrix}, \begin{pmatrix} 1/\sqrt{6} \\ -2/\sqrt{6} \\ 1/\sqrt{6} \end{pmatrix}, \begin{pmatrix} 1/\sqrt{2} \\ 0 \\ -1/\sqrt{2} \end{pmatrix} \right\}$.

(4) $\left\{ \begin{pmatrix} 1/\sqrt{2} \\ 1/\sqrt{2} \\ 0 \\ 0 \end{pmatrix}, \begin{pmatrix} -1/\sqrt{6} \\ 1/\sqrt{6} \\ 2/\sqrt{6} \\ 0 \end{pmatrix}, \begin{pmatrix} 1/\sqrt{12} \\ -1/\sqrt{12} \\ 1/\sqrt{12} \\ 3/\sqrt{12} \end{pmatrix}, \begin{pmatrix} -1/2 \\ 1/2 \\ -1/2 \\ 1/2 \end{pmatrix} \right\}$.

2 (1) $\{ (\sqrt{5}/\sqrt{2})x^2, (\sqrt{3}/\sqrt{2})x, (3/2\sqrt{2})(1 - (5/3)x^2) \}$.

(2) $\{ (\sqrt{6}/4)(1 + x), (\sqrt{10}/4)(-1 + x + 2x^2), (\sqrt{2}/4)(-1 - 2x + 5x^2) \}$.

3 We have only to verify $^tPP = E_3$.

4 (1) $a = \pm(\sqrt{3}/3)$, $b = \pm(\sqrt{2}/2)$, $c = \pm(\sqrt{6}/6)$.

(2) $a = \pm(\sqrt{6}/6)$, $b = \pm(\sqrt{2}/2)$, $c = \pm(\sqrt{3}/3)$.

5 (1) We note that $\mathbf{0}_V \in W \cap W^\perp$. Let $\mathbf{u} \in W \cap W^\perp$. Since $\mathbf{u} \in W^\perp$, $\mathbf{u}$ is orthogonal to any vector in W. In particular, $\mathbf{u}$ is orthogonal to $\mathbf{u}$ itself. Thus we have $\|\mathbf{u}\|^2 = (\mathbf{u}, \mathbf{u}) = 0$. Therefore $\mathbf{u} = \mathbf{0}_V$.

(2) Let $\mathbf{u} \in V$ and $\{\mathbf{u}_1, \ldots, \mathbf{u}_r\}$ be an orthonormal basis of W. We let $\mathbf{v} = (\mathbf{u}, \mathbf{u}_1)\mathbf{u}_1 + \cdots + (\mathbf{u}, \mathbf{u}_r)\mathbf{u}_r \in W$ and $\mathbf{w} = \mathbf{u} - \mathbf{v}$. Then $(\mathbf{w}, \mathbf{u}_i) = 0$ for any $i = 1, \ldots, r$. Thus we see that $\mathbf{w} \in W^\perp$ and obtain $\mathbf{u} = \mathbf{v} + \mathbf{w} \in W + W^\perp$. Therefore $V = W + W^\perp$. Since $W \cap W^\perp = \{\mathbf{0}_V\}$ by (1), we have $V = W \oplus W^\perp$ by Theorem 5.4.1.

(3) If $\mathbf{u} \in W$, then $(\mathbf{u}, \mathbf{v}) = 0$ for any $\mathbf{v} \in W^{\perp}$. Therefore we have $W \subset (W^{\perp})^{\perp}$. Let $\mathbf{u} \in (W^{\perp})^{\perp}$. Then $\mathbf{u} = \mathbf{v} + \mathbf{w}$ $(\mathbf{v} \in W, \mathbf{w} \in W^{\perp})$ by (2). Since $\mathbf{u} \in (W^{\perp})^{\perp}$, we have $(\mathbf{u}, \mathbf{w}) = 0$. On the other hand, we see that $0 = (\mathbf{u}, \mathbf{w}) = (\mathbf{v} + \mathbf{w}, \mathbf{w}) = \|\mathbf{w}\|^2$. Then $\mathbf{w} = \mathbf{0}_V$. This implies that $\mathbf{u} = \mathbf{v} \in W$ and $(W^{\perp})^{\perp} \subset W$. Thus we have $(W^{\perp})^{\perp} = W$.

(4) $(\Rightarrow)$ Let $\mathbf{u} \in W_2^{\perp}$. Then $\mathbf{u}$ is orthogonal to any vector in W_2. In particular, $\mathbf{u}$ is orthogonal to any vector in W_1, since $W_1 \subset W_2$. Therefore $\mathbf{u} \in W_1^{\perp}$. This implies $W_2^{\perp} \subset W_1^{\perp}$. $(\Leftarrow)$ Assume $W_1^{\perp} \supset W_2^{\perp}$. By (3) and $(\Rightarrow)$, we have $W_1 = (W_1^{\perp})^{\perp} \subset (W_2^{\perp})^{\perp} = W_2$.

(5) is clear by (2).

6 (1) Since ${}^tPP = E_n$, we have ${}^tP = P^{-1}$. Then $P{}^tP = PP^{-1} = E_n$, or ${}^t({}^tP){}^tP = E_n$. Therefore ${}^tP$ is an orthogonal matrix. Since $P^{-1} = {}^tP$, P^{-1} is an orthogonal matrix.

(2) Since ${}^t(PQ)PQ = {}^tQ{}^tPPQ = {}^tQQ = E_n$, PQ is an orthogonal matrix.

7 $(\Rightarrow)$ is obvious. $(\Leftarrow)$ By Exercises $(\S 6.1)5(4)$, we have $(T(\mathbf{u}), T(\mathbf{v})) = (1/2)(\|T(\mathbf{u} + \mathbf{v})\|^2 - \|T(\mathbf{u})\|^2 - \|T(\mathbf{v})\|^2) = (1/2)(\|\mathbf{u} + \mathbf{v}\|^2 - \|\mathbf{u}\|^2 - \|\mathbf{v}\|^2) = (\mathbf{u}, \mathbf{v})$.

6.3

1 Answers are not uniquely determined. We denote the given matrices by A and orthogonal matrices by P.

(1) $P = \begin{pmatrix} 1/\sqrt{2} & -1/\sqrt{2} \\ 1/\sqrt{2} & 1/\sqrt{2} \end{pmatrix}$, $P^{-1}AP = \begin{pmatrix} 1 & 0 \\ 0 & -1 \end{pmatrix}$.

(2) $P = \begin{pmatrix} 1/\sqrt{2} & -1/\sqrt{2} \\ 1/\sqrt{2} & 1/\sqrt{2} \end{pmatrix}$, $P^{-1}AP = \begin{pmatrix} -1 & 0 \\ 0 & 3 \end{pmatrix}$.

(3) $P = \begin{pmatrix} -1/\sqrt{2} & 1/\sqrt{2} \\ 1/\sqrt{2} & 1/\sqrt{2} \end{pmatrix}$, $P^{-1}AP = \begin{pmatrix} -1 & 0 \\ 0 & 5 \end{pmatrix}$.

(4) $P = \begin{pmatrix} 0 & 1/\sqrt{2} & -1/\sqrt{2} \\ 1 & 0 & 0 \\ 0 & 1/\sqrt{2} & 1/\sqrt{2} \end{pmatrix}$, $P^{-1}AP = \begin{pmatrix} 1 & 0 & 0 \\ 0 & 1 & 0 \\ 0 & 0 & -1 \end{pmatrix}$.

(5) $P = \begin{pmatrix} -1/\sqrt{2} & 0 & 1/\sqrt{2} \\ 0 & 1 & 0 \\ 1/\sqrt{2} & 0 & 1/\sqrt{2} \end{pmatrix}$, $P^{-1}AP = \begin{pmatrix} 0 & 0 & 0 \\ 0 & 1 & 0 \\ 0 & 0 & 2 \end{pmatrix}$.

(6) $P = \begin{pmatrix} -3/\sqrt{10} & 1/\sqrt{14} & 1/\sqrt{35} \\ 0 & -2/\sqrt{14} & 5/\sqrt{35} \\ 1/\sqrt{10} & 3/\sqrt{14} & 3/\sqrt{35} \end{pmatrix}$, $P^{-1}AP = \begin{pmatrix} 1 & 0 & 0 \\ 0 & -1 & 0 \\ 0 & 0 & 6 \end{pmatrix}$.

2 Answers are not uniquely determined. We denote the given matrices by A and orthogonal matrices by P.

(1) We have only to orthonormalize the eigenvectors for the eigenvalues $-1, 1, 2$

of A. $P = \begin{pmatrix} 0 & -1 & 0 \\ 0 & 0 & 1 \\ 1 & 0 & 0 \end{pmatrix}$, $P^{-1}AP = \begin{pmatrix} -1 & 2 & 4 \\ 0 & 1 & -2 \\ 0 & 0 & 2 \end{pmatrix}$.

(2) We have only to orthonormalize the eigenvectors for the eigenvalues $-1, 2, 3$ of A.

$$P = \begin{pmatrix} -2/3 & 1/\sqrt{18} & -1/\sqrt{2} \\ -2/3 & 1/\sqrt{18} & 1/\sqrt{2} \\ 1/3 & 4/\sqrt{18} & 0 \end{pmatrix}, \quad P^{-1}AP = \begin{pmatrix} -1 & -6\sqrt{2} & 14\sqrt{2}/3 \\ 0 & 2 & 11/3 \\ 0 & 0 & 3 \end{pmatrix}.$$

3 ("If" part.) Any real symmetric matrix is diagonalizable by an orthogonal matrix by Theorem 6.3.7. ("Only if" part.) Assume that a real square matrix A is diagonalizable by an orthogonal matrix P and $P^{-1}AP = D$ (a diagonal matrix). Since $A = PDP^{-1}$, we have ${}^tA = {}^t(PDP^{-1}) = ({}^tP^{-1}){}^tD{}^tP = PDP^{-1} = A$. Therefore A is a symmetric matrix.

4 Let $f(x) = a_0 + a_1x$ and $g(x) = b_0 + b_1x$. Then $(T(f(x)), g(x)) = (a_1 + 3a_0x, b_0 + b_1x) = \int_{-1}^{1} (a_1 + 3a_0x)(b_0 + b_1x)dx = \int_{-1}^{1} (a_1b_0 + (3a_0b_0 + a_1b_1x) + 3a_0b_1x^2)\,dx = \left[a_1b_0x + \dfrac{3a_0b_0 + a_1b_1}{2}x^2 + a_0b_1x^3 \right]_{-1}^{1} = 2(a_1b_0 + a_0b_1)$.

On the other hand, $(f(x), T(g(x))) = (T(g(x)), f(x)) = 2(b_1a_0 + b_0a_1) = (T(f(x)), g(x))$. Namely, $(T(f(x)), g(x)) = (f(x), T(g(x)))$. Therefore $T^* = T$.

5 Since $\dim(V^*) = \dim(V) = n$, we have only to prove that $p_{u_i}(u_j) = \delta_{i,j}$. This is true since $p_{u_i}(u_j) = (u_j, u_i) = \delta_{i,j}$ by definition.

6 Let $a, b \in R^n$. Then $(a, (T_A)^*(b)) = (T_A(a), b) = {}^t(Aa)b = ({}^ta\,{}^tA)b = {}^ta({}^tAb) = (a, T_{{}^tA}(b))$. Therefore $(T_A)^* = T_{{}^tA}$.

6.4

1 We denote by A the coefficient matrices of quadratic forms.

(1) $q(x_1, x_2) = A[\mathbf{x}] = D[\mathbf{y}] = 3y_1^2 - y_2^2$. $A = \begin{pmatrix} 1 & 2 \\ 2 & 1 \end{pmatrix}$, $P = \dfrac{\sqrt{2}}{2}\begin{pmatrix} 1 & -1 \\ 1 & 1 \end{pmatrix}$,

$P^{-1} = {}^tP = \dfrac{\sqrt{2}}{2}\begin{pmatrix} 1 & 1 \\ -1 & 1 \end{pmatrix}$, $\mathbf{y} = P^{-1}\mathbf{x}$, $D = {}^tPAP = \begin{pmatrix} 3 & 0 \\ 0 & -1 \end{pmatrix}$.

(2) $q(x_1, x_2, x_3) = A[\mathbf{x}] = D[\mathbf{y}] = \sqrt{2}y_1^2 - \sqrt{2}y_2^2$. $A = \begin{pmatrix} 0 & 1 & 0 \\ 1 & 0 & 1 \\ 0 & 1 & 0 \end{pmatrix}$,

$P = \dfrac{1}{2}\begin{pmatrix} 1 & 1 & -\sqrt{2} \\ \sqrt{2} & -\sqrt{2} & 0 \\ 1 & 1 & \sqrt{2} \end{pmatrix}$, $P^{-1} = {}^tP = \dfrac{1}{2}\begin{pmatrix} 1 & \sqrt{2} & 1 \\ 1 & -\sqrt{2} & 1 \\ -\sqrt{2} & 0 & \sqrt{2} \end{pmatrix}$,

$\mathbf{y} = P^{-1}\mathbf{x}$, $D = {}^tPAP = \begin{pmatrix} \sqrt{2} & 0 & 0 \\ 0 & -\sqrt{2} & 0 \\ 0 & 0 & 0 \end{pmatrix}$.

(3) $q(x_1, x_2, x_3) = A[\mathbf{x}] = D[\mathbf{y}] = y_1^2 + 3y_2^2 - y_3^2.$

$$A = \begin{pmatrix} 1 & \sqrt{2} & 0 \\ \sqrt{2} & 1 & \sqrt{2} \\ 0 & \sqrt{2} & 1 \end{pmatrix}, \quad P = \frac{1}{2}\begin{pmatrix} -\sqrt{2} & 1 & 1 \\ 0 & \sqrt{2} & -\sqrt{2} \\ \sqrt{2} & 1 & 1 \end{pmatrix},$$

$$P^{-1} = {}^tP = \frac{1}{2}\begin{pmatrix} -\sqrt{2} & 0 & \sqrt{2} \\ 1 & \sqrt{2} & 1 \\ 1 & -\sqrt{2} & 1 \end{pmatrix}, \quad \mathbf{y} = P^{-1}\mathbf{x},$$

$$D = {}^tPAP = \begin{pmatrix} 1 & 0 & 0 \\ 0 & 3 & 0 \\ 0 & 0 & -1 \end{pmatrix}.$$

2   We denote by A the coefficient matrices of quadratic forms.

(1) $q(x_1, x_2, x_3) = (x_1 + x_2)^2 - x_3^2 = y_1^2 - y_2^2, \quad y_1 = x_1 + x_2, \; y_2 = x_3.$
$\operatorname{sgn}(q(x_1, x_2, x_3)) = (1, 1). \; \mathbf{y} = P^{-1}\mathbf{x}.$

$$A = \begin{pmatrix} 1 & 1 & 0 \\ 1 & 1 & 0 \\ 0 & 0 & -1 \end{pmatrix}, \quad P^{-1} = \begin{pmatrix} 1 & 1 & 0 \\ 0 & 0 & 1 \\ 0 & 1 & 0 \end{pmatrix}, \quad P = \begin{pmatrix} 1 & 0 & -1 \\ 0 & 0 & 1 \\ 0 & 1 & 0 \end{pmatrix},$$

$${}^tPAP = \begin{pmatrix} 1 & 0 & 0 \\ 0 & -1 & 0 \\ 0 & 0 & 0 \end{pmatrix}.$$

The third row of P^{-1} can be taken to be any row so that P^{-1} is a regular matrix.

(2) $q(x_1, x_2, x_3, x_4) = (x_1 + x_2)^2 + x_2^2 + (x_3 + x_4)^2 - (x_3 - x_4)^2 = y_1^2 + y_2^2 + y_3^2 - y_4^2, \; y_1 = x_1 + x_2, \; y_2 = x_2, \; y_3 = x_3 + x_4, \; y_4 = x_3 - x_4. \; \operatorname{sgn}(q(x_1, x_2, x_3, x_4)) = (3, 1).$

$$A = \begin{pmatrix} 1 & 1 & 0 & 0 \\ 1 & 2 & 0 & 0 \\ 0 & 0 & 0 & 2 \\ 0 & 0 & 2 & 0 \end{pmatrix}, \quad P = \begin{pmatrix} 1 & -1 & 0 & 0 \\ 0 & 1 & 0 & 0 \\ 0 & 0 & 1/2 & 1/2 \\ 0 & 0 & 1/2 & -1/2 \end{pmatrix}, \quad {}^tPAP = \begin{pmatrix} 1 & 0 & 0 & 0 \\ 0 & 1 & 0 & 0 \\ 0 & 0 & 1 & 0 \\ 0 & 0 & 0 & -1 \end{pmatrix}.$$

(3) $q(x_1, x_2, x_3, x_4) = (x_1 + 2x_3)^2 - x_3^2 + (2x_2 - x_4)^2 - (x_2 + x_4)^2 = y_1^2 + y_2^2 - y_3^2 - y_4^2, \; y_1 = x_1 + 2x_3, \; y_2 = 2x_2 - x_4, \; y_3 = x_3, \; y_4 = x_2 + x_4. \; \operatorname{sgn}(q(x_1, x_2, x_3, x_4)) = (2, 2).$

$$A = \begin{pmatrix} 1 & 0 & 2 & 0 \\ 0 & 3 & 0 & -3 \\ 2 & 0 & 3 & 0 \\ 0 & -3 & 0 & 0 \end{pmatrix}, \quad P = \begin{pmatrix} 1 & 0 & -2 & 0 \\ 0 & 1/3 & 0 & 1/3 \\ 0 & 0 & 1 & 0 \\ 0 & -1/3 & 0 & 2/3 \end{pmatrix}, \quad {}^tPAP = \begin{pmatrix} 1 & 0 & 0 & 0 \\ 0 & 1 & 0 & 0 \\ 0 & 0 & -1 & 0 \\ 0 & 0 & 0 & -1 \end{pmatrix}.$$

3   (1) Since $|A_3| = \begin{vmatrix} 2 & 0 & -1 \\ 0 & 2 & 0 \\ -1 & 0 & 1 \end{vmatrix} = 2 > 0, \; |A_2| = \begin{vmatrix} 2 & 0 \\ 0 & 2 \end{vmatrix} = 4 > 0$ and $|A_1| = |2| = 2 > 0, \; q(x_1, x_2, x_3)$ is positive definite by Theorem 6.4.3.

(2) Since $\left| A_3 \right| = \begin{vmatrix} 1 & 1 & 1 \\ 1 & 3 & -1 \\ 1 & -1 & 4 \end{vmatrix} = 2 > 0$, $\left| A_2 \right| = \begin{vmatrix} 1 & 1 \\ 1 & 3 \end{vmatrix} = 2 > 0$ and $\left| A_1 \right| =$

$\left| 1 \right| = 1 > 0$, $q(x_1, x_2, x_3)$ is positive definite by Theorem 6.4.3.

4 We know that $q(x_1, \ldots, x_n)$ is negative definite if and only if the quadratic form $q_1(x_1, \ldots, x_n) = -A[\mathbf{x}]$ is positive definite. Let $B = -A$, and B_k the primary submatrix of B of degree k for $k = 1, \ldots, n$. Then by Theorem 6.4.3, $q_1(x_1, \ldots, x_n)$ is positive definite if and only if $\left| B_k \right| = (-1)^k \left| A_k \right| > 0$ for $k = 1, \ldots, n$.

5 (1) Let $A = \begin{pmatrix} -4 & 1 \\ 1 & -1 \end{pmatrix}$. Since $(-1)^2 \left| A_2 \right| = \begin{vmatrix} -4 & 1 \\ 1 & -1 \end{vmatrix} = 3 > 0$ and

$(-1) \left| A_1 \right| = - \left| -4 \right| = 4 > 0$, $q(x_1, x_2)$ is negative definite by Exercise 4 above.

(2) Let $A = \begin{pmatrix} -1 & -1 & 1 \\ -1 & -3 & 5 \\ 1 & 5 & -12 \end{pmatrix}$. Since $(-1)^3 \left| A_3 \right| = - \begin{vmatrix} -1 & -1 & 1 \\ -1 & -3 & 5 \\ 1 & 5 & -12 \end{vmatrix} =$

$6 > 0$, $(-1)^2 \left| A_2 \right| = \begin{vmatrix} -1 & -1 \\ -1 & -3 \end{vmatrix} = 2 > 0$ and $(-1) \left| A_1 \right| = - \left| -1 \right| = 1 > 0$,

the quadratic form $q(x_1, x_2, x_3)$ is negative definite by Exercise 4 above.

7.1

1 (1) 4.   (2) 6.

2 (1) $-2 + i$.   (2) $15 + 5i$.   (3) $-9 - 4i$.   (4) 0.

3 (1) Orthogonal.   (2) Not orthogonal.

4 (1) Solving the linear equation $x_1 - ix_2 + x_3 = 0$, we have

$$W^\perp = \left\{ a_1 \begin{pmatrix} i \\ 1 \\ 0 \end{pmatrix} + a_2 \begin{pmatrix} -1 \\ 0 \\ 1 \end{pmatrix} \,\middle|\, a_1, a_2 \in \mathbf{C} \right\}.$$

(2) Solving the linear equation $\begin{cases} ix_1 + (2 + i)x_2 + & 2x_3 = 0, \\ x_1 + (1 - i)x_2 + (1 + 2i)x_3 = 0, \end{cases}$ we have

$$W^\perp = \left\{ a \begin{pmatrix} 2 - 7i \\ -4 + i \\ 1 \end{pmatrix} \,\middle|\, a \in \mathbf{C} \right\}.$$

5 (1) Take $\mathbf{a} = \begin{pmatrix} 1 \\ 0 \\ 0 \end{pmatrix} \neq \mathbf{0}_3$. Then $(\mathbf{a}, \mathbf{a}) = 1 \cdot 0 + 0 \cdot 0 + 0 \cdot 1 = 0$. Therefore

$(\ , \ )$ is not a Hermitian inner product of $\mathbf{C}^3$.

(2) The linearity and $\overline{(\mathbf{a}, \mathbf{b})} = (\mathbf{b}, \mathbf{a})$ are obvious. We show that if $\mathbf{a} \in \mathbf{C}^3$ then

$(\mathbf{a}, \mathbf{a}) > 0$. In fact, if $\mathbf{a} = \begin{pmatrix} a_1 \\ a_2 \\ a_3 \end{pmatrix} \neq \mathbf{0}_3$, then $(\mathbf{a}, \mathbf{a}) = 3|a_1|^2 + 2|a_2|^2 + |a_3|^2 > 0$,

and therefore $(\ ,\ )$ is a Hermitian inner product. We note that this Hermitian inner product is different from the standard Hermitian inner product of $\mathbf{C}^3$.

6 If $\mathbf{v} = a_1\mathbf{u}_1 + \cdots + a_n\mathbf{u}_n$, then $(\mathbf{v}, \mathbf{u}_k) = a_k$ for $k = 1, \ldots, n$. Therefore $\mathbf{v}$ is determined by the values $(\mathbf{v}, \mathbf{u}_k)$ for $k = 1, \ldots, n$.

7 (1) We assume that $\mathbf{u} \in (W_1 + W_2)^\perp$. Then $\mathbf{u} \in W_1^\perp$ and $\mathbf{u} \in W_2^\perp$. Therefore we see that $(W_1 + W_2)^\perp \subset W_1^\perp \cap W_2^\perp$. Conversely, assume $\mathbf{u} \in W_1^\perp \cap W_2^\perp$. Then $\mathbf{u} \in W_1^\perp$ and $\mathbf{u} \in W_2^\perp$. Therefore $\mathbf{u}$ is orthogonal to vectors in $W_1 + W_2$. Thus $\mathbf{u} \in (W_1 + W_2)^\perp$. In other words, $W_1^\perp \cap W_2^\perp \subset (W_1 + W_2)^\perp$.

(2) By (1) and $(W^\perp)^\perp = W$, we have $(W_1^\perp + W_2^\perp)^\perp = (W_1^\perp)^\perp \cap (W_2^\perp)^\perp = W_1 \cap W_2$. Then we see that $(W_1 \cap W_2)^\perp = ((W_1^\perp + W_2^\perp)^\perp)^\perp = W_1^\perp + W_2^\perp$.

8 ($\Rightarrow$) Let $\mathbf{u} = \mathbf{u}_1 + \mathbf{u}_2$ ($\mathbf{u}_1 \in W$, $\mathbf{u}_2 \in W^\perp$). If $J = P_{V/W}$, then $J(\mathbf{u}) = \mathbf{u}_1$. Therefore we have $J(J(\mathbf{u})) = J(\mathbf{u}_1) = \mathbf{u}_1 = J(\mathbf{u})$ and $J^2 = J$, which is (i). Next, let $\mathbf{v} = \mathbf{v}_1 + \mathbf{v}_2$ ($\mathbf{v}_1 \in W$, $\mathbf{v}_2 \in W^\perp$) be another vector in V. Then $(J(\mathbf{u}), \mathbf{v}) = (\mathbf{u}_1, \mathbf{v}_1 + \mathbf{v}_2) = (\mathbf{u}_1, \mathbf{v}_1) = (\mathbf{u}_1 + \mathbf{u}_2, \mathbf{v}_1) = (\mathbf{u}, J(\mathbf{v}))$. Since $(J(\mathbf{u}), \mathbf{v}) = (\mathbf{u}, J^*(\mathbf{v}))$ by the definition of the adjoint transformation, we have $(\mathbf{u}, J(\mathbf{v})) = (\mathbf{u}, J^*(\mathbf{v}))$. As this equality holds for any vectors $\mathbf{u}, \mathbf{v} \in V$, we have $J^* = J$, which is (ii).

($\Leftarrow$) Assume that J is a linear transformation satisfying (i) and (ii). Let $W = \{\mathbf{u} \in V \mid J(\mathbf{u}) = \mathbf{u}\}$ and $W' = \{\mathbf{u} \in V \mid J(\mathbf{u}) = \mathbf{0}_V\}$. Then $W = J(V)$, $W' = (I_V - J)(V)$ and $V = W \oplus W'$. In fact, $W \cap W' = \{\mathbf{0}_V\}$ and $\mathbf{u} = J(\mathbf{u}) + (\mathbf{u} - J(\mathbf{u}))$ ($J(\mathbf{u}) \in W$, $\mathbf{u} - J(\mathbf{u}) \in W'$ for any $\mathbf{u} \in V$). Furthermore, for $\mathbf{w} \in W$ and $\mathbf{w}' \in W'$, we have $(\mathbf{w}, \mathbf{w}') = (J(\mathbf{w}), \mathbf{w}') = (\mathbf{w}, J(\mathbf{w}')) = (\mathbf{w}, \mathbf{0}_V) = 0$. Then $W' \subset W^\perp$. Since $\dim(W') = n - \dim(W) = \dim(W^\perp)$, we have $W' = W^\perp$ and $J = P_{V/W}$

9 The spaces W_1 and W_2 are orthogonal $\Leftrightarrow$ $W_2 \subset W_1^\perp$ $\Leftrightarrow$ $I_1 I_2 = O_V$.

10 (1) For any vectors $\mathbf{u}, \mathbf{v} \in V$, we see that $((T_1 T_2)(\mathbf{u}), \mathbf{v}) = (\mathbf{u}, (T_1 T_2)^*(\mathbf{v}))$ and $((T_1 T_2)(\mathbf{u}), \mathbf{v}) = (T_2(\mathbf{u}), T^* T_1^*(\mathbf{v})) = (\mathbf{u}, T_2^*(T_1^*(\mathbf{v}))) = (\mathbf{u}, (T_2^* T_1^*)(\mathbf{v})))$. Therefore $(T_1 T_2)^* = T_2^* T_1^*$.

(2) Since $(\mathbf{u}, \mathbf{v}) = (T^{-1}(T(\mathbf{u})), \mathbf{v}) = (T(\mathbf{u}), (T^{-1})^*(\mathbf{v})) = (\mathbf{u}, T^*(T^{-1})^*(\mathbf{v}))$ for any $\mathbf{u}, \mathbf{v} \in V$, we have $T^*(T^{-1})^* = I_V$. Therefore $(T^*)^{-1} = (T^{-1})^*$.

11 (1) $(A_1 A_2)^* = {}^t\overline{(A_1 A_2)} = {}^t(\overline{A_1} \overline{A_2}) = {}^t\overline{A_2} \, {}^t\overline{A_1} = A_2^* A_1^*$.

(2) $(A^*)^{-1} = ({}^t\overline{A})^{-1} = {}^t(\overline{A}^{-1}) = {}^t(\overline{A^{-1}}) = (A^{-1})^*$.

7.2 Answers to 1 through 4 are not necessarily uniquely determined.

1 To show that A is a Hermitian matrix, we have only to verify the equality $A^* = A$, or ${}^t\overline{A} = A$.

(1) The equality $A^* = A$ can be easily seen. Since $g_A(t) = (t - 2)^2 t$, the eigenvalues of A are $2, 0$. Solving the linear equation $A\mathbf{x} = 2\mathbf{x}$, we

obtain $\mathbf{x} = a_1 \begin{pmatrix} i \\ 1 \\ 0 \end{pmatrix} + a_2 \begin{pmatrix} 0 \\ 0 \\ 1 \end{pmatrix}$ $(a_1, a_2 \in \mathbf{C})$. Solving the linear equation

$A\mathbf{x} = \mathbf{0}_3$, we obtain $\mathbf{x} = a_3 \begin{pmatrix} -i \\ 1 \\ 0 \end{pmatrix}$ $(a_3 \in \mathbf{C})$. Thus we have the basis

$\left\{ \begin{pmatrix} i \\ 1 \\ 0 \end{pmatrix}, \begin{pmatrix} 0 \\ 0 \\ 1 \end{pmatrix}, \begin{pmatrix} -i \\ 1 \\ 0 \end{pmatrix} \right\}$ of $\mathbf{C}^3$ consisting of eigenvectors of A. Orthonormal-

izing the basis, we obtain an orthonormal basis

$\left\{ \begin{pmatrix} i/\sqrt{2} \\ 1/\sqrt{2} \\ 0 \end{pmatrix}, \begin{pmatrix} 0 \\ 0 \\ 1 \end{pmatrix}, \begin{pmatrix} -i/\sqrt{2} \\ 1/\sqrt{2} \\ 0 \end{pmatrix} \right\}$. Then the matrix $U = \begin{pmatrix} i/\sqrt{2} & 0 & -i/\sqrt{2} \\ 1/\sqrt{2} & 0 & 1/\sqrt{2} \\ 0 & 1 & 0 \end{pmatrix}$

is a unitary matrix and we have $U^{-1}AU = U^*AU = \begin{pmatrix} 2 & 0 & 0 \\ 0 & 2 & 0 \\ 0 & 0 & 0 \end{pmatrix}$.

(2) The equality $A^* = A$ can be easily seen. Since $g_A(t) = (t - 2 - \sqrt{2})(t - 2 + \sqrt{2})(t - 2)$, the eigenvalues of A are $2 \pm \sqrt{2}, 2$. By a similar calculation to (1), we have a unitary matrix $U = \begin{pmatrix} (1 + \sqrt{2})i/\sqrt{4 + 2\sqrt{2}} & (1 - \sqrt{2})i/\sqrt{4 - 2\sqrt{2}} & 0 \\ 0 & 0 & 1 \\ 1/\sqrt{4 + 2\sqrt{2}} & 1/\sqrt{4 - 2\sqrt{2}} & 0 \end{pmatrix}$. Then $U^{-1}AU =$

$U^*AU = \begin{pmatrix} 2 + \sqrt{2} & 0 & 0 \\ 0 & 2 - \sqrt{2} & 0 \\ 0 & 0 & 2 \end{pmatrix}$.

2   To show that A is a unitary matrix, we have only to verify the equality $A^*A = E_3$, or ${}^t\overline{A}A = E_3$.

(1) (i) Since $A^*A = \begin{pmatrix} \frac{1-i}{2} & \frac{-1-i}{2} & 0 \\ \frac{-1-i}{2} & \frac{1-i}{2} & 0 \\ 0 & 0 & i \end{pmatrix} \begin{pmatrix} \frac{1+i}{2} & \frac{-1+i}{2} & 0 \\ \frac{-1+i}{2} & \frac{1+i}{2} & 0 \\ 0 & 0 & -i \end{pmatrix} = E_3$, A is a unitary

matrix.

(ii) Since $g_A(t) = (t - i)(t + i)(t - 1)$, the eigenvalues of A are $\pm i, 1$. Solving linear equations $A\mathbf{x} = i\mathbf{x}$, $A\mathbf{x} = -i\mathbf{x}$ and $A\mathbf{x} = \mathbf{x}$, we have a basis

$\left\{ \begin{pmatrix} 1 \\ 1 \\ 0 \end{pmatrix}, \begin{pmatrix} 0 \\ 0 \\ 1 \end{pmatrix}, \begin{pmatrix} -1 \\ 1 \\ 0 \end{pmatrix} \right\}$ of $\mathbf{C}^3$ consisting of eigenvectors of A. Orthonor-

malizing the basis, we have a unitary matrix $U = \begin{pmatrix} 1/\sqrt{2} & 0 & -1/\sqrt{2} \\ 1/\sqrt{2} & 0 & 1/\sqrt{2} \\ 0 & 1 & 0 \end{pmatrix}$ and

$$U^{-1}AU = U^*AU = \begin{pmatrix} i & 0 & 0 \\ 0 & -i & 0 \\ 0 & 0 & 1 \end{pmatrix}.$$

(2) (i) The equality $A^*A = E_3$ can be easily seen.

(ii) Since $g_A(t) = (t - (1+i)/\sqrt{2})(t - i)(t - 1)$, the eigenvalues of A are $(1+i)/\sqrt{2}, i, 1$. Solving linear equations $Ax = \frac{1+i}{\sqrt{2}}x$, $Ax = ix$ and $Ax = x$, we have a basis $\left\{ \begin{pmatrix} 0 \\ 1 \\ 0 \end{pmatrix}, \begin{pmatrix} \sqrt{3} \\ 0 \\ 1 \end{pmatrix}, \begin{pmatrix} -1/\sqrt{3} \\ 0 \\ 1 \end{pmatrix} \right\}$ of $\mathbf{C}^3$ consisting of eigenvectors of A. Orthonormalizing this basis, we have a unitary matrix

$$U = \begin{pmatrix} 0 & \sqrt{3}/2 & -1/2 \\ 1 & 0 & 0 \\ 0 & 1/2 & \sqrt{3}/2 \end{pmatrix} \text{ and } U^{-1}AU = U^*AU = \begin{pmatrix} (1+i)/\sqrt{2} & 0 & 0 \\ 0 & i & 0 \\ 0 & 0 & 1 \end{pmatrix}.$$

3 To show that A is a normal matrix, we have only to verify the equality $A^*A = AA^*$.

(1) (i) Since $A^*A = AA^* = \begin{pmatrix} 2 & 0 \\ 0 & 2 \end{pmatrix}$, A is a normal matrix.

(ii) Since $g_A(t) = (t - (1+i))(t + (1+i))$, the eigenvalues of A are $\pm(1+i)$. We solve the linear equations $Ax = (1+i)x$ and $Ax = -(1+i)x$ and obtain a basis $\left\{ \begin{pmatrix} (1+i)/\sqrt{2} \\ 1 \end{pmatrix}, \begin{pmatrix} -(1+i)/\sqrt{2} \\ 1 \end{pmatrix} \right\}$ of $\mathbf{C}^2$ consisting of eigenvectors of A. Orthonormalize the basis and we have an orthonormal basis $\left\{ \begin{pmatrix} (1+i)/2 \\ 1/\sqrt{2} \end{pmatrix}, \begin{pmatrix} -(1+i)/2 \\ 1/\sqrt{2} \end{pmatrix} \right\}$. Then the matrix $U = \begin{pmatrix} (1+i)/2 & -(1+i)/2 \\ 1/\sqrt{2} & 1/\sqrt{2} \end{pmatrix}$ is a unitary matrix and we have $U^{-1}AU = U^*AU = \begin{pmatrix} 1+i & 0 \\ 0 & -(1+i) \end{pmatrix}$.

(2) (i) Since $A^*A = AA^* = \begin{pmatrix} 5 & 0 & -4i \\ 0 & 4 & 0 \\ 4i & 0 & 5 \end{pmatrix}$, A is a normal matrix.

(ii) Since $g_A(t) = (t + i)(t - 2i)(t - 3i)$, the eigenvalues of A are $-i, 2i, 3i$. Solving the linear equations $Ax = -ix$, $Ax = 2ix$ and $Ax = 3ix$, we obtain a basis $\left\{ \begin{pmatrix} i \\ 0 \\ 1 \end{pmatrix}, \begin{pmatrix} 0 \\ 1 \\ 0 \end{pmatrix}, \begin{pmatrix} -i \\ 0 \\ 1 \end{pmatrix} \right\}$ of $\mathbf{C}^3$ consisting of eigenvectors of A. Orthonormalizing the basis, we obtain an orthonormal basis $\left\{ \begin{pmatrix} i/\sqrt{2} \\ 0 \\ 1/\sqrt{2} \end{pmatrix}, \begin{pmatrix} 0 \\ 1 \\ 0 \end{pmatrix}, \begin{pmatrix} -i/\sqrt{2} \\ 0 \\ 1/\sqrt{2} \end{pmatrix} \right\}$. Then the matrix $U = \begin{pmatrix} i/\sqrt{2} & 0 & -i/\sqrt{2} \\ 0 & 1 & 0 \\ 1/\sqrt{2} & 0 & 1/\sqrt{2} \end{pmatrix}$

is a unitary matrix and $U^{-1}AU = U^*AU = \begin{pmatrix} -i & 0 & 0 \\ 0 & 2i & 0 \\ 0 & 0 & 3i \end{pmatrix}$.

4 To see that T_A is a normal transformation, we have only to show that the matrix A is a normal matrix by Theorem 7.2.9 (2).

(1) (i) $AA^* = \begin{pmatrix} 3 & 0 & -\sqrt{2}i \\ 0 & 2i & 0 \\ \sqrt{2}i & 0 & 4 \end{pmatrix} \begin{pmatrix} 3 & 0 & -\sqrt{2}i \\ 0 & -2i & 0 \\ \sqrt{2}i & 0 & 4 \end{pmatrix} = \begin{pmatrix} 11 & 0 & -7\sqrt{2}i \\ 0 & 4 & 0 \\ 7\sqrt{2}i & 0 & 18 \end{pmatrix}$.

$A^*A = \begin{pmatrix} 3 & 0 & -\sqrt{2}i \\ 0 & -2i & 0 \\ \sqrt{2}i & 0 & 4 \end{pmatrix} \begin{pmatrix} 3 & 0 & -\sqrt{2}i \\ 0 & 2i & 0 \\ \sqrt{2}i & 0 & 4 \end{pmatrix} = \begin{pmatrix} 11 & 0 & -7\sqrt{2}i \\ 0 & 4 & 0 \\ 7\sqrt{2}i & 0 & 18 \end{pmatrix}$.

Therefore $AA^* = A^*A$.

(ii) Since $g_A(t) = (t - 2i)(t - 5)(t - 2)$, the eigenvalues of A are $2i$, 5, 2.

Solving $Ax = 2ix$, we have $W_1 = W(2i\,;A) = \left\{ a_1 \begin{pmatrix} 0 \\ 1 \\ 0 \end{pmatrix} \middle| a_1 \in \mathbf{C} \right\}$.

Solving $Ax = 5x$, we have $W_2 = W(5\,;A) = \left\{ a_2 \begin{pmatrix} -i/\sqrt{2} \\ 0 \\ 1 \end{pmatrix} \middle| a_1 \in \mathbf{C} \right\}$.

Solving $Ax = 2x$, we have $W_3 = W(2\,;A) = \left\{ a_3 \begin{pmatrix} \sqrt{2}i \\ 0 \\ 1 \end{pmatrix} \middle| a_3 \in \mathbf{C} \right\}$.

Thus we obtain three subspaces of $\mathbf{C}^3$ which are orthogonal to each other and satisfy $\mathbf{C}^3 = W_1 \oplus W_2 \oplus W_3$. If we denote by I_i the orthogonal projection of $\mathbf{C}^3$ into W_i for $i = 1, 2, 3$, then we have the spectral resolution $T_A = 2iI_1 + 5I_2 + 2I_3$. The orthogonal projections I_1, I_2, I_3 are given by

$I_1 \left(\begin{pmatrix} x_1 \\ x_2 \\ x_3 \end{pmatrix} \right) = x_2 \begin{pmatrix} 0 \\ 1 \\ 0 \end{pmatrix}$, $I_2 \left(\begin{pmatrix} x_1 \\ x_2 \\ x_3 \end{pmatrix} \right) = \frac{1}{3}(\sqrt{2}ix_1 + 2x_3) \begin{pmatrix} -i/\sqrt{2} \\ 0 \\ 1 \end{pmatrix}$ and

$I_3 \left(\begin{pmatrix} x_1 \\ x_2 \\ x_3 \end{pmatrix} \right) = \frac{1}{3}(-\sqrt{2}ix_1 + x_3) \begin{pmatrix} \sqrt{2}i \\ 0 \\ 1 \end{pmatrix}$.

(2) (i) Since $A = \begin{pmatrix} 6 & -\sqrt{2}i & -\sqrt{2}i \\ \sqrt{2}i & 5 & 1 \\ \sqrt{2}i & 1 & 5 \end{pmatrix} = A^*$, A is a Hermitian matrix.

Therefore A is a normal matrix.

(ii) Since $g_A(t) = (t - 4)^2(t - 8)$, the eigenvalues of A are $4, 8$. Solving $Ax = 4x$, we have $W_1 = W(4\,;A) =$ $\left\{ x = a_1 \begin{pmatrix} i/\sqrt{2} \\ 1 \\ 0 \end{pmatrix} + a_2 \begin{pmatrix} i/\sqrt{2} \\ 0 \\ 1 \end{pmatrix} \middle| a_1, a_2 \in \mathbf{C} \right\}$. Solving $Ax = 8x$, we have $\ W_2 = W(8\,;A) = \left\{ a_3 \begin{pmatrix} -\sqrt{2}i \\ 1 \\ 1 \end{pmatrix} \,\middle|\, a_3 \in \mathbf{C} \right\}.$ Thus we obtain two

subspaces W_1 and W_2 of $\mathbf{C}^3$ which are orthogonal to each other and satisfy $\mathbf{C}^3 = W_1 \oplus W_2$. If we denote by I_i the orthogonal projection of $\mathbf{C}^3$ into W_i for $i = 1, 2$, then we have the spectral resolution $T_A = 4I_1 + 8I_2$.

The orthogonal projections I_1 and I_2 are given by $I_1 \left(\begin{pmatrix} x_1 \\ x_2 \\ x_3 \end{pmatrix} \right) =$

$$\left(-\frac{\sqrt{2}}{4}ix_1 + \frac{3}{4}x_2 - \frac{1}{4}x_3 \right) \begin{pmatrix} i/\sqrt{2} \\ 1 \\ 0 \end{pmatrix} + \left(-\frac{\sqrt{2}}{4}ix_1 - \frac{1}{4}x_2 + \frac{3}{4}x_3 \right) \begin{pmatrix} i/\sqrt{2} \\ 0 \\ 1 \end{pmatrix}$$

and $I_2 \left(\begin{pmatrix} x_1 \\ x_2 \\ x_3 \end{pmatrix} \right) = \left(\frac{\sqrt{2}}{4}ix_1 + \frac{1}{4}x_2 + \frac{1}{4}x_3 \right) \begin{pmatrix} -\sqrt{2}i \\ 1 \\ 1 \end{pmatrix}.$

5 If T is a normal transformation, then there is an orthogonal basis consisting of eigenvectors of T. Let λ be an eigenvalue of T_1 and $W(\lambda\,;T_1)$ the eigenspace of T_1 with the eigenvalue λ. Then $V = \bigoplus_\lambda W(\lambda\,;T_1)$. Since T_1 and T_2 are commutative, T_2 maps $W(\lambda\,;T_1)$ into itself. We also know that since T_1^* and T_2^* are commutative, T_2^* maps $W(\bar{\lambda}\,;T_1^*)$ into itself. Noting that $W(\bar{\lambda}\,;T_1^*) = W(\lambda\,;T_1)$, the restriction of T_2 on $W(\lambda\,;T_1)$ is a normal transformation of $W(\lambda\,;T_1)$. Take an orthonormal basis of $W(\lambda\,;T_1)$ consisting of eigenvectors of T_2. Then the union of such bases for all eigenvalues λ of T_1 is an orthonormal basis of V consisting of eigenvectors of both T_1 and T_2.

6 (1) ($\Rightarrow$) Theorem 7.2.2. ($\Leftarrow$) We assume that the different eigenvalues $\lambda_1, \ldots, \lambda_r$ are all real numbers. Let $T = \lambda_1 I_1 + \cdots + \lambda_r I_r$ be the spectral resolution of T. Since $I_i^* = I_i$, we have $T^* = \bar{\lambda}_1 I_1 + \cdots + \bar{\lambda}_r I_r = \lambda_1 I_1 + \cdots + \lambda_r I_r = T$.

(2) ($\Rightarrow$) Let T be a unitary transformation. Let λ be an eigenvalue of T and $\mathbf{u}$ an eigenvector of T with the eigenvalue λ. Since T is a unitary transformation, we have $\|T(\mathbf{u})\| = \|\mathbf{u}\|$ by Theorem 7.2.3. On the other hand, we see that

$$\|T(\mathbf{u})\| = \|\lambda\mathbf{u}\| = |\lambda| \cdot \|\mathbf{u}\|$$

by definition. Since $\|\mathbf{u}\| \neq 0$, we have $|\lambda| = 1$.

($\Leftarrow$) Let $\{\mathbf{u}_1, \ldots, \mathbf{u}_n\}$ be an orthonormal basis consisting of eigenvectors of T and λ_i the eigenvalue of T corresponding to $\mathbf{u}_i$ for $i = 1, \ldots, n$, respectively. We let $\mathbf{u} = a_1 \mathbf{u}_1 + \cdots + a_n \mathbf{u}_n$ be a vector in V. Since $|\lambda_i| = 1$ for $i = 1, \ldots, n$, we have

$$\|T(\mathbf{u})\|^2 = (T(\mathbf{u}), T(\mathbf{u})) = \sum_{i,j=1}^n a_i \lambda_i \overline{a_j \lambda_j}(\mathbf{u}_i, \mathbf{u}_j) = \sum_{i=1}^n |a_i|^2 |\lambda_i|^2 = \sum_{i=1}^n |a_i|^2 = \|\mathbf{u}\|^2.$$

Therefore by Theorem 7.2.3, T is a unitary transformation.

7 (1) We have $AA^* = \begin{pmatrix} 1 & 0 \\ 0 & 4 \end{pmatrix}$. Since AA^* is a positive definite Hermitian matrix,

we let $H = \sqrt{AA^*} = \begin{pmatrix} 1 & 0 \\ 0 & 2 \end{pmatrix}$, which is also a positive definite Hermitian matrix.

We put $U = H^{-1}A = \begin{pmatrix} 1 & 0 \\ 0 & 1/2 \end{pmatrix}\begin{pmatrix} 0 & i \\ 2 & 0 \end{pmatrix} = \begin{pmatrix} 0 & i \\ 1 & 0 \end{pmatrix}$. Then U is a unitary matrix.

Therefore we have $A = HU = \begin{pmatrix} 1 & 0 \\ 0 & 2 \end{pmatrix}\begin{pmatrix} 0 & i \\ 1 & 0 \end{pmatrix}$, which is a product of a positive
definite Hermitian matrix H and a unitary matrix U.

(2) We have $AA^* = \begin{pmatrix} 4 & \sqrt{2} \\ \sqrt{2} & 3 \end{pmatrix}$. The characteristic polynomial is $g_{AA^*}(t) =$
$(t - 5)(t - 2)$ and the eigenvalues of AA^* are 2 and 5. Therefore AA^* is a positive definite Hermitian matrix and we can consider a square root $\sqrt{AA^*}$. Solving
the linear equation $AA^*\mathbf{x} = 2\mathbf{x}$, we get solutions $\mathbf{x} = a_1 \begin{pmatrix} -1/\sqrt{2} \\ 1 \end{pmatrix}$ $(a_1 \in \mathbf{C})$.

Solving the linear equation $AA^*\mathbf{x} = 5\mathbf{x}$, we get solutions $\mathbf{x} = a_2 \begin{pmatrix} \sqrt{2} \\ 1 \end{pmatrix}$ $(a_2 \in$

$\mathbf{C})$. Orthonormalizing the basis $\left\{ \begin{pmatrix} -1/\sqrt{2} \\ 1 \end{pmatrix}, \begin{pmatrix} \sqrt{2} \\ 1 \end{pmatrix} \right\}$ of $\mathbf{C}^2$, we obtain

the orthonormal basis $\left\{ \begin{pmatrix} -1/\sqrt{3} \\ \sqrt{2}/\sqrt{3} \end{pmatrix}, \begin{pmatrix} \sqrt{2}/\sqrt{3} \\ 1/\sqrt{3} \end{pmatrix} \right\}$. Then the matrix $U_0 =$

$\begin{pmatrix} -1/\sqrt{3} & \sqrt{2}/\sqrt{3} \\ \sqrt{2}/\sqrt{3} & 1/\sqrt{3} \end{pmatrix}$ is a unitary matrix and $U_0^{-1}AA^*U_0 = \begin{pmatrix} 2 & 0 \\ 0 & 5 \end{pmatrix}$. Then

$$\sqrt{AA^*} = U_0 \begin{pmatrix} \sqrt{2} & 0 \\ 0 & \sqrt{5} \end{pmatrix} U_0^{-1} = \frac{1}{3}\begin{pmatrix} \sqrt{2}+2\sqrt{5} & -2+\sqrt{10} \\ -2+\sqrt{10} & 2\sqrt{2}+\sqrt{5} \end{pmatrix}.$$

Let $H = \sqrt{AA^*}$ and

$$U = H^{-1}A = \frac{1}{3\sqrt{10}}\begin{pmatrix} 2\sqrt{2}+\sqrt{5} & 2-\sqrt{10} \\ 2-\sqrt{10} & \sqrt{2}+2\sqrt{5} \end{pmatrix}\begin{pmatrix} 1-i & \sqrt{2} \\ \sqrt{2} & -i \end{pmatrix}$$

$$= \frac{1}{3\sqrt{10}}\begin{pmatrix} 4\sqrt{2}-\sqrt{5}-(2\sqrt{2}+\sqrt{5})i & 4+\sqrt{10}-(2-\sqrt{10})i \\ 4+\sqrt{10}-(2-\sqrt{10})i & 2\sqrt{2}-2\sqrt{5}-(\sqrt{2}+2\sqrt{5})i \end{pmatrix}.$$

Then U is a unitary matrix. Thus the matrix $A = HU$ is a product of a positive definite Hermitian matrix H and a unitary matrix U.

8 (1) Let $H = \frac{1}{2}(A + A^*)$ and $K = -\frac{i}{2}(A - A^*)$. Then $A = H + iK$. Since
$H^* = \frac{1}{2}(A^* + A) = H$ and $K^* = \frac{i}{2}(A^* - A) = -\frac{i}{2}(A - A^*) = K$, both H
and K are Hermitian matrices. To see the uniqueness, we assume that $A = H +$
iK (H, K: Hermitian matrices). Then $A^* = H^* + (iK)^* = H - iK$. Therefore
$H = \frac{1}{2}(A + A^*)$ and $K = \frac{1}{2i}(A - A^*) = -\frac{i}{2}(A - A^*)$.

(2) We see that $A = H + iK$ and $A^* = H - iK$ by (1). Then $A = A^* \Leftrightarrow iK = -iK$, namely, $K = O_n$.

(3) We see that $A^*A = (H - iK)(H + iK) = H^2 + K^2 + i(HK - KH)$ and $AA^* = (H + iK)(H - iK) = H^2 + K^2 + i(KH - HK)$. Then $A^*A = AA^* \iff HK - KH = KH - HK$, namely, $HK = KH$.

(4) Assume that A is a unitary matrix. Then $A^*A = (H - iK)(H + iK) = H^2 + K^2 + i(HK - KH) = E_n$ and $AA^* = (H + iK)(H - iK) = H^2 + K^2 + i(KH - HK) = E_n$ by Theorem 7.2.5. Therefore we have $2(H^2 + K^2) = 2E_n$, namely, $H^2 + K^2 = E_n$. The equality $HK = KH$ is shown by (3). Conversely, if A satisfies $H^2 + K^2 = E_n$ and $HK = KH$, it is easy to see that $A^*A = E_n$, namely, A is a unitary matrix.

8.1

1 (1) $g_A(t) = |tE_3 - A| = (t - 1)^3$. The eigenvalue of A is 1 and the generalized eigenspace is $\tilde{W}(1 ; A) = \mathbf{C}^3$.

(2) $g_A(t) = |tE_3 - A| = (t - 2)^2(t + 1)$. The eigenvalues of A are 2 and -1. The generalized eigenspaces are

$$\tilde{W}(2 ; A) = \mathrm{Ker}\,((A - 2E_3)^2) = \left\{ a\begin{pmatrix} 3 \\ 2 \\ 0 \end{pmatrix} + b\begin{pmatrix} 3 \\ 0 \\ 2 \end{pmatrix} \middle| a, b \in \mathbf{C} \right\} \text{ and}$$

$$\tilde{W}(-1 ; A) = \mathrm{Ker}\,(A + E_3) = \left\{ c\begin{pmatrix} 0 \\ 1 \\ -2 \end{pmatrix} \middle| c \in \mathbf{C} \right\}.$$

(3) $g_A(t) = |tE_3 - A| = (t - 1)^2(t + 2)$. The eigenvalues of A are 1 and -2. The generalized eigenspaces are

$$\tilde{W}(1 ; A) = \mathrm{Ker}\,((A - E_3)^2) = \left\{ a\begin{pmatrix} -2 \\ 0 \\ 1 \end{pmatrix} + b\begin{pmatrix} 0 \\ 1 \\ 0 \end{pmatrix} \middle| a, b \in \mathbf{C} \right\} \text{ and}$$

$$\tilde{W}(-2 ; A) = \mathrm{Ker}\,(A + 2E_3) = \left\{ c\begin{pmatrix} -1 \\ 0 \\ 1 \end{pmatrix} \middle| c \in \mathbf{C} \right\}.$$

(4) $g_A(t) = |tE_3 - A| = (t - 1)(t - i)^2$. The eigenvalues of A are 1 and i. The generalized eigenspaces are

$$\tilde{W}(1 ; A) = \mathrm{Ker}\,(A - E_3) = \left\{ a\begin{pmatrix} 0 \\ 1 \\ 0 \end{pmatrix} \middle| a \in \mathbf{C} \right\} \text{ and}$$

$$\tilde{W}(i ; A) = \mathrm{Ker}\,((A - iE_3)^2) = \left\{ b\begin{pmatrix} 1 \\ 1 \\ 0 \end{pmatrix} + c\begin{pmatrix} 1 \\ 0 \\ 1 \end{pmatrix} \middle| b, c \in \mathbf{C} \right\}.$$

(5) $g_A(t) = |tE_3 - A| = (t - 2)^2(t - i)$. The eigenvalues of A are 2 and i. The generalized eigenspaces are

$$\tilde{W}(2\,;A) = \mathrm{Ker}\,((A - 2E_3)^2) = \left\{ a \begin{pmatrix} -1 \\ 1 \\ 0 \end{pmatrix} + b \begin{pmatrix} 0 \\ 0 \\ 1 \end{pmatrix} \,\middle|\, a, b \in \mathbf{C} \right\} \text{ and }$$

$$\tilde{W}(i\,;A) = \mathrm{Ker}\,(A - iE_3) = \left\{ c \begin{pmatrix} 1 \\ 0 \\ 1 \end{pmatrix} \,\middle|\, c \in \mathbf{C} \right\}.$$

(6) $g_A(t) = \left| tE_4 - A \right| = (t - 2)^2(t + 1)^2$. The eigenvalues of A are 2 and -1. The generalized eigenspaces are

$$\tilde{W}(2\,;A) = \mathrm{Ker}\,((A - 2E_4)^2) = \left\{ a \begin{pmatrix} 1 \\ 0 \\ 0 \\ 0 \end{pmatrix} + b \begin{pmatrix} 0 \\ -1 \\ 1 \\ 0 \end{pmatrix} \,\middle|\, a, b \in \mathbf{C} \right\} \text{ and }$$

$$\tilde{W}(-1\,;A) = \mathrm{Ker}\,((A + E_4)^2) = \left\{ c \begin{pmatrix} 0 \\ -1 \\ 2 \\ 0 \end{pmatrix} + d \begin{pmatrix} 0 \\ 0 \\ 0 \\ 1 \end{pmatrix} \,\middle|\, c, d \in \mathbf{C} \right\}.$$

(7) $g_A(t) = \left| tE_4 - A \right| = (t - 1)(t - i)^3$. The eigenvalues of A are 1 and i. The generalized eigenspaces are

$$\tilde{W}(1\,;A) = \mathrm{Ker}\,(A - E_4) = \left\{ a \begin{pmatrix} i \\ 0 \\ 0 \\ 1 \end{pmatrix} \,\middle|\, a \in \mathbf{C} \right\} \text{ and }$$

$$\tilde{W}(i\,;A) = \mathrm{Ker}\,((A - iE_4)^3) = \left\{ b \begin{pmatrix} 1 \\ 0 \\ 0 \\ 0 \end{pmatrix} + c \begin{pmatrix} 0 \\ 1 \\ 0 \\ 0 \end{pmatrix} + d \begin{pmatrix} 0 \\ 0 \\ 1 \\ 0 \end{pmatrix} \,\middle|\, b, c, d \in \mathbf{C} \right\}.$$

(8) $g_A(t) = \left| tE_4 - A \right| = (t - 1)^2(t - i)^2$. The eigenvalues of A are 1 and i. The generalized eigenspaces are

$$\tilde{W}(1\,;A) = \mathrm{Ker}\,((A - E_4)^2) = \left\{ a \begin{pmatrix} 0 \\ 0 \\ 1 \\ 0 \end{pmatrix} + b \begin{pmatrix} 0 \\ -1 \\ 0 \\ 1 \end{pmatrix} \,\middle|\, a, b \in \mathbf{C} \right\} \text{ and }$$

$$\tilde{W}(i\,;A) = \mathrm{Ker}\,((A - iE_4)^2) = \left\{ c \begin{pmatrix} 2 \\ -1 \\ 1 \\ 0 \end{pmatrix} + d \begin{pmatrix} 1 \\ -1 \\ 0 \\ 1 \end{pmatrix} \,\middle|\, c, d \in \mathbf{C} \right\}.$$

2   Answers are not unique.

(1) Let $f_1(t) = (t + 1)^2$ and $f_2(t) = (t - 2)^2$. Dividing $f_1(t)$ by $f_2(t)$, we obtain $f_1(t) - f_2(t) = 6t - 3$. Next, dividing $f_2(t)$ by $6t - 3$, we have $f_2(t) = (\frac{1}{6}t - \frac{7}{12})(6t - 3) + \frac{9}{4}$. Therefore $\frac{9}{4} = -(\frac{1}{6}t - \frac{7}{12})f_1(t) + (\frac{1}{6}t + \frac{5}{12})f_2(t)$. Divide both sides by $\frac{9}{4}$, and put $g_1(t) = -\frac{2}{27}t + \frac{7}{27}$, $g_2(t) = \frac{2}{27}t + \frac{5}{27}$. Then we

have $g_1(t)f_1(t) + g_2(t)f_2(t) = 1$.

(2) Let $g_1(t) = -\frac{1}{4}t + \frac{1}{4}$ and $g_2 = \frac{1}{4}$. Then $g_1(t)f_1(t) + g_2(t)f_2(t) = 1$.

(3) Let $g_1(t) = \frac{7}{27}t + \frac{20}{27}$ and $g_2(t) = -\left(\frac{7}{27}t^2 + \frac{2}{9}t + \frac{1}{3}\right)$. Then $g_1(t)f_1(t) + g_2(t)f_2(t) = 1$.

3   Let $n_i = \deg(A_i)$ for $i = 1, \ldots, r$. We show by induction on r. Let $r = 2$. Since the characteristic polynomials $g_{A_1}(t)$ and $g_{A_2}(t)$ have no common divisor, there exist polynomials $q_1(t)$ and $q_2(t)$ satisfying $1 = q_1(t)g_{A_1}(t) + q_2(t)g_{A_2}(t)$ by Theorem 5.4.5. Then we have $E = q_1(C)g_{A_1}(C) + q_2(C)g_{A_2}(C)$ for any square matrix C. Let $f(t) = f_1(t)q_2(t)g_{A_2}(t) + f_2(t)q_1(t)g_{A_1}(t)$. We note $g_{A_i}(A_i) = O_{n_i}$ for $i = 1, 2$. Then we have $f(A_1) = f_1(A_1)q_2(A_1)g_{A_2}(A_1) + f_2(A_1)q_1(A_1)g_{A_1}(A_1) = f_1(A_1)q_2(A_1)g_{A_2}(A_1) = f_1(A_1)(E_{n_1} - q_1(A_1)g_{A_1}(A_1)) = f_1(A_1)E_{n_1} = f_1(A_1) = B_1$. Similarly, we have $f(A_2) = B_2$. Therefore $f(A) = f(A_1 \oplus A_2) = f(A_1) \oplus f(A_2) = B_1 \oplus B_2 = B$. We assume that $r > 2$. Let $A_0 = A_1 \oplus \cdots \oplus A_{r-1}$ and $B_0 = B_1 \oplus \cdots \oplus B_{r-1}$. Then $A = A_0 \oplus A_r$ and $B = B_0 \oplus B_r$. Now, there exists a polynomial $f_0(t)$ satisfying $f_0(A_0) = B_0$ by the induction assumption. Since the characteristic polynomial $g_{A_0}(t)$ is the product $g_{A_1}(t) \cdots g_{A_{r-1}}(t)$, the polynomials $g_{A_0}(t)$ and $g_{A_r}(t)$ have no common divisor. Therefore by the result on $r = 2$, there exists a polynomial $f(t)$ satisfying $f(A) = B$.

8.2

1   We denote by B the Jordan normal form of A.

(1) We have $g_A(t) = t^4$. Since $A^2 \neq O_4$ and $A^3 = O_4$, A is a nilpotent matrix and $p_A(t) = t^3$. Therefore $B = J(0; 3) \oplus J(0; 1)$.

(2) We have $g_A(t) = t^4$. Since $A^2 = O_4$, A is a nilpotent matrix and $p_A(t) = t^2$. Then B is either $B = J(0; 2) \oplus J(0; 2)$ or $B = J(0; 2) \oplus J(0; 1) \oplus J(0; 1)$. Since rank $(A) = 2$, we have $B = J(0; 2) \oplus J(0; 2)$ by Theorem 8.2.5.

(3) We have $g_A(t) = t^6$. Since $A^3 \neq O_6$ and $A^4 = O_6$. Thus A is a nilpotent matrix and $p_A(t) = t^4$. Then B is either $J(0; 4) \oplus J(0; 1) \oplus J(0; 1)$ or $J(0; 4) \oplus J(0; 2)$. Since rank $(A) = 4$, we have $B = J(0; 4) \oplus J(0; 2)$ by Theorem 8.2.5.

(4) We have $g_A(t) = t^6$. Since $A^3 \neq O_6$ and $A^4 = O_6$, A is a nilpotent matrix and $p_A(t) = t^4$. Then B is either $J(0; 4) \oplus J(0; 2)$ or $J(0; 4) \oplus J(0; 1) \oplus J(0; 1)$. Since rank $(A) = 3$, we have $B = J(0; 4) \oplus J(0; 1) \oplus J(0; 1)$ by Theorem 8.2.5.

2   (1) Since $g_A(t) = (t - 2)(t - 3)^2$, the minimal polynomial $p_A(t)$ is either $g_A(t)$ or $(t - 2)(t - 3)$ by Theorem 5.4.6. Since $(A - 2E_3)(A - 3E_3) = O_3$, we have $p_A(t) = (t - 2)(t - 3)$. Then A is diagonalizable by Theorem 5.5.4 and the Jordan normal form of A is $J(2; 1) \oplus J(3; 1) \oplus J(3; 1)$.

(2) Since $g_A(t) = (t-3)^3$, the minimal polynomial $p_A(t)$ is $g_A(t)$, $(t-3)$ or $(t-3)^2$ by Theorem 5.4.6. Since $A - 3E_3 \neq O_3$ and $(A - 3E_3)^2 \neq O_3$, we have $p_A(t) = g_A(t) = (t-3)^3$. Then the Jordan normal form of A is $J(3; 3)$.

(3) Since $g_A(t) = (t^2+1)(t-i) = (t-i)^2(t+i)$, the minimal polynomial $p_A(t)$ is either $g_A(t)$ or $(t-i)(t+i) = t^2+1$ by Theorem 5.4.6. Since $A^2 + E_3 = O_3$, we have $p_A(t) = t^2+1$. Then applying Theorem 5.5.4, A is diagonalizable and the Jordan normal form of A is $J(i; 1) \oplus J(i; 1) \oplus J(-i; 1)$.

(4) Since $g_A(t) = (t+1)(t-2)^3$, the minimal polynomial $p_A(t)$ is $g_A(t)$, $(t+1)(t-2)^2$ or $(t+1)(t-2)$ by Theorem 5.4.6. Since $(A + E_4)(A - 2E_4) \neq O_4$ and $(A + E_4)(A - 2E_4)^2 = O_4$, we have $p_A(t) = (t+1)(t-2)^2$. Therefore the Jordan normal form of A is $J(2; 1) \oplus J(2; 2) \oplus J(-1; 1)$.

(5) The characteristic polynomial of A is $g_A(t) = (t-1)^3(t-2)^2$. Since rank $(A - E_5) = 3$, we have dim $(W(1; A)) = 2$. Since rank $(A - 2E_5) = 4$, we have dim $(W(2; A)) = 1$. Then $p_A(t) = (t-1)^2(t-2)^2$ and the Jordan normal form of A is $J(1; 1) \oplus J(1; 2) \oplus J(2; 2)$.

(6) The characteristic polynomial of A is $g_A(t) = (t-1)(t+1)^2(t-2)^2$. Since rank $(A + E_5) = 4$, we have dim $(W(-1; A)) = 1$. Since rank $(A - 2E_5) = 4$, we have dim $(W(2; A)) = 1$. Then $p_A(t) = g_A(t) = (t-1)(t+1)^2(t-2)^2$ and the Jordan normal form of A is $J(1; 1) \oplus J(-1; 2) \oplus J(2; 2)$.

(7) The characteristic polynomial of A is $g_A(t) = (t-i)(t-1)^2(t-3)^2$. Since rank $(A - E_5) = 4$, we have dim $(W(1; A)) = 1$. Since rank $(A - 3E_5) = 3$, we have dim $(W(3; A)) = 2$. Then $p_A(t) = (t-i)(t-1)^2(t-3)$ and the Jordan normal form of A is $J(i; 1) \oplus J(1; 2) \oplus J(3; 1) \oplus J(3; 1)$.

3 (1) rank $(A) = 8$ and $p_A(t) = t^3(t-2)^3$.

(2) rank $(A) = 7$ and $p_A(t) = (t-3)^2(t-i)^2$.

(3) rank $(A) = 9$ and $p_A(t) = t^2(t-1)^3(t+2)^3$.

(4) rank $(A) = 6$ and $p_A(t) = t(t-2)^3(t-i)^2(t+i)$.

4 Let B be the Jordan normal form of A and λ an eigenvalue of A. Then the direct sum of all Jordan cells with the eigenvalue λ appearing in B is $J(\lambda; k_1) \oplus \cdots \oplus J(\lambda; k_l)$ $(k_1 \geq \cdots \geq k_l \geq 1,\ l = l_\lambda)$. We easily see that the minimal polynomial of this matrix is equal to $(t-\lambda)^{k_1}$ and the characteristic polynomial is $(t-\lambda)^k$ $(k = k_1 + \cdots + k_l)$. Therefore the minimal polynomial is equal to the characteristic polynomial if and only if $l = 1$ for all eigenvalues λ.

5 Let A be a nilpotent and semisimple matrix. Since A is a semisimple matrix, A is similar to a diagonal matrix. As A is a nilpotent matrix, all the eigenvalues of A are 0. Then we have $A = O_n$.

6 (1) There exists a regular matrix P such that $A_0 = P^{-1}AP$ is the Jordan normal form of A, i.e., $A_0 = J(\lambda_1) \oplus \cdots \oplus J(\lambda_r)$. Here $\lambda_1, \ldots, \lambda_r$ are all the different eigenvalues of A and $J(\lambda_i) = J(\lambda_i; k_{i,1}) \oplus \cdots \oplus J(\lambda_i; k_{i,m_i})$ for

$i = 1, \ldots, r$ as in the Theorem 8.2.3. We let $k_i = k_{i,1} + \cdots + k_{i,m_i}$, $H_i = \lambda_i E_{k_i}$ and $N_i = J(\lambda_i) - H_i = J(\lambda_i) - \lambda_i E_{k_i}$ for $i = 1, \ldots, r$. We also let $H_0 = \lambda_1 E_{k_1} \oplus \cdots \oplus \lambda_r E_{k_r}$ and $N_0 = N_1 \oplus \cdots \oplus N_r$. Then we have $A_0 = H_0 + N_0$. By the definition of the Jordan normal form, H_0 is a diagonal matrix and N_0 is a nilpotent matrix. Let $H = P H_0 P^{-1}$ and $N = P N_0 P^{-1}$. Then H is a semisimple matrix and N is a nilpotent matrix. Since $A = P A_0 P^{-1} = P(H_0 + N_0)P^{-1} = P H_0 P^{-1} + P N_0 P^{-1} = H + N$, we see that A is a sum of a semisimple matrix H and a nilpotent matrix N. As H_i and N_i are commutative for $i = 1, \ldots, r$, H_0 and N_0 are also commutative. Therefore $H = P H_0 P^{-1}$ and $N = P N_0 P^{-1}$ are commutative. We note that $N_i = J(\lambda_i) - \lambda_i E_{k_i}$ is a polynomial of $J(\lambda_i)$. Since the characteristic polynomial $g_{J(\lambda_i)}(t)$ is a power of $t - \lambda_i$ for $i = 1, \ldots, r$ and $\lambda_i \neq \lambda_j$ for $i \neq j$, we see that any pair of the characteristic polynomials $g_{J(\lambda_i)}(t)$ has no common divisor. So, we can apply Exercises (§ 8.1) 3 to $A_0 = J(\lambda_1) \oplus \cdots \oplus J(\lambda_r)$ and $N_0 = N_1 \oplus \cdots \oplus N_r$, and see that N_0 is a polynomial of A_0. Since $H_0 = A_0 - N_0$, H_0 is also a polynomial of A_0. Therefore the matrices $H = P H_0 P^{-1}$ and $N = P N_0 P^{-1}$ are polynomials of $P A_0 P^{-1} = A$.

We shall show the uniqueness of H and N. Assume that $A = H' + N'$ is another sum of A with a semisimple matrix H' and a nilpotent matrix N' satisfying $H'N' = N'H'$. Then H' and N' are commutative with the matrix $A = H' + N'$. Since the matrices H and N are polynomials of A, the matrices H, N, H', N' are commutative to each other. Since H and H' are semisimple (i.e. diagonalizable) matrices, there exists a regular matrix Q such that both $Q^{-1} H Q$ and $Q^{-1} H' Q$ are diagonal matrices by Theorem 5.5.6. Therefore $H - H'$ is a semisimple matrix. Since $N' - N$ is a nilpotent matrix by Exercises (§ 1.2) 8, the matrix $H - H' = N' - N$ is a semisimple and nilpotent matrix. Therefore we have $H' = H$ and $N' = N$ by Exercise 5 above.

(2) Assume that A is a real square matrix. Let $A = H + N$ with a semisimple matrix H and a nilpotent matrix N satisfying $HN = NH$. Then $\overline{A} = \overline{H} + \overline{N}$. We easily see that $\overline{H}$ is a semisimple matrix and $\overline{N}$ is a nilpotent matrix satisfying $\overline{H}\,\overline{N} = \overline{N}\,\overline{H}$. Since $A = \overline{A}$, we have $H + N = \overline{H} + \overline{N}$. As the pair of H and N is uniquely determined by (1), we have $H = \overline{H}$ and $N = \overline{N}$. Therefore both H and N are real matrices.

References

Bourbaki, N., Eléments de mathématique, Livre II Algèbre, Ch. 2 Algèbre linéaire, Hermann (1967).

Chatelin, F., Eigenvalues of Matrices, John Wiley & Sons, Ltd. (1993).

Jacobson, N., Basic Algebra I 2nd ed., W.H. Freeman (1985).

Lang, S., Linear Algebra 3rd ed., Springer–Verlag (1987).

Mostow, G. and Sampson, J., Linear Algebra, MacGraw-Hill (1969).

Rosen, S., Advanced Linear Algebra 3rd ed., Springer (2007).

Satake, I., Linear Algebra, M. Dekker (1975).

Strang, G., Linear Algebra and its Applications, Academic Press (1980).

Lake Kanayama, Furano, Japan

Index of Theorems

Index

Printed in the United States
by Baker & Taylor Publisher Services

Printed in the United States
by Baker & Taylor Publisher Services